Lecture Notes in Computer Science 2227

Edited by G. Goos, J. Hartmanis and J. van Leeuwen

Springer

Berlin
Heidelberg
New York
Barcelona
Hong Kong
London
Milan
Paris
Tokyo

Serdar Boztaş   Igor E. Shparlinski (Eds.)

Applied Algebra, Algebraic Algorithms and Error-Correcting Codes

14th International Symposium, AAECC-14
Melbourne, Australia, November 26-30, 2001
Proceedings

Springer

Series Editors

Gerhard Goos, Karlsruhe University, Germany
Juris Hartmanis, Cornell University, NY, USA
Jan van Leeuwen, Utrecht University, The Netherlands

Volume Editors

Serdar Boztaş
RMIT University, Department of Mathematics
GPO Box 2476V, Melbourne 3001, Australia
E-mail: serdar@rmit.edu.au

Igor E. Shparlinski
Macquarie University, Department of Computing
NSW 2109, Australia
E-mail: igor@comp.mq.edu.au

Cataloging-in-Publication Data applied for

Die Deutsche Bibliothek - CIP-Einheitsaufnahme

Applied algebra, algebraic algorithms and error correcting codes : 14th
international symposium ; proceedings / AAECC 14, Melbourne, Australia,
November 26 - 30, 2001. Serdar Boztaş ; Igor E. Shparlinski (ed.). - Berlin ;
Heidelberg ; New York ; Barcelona ; Hong Kong ; London ; Milan ; Paris ; Tokyo :
Springer, 2002
  (Lecture notes in computer science ; Vol. 2227)
  ISBN 3-540-42911-5

CR Subject Classification (1998): E.4, I.1, E.3, G.2, F.2

ISSN 0302-9743
ISBN 3-540-42911-5 Springer-Verlag Berlin Heidelberg New York

Springer-Verlag Berlin Heidelberg New York
a member of BertelsmannSpringer Science+Business Media GmbH

http://www.springer.de

© Springer-Verlag Berlin Heidelberg 2001
Printed in Germany

Typesetting: Camera-ready by author, data conversion by Steingräber Satztechnik GmbH, Heidelberg
Printed on acid-free paper   SPIN: 10840981   06/3142   5 4 3 2 1 0

Preface

The AAECC Symposia Series was started in 1983 by Alain Poli (Toulouse), who, together with R. Desq, D. Lazard, and P. Camion, organized the first conference. Originally the acronym AAECC meant "Applied Algebra and Error-Correcting Codes". Over the years its meaning has shifted to "Applied Algebra, Algebraic Algorithms, and Error-Correcting Codes", reflecting the growing importance of complexity in both decoding algorithms and computational algebra.

AAECC aims to encourage cross-fertilization between algebraic methods and their applications in computing and communications. The algebraic orientation is towards finite fields, complexity, polynomials, and graphs. The applications orientation is towards both theoretical and practical error-correction coding, and, since AAECC 13 (Hawaii, 1999), towards cryptography. AAECC was the first symposium with papers connecting Gröbner bases with E-C codes. The balance between theoretical and practical is intended to shift regularly; at AAECC-14 the focus was on the theoretical side.

The main subjects covered were:

- Codes: iterative decoding, decoding methods, block codes, code construction.
- Codes and algebra: algebraic curves, Gröbner bases, and AG codes.
- Algebra: rings and fields, polynomials.
- Codes and combinatorics: graphs and matrices, designs, arithmetic.
- Cryptography.
- Computational algebra: algebraic algorithms.
- Sequences for communications.

Six invited speakers covered the areas outlined:

- Robert Calderbank, "Combinatorics, Quantum Computers, and Cellular Phones"
- James Massey, "The Ubiquity of Reed-Muller Codes"
- Graham Norton, "Gröbner Bases over a Principal Ideal Ring"
- Vera Pless, "Self-dual Codes – Theme and Variations"
- Amin Shokrollahi, "Design of Differential Space-Time Codes Using Group Theory"
- Madhu Sudan, "Ideal Error-Correcting Codes: Unifying Algebraic and Number-Theoretic Algorithms".

Except for AAECC-1 (*Discrete Mathematics* 56, 1985) and AAECC-7 (*Discrete Applied Mathematics* 33, 1991), the proceedings of all the symposia have been published in Springer-Verlag's *Lecture Notes in Computer Science* (Vols. 228, 229, 307, 356, 357, 508, 539, 673, 948, 1255, 1719).

It is a policy of AAECC to maintain a high scientific standard, comparable to that of a journal. This has been made possible thanks to the many referees involved. Each submitted paper was evaluated by at least two international researchers.

AAECC-14 received and refereed 61 submissions. Of these, 1 was withdrawn, 36 were selected for publication in these proceedings, while 7 additional works contributed to the symposium as oral presentations. Unrefereed talks were presented in a "Recent Results" session.

The symposium was organized by Serdar Boztaş, Tom Høholdt, Kathy Horadam, Igor E. Shparlinski, and Branka Vucetic, with the help of Asha Baliga, Pride Conference Management (Juliann Smith), and the Department of Mathematics, RMIT University. It was sponsored by the Australian Mathematical Society.

We express our thanks to the staff of Springer-Verlag, especially Alfred Hofmann and Anna Kramer, for their help in the preparation of these proceedings.

August 2001   Serdar Boztaş and Igor E. Shparlinski

Organization

Steering Committee

General Chair: Kathy Horadam (RMIT Univ., AUS)
Conference Co-chair: Tom Høholdt (Technical Univ. of Denmark, DK)
Program Chair: Igor Shparlinski (Macquarie Univ., AUS)
Program Co-chair: Branka Vucetic (Sydney Univ., AUS)
Publication: Serdar Boztaş (RMIT Univ., AUS)

Conference Committee

J. Calmet	T. Høholdt	S. Lin
M. Clausen	K. Horadam	O. Moreno
G. Cohen	H. Imai	H. Niederreiter
P.G. Farrell	H. Janwa	A. Poli
G.L. Feng	J.M. Jensen	T.R.N. Rao
M. Giusti	R. Kohno	S. Sakata
J. Heintz	H.W. Lenstra Jr.	P. Solé

Program Committee

I.F. Blake	M. Giusti	S. Litsyn
J. Calmet	J. Gutierrez	A. Nechaev
C. Carlet	J. Heintz	H. Niederreiter
P. Charpin	T. Helleseth	D. Panario
M. Clausen	H. Imai	S. Sakata
P.G. Farrell	E. Kaltofen	P. Solé
M. Fossorier	T. Kasami	H. van Tilborg
M. Giesbrecht	L. Knudsen	C. Xing

Local Organizing Committee

Asha Baliga	Serdar Boztaş	Kathy Horadam

Referees

D. Augot	N. Boston	C. Carlet
A. Baliga	F. Boulier	P. Charpin
I.F. Blake	S. Boztaş	M. Clausen
A. Bonnecaze	J. Calmet	G. Cohen

R. Cramer
I. Damgård
M. Dichtl
C. Ding
I. Duursma
P.G. Farrell
G-L. Feng
H.C. Ferreira
M. Fossorier
T. Fujiwara
P. Gaborit
J. Galati
S. Galbraith
S. Gao
V.P. Gerdt
M. Giesbrecht
M. Giusti
F. Griffin
J. Gutierrez
Y.S. Han

C. Hao
T. Hashimoto
J. Heintz
T. Helleseth
K. Horadam
X-D. Hou
H. Imai
J. Jensen
G. Kabatiansky
E. Kaltofen
T. Kasami
F. Keqin
T. Kløve
L. Knudsen
L. Kulesz
T. Laihonen
S. Ling
S. Litsyn
F. Morain
R. Morelos-Zaragoza

S. Murphy
V.K. Murty
A. Nechaev
H. Niederreiter
D. Panario
L. Pecquet
V. Rijmen
S. Sakata
P. Sarkar
H.G. Schaathun
I. Shparlinski
B. Shung
A. Silverberg
P. Solé
B. Stevens
H. van Tilborg
B. Vucetic
J.L. Walker
K. Yang
C. Xing

Sponsoring Institutions

Australian Mathematical Society

Table of Contents

Invited Contributions

Block Codes

Code Constructions

Codes and Algebra: Rings and Fields

Codes and Algebra: Algebraic Geometry Codes

Sequences

Cryptography

Algorithms

Algorithms: Decoding

Algebraic Constructions

The Ubiquity of Reed-Muller Codes

James L. Massey

ETH-Zürich and Lund University
Trondhjemsgade 3 2TH, DK-2100 Copenhagen East
JamesMassey@compuserve.com

Abstract. It is argued that the nearly fifty-year-old Reed-Muller codes underlie a surprisingly large number of algebraic problems in coding and cryptography. This thesis is supported by examples that include some new results such as the construction of a new class of constant-weight cyclic codes with a remarkably simple decoding algorithm and a much simplified derivation of the well-known upper bound on the linear complexity of the running key produced by a nonlinearly filtered maximal-length shift-register.

1 Introduction

The Reed-Muller codes, which were actually discovered by Muller [1], were the first nontrivial class of multiple-error-correcting codes. Reed [2] gave a simple majority-logic decoding algorithm for these binary codes that corrects all errors guaranteed correctable by their minimum distance; he also gave an insightful description of these codes that has been adopted by most later researchers and that we will also follow here.

Nearly 50 years have passed since the discovery of the Reed-Muller codes. It is our belief that when one digs deeply into almost any algebraic problem in coding theory or cryptography, one finds these venerable codes (or closely related codes) lying at the bottom. We illustrate this "ubiquity" of the Reed-Muller codes in what follows with a number of examples that include some new results.

In Section 2, we describe the two matrices whose properties underlie the construction and theory of the Reed-Muller codes. The codes themselves are introduced in Section 3. In Section 4 we show how the Reed-Muller codes have been used in a natural way to measure the nonlinearity of a binary function of m binary variables, a problem that arises frequently in cryptography. In Section 5 we use Reed-Muller coding concepts to construct a new class of constant-weight cyclic codes that have an astonishingly simple decoding algorithm. The cyclic Reed-Muller codes are introduced in Section 6 where we also describe an "unconventional" encoder for these codes. This encoder is seen in Section 7 to be the same as the running-key generator for a stream cipher of the type called a nonlinearly filtered maximal-length shift register, which leads to an extremely simple derivation of a well-known upper bound on the linear complexity of the resulting running key. We conclude with some remarks in Section 8.

S. Boztaş and I.E. Shparlinski (Eds.): AAECC-14, LNCS 2227, pp. 1–12, 2001.
© Springer-Verlag Berlin Heidelberg 2001

2   Two Useful Matrices

In this section we describe two matrices whose properties will be exploited in the sequel.

Let $\mathbf{M}_m$ denote the $2^m \times 2^m$ binary matrix in which the entries in row $i + 1$ are the coefficients of $(1 + x)^i$ in order of ascending powers of x for $i = 0, 1, 2, \ldots, 2^m - 1$. For $m = 3$, this matrix is

$$\mathbf{M}_3 = \begin{bmatrix} 1\,0\,0\,0\,0\,0\,0\,0 \\ 1\,1\,0\,0\,0\,0\,0\,0 \\ 1\,0\,1\,0\,0\,0\,0\,0 \\ 1\,1\,1\,1\,0\,0\,0\,0 \\ 1\,0\,0\,0\,1\,0\,0\,0 \\ 1\,1\,0\,0\,1\,1\,0\,0 \\ 1\,0\,1\,0\,1\,0\,1\,0 \\ 1\,1\,1\,1\,1\,1\,1\,1 \end{bmatrix} .$$

Some Properties of $\mathbf{M}_m$:

1. The i-th row of $\mathbf{M}_m$ is the i-th row of Pascal's triangle with entries reduced modulo 2. Equivalently, each row after the first is obtained by adding the previous row to its own shift right by one position.
2. The Hamming weight of row $i + 1$, i.e., the number of nonzero coefficients in $(1 + x)^i$, is equal to the Hamming weight $W_2(i)$ of the radix-two representation of the integer i for $i = 0, 1, 2, \ldots, 2^m - 1$, cf. Lemma 1 in [3].
3. The matrix $\mathbf{M}_m$ is its own inverse, cf. [4].
4. The sum of any selection of rows of the matrix $\mathbf{M}_m$ has Hamming weight at least that of the uppermost row included in the sum, cf. Theorem 1.1 in [3].

Of special interest to us here will be the submatrix $\mathbf{A}_m$ of $\mathbf{M}_m$ consisting of the m rows with Hamming weight 2^{m-1}. For $m = 3$, this matrix is

$$\mathbf{A}_3 = \begin{bmatrix} \mathbf{a}_1 \\ \mathbf{a}_2 \\ \mathbf{a}_3 \end{bmatrix} = \begin{bmatrix} 1\,1\,1\,1\,0\,0\,0\,0 \\ 1\,1\,0\,0\,1\,1\,0\,0 \\ 1\,0\,1\,0\,1\,0\,1\,0 \end{bmatrix}$$

where here and hereafter we denote the rows of $\mathbf{A}_m$ as $\mathbf{a}_1, \mathbf{a}_2, \ldots, \mathbf{a}_m$.

Some Properties of $\mathbf{A}_m$:

1. The j-th column of $\mathbf{A}_m$, when read downwards with its entries considered as integers, contains the radix-two representation of the integer $2^m - j$ for $j = 1, 2, \ldots, 2^m$.
2. The ith row $\mathbf{a}_i$ of $\mathbf{A}_m$, when treated as the function table of a binary-valued function of m binary variables in the manner that the entry in the jth column is the value of the function $f(x_1, x_2, \ldots, x_m)$ when $x_1, x_2, \ldots, x_m$ considered as integers is the radix-two representation of the integer $2^m - j$, corresponds to the function $f(x_1, x_2, \ldots, x_m) = x_i$ for $i = 1, 2, \ldots, m$ and $j = 1, 2, \ldots, 2^m$.
3. Cyclic shifting the rows of $\mathbf{A}_m$ in any way (i.e., allowing different numbers of shifts for each row) is equivalent to a permutation of the columns of $\mathbf{A}_m$.

4. Every $m \times 2^m$ binary matrix whose columns are all different can be obtained from $\mathbf{A}_m$ by a permutation of the columns.

3   Reed-Muller Codes

Following Reed's notation [2] for the Reed-Muller codes, we use juxtaposition of row vectors to denote their term-by-term product, which we will refer to as the *Hadamard product* of these row vectors. For instance, for $m = 3$, $\mathbf{a}_1\mathbf{a}_3 = \begin{bmatrix} 1\,0\,1\,0\,0\,0\,0\,0 \end{bmatrix}$ and $\mathbf{a}_1\mathbf{a}_2\mathbf{a}_3 = \begin{bmatrix} 1\,0\,0\,0\,0\,0\,0\,0 \end{bmatrix}$. We also write $\mathbf{a}_0$ to denote the all-one row vector of length 2^m.

Let $\mathrm{RM}(m, \mu)$, where $1 \leq \mu < m$, denote the μth-order Reed-Muller code of length $n = 2^m$. $\mathrm{RM}(m, \mu)$ can be defined as the linear binary code for which the matrix $\mathbf{G}_m^\mu$, which has as rows $\mathbf{a}_0$, $\mathbf{a}_1$, ... $\mathbf{a}_m$ together with all Hadamard products of $\mathbf{a}_1$, $\mathbf{a}_2$, ... $\mathbf{a}_m$ taken μ or fewer at a time, is a generator matrix. For instance, the second-order Reed-Muller code $\mathrm{RM}(3, 2)$ has the generator matrix

$$
\mathbf{G}_3^2 = \begin{bmatrix} \mathbf{a}_0 \\ \mathbf{a}_1 \\ \mathbf{a}_2 \\ \mathbf{a}_3 \\ \mathbf{a}_1\mathbf{a}_2 \\ \mathbf{a}_1\mathbf{a}_3 \\ \mathbf{a}_2\mathbf{a}_3 \end{bmatrix} = \begin{bmatrix} 1\,1\,1\,1\,1\,1\,1\,1 \\ 1\,1\,1\,1\,0\,0\,0\,0 \\ 1\,1\,0\,0\,1\,1\,0\,0 \\ 1\,0\,1\,0\,1\,0\,1\,0 \\ 1\,1\,0\,0\,0\,0\,0\,0 \\ 1\,0\,1\,0\,0\,0\,0\,0 \\ 1\,0\,0\,0\,1\,0\,0\,0 \end{bmatrix}.
$$

It is also convenient to define the 0th-order Reed-Muller code $\mathrm{RM}(m, 0)$ as the binary linear code with generator matrix $\mathbf{G}_m^0 = [\mathbf{a}_0]$. For instance for $m = 3$,

$$
\mathbf{G}_3^0 = \begin{bmatrix} 1\,1\,1\,1\,1\,1\,1\,1 \end{bmatrix}.
$$

The following proposition is a direct consequence of Properties 3 and 4 of the matrix $\mathbf{M}_m$.

Proposition 1. *The Reed-Muller code RM(m, μ) of length $n = 2^m$, where $0 < \mu < m$, has dimension $k = \sum_{i=0}^{\mu} \binom{m}{i}$ and minimum distance $d = 2^{m-\mu}$. Moreover, its dual code is the Reed-Muller code RM($m, m - 1 - \mu$).*

4   Measuring Nonlinearity

It is often the case in cryptography that one wishes to find a binary-valued function $f(x_1, x_2, \ldots, x_m)$ of m binary variables that is "highly nonlinear". Rueppel [5] showed that the Reed-Muller codes can be used to measure the amount of nonlinearity in a very natural way. His approach is based on the following proposition, which is an immediate consequence of Property 2 of the matrix $\mathbf{A}_m$ and of the facts that

$$
\mathbf{G}_m^1 = \begin{bmatrix} \mathbf{a}_0 \\ \mathbf{A}_m \end{bmatrix}
$$

and that $\mathbf{a}_0$ is the function table of the constant function 1.

Proposition 2. *The codewords in the first-order Reed-Muller code of length 2^m, $RM(m,1)$, correspond to the function tables of all linear and affine functions of m binary variables when the entry in the jth position is considered as the value of the function $f(x_1, x_2, \ldots, x_m)$ where $x_1, x_2, \ldots, x_m$ give the radix-two representation of the integer $2^m - j$ for $j = 1, 2, \ldots, 2^m$.*

Rueppel, cf. pp. 127–129 in [5], exploited the content of Proposition 2 to assert that the best linear or affine approximation to a binary function $f(x_1, x_2, \ldots, x_m)$ with function table $\mathbf{y} = \begin{bmatrix} y_1 \ y_2 \ \cdots \ y_{2^m} \end{bmatrix}$ has as its function table the codeword in $RM(m,1)$ closest (in the Hamming metric) to $\mathbf{y}$. If e is the number of errors in this best approximation, *i.e.*, the Hamming distance from this closest codeword to $\mathbf{y}$, then $e/2^m$ is the error rate of this best linear or affine approximation to $f(x_1, x_2, \ldots, x_m)$.

Sometimes in cryptography one knows only that the function to be approximated is one of a set of t functions. In this case, Rueppel suggested taking the best linear or affine approximation to be the function corresponding to the codeword in $RM(m,1)$ at the smallest average Hamming distance to the function tables $\mathbf{y}_1, \mathbf{y}_2, \ldots, \mathbf{y}_t$ and to use the smallness of the average error rate as the measure of goodness. As an example of this method, Rueppel showed that the best linear or affine approximation to the most significant input bit of "S-box" S_5 of the Data Encryption Standard (DES) [6] from the four different output functions $f(x_1, x_2, \ldots, x_{16})$ determined by the two "control bits" for this S-Box is the affine function $1 + x_1 + x_2 + x_3 + x_4$ and has an error rate of only $12/64$ or 18.8%. It is hardly surprising that, seven years later, Matsui [7] built his "linear cryptanalysis" attack against DES on this "linear weakness" in S-box S_5.

5   Easily Decodable Constant-Weight Cyclic Codes

There are many ways to combine binary vectors to obtain another binary vector in addition to summing and to taking their Hadamard product. One of the most interesting ways when the number of vectors is odd is by *majority combining* in each bit position. For instance, majority combining of the three rows in

$$\mathbf{A}_3 = \begin{bmatrix} \mathbf{a}_1 \\ \mathbf{a}_2 \\ \mathbf{a}_3 \end{bmatrix} = \begin{bmatrix} 1\,1\,1\,1\,0\,0\,0\,0 \\ 1\,1\,0\,0\,1\,1\,0\,0 \\ 1\,0\,1\,0\,1\,0\,1\,0 \end{bmatrix}$$

gives the row vector

$$\mathbf{v}_3 = \begin{bmatrix} 1\,1\,1\,0\,1\,0\,0\,0 \end{bmatrix}.$$

The sequence $\mathbf{v}_m$ obtained by majority combining the rows of $\mathbf{A}_m$ was introduced by Stiffler [8] as one period of a periodic "ranging sequence" with the property that, when corrupted by additive noise, it could be synchronized by serial processing with a single correlator much faster than could any previously proposed ranging sequence of the same period. We adopt a coding viewpoint here and, for odd m at least 3, take $\mathbf{v}_m$ and its $2^m - 1$ cyclic shifts to be the codewords in a binary cyclic constant–weight code, which we denote by S_m and call a *Stiffler code*.

Proposition 3. *For every odd m at least 3, the Stiffler code S_m is a cyclic constant-weight binary code with length $n = 2^m$ having n codewords of weight $w = 2^{m-1}$ and minimum distance $d = 2\binom{m-1}{(m-1)/2}$.*

For instance, the $n = 8$ codewords

$$[1\,1\,1\,0\,1\,0\,0\,0], [0\,1\,1\,1\,0\,1\,0\,0], [0\,0\,1\,1\,1\,0\,1\,0], [0\,0\,0\,1\,1\,1\,0\,1],$$
$$[1\,0\,0\,0\,1\,1\,1\,0], [0\,1\,0\,0\,0\,1\,1\,1], [1\,0\,1\,0\,0\,0\,1\,1], [1\,1\,0\,1\,0\,0\,0\,1]$$

in S_3 have weight $w = 4$ and are easily checked to have minimum distance $d = 2\binom{2}{1} = 4$. Because the codewords in S_m form a single cyclic equivalence class, the code has a well-defined distance distribution. The distance distribution for S_3 is $D_0 = 1$, $D_4 = 5$ and $D_6 = 2$ where D_i is the number of codewords at distance i from a fixed codeword.

Before proving Proposition 3, it behooves us to say why the Stiffler codes are interesting. From a distance viewpoint, they are certainly much inferior to the first-order Reed-Muller code $\mathrm{RM}(m,1)$ which have $n = 2^m$, dimension $k = m+1$ (and thus $2n$ codewords), and minimum distance $d = 2^{m-1}$. The saving grace of the Stiffler codes is that they can be decoded up to their minimum distance much more simply than even the first-order Reed-Muller codes.

To prove Proposition 3, we first note that row $\mathbf{a}_1$ of $\mathbf{A}_m$ affects the majority combining that produces $\mathbf{v}_m$ only in those $2\binom{m-1}{(m-1)/2}$ columns where the remaining $m-1$ rows of $\mathbf{A}_m$ contain an equal number of zeroes and ones. Complementing row $\mathbf{a}_1$ of $\mathbf{A}_m$ and then majority combining with the remaining rows would thus produce a new row vector at distance $2\binom{m-1}{(m-1)/2}$ from $\mathbf{v}_m$–but this complementing of the first row of $\mathbf{A}_m$ without changing the remaining rows is equivalent to cyclic shifting *all* rows of $\mathbf{A}_m$ by 2^{m-1} positions so that this new row vector is the cyclic shift of $\mathbf{v}_m$ by 2^{m-1} positions and is thus also a codeword in S_m. It follows that the minimum distance of S_m cannot exceed $2\binom{m-1}{(m-1)/2}$.

We complete the proof of Proposition 3 by showing that the following decoding algorithm for S_m corrects all patterns of $\binom{m-1}{(m-1)/2} - 1$ or fewer errors and either corrects or detects every pattern of $\binom{m-1}{(m-1)/2}$ errors, which implies that the minimum distance cannot be less than $2\binom{m-1}{(m-1)/2}$. We first note, however, that every row of $\mathbf{A}_m$, say row $\mathbf{a}_i$, agrees with $\mathbf{v}_m$ in exactly $2^{m-1} + \binom{m-1}{(m-1)/2}$ positions, *i.e.*, in all $2\binom{m-1}{(m-1)/2}$ positions where $\mathbf{a}_i$ affects the majority combining and in exactly half of the remaining $2^m - 2\binom{m-1}{(m-1)/2}$ positions. We note also that by the *decimation by 2* of a vector of even length, say $\mathbf{r} = [r_1\,r_2\,r_3\,r_4\,\dots\,r_{2L-1}\,r_{2L}]$, is meant the vector $[r_1\,r_3\,\dots\,r_{2L-1}\,r_2\,r_4\,\dots\,r_{2L}]$ whose two subvectors $[r_1\,r_3\,\dots\,r_{2L-1}]$ and $[r_2\,r_4\,\dots\,r_{2L}]$ are called the *phases* of this decimation by 2 of $\mathbf{r}$.

Decoding algorithm for S_m:
Let $\mathbf{r} = \begin{bmatrix} r_1 \ r_2 \ldots r_{2^m} \end{bmatrix}$ be the binary received vector.
Step 0: Set $i = m$ and $\tilde{\mathbf{r}} = \mathbf{r}$.
Step 1: If the Hamming distance from $\mathbf{a}_m = \begin{bmatrix} 1\,0\,1\,0 \ldots 1\,0 \end{bmatrix}$ to $\tilde{\mathbf{r}}$ is less than $n/2 = 2^{m-1}$ or greater than $n/2 = 2^{m-1}$, set δ_i to 0 or 1, respectively. If this distance is equal to $n/2 = 2^{m-1}$, announce a detected error and stop.
Step 2: If $i = 1$, stop and announce the decoding decision as the right cyclic shift of $\mathbf{v}_m$ by $\delta = \delta_1 2^{m-1} + \delta_2 2^{m-2} + \ldots + \delta_m$ positions.
Step 3: If $\delta_i = 1$, shift $\tilde{\mathbf{r}}$ cyclically left by one position.
Step 4: Replace $\tilde{\mathbf{r}}$ by its decimation by 2, decrease i by 1, then return to Step 1.

Example of Decoding for S_3 :
Suppose that $\mathbf{r} = \begin{bmatrix} 1\,1\,0\,0\,0\,1\,1\,1 \end{bmatrix}$ is the received vector.
    We begin by setting $i = 3$ and $\tilde{\mathbf{r}} = \begin{bmatrix} 1\,1\,0\,0\,0\,1\,1\,1 \end{bmatrix}$.
    Because the Hamming distance from $\mathbf{a}_3 = \begin{bmatrix} 1\,0\,1\,0\,1\,0\,1\,0 \end{bmatrix}$ to $\tilde{\mathbf{r}}$ is 5, which exceeds $n/2 = 4$, we set $\delta_3 = 1$ and then shift $\tilde{\mathbf{r}}$ cyclically to the left by one position to obtain $\tilde{\mathbf{r}} = \begin{bmatrix} 1\,0\,0\,0\,1\,1\,1\,1 \end{bmatrix}$. We then decimate $\tilde{\mathbf{r}}$ by 2 to obtain $\tilde{\mathbf{r}} = \begin{bmatrix} 1\,0\,1\,1\,0\,0\,1\,1 \end{bmatrix}$, after which we decrease i to 2.
    Because the Hamming distance from $\mathbf{a}_3$ to $\tilde{\mathbf{r}}$ is 3, which is less than $n/2 = 4$, we set $\delta_2 = 0$. We then decimate $\tilde{\mathbf{r}}$ by 2 to obtain $\tilde{\mathbf{r}} = \begin{bmatrix} 1\,1\,0\,1\,0\,1\,0\,1 \end{bmatrix}$, after which we decrease i to 1.
    Because the Hamming distance from $\mathbf{a}_3$ to $\tilde{\mathbf{r}}$ is 7, which exceeds $n/2 = 4$, we set $\delta_1 = 1$.
    We now announce the decoding decision as the right shift of $\mathbf{v}_m$ by $\delta = 4\delta_1 + 2\delta_2 + \delta_3 = 5$ positions, *i.e.*, as the codeword $\begin{bmatrix} 0\,1\,0\,0\,0\,1\,1\,1 \end{bmatrix}$, which we note is at Hamming distance 1 from the received word so that we have corrected an apparent single error.

To justify this decoding algorithm, which is an adaptation to the decoding problem for S_m of the algorithm given by Stiffler [8] for synchronization of the periodic ranging sequence with pattern $\mathbf{v}_m$ within one period, we argue as follows:
    Suppose that the transmitted codeword is the right cyclic shift of $\mathbf{v}_m$ by an *even* number of bit positions. Because $\mathbf{a}_m$ is unchanged by a right cyclic shift by an *even* number of bit positions, $\mathbf{a}_m$ will agree with the transmitted codeword in the same number of bit positions as it agrees with $\mathbf{v}_m$, *i.e.*, in $2^{m-1} + \binom{m-1}{(m-1)/2}$ bit positions. Thus, if $\binom{m-1}{(m-1)/2} - 1$ or fewer errors occur, $\mathbf{a}_m$ will agree with the transmitted codeword in more than $n/2 = 2^{m-1}$ positions–and in at least $n/2 = 2^{m-1}$ positions if exactly $\binom{m-1}{(m-1)/2}$ errors occur. Suppose conversely that the transmitted codeword is the right cyclic shift of $\mathbf{v}_m$ by an *odd* number of bit positions. Because $\mathbf{a}_m$ is complemented by a right cyclic shift by an *odd* number of bit positions, $\mathbf{a}_m$ will disagree with the transmitted codeword in the same number of bit positions as it agrees with $\mathbf{v}_m$, *i.e.*, in $2^{m-1} + \binom{m-1}{(m-1)/2}$ bit positions. Thus, if $\binom{m-1}{(m-1)/2} - 1$ or fewer errors occur, $\mathbf{a}_m$ will disagree with the transmitted codeword in more than $n/2 = 2^{m-1}$ positions–and in at least $n/2 = 2^{m-1}$ positions if exactly $\binom{m-1}{(m-1)/2}$ errors occur. It follows that the value

of δ_m produced by the decoding algorithm in Step 1 will be 0 or 1 according as the transmitted codeword is the right cyclic shift of $\mathbf{v}_m$ by an even or odd number of bit positions, respectively, when either $\binom{m-1}{(m-1)/2} - 1$ or fewer errors occur or when exactly $\binom{m-1}{(m-1)/2}$ errors occur but a detected error is not announced before δ_m has been determined by the decoding algorithm.

Suppose we correctly find in Step 1 that $\delta_m = 0$, i.e., that the transmitted codeword is the right cyclic shift of $\mathbf{v}_m$ by an *even* number of bit positions, say by $\delta = 2j$ positions. We note that shifting each row of $\mathbf{A}_m$ cyclically rightwards by $\delta = 2j$ positions then decimating by 2 each of the resulting rows gives exactly the same result as first decimating by 2 each row of $\mathbf{A}_m$ then shifting each of the resulting rows cyclically rightwards by $\delta/2 = j$ positions. Moreover, the row $\mathbf{a}_{m-1}$ is converted in this manner to the row $\mathbf{a}_m$. It follows that if we first decimate the received word by 2, then we can determine the parity δ_{m-1} of $\delta/2 = j$ by using exactly the same procedure that was used to determine the parity δ_m of $\delta = 2j$. Conversely, suppose we correctly find in Step 1 that $\delta_m = 1$, i.e., that the transmitted codeword is the right cyclic shift of $\mathbf{v}_m$ by an *odd* number of bit positions, say $\delta = 2j + 1$ positions. In this case we can perform a *left* cyclic shift of the received word by 1 position to arrive again at the case where the transmitted codeword is the right cyclic shift of $\mathbf{v}_m$ by an *even* number $\delta - 1 = 2j$ of bit positions. This is the purpose of Step 3 of the decoding algorithm.

It follows from the above that the decoding algorithm will correctly decide δ_{m-1} when either $\binom{m-1}{(m-1)/2} - 1$ or fewer errors occur or exactly $\binom{m-1}{(m-1)/2}$ errors occur but a detected error is not announced before δ_{m-1} has been determined by the decoding algorithm. A simple induction establishes that this is also true for the decisions on $\delta_{m-2}, \delta_{m-3}, \ldots, \delta_1$ and hence that the received word will always be correctly decoded when either $\binom{m-1}{(m-1)/2} - 1$ or fewer errors occur or when exactly $\binom{m-1}{(m-1)/2}$ errors occur but a detected error is not announced.

Note that the above decoding algorithm for the Stiffler code S_m requires the operation of comparing the distance between two binary words to the fixed "threshold" $n/2$ to be performed only m times, which is the least number possible to determine the codeword in a code with 2^m codewords. Note also that this decoding algorithm also corrects many error patterns of weight greater than half the minimum distance of the code.

We remark that one can also define Stiffler codes of length $n = 2^m$, m *even*, by choosing the sequence $\mathbf{v}_m$ as the result of majority combining of the $m + 1$ rows of the generator matrix $\mathbf{G}_m^1$ of the first-order Reed-Muller code. Most of the above discussion goes through virtually unchanged for these codes–the details are left to the reader.

6   Cyclic Reed-Muller Codes

The Reed-Muller codes are not cyclic, nor are they equivalent to cyclic codes. However, if the last digit is removed from each codeword, they become equivalent to cyclic codes as was first noted by Kasami, Lin and Peterson [9] and by Goethals and Delsarte [10]. We follow here the approach of [9] to the cyclic Reed-Muller codes.

A binary *maximum-length sequence* of length $2^m - 1$ can be defined as the first period of the output sequence produced by an m-stage binary linear-feedback shift register as in Fig. 1, whose feedback polynomial $h(X) = 1 + h_1 X + \ldots + h_{m-1} X^{m-1} + X^m$ is a primitive polynomial, when the initial state of the shift register is not all zero, cf. §7.4 in [11]. The period has length $2^m - 1$, which means that every non-zero state occurs exactly once as the maximum-length sequence is produced by the m-stage binary linear-feedback shift register. For instance, the sequence $\begin{bmatrix} 1\,0\,0\,1\,1\,1\,0 \end{bmatrix}$ is the binary maximum-length sequence produced by the 3-stage LFSR with the primitive feedback polynomial $h(X) = 1 + X^2 + X^3$ when the initial state is $\begin{bmatrix} 1\,0\,0 \end{bmatrix}$.

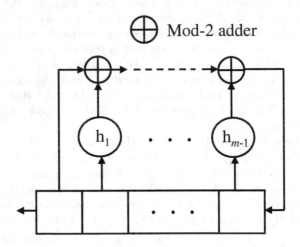

Fig. 1. An m-stage binary linear-feedback shift register.

Suppose that $\mathbf{b} = \begin{bmatrix} b_1\, b_2\, \ldots\, b_{2^m-1} \end{bmatrix}$ is a binary maximum-length sequence. If T denotes the left cyclic shift operator, then the matrix

$$\mathbf{B}_m = \begin{bmatrix} \mathbf{b}_1 \\ \mathbf{b}_2 \\ \ldots \\ \mathbf{b}_m \end{bmatrix} = \begin{bmatrix} \mathbf{b} \\ T\mathbf{b} \\ \ldots \\ T^{m-1}\mathbf{a} \end{bmatrix}$$

has the property that its $2^m - 1$ columns contain every non-zero m-tuple. (Here and hereafter we denote the rows of $\mathbf{B}_m$ as $\mathbf{b}_1, \mathbf{b}_2, \ldots, \mathbf{b}_m$. We will also write $\mathbf{b}_0$ to denote the all-one row of length $2^m - 1$.) For instance with $m = 3$, the maximum-length sequence $\mathbf{b} = \begin{bmatrix} 1\,0\,0\,1\,1\,1\,0 \end{bmatrix}$ specifies the matrix

$$\mathbf{B}_3 = \begin{bmatrix} \mathbf{b}_1 \\ \mathbf{b}_2 \\ \mathbf{b}_3 \end{bmatrix} = \begin{bmatrix} 1\,0\,0\,1\,1\,1\,0 \\ 0\,0\,1\,1\,1\,0\,1 \\ 0\,1\,1\,1\,0\,1\,0 \end{bmatrix}.$$

The row space of the matrix $\mathbf{B}_m$ is a cyclic code as follows from the fact that any linear combination of its rows is equal to the first $2^m - 1$ output digits from the linear-feedback shift register when the first m digits of the linear combination are used as the initial state. Thus the result is either some cyclic shift of $\mathbf{b}$ or (for the trivial linear combination of rows) the all-zero sequence.

It follows from Property 4 of the matrix $\mathbf{A}_m$ that the matrix $\mathbf{B}_m$ is just a column permutation of $\mathbf{A}_m$ when the last column of $\mathbf{A}_m$ (i.e., the all-zero column) is removed before permuting. It follows that the matrix $\widetilde{\mathbf{G}}_m^\mu$, which has as rows $\mathbf{b}_0$, $\mathbf{b}_1$, ... $\mathbf{b}_m$ together with all Hadamard products of $\mathbf{b}_1$, $\mathbf{b}_2$, ... , $\mathbf{b}_m$ taken μ or fewer at a time, is a generator matrix for a linear binary code equivalent to the the μth-order Reed-Muller code $\mathrm{RM}(m, \mu)$ with the last digit of each codeword removed. Moreover, the code generated by $\widetilde{\mathbf{G}}_m^\mu$ is *cyclic*. This can be seen by noting that the left cyclic shift of each codeword is contained in the linear code generated by the matrix whose rows are $T\mathbf{b}_0 = \mathbf{b}_0$, $T\mathbf{b}_1$, ..., $T\mathbf{b}_m$ together with all Hadamard products of $T\mathbf{b}_1$, $T\mathbf{b}_2$, ... , $T\mathbf{b}_m$ taken μ or fewer at a time. But $T\mathbf{b}_1$, ..., $T\mathbf{b}_m$ are all codewords in the code generated by $\widetilde{\mathbf{G}}_m^\mu$ since they are in the row space of its submatrix $\mathbf{B}_m$. Moreover, $T\mathbf{b}_1 T\mathbf{b}_2 = T(\mathbf{b}_1\mathbf{b}_2)$, $T\mathbf{b}_1 T\mathbf{b}_2 T\mathbf{b}_3 = T(\mathbf{b}_1\mathbf{b}_2\mathbf{b}_3)$, etc. up to Hadamard products of order μ. It follows that $T(\mathbf{b}_1\mathbf{b}_2)$, $T(\mathbf{b}_1\mathbf{b}_2\mathbf{b}_3)$, etc. up to Hadamard products of order μ, are al so codewords in the code generated by $\widetilde{\mathbf{G}}_m^\mu$. Hence the code generated by $\widetilde{\mathbf{G}}_m^\mu$ is indeed cyclic. This code will be called the μth-order *cyclic Reed-Muller code* of length $n = 2^m - 1$.

The above description of the cyclic Reed-Muller codes suggests that one can build an "unconventional" encoder for these codes as we now describe. Every codeword in the μth-order cyclic Reed-Muller code of length $n = 2^m - 1$ can be written as

$$u_0\mathbf{b}_0 + u_1\mathbf{b}_1 + \ldots + u_m\mathbf{b}_m + u_{1,2}\mathbf{b}_1\mathbf{b}_2 + \ldots + u_{m-1,m}\mathbf{b}_{m-1}\mathbf{b}_m$$
$$+ \ldots + u_{m-\mu+1,m-\mu+2,\ldots,m}\mathbf{b}_{m-\mu+1}\mathbf{b}_{m-\mu+2}\cdots\mathbf{b}_m$$

and hence u_0, u_1, ... , u_m, $u_{1,2}$, ... , $u_{m-1,m}$, ... , $u_{m-\mu+1,m-\mu+2,\ldots,m}$ can be taken as the k information bits of the cyclic Reed-Muller code. We next note that the sequence seen in the leftmost stage of the linear-feedback shift register that generates the maximal-length sequence $\mathbf{b}$ is just the sequence $\mathbf{b}_1 = \mathbf{b}$. Moreover, $\mathbf{b}_2 = T\mathbf{b}$ is the sequence seen in the second stage of this register, *etc.* It follows that one can determine the codeword corresponding to a particular choice of the information bits just by applying the logical circuitry that creates and adds the necessary Hadamard products of the contents of the various stages of this linear-feedback shift register. Fig. 2 shows the logic corresponding to the choice $u_0 = 1$, $u_1 = 1$, $u_2 = u_3 = 0$, $u_{1,2} = u_{1,3} = 0$, $u_{2,3} = 1$ of the information bits in the second-order cyclic Reed-Muller code of length $n = 7$ whose underlying maximal-length sequence $\mathbf{b} = \begin{bmatrix} 1\,0\,0\,1\,1\,1\,0 \end{bmatrix}$ is generated by the linear-feedback shift register in Fig. 2 when its initial state is $\begin{bmatrix} 1\,0\,0 \end{bmatrix}$. The AND gate produces the Hadamard product of the sequences $\mathbf{b}_2$ and $\mathbf{b}_3$. The constant 1 input to the mod-2 adder is equivalent to having the all-one sequence $\mathbf{b}_0$ at this input to the adder. The resulting codeword is $\mathbf{b}_0 + \mathbf{b}_1 + \mathbf{b}_2\mathbf{b}_3 = \begin{bmatrix} 0\,1\,0\,1\,0\,0\,1 \end{bmatrix}$.

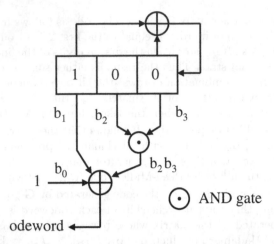

Fig. 2. An unconventional encoder for the 2nd order cyclic Reed-Muller Code of length $n = 7$, illustrated for the choice of information bits $u_0 = 1$, $u_1 = 1$, $u_2 = u_3 = 0$, $u_{1,2} = u_{1,3} = 0$, $u_{2,3} = 1$. The resulting codeword is $b_0 + b_1 + b_2 b_3 = [0\ 1\ 0\ 1\ 0\ 0\ 1]$. (The contents shown are the initial state of the register).

7   Nonlinearly Filtered Maximal-Length Shift Registers

In an additive binary stream cipher, a "running key" is added modulo-2 to the plaintext sequence to produce the ciphertext sequence. The running key is determined by the secret key in a manner that the future of the running key should be difficult to predict by an attacker who has observed some past segment of it. A structure often used to generate the "running-key" in an additive stream cipher is what Rueppel [5] has called a *nonlinearly filtered maximal-length shift register*. In this structure, the initial state of a maximal-length shift register (*i.e.*, a linear-feedback shift register with a primitive feedback polynomial) is determined in some manner by the secret key. The running key is the sequence produced by a fixed nonlinear binary function applied to the stages of this maximal-length linear-feedback shift register. This binary function $f(x_1, x_2, \ldots, x_m)$ is usually characterized by its *algebraic normal form*, namely by its expression in the form

$$f(x_1, x_2, \ldots, x_m) = u_0 + u_1 x_1 + \ldots + u_m x_m + u_{1,2} x_1 x_2 + \ldots + u_{m-1,m} x_{m-1} x_m$$
$$+ \ldots + u_{m-\mu+1,m-\mu+2,\ldots,m} x_{m-\mu+1} x_{m-\mu+2} \cdots x_m$$

where u_0, u_1, $\ldots$, u_m, $u_{1,2}$, $\ldots$, $u_{m-1,m}$, $\ldots$, $u_{m-\mu+1,m-\mu+2,\ldots,m}$ are binary coefficients uniquely determined by the function. The function $f(x_1, x_2, \ldots, x_m)$ is said to have *nonlinear order* μ if μ is the maximum number of variables appearing in a product with a nonzero coefficient in its algebraic normal form. For instance, $f(x_1, x_2, \ldots, x_3) = 1 + x_1 + x_2 x_3$ has nonlinear order 2.

The above formulation of nonlinearly filtered maximal-length shift registers makes the following characterization immediate.

Proposition 4. *The running key produced by a nonlinearly filtered m-stage maximal-length shift register in which the filtering logic has nonlinear order μ is the periodic repetition of a codeword in the μth-order cyclic Reed-Muller Code of length $n = 2^m - 1$ whose underlying maximal-length sequence $\mathbf{b}$ is generated by this maximal-length shift register.*

An important measure of the cryptographic quality of a running key is its *linear complexity, i.e.,* the number of stages in the shortest linear-feedback shift register (cf. Fig. 1) that for some choice of the initial state can generate the sequence as its output. But it is well known (cf. §8.7 in [11]) that every codeword (as well as its periodic repetition) in a cyclic code of dimension k can be generated by a k-stage linear-feedback shift register as in Fig. 1 where the feedback polynomial is the so-called parity-check polynomial of the cyclic code; moreover, the periodic repetitions of some codewords cannot be generated by a shorter linear-feedback shift register. Proposition 4 thus implies the following upper bound on the linear complexity of the running key produced by a nonlinearly filtered m-stage maximal-length shift register.

Corollary 1. *The linear complexity of the running key produced by a nonlinearly filtered m-stage maximal-length shift register in which the filtering logic has nonlinear order μ is at most $\sum_{i=0}^{\mu} \binom{m}{i}$. Equality holds for some choices of the nonlinear-order-μ filtering logic.*

The bound of Corollary 1 is essentially the bound on linear complexity proved by Key [12]. However, Key required $u_0 = 0$ in the algebraic normal form of his filtering function, which implies that the running key will lie in the subcode of the μth-order cyclic Reed-Muller Code generated by the matrix obtained from $\tilde{\mathbf{G}}_m^{\mu}$ by removing the all-one row $\mathbf{b}_0$. This subcode has dimension one less than does the parent code and hence Key's bound on the linear complexity of the running key is $\sum_{i=1}^{\mu} \binom{m}{i}$.

8   Concluding Remarks

We have only touched on a very few of the many places in which the Reed-Muller codes can be shown to underlie an algebraic problem in coding or cryptography. One can easily give many more such examples. For instance, one can show that Meier and Staffelbach's iterative algorithm for attacking an additive stream cipher [13] is an adaptation of Reed's majority-logic decoding algorithm [2] applied to the first-order cyclic Reed-Muller codes. The construction used to obtain the Stiffler codection 5 can be applied to the ranging sequences recently proposed [14] as an improvement for long periods on the Stiffler ranging sequences to obtain a new class of cyclic constant-weight codes that are even simpler to decode than the Stiffler codes. We will not go into these and other examples here, but we do wish to encourage the reader to be on the alert for exploiting the Reed-Muller codes whenever he or she investigates an algebraic problem in coding or cryptography.

References

1. Muller, D. E.: Application of Boolean Algebra to Switching Circuit Design and to Error Detecting. IRE Trans. Electronic Computers **EC-3** (1954) 6-12.
2. Reed, I. S.: A Class of Multiple-Error-Correcting Codes and The Decoding Scheme. IRE Trans. Inform. Th. **IT-4** (1954) 38-49.
3. Massey, J. L., Costello, Jr., D. J., Justesen, J.: Polynomial Weights and Code Constructions. IEEE Trans. Inform. Th. **IT-19** (1973) 101-110.
4. Preparata, F. P.: State-Logic Relations for Autonomous Sequential Networks. IRE Trans. Electronic Computers **EC-13** (1964) 542-548.
5. Rueppel, R. A.: Analysis and Design of Stream Ciphers. Springer-Verlag, Berlin Heidelberg New York (1986).
6. U.S. Department of Commerce, National Bureau of Standards: Data Encryption Standard, FIPS Pub 46 (1977).
7. Matsui, M.: Linear Cryptanalysis Method for DES Cipher. Advances in Cryptology-EUROCRYPT '93 (Ed. T. Helleseth), Lecture Notes in Computer Science, Vol. 765. Springer-Verlag, Berlin Heidelberg New York (1994) 386-397.
8. Stiffler, J. J.: Rapid Acquisition Sequences. IEEE Trans. Inform. Th. **IT-14** (1968) 221-225.
9. Kasami, T., Lin, S., Peterson, W. W.: New Generalizations of the Reed-Muller Codes; Part I: Primitive Codes. IEEE Trans. Inform. Th. **IT-14** (1968) 189-199.
10. Goethals, J. M., Delsarte, P.: On a Class of Majority-Logic Decodable Cyclic Codes. IEEE Trans. Inform. Th. **IT-14** (1968) 182-188.
11. Peterson, W. W., Weldon, Jr., E. J.: Error-Correcting Codes (2nd Ed.). M.I.T. Press, Cambridge, Mass. (1972).
12. Key, E.L.: An Analysis of the Structure and Complexity of Nonlinear Binary Sequence Generators. IEEE Trans. Inform. Th. **IT-22** (1976) 732-736.
13. Meier, W., Staffelbach, O.: Fast Correlation Attacks on Certain Stream Ciphers. J. of Cryptology **1** (1989) 159-176.
14. Ganz, J., Hiltgen, A. P., Massey, J. L.: Fast Acquisition Sequences. Proc. 6th Int. Symp. on Comm. Theory and Appl., Ambleside, England (2001) 471-476.

Self-dual Codes-Theme and Variations

Vera Pless

University of Illinois at Chicago
Department of Mathematics, Statistics,
and Computer Science (MC 249)
851 South Morgan Street
Chicago, Illinois 60607-7045
pless@math.uic.edu

Abstract. Self-dual codes over $GF(2)$, $GF(3)$ and $GF(4)$ were classified from the early 70's until the early 80's. A method for how to do this and efficient descriptions of the codes were developed [3, 4, 17, 20, 21]. New results related to the binary classifications have recently appeared. New formats and classifications have also recently occurred. These events, their relations to the old classifications and open problems will be given.

1 Basics

An $[n, k, d]$ *code* over F_q, for q a prime power, is a k dimensional subspace of F_q^n. The *weight* of a vector is the number of non-zero components it has. The smallest non-zero weight of a code, called the *minimum weight* and denoted by d, if known, is the third parameter in the description of a code. For error-correcting and design purposes, the larger d is the better the code is.

A basis of a code C is called a *generator matrix*. If $x = (x_1, \ldots, x_n)$ and $y = (y_1, \ldots, y_n)$, the inner product of x and y, $x \cdot y = \sum_{i=1}^{n} x_i y_i$. If C is an $[n, k]$ code, the set of all vectors orthogonal to C, with respect to this inner product, is an $[n, n-k]$ code, $C^\perp$, called the *dual* of C. C is called *self-orthogonal*, s.o., if $C \subset C^\perp$. If $C = C^\perp$, C is *self-dual*, s.d. If the field is F_4, the Hermitian inner product, $x \cdot y = \sum_{i=1}^{n} x_i y_i^2$ is used. If C is a s.o. binary code, every vector has even weight as it must be orthogonal to itself. Some s.o. binary codes have all weights divisible by 4. Similarly the weight of a s.o. vector over $GF(3)$ is divisible by 3. It can be shown that vectors in a Hermitian s.o. F_4 code also have even weight. We call a code divisible if the weight of every vector is divisible by a fixed integer $\Delta > 1$. The Gleason-Pierce-Ward Theorem states that divisible $[n, n/2]$ codes exist only for the values of q and Δ mentioned above, except in one other trivial situation, and that the codes are always self-dual except possibly when $q = \Delta = 2$. Many combinatorially interesting self-dual codes with high minimum weights have been found.

Two binary codes are *equivalent* if one can be gotten from the other by a permutation of coordinate indices. One also allows multiplication of coordinates

S. Boztaş and I.E. Shparlinski (Eds.): AAECC-14, LNCS 2227, pp. 13–21, 2001.

by a fixed constant in a non-binary field. When codes of a fixed length are classified, they are classified up to equivalence. If a code is the only code with certain parameters $[n, k, d]$ up to equivalence, then the code is called *unique*.

The *weight distribution* of a code is the number of vectors of any fixed weight in the code. It is well known that the weight distribution of a code is related to the weight distribution of its dual code [18]. When C is self-dual, this gives greater constraints on the possible weight distribution of a self-dual code. In particular bounds on the highest minimum weight are available. When a weight distribution is expressed as a polynomial, in two variables, it is called a *weight enumerator*. The weight enumerators of self-dual codes over F_2, F_3 and F_4 are combinations of weight enumerators of self-dual codes of small length called the *Gleason polynomials*.

As we saw, all vectors in a binary s.d. code have even weights. The binary s.o. codes where all the weights are divisible by 4 are called *doubly-even*, d.e., or Type II if they are self-dual. A s.d. code where it is not necessarily so that all weights are divisible by 4 is of Type I. If a binary code has all weights divisible by 4 it must be s.o. If it contains only even weight vectors, hence called *even*, it need not be s.o.

A famous example of a Type II code is the $[8, 4, 4]$ Hamming code H whose generator matrix $G(H)$ follows

$$G(H) = \begin{pmatrix} 1\,1\,1\,1\,0\,0\,0\,0 \\ 1\,1\,0\,0\,1\,1\,0\,0 \\ 1\,1\,0\,0\,0\,0\,1\,1 \\ 1\,0\,1\,0\,1\,0\,1\,0 \end{pmatrix}.$$

This is the unique $[8, 4, 4]$ binary code. The weight enumerator of a code can be written as a homogeneous polynomial in two variables. For H this is $x^8 + 14x^4y^4 + x^8$. This says that H contains 1 vector of weight 0, 14 vectors of weight 4 and 1 vector of weight 8. The Gleason polynomials for Type II codes are the weight enumerator of H and the weight enumerator of the unique $[24, 12, 8]$ Golay code; $y^{24} + 759x^8y^{16} + 2576x^{12}y^{12} + 759x^{16}y^8 + x^{24}$ [18]. Clearly the length n of any Type II code is divisible by 8. The Gleason polynomials for Type I codes are the weight enumerator of the Hamming $[8, 4, 4]$ code and the weight enumerator of the unique $[2, 1, 2]$ code; $y^2 + x^2$. This is the only case where the weight enumerator of a code which is not s.d. can be a combination of Gleason polynomials. Such a code is an even $[n, n/2]$ code C where C has the same weight enumerator as its dual code $C^\perp$. These codes are called even *formally self-dual*, f.s.d.

Consider the codes C_1 and C_2 with generator matrices G_1 and G_2.

$$G_1 = \begin{pmatrix} 1\,1\,0\,0\,0\,0 \\ 0\,0\,1\,1\,0\,0 \\ 0\,0\,0\,0\,1\,1 \end{pmatrix} \qquad G_2 = \begin{pmatrix} 1\,1\,0\,0\,0\,0 \\ 1\,0\,1\,0\,0\,0 \\ 1\,1\,1\,1\,1\,1 \end{pmatrix}$$

C_1 is s.d., C_2 is f.s.d. The weight enumerator of both is $(y^2+x^2)^3 = y^6 + 3x^2y^4 + 3x^4y^2 + x^6$. Clearly these codes are not equivalent.

If C an $[n_1, k_1, d_1]$ code, and C' an $[n_2, k_2, d_2]$ code, have generator matrices G and G', the *direct sum* of C and C' has generator matrix $\begin{pmatrix} G & 0 \\ 0 & G' \end{pmatrix}$ and is an $[n_1 + n_2, k_1 + k_2, \min(d_1, d_2)]$ code. The code C_1 in the example above is a direct sum of three copies of the $[2, 1, 2]$ code. A code which is not a direct sum is called *indecomposable*. The direct sum of a code C with itself n times is denoted by nC.

2 Binary Self-dual Codes

From the "combinations" of Gleason polynomials one can determine bounds on the minimum weights d of Type I (or even f.s.d.) codes and Type II codes of length n

Type I bound: $d \overset{\leq}{=} 2\lfloor \frac{n}{8} \rfloor + 2$

Type II bound: $d \overset{\leq}{=} 4\lfloor \frac{n}{24} \rfloor + 4$.

A code whose minimum weight meets these bounds is called *extremal*. An extremal code has a unique weight enumerator even though the code need not be unique. Some of these unique weight enumerators have a negative coefficient, hence there can be no extremal codes when this occurs. Many authors have investigated this and for this reason, extremal Type I codes cannot exist for $n = 32, 40, 42, 48, 50, 52$ and $n \overset{\geq}{=} 56$. Other reasons [8] show that extremal Type I or even f.s.d. codes can only exist for $n \overset{\leq}{=} 30$, $n \neq 16$, $n \neq 26$. It can be seen from the classifications [8] of binary s.d. codes and divisibility conditions [25], that the only s.d. codes meeting the Type I bound are of lengths $n = 2, 4, 6, 8, 12, 14, 22$ and 24. For $n = 8$ and 24 the codes are Type II, the unique Hamming and Golay codes. There are also even f.s.d. extremal codes of lengths $n = 10, 18, 20, 28$ and 30 [1, 9, 10, 15]. A $[10, 5, 4]$ code is unique if you assume the all one vector is in it. An $[18, 9, 6]$ code is unique [24]. The weight enumerators for all Type II extremal codes have a negative coefficient for $n > 3928$ and some do for various values of $n \overset{\geq}{=} 3696$. For n divisible by 24 extremal codes are known to exist at lengths 24 and 48. It is an open problem whether the length 48 code is unique and whether there are any extremal codes at lengths 72 and 96 [18]. Extremal Type II codes exist at lengths 8, 16, 24, 32, 40, 48, 56, 64, 80, 88, 104, 136 [3, 5, 14, 17, 20, 23] and maybe at other lengths.

A $t - (v, k, \lambda)$ *design* is a set of v points and a set of blocks of these points of size k such that every t points is contained in exactly λ blocks. The larger t is and the smaller λ is, the more interesting is the design, particularly if $\lambda = 1$.

It is quite surprising that, on occasion, the set of all vectors of a fixed weight in a code form the blocks of some t-design. When this happens we say the vectors "hold" the design. This happens for extremal codes. It is known [18] that the vectors of any fixed weight in a Type II extremal code "hold" a 5-design if $n = 24r$, a 3-design if $n = 24r + 8$, and a 1-design if $n = 24r + 16$. Thus the vectors of weight 4 in the Hamming code form a 3-(8,4,1) design and the vectors of weight 8 in the Golay code form the well-known 5-(24,8,1) design. It is interesting that in both the lengths 10 and 18 even f.s.d. codes, the union of vectors of a fixed weight in the code and their duals "hold" 3-designs [15].

If C is a Type I s.d. code, then the set of d.e. vectors in C forms a subcode, C_0, of codimension 1. From the weight distribution of C, the weight distribution of C_0 can be easily determined, hence the weight distribution of $C_0^{\perp}$. Using this and linear programming, Rains [22] was able to show that there is something wrong with the weight distribution of Type I extremal codes of length > 24 and that such codes of length $n \neq 22 (\mathrm{mod}\ 24)$ must meet the Type II bound.

The following is the new bound for binary self-dual codes.

Theorem 1. *Let C be an $[n, n/2, d]$ binary self-dual code.*

$$\text{Then} \quad d \overset{\leq}{=} 4 \left\lfloor \frac{n}{24} \right\rfloor + 4 \text{ if } n \neq 22 \ (\mathrm{mod}\ 24).$$

$$\text{If } n \equiv 22 (\mathrm{mod}\ 24), \text{ then } d \overset{\leq}{=} 4 \left\lfloor \frac{n}{24} \right\rfloor + 6.$$

Further if C is an extremal code of length $n \equiv 22 (\mathrm{mod}\ 24)$, then C is a "child" of an extremal Type II code C' of length $n + 2$. That is C consists of all vectors in C' with either 11 or 00 in 2 fixed positions with those positions removed.

If C is an extremal code of length divisible by 24, then C is Type II.

The last statement is not true for other lengths n; for example there are Type I [32,16,8] codes [5]. The above argument does not hold for even f.s.d. codes. Hence these codes satisfy the following weak Type I bound

$$d \overset{\leq}{=} 2 \left\lfloor \frac{n}{8} \right\rfloor \quad \text{if } n \overset{\geq}{=} 32.$$

A much better bound is needed. Another open question is whether there can be a f.s.d. code of length $24r$ which satisfies the Type II bound. There are even f.s.d. [32,16,8] codes [8].

The classification of self-dual codes began in the seventies [17]. The method used in the beginning remained essentially the same throughout the succeeding classifications. The classification proceeds from smaller n to larger n and codes are classified up to equivalence. The process begins with the formula for the number of self-dual codes of length n of the type one is trying to classify. The formulas for the numbers of ternary and quaternary self-dual codes are related to the number of totally isotropic subspaces in a finite geometry and are demonstrated in [16] as are the number of Type I binary codes. The number of Type II binary codes are in [23].

The number of self-dual binary codes of even length n is $\displaystyle\prod_{i=1}^{n/2-1} (2^i + 1)$.

If 8 divides n, the number of Type II binary codes is $\displaystyle\prod_{i=0}^{n/2-2} (2^i + 1)$.

The group of a binary code C is the set of all coordinate permutations sending the code onto itself. The group of the Hamming code has order 1,344. If C has length n, then the number of codes equivalent to C is $\frac{n!}{|\mathcal{G}(C)|}$ where $|\mathcal{G}(C)|$ is the order of $\mathcal{G}(C)$ the group of C. Hence to classify self-dual codes of length n, it is necessary to find inequivalent s.d. codes $C_1, \ldots, C_r$ so that

$$\sum_{i=1}^{r} \frac{n!}{|\mathcal{G}(C_i)|} = N$$

the number of self-dual codes of length n. This is called the *mass-formula*. Note that $\frac{8!}{1,344} = 30$, the number of Type II codes, so the Hamming code is the only d.e. code of length 8.

Direct sums can be found directly, indecomposable codes are harder. Some of these can be constructed by "glueing" shorter indecomposable s.o. component codes together. This is a process of taking a direct sum of component codes and adjoining even weight, s.o. vectors consisting of portions in the duals of the component codes. This is called the "glue" space. Often the component codes have minimum weight 4 and then one wants no additional weight 4's in the glue space and also to have it large enough to get, together with the component codes, a self-dual code. The resulting code is then labeled by the labels of the component codes. Let d_6 be the $[6, 2, 4]$ code with the following generator matrix $\begin{pmatrix} 1 1 1 1 0 0 \\ 1 1 0 0 1 1 \end{pmatrix}$. Similarly d_{2m} is a $[2m, m-1, 4]$ code. Thus 4 d_6 is the label of a $[24,12,4]$ Type I code. The component codes have dimension 8 and the "glue" space has dimension 4. Some components have minimum weight greater than 4 [20].

The Type II codes have been classified until length 32 [3] where it was found that there are exactly 5 inequivalent extremal $[32, 16, 8]$ d.e. codes. From this classification, it is possible to determine the number of inequivalent Type I codes of lengths 26 through 30. Both the Type I and Type II codes of lengths 24 and less were previously classified [17, 20]. Following are the numbers of inequivalent self-dual binary codes.

length	2	4	6	8	10	12	14	16	18	20	22	24	26	28	30
number s.d.	1	1	1	2	2	3	4	7	9	16	25	55	103	261	731

length	8	16	24	32
number of Type II	1	2	9	85

It can be shown [3] that there are more than 17,000 inequivalent Type II codes of length 40 so the codes of this and larger lengths are too numerous to classify.

The ternary s.d. codes were classified until length 20 [21] and the F_4 codes until length 16 [4]. The methods were similar to the binary methods.

Actually the extremal codes are the ones of interest. Many have been found but it is difficult to determine all the inequivalent ones of any given length. Much work has been done on length 48. It is also interesting to find *optimal* self-dual codes, that is s.d. or f.s.d. codes of the largest minimum weight which can exist for a specific length n where extremal codes cannot exist. The weight distribution of an optimal code need not be unique.

Extremal F_4 codes exist at lengths 2,4,6,8,10,14,16,18,20,28 and 30. They do not exist for $n = 12, 24, 102, 108, 114, 120, 122$ and all $n \geq 126$ (because of negative coefficients in their weight enumerators). The other lengths $n = 26, 32, \ldots$ are as yet undecided [23].

3   Even Formally Self-dual

There is no mass formula for even, f.s.d. codes as the number of such codes of a fixed length is not known. However the extremal f.s.d. codes were classified until length 20[1, 9, 10, 15]. For $n = 28$, a f.s.d. code is unique as a $[28, 14, 8]$ code is unique. A table of these, optimal f.s.d., and extremal and optimal s.d. codes until length 48 are in [8]. The extremal f.s.d. codes of lengths 10, 18, 20, 28 have higher minimum weights then the extremal or optimal s.d. codes of those lengths. It is not known whether there is a f.s.d. (or any) $[40, 20, 10]$ code. If so, it would have higher minimum weight than the extremal $[40, 20, 8]$ Type II code. Optimal f.s.d. codes of lengths 34, 42 and 44 have higher minimum weights than the optimal Type I codes of those lengths.

The tool used to classify the extremal f.s.d. codes of lengths 10 and 18 were the 3-design property they had [15]. At lengths 14, 20 and 22, [9, 10] the main tool used was the following Balance Principle. Even though this has a simpler analogue for s.d. codes, it was not used as codes of these lengths were completely classified. Let $\{x\}$ denote the code with generator matrix X and k_x the dimension of $\{x\}$.

Theorem 2. *Balance Principle: Let C be a binary code of length $n_1 + n_2$ with $\dim C = \dim C^{\perp}$. Assume A and B (F and J) generate subcodes of $C(C^{\perp})$ of the largest dimension with support under the first n_1 and last n_2 coordinates. Then*

$$G(C) = \begin{pmatrix} A & O \\ O & B \\ D & E \end{pmatrix} \qquad \text{and} \qquad G(C^{\perp}) = \begin{pmatrix} F & O \\ O & J \\ L & M \end{pmatrix},$$

a) $k_D = k_E = k_L = k_M$,
b) $\{A\}^{\perp} = \{F \cup L\}$, $\{B\}^{\perp} = \{J \cup M\}$, $\{F\}^{\perp} = \{A \cup D\}$, $\{J\}^{\perp} = \{B \cup E\}$
   and
c) $n_1 - 2k_A = n_2 - 2k_J$, $n_1 - 2k_F = n_2 - 2k_B$.

Using this theorem, a computer and the desired extremal or optimal minimum weight it is possible to find such codes of length ≤ 22. It gets much harder as the length increases.

4 Additive F_4 Codes

Further classes of s.d. codes related to characteristic 2 and concepts of Type I and Type II have recently been investigated. These codes, however, are not vector spaces over fields. One of these classes are additive F_4 codes which are of interest because of their relation to quantum computing [2]. We can describe an additive code by means of a generator matrix but it should be noted that code words are sums of the vectors in this matrix (not scalar multiples). The inner product used here, called the trace inner product, is defined on components. If the components are equal, it is zero, if unequal and non-zero, it is one. The inner product of two vectors is the sum of the inner products of corresponding components. If two vectors are orthogonal *wrt* the Hermitian inner product, they are also orthogonal *wrt* the trace inner product, not conversely. Hence a s.d. F_4 code gives a s.d. additive code. A code of length n with k generators and minimum weight d is denoted as an $(n, 2^k, d)$ code. It contains 2^k vectors. An F_4 linear $[n, k]$ code is also an additive $(n, 2^{2k})$ code. Its generator matrix as an additive code would consist of its generator matrix as an F_4 code and w times this matrix. The *hexacode* is the very interesting unique linear s.d. [6,3,4] code with the following generator matrix

$$G = \begin{pmatrix} 1 & 0 & 0 & 1 & 1 & 1 \\ 0 & 1 & 0 & 1 & w & \overline{w} \\ 0 & 0 & 1 & 1 & \overline{w} & w \end{pmatrix}.$$

As an additive code its generator matrix would consist of G and wG.

A s.d. additive code can include vectors of odd weight. If so the code is called Type I. If all weights are even, the code is Type II. It can be shown that Type II codes exist only if n is even. A mass formula holds for these codes as do the following bounds on the minimum weights of Type I and Type II codes [13]

$$d_I \overset{\leq}{=} \begin{cases} 2\left\lfloor \frac{n}{6} \right\rfloor + 1 \text{ if } n \equiv 0 \ (\text{mod } 6) \\[2mm] 2\left\lfloor \frac{n}{6} \right\rfloor + 3 \text{ if } n \equiv 5 \ (\text{mod } 6) \\[2mm] 2\left\lfloor \frac{n}{6} \right\rfloor + 2 \text{ otherwise} \end{cases}$$

$$d_{II} \overset{\leq}{=} 2\left\lfloor \frac{n}{6} \right\rfloor + 2.$$

Again a code that meets the bound is called extremal. Hohn [13] classified Type I codes, up to equivalence, until length 7 and Type II codes of length 8. There is an interesting unique extremal $(12, 2^{12}, 6)$ code, the dodecacode [2]. An investigation of extremal and optimal additive s.d. codes was extended to length 16 [12]. This latter used an appropriate variant of the Balance Principle.

Self-dual additive codes of length n are related to binary self-dual codes of length $4n$. If C is an additive s.d. code of length n, the map

$$\rho : 0 \longrightarrow 0000, \quad 1 \longrightarrow 0110, \quad w \longrightarrow 1010, \quad \overline{w} \longrightarrow 1100$$

sends C onto a binary s.o. code $\rho(C)$ of length $4n$. Let $(nd_4)_0$ be the $[4n, n-1]$ binary linear code consisting of all code words of weights divisible by 8 from the $[4n, n]$ code nd_4. If n is even, we can construct the d.e. code $\rho_B(C) = \rho(C) + (nd_4)_0 + e$ where e is the length $4n$ vector

$$0001\ 0001\ \ldots\ 0001 \text{ if } n \equiv 0 \pmod 4$$
$$\text{or } 0001\ 0001\ \ldots\ 1110 \text{ if } n \equiv 2 \pmod 4 .$$

If C is Type I, we can construct a Type I binary code in a similar fashion. The construction of $\rho_B(C)$ explains why a Type II additive code must have even weight n as $\rho_B(C)$ of Type II implies that 8 divides $4n$. Also this map explains the hexacode construction of the Golay code [18].

An interesting open question is the existence of a $(24, 2^{24}, 10)$ additive F_4 Type II code as a $[24,12,10]$ Hermitian self-dual code cannot exist.

5 Z_4 Codes

Another interesting family of self-dual codes are the codes over Z_4, the integers mod 4. Here one can take all linear combinations of a generator matrix. The usual inner product is used. Up to equivalence every s.d. Z_4 code has a generator matrix of the following form

$$G = \begin{pmatrix} A & (I_k + 2B) \\ & 2D \end{pmatrix}$$

where A, B and D are binary matrices [18]. Further $G_1 = [A, I_k]$ generates a doubly-even binary code and $G_2 = \begin{bmatrix} A & I_k \\ D & \end{bmatrix}$ generates $C_2 = C_1^\perp$. In this situation C is said to have type $4^k 2^{n-2k}$ and indeed contains this many codewords. The *Lee weight* of a vector counts 1 and 3 as 1 and 2 as 2. The *Euclidean* weight counts 1 and 3 as 1 and 2 as 4. A s.d. code has Type II if it contains a vector equivalent to the all-one vector and all Euclidean weights are divisible by 8. It is known that Type II codes can only exist at lengths divisible by 8. Other s.d. codes are Type I.

Conway and Sloane [5] classified the self-dual, Z_4 codes until length 9 without a mass formula. Gaborit [11] found such a formula for both Type I and Type II s.d. codes. Using this we [19] classified the Type II codes of length 16. If we take a coordinate position in a Type II code of length n and eliminate all codewords with either 0 or 2 in that position, and then remove that position we get a s.d. code of length $n-1$ called the *shortened* code. Using shortened codes we [7] were able to classify the s.d. Z_4 codes of length 15 or less. A modification of the notation for self-orthogonal, d.e. binary codes was useful in both of these classifications as such codes were used for the matrix G_1 in the canonical form described above. This notation then described the s.d. codes of length 16 and lower. The notation e_8 denotes the Hamming $[8,4,4]$ code and f denotes a code with no weight 4 vectors. For example, the Type II codes of length 8 have the labels $4 - e_8$, $3 - d_8$, $2 - 2d_4$ and $1 - f$. They have types 4^4, $4^3 \cdot 2^2$, $4^2 \cdot 2^4$ and 4.2^6.

References

1. Bachoc, C.: On harmonic weight enumerators of binary codes. DESI (1999) 11–28.
2. Calderbank, A.R., Rains, E.M., Shor, P.W., Sloane, N.J.A.: Quantum error correction via codes over $GF(4)$. IEEE Trans. Inform. Th. IT-44 (1998) 1369–1387.
3. Conway, J.H., Pless, V.: An enumeration of self-dual codes. JCT(A) **28** (1980) 26–53.
4. Conway, J.H., Pless, V., Sloane, N.J.A.: Self-dual codes over $GF(3)$ and $GF(4)$ of length not exceeding 16. IEEE Trans. Inform. Th. IT-25 (1979) 312–322.
5. Conway, J.H., Sloane, N.J.A.: A new upper bound on the minimum distance of self-dual codes. IEEE Trans. Inform. Th. IT-36 (1990) 1319–1333.
6. Conway, J.H., Sloane, N.J.A.: Self-dual codes over the integers modulo 4. JCT(A) **62** (1993) 31–45.
7. Fields, J.E., Gaborit, P., Huffman, W.C., Pless, V.: All Self-Dual Z_4 Codes of Length 15 or Less are Known. IEEE Trans. Inform. Th. **44** (1998) 311–322.
8. Fields, J.E., Gaborit, P., Huffman, W.C., Pless, V.: On the classification of formally self-dual codes. Proceedings of thirty-sixth Allerton Conference, UIUC, (1998) 566–575.
9. Fields, J.E., Gaborit, P., Huffman, W.C., Pless, V.: On the classification of extremal, even formally self-dual codes. DESI **18** (1999) 125–148.
10. Fields, J.E., Gaborit, P., Huffman, W.C., Pless, V.: On the classification of extremal, even, formally self-dual codes of lengths 20 and 22 (preprint).
11. Gaborit, P.: Mass formulas for self-dual codes over Z_4 and $F_q + uF_q$ rings. IEEE Trans. Inform. Th. **42** (1996) 1222–1228.
12. Gaborit, P., Huffman, W.C., Kim, J.L., Pless, V.: On Additive $GF(4)$ Codes. DIMACS Series in Discrete Mathematics and Theoretical Computer Science, **56** (2001) 135–149.
13. Hohn, G.: Self-dual codes over Kleinian four group. (preprint) 1996.
14. Gulliver, T.A., Harada, M.: Classification of extremal double circulant formally self-dual even codes. DESI **11** (1997) 25–35.
15. Kennedy, G.T., Pless, V.: On designs and formally self-dual codes. DESI **4** (1994) 43–55.
16. Pless, V.: On the uniqueness of the Golay codes. JCT(A) **5** (1968) 215–228.
17. Pless, V.: A classification of self-orthogonal codes over $GF(2)$. Discrete Math. **3** (1972) 209–246.
18. Pless, V.: Introduction to the Theory of Error Correcting Codes, third edition, New York, Wiley, 1998.
19. Pless, V., Leon, J.S., Fields, J.: All Z_4 Codes of Type II and Length 16 are known. JCT(A) **78** (1997) 32–50.
20. Pless, V., Sloane, N.J.A.: On the classification and enumeration of self-dual codes. JCT(A) **18** (1975) 313–333.
21. Pless, V., Sloane, N.J.A., Ward, H.N.: Ternary codes of minimum weight 6 and the classification of self-dual codes of length 20. IEEE Trans. Inform. Th. **26** (1980) 305–316.
22. Rains, E.M.: Shadow bounds for self-dual codes. IEEE Trans. Inform. Th. **44** (1998) 134–139.
23. Rains, E.M., Sloane, N.J.A.: Self-dual codes. Handbook of Coding Theory. V. Pless and W.C. Huffman, North Holland, eds., Amsterdam, 1998.
24. Simonis, J.: The [18,9,6] code is unique. Discrete Math. 106/107 (1992) 439–448.
25. Ward, H.N.: A bound for divisible codes. IEEE Trans. Inform. Th. **38** (1992) 191–194.

Design of Differential Space-Time Codes
Using Group Theory

Amin Shokrollahi

Digital Fountain
39141 Civic Center Drive
Fremont, CA 94538, USA
amin@digitalfountain.com

Abstract. It is well-known that multiple transmit and receiving antennas can significantly improve the performance of wireless networks. The design of good modulation schemes for the model of multiple antenna wireless transmission in a fast fading environment (e.g., mobile communication) leads to an interesting packing problem for unitary matrices. Surprisingly, the latter problem is related to certain aspects of finite (and infinite) group theory. In this paper we will give a brief survey of some of these connections.

1 Introduction

Multiple-antenna wireless communication links promise very high data rates with low error probabilities, especially when the channel is known at the receiver [19,5]. The channel model adapted in this scenario is that of multiple-input multiple-output Rayleigh flat fading channel. In cases where the fading coefficients are neither known to the sender nor to the receiver (a case particularly interesting for mobile communication), the design of modulation schemes for the transmission leads to a non-standard packing problem for unitary matrices. Interestingly, some good solutions to the latter problem are intimately related to certain questions from the representation theory of finite and Lie groups. This paper surveys some of these connections and poses several open questions.

A unitary space-time code (constellation) S of rate R is a collection of 2^{RM} unitary $M \times M$-matrices. We will discuss in the next section how these matrices can be used for modulation of information in a mobile wireless setting. As it turns out, in such a transmission the probability of mistaking a matrix V with a matrix W decreases with the quantity

$$\zeta(S) := \frac{1}{2} \min_{A,B \in S, A \neq B} |\det(A - B)|^{1/M}. \tag{1}$$

This quantity is called the *minimum diversity distance* of the code S. The code is said to be *fully diverse* if it has positive minimum diversity distance, i.e., if for all $A, B \in S$ with $A \neq B$ the eigenvalues of $A - B$ are nonzero.

These notions lead to an interesting and packing problem: finding large finite subsets of unitary matrices that have a large minimum diversity distance. More precisely, let

$$A(M, L) := \sup\{\epsilon \mid \exists S \in U(M), \zeta(S) = \epsilon\}. \tag{2}$$

S. Boztaş and I.E. Shparlinski (Eds.): AAECC-14, LNCS 2227, pp. 22–35, 2001.

At a fundamental level the determination of lower and upper bounds for $A(M, L)$ is complicated by the fact that the diversity distance is not a distance function: given two matrices A and B, it is quite possible that $\det(A - B)$ is zero while $A \neq B$.

In this paper we will discuss some upper lower and upper bounds for the function $A(M, L)$ for various M and L. Upper bound results are rather poor and close to trivial. We will give a brief survey on how to relate lower bounds for this function (i.e., construction of good space-time codes) to the theory of finite groups.

The paper is organized as follows. In the next section we will introduce the transmission model and provide formulas for the pairwise probability of error of the maximum likelihood decoder. In Section 3 we derive some upper and lower bounds for the function $A(M, L)$. Section 4 introduces the concept of fixed-point-free groups and reviews some basic facts from representation theory. Section 5 discusses space-time codes of positive minimum diversity distance that form an abelian group under matrix multiplication, while the following section investigates such groups whose order is a power of a prime. Section 7 briefly discusses the classification of fixed-point-free groups. The last two sections deal with construction of codes from compact Lie groups and construction of good codes with zero minimum diversity distance.

For reasons of space, we have omitted in this note the discussion of decoding algorithms for space-time codes. For decoding of space-time group codes we refer the reader to [4,16].

2 Multiple Antenna Space-Time Modulation

2.1 The Rayleigh Flat Fading Channel

Consider a communication link with M transmitter antennas and N receiver antennas operating in a Rayleigh flat-fading environment. The nth receiver antenna responds to the symbol sent on the mth transmitter antenna through statistically independent multiplicative complex-Gaussian fading coefficients h_{mn}. The received signal at the nth antenna is corrupted at time t by additive complex-Gaussian noise w_{tn} that is statistically independent among the receiver antennas and also independent from one symbol to the next. We assume that time is discrete, $t = 0, 1, \ldots$.

It is convenient to group the symbols transmitted over the M antennas in blocks of M channel uses. We use $\tau = 0, 1, \ldots$ to index these blocks; within the τth block, $t = \tau M, \ldots, \tau M + M - 1$. The transmitted signal is written as an $M \times M$ matrix S_τ whose mth column contains the symbols transmitted on the mth antenna as a function of time; equivalently, the rows contain the symbols transmitted on the M antennas at any given time. The fading coefficients h_{mn} are assumed to be constant over these M channel uses.

Similarly, the received signals are organized in $M \times N$ matrices X_τ. Since we have assumed that the fading coefficients are constant within the block of M symbols, the action of the channel is given by the simple matrix equation

$$X_\tau = \sqrt{\rho}\, S_\tau\, H_\tau + W_\tau \quad \text{for} \quad \tau = 0, 1, \ldots. \tag{3}$$

Here $H_\tau = (h_{mn})$ and $W_\tau = (w_{tn})$ are $M \times N$ matrices of independent $\mathcal{CN}(0,1)$-distributed random variables. Because of the power normalization, ρ is the expected SNR at each receiver antenna.

2.2 Known Channel Modulation

We first discuss the case where the receiver knows the channel H_τ. Typically, this could be the case in a fixed wireless environment. We assume that the data to be transmitted is a sequence $z_0, z_1, \ldots$ with $z_\tau \in \{0, \ldots, L-1\}$. Each transmitted matrix occupies M time samples of the channel, implying that transmitting at a rate of R bits per channel use requires a constellation $\mathcal{S} = \{S_1, \ldots, S_L\}$ of $L = 2^{RM}$ unitary signal matrices.

The quality of a constellation $\mathcal{S}$ is determined by the probability of error of mistaking one symbol of $\mathcal{S}$ for another, using the Maximum Likelihood Decoding. In [18,9] it is shown that pairwise probability of mistaking A for B in case of a known channel is given by

$$P(A, B) \leq \frac{1}{2} \prod_{m=1}^{M} \left[1 + \frac{\rho}{4} \sigma_m^2 (A - B) \right]^{-N},$$

where $\sigma_m(A - B)$ is the mth singular value of the $M \times M$ matrix $A - B$ (in some ordering).

2.3 Differential Modulation

When the receiver does not know the channel, or when the channel changes rather rapidly, one can communicate using multiple-antenna differential modulation [8]. Here, we transmit an $M \times M$ unitary matrix that is the product of the previously transmitted matrix and a unitary data matrix taken from the constellation. In other words, $S_\tau = V_{z_\tau} S_{\tau-1}$ for $\tau = 1, 2, \ldots$, with $S_0 = I_M$. In [8] the pairwise probability of error under the Maximum Likelihood Decoding was shown to satisfy

$$P(S_\ell, S_{\ell'}) \leq \frac{1}{2} \prod_{m=1}^{M} \left[1 + \frac{\rho^2}{4(1 + 2\rho)} \sigma_m^2 (S_\ell - S_{\ell'}) \right]^{-N}.$$

At high SNR, the bounds for the known and the unknown channel depend primarily on the product of the nonzero singular values of $S_\ell - S_{\ell'}$. Let $\text{Sing}^*(A)$ denote the multiset of nonzero singular values of the matrix A, counted with multiplicities. The size of $\text{Sing}^*(A)$ is thus equal to the rank $\text{rk}(A)$ of A, if A is a square matrix. Then, for high SNR we may write

$$P(S_\ell, S_{\ell'}) \leq \frac{1}{2} \left(\frac{\alpha}{\rho} \right)^{N\text{rk}(S_\ell - S_{\ell'})} \cdot \prod_{\lambda \in \text{Sing}^*(S_\ell - S_{\ell'})} \lambda^{-2N}, \tag{4}$$

where $\alpha = 4$ for the unknown channel case and $\alpha = 8$ for the known channel case.

A constellation $\mathcal{S}$ is called *fully diverse* if for any two matrices S_ℓ and $S_{\ell'}$ the difference $S_\ell - S_{\ell'}$ has full rank. In other words, $\mathcal{S}$ is fully diverse of $\zeta(\mathcal{S}) \neq 0$. It is clear from (4) that for two fully diverse constellations $\mathcal{S}$ and $\mathcal{S}'$ that have the same number of transmit/receive antennas, the upper bound of the pairwise probability of error is smaller for the constellation with the larger diversity distance.

3 A Packing Problem for Unitary Matrices

Construction of constellations with large minimum diversity distance resembles at first sight the problem of packing points on the sphere, or the problem of constructing good error-correcting codes. However, there is a major difference between the first and the latter two problems: the latter two problems are concerned with packing points with respect to a metric, while the first is not.

Similar to the theory of error-correcting codes, we define the function $A(M, L)$, see 2. Let us first discuss some elementary properties of this function.

Proposition 1. (1) *For all M and L we have* $0 \leq A(M, L) \leq 1$.
(2) *For any K, M, and L we have* $A(M, L)^M A(K, L)^K \leq A(M + K, L)^{M+K}$.
(3) $A(M, L) \leq A(2M, L)$.
(4) $A(1, L) \leq A(M, L)$.
(5) $A(M, L) \geq A(M, L + 1)$.

Proof. (1) First, note that for two unitary matrices A and B we have $|\det(A - B)| = |\det(I - AB^*)|$ where B^* is the Hermitian transpose of B (which is its inverse sine B is unitary), and I denotes the identity matrix of appropriate size. So, we need to show that $0 \leq |\det(I - C)| \leq 2^M$ for any $M \times M$-unitary matrix C. Note that $|\det(I-C)| = \prod_{i=1}^{M} |1-\lambda_i|$, where the λ_i are the eigenvalues of C. Since $|1-\lambda_i| \leq 2$, the result follows.

(2) Let $\mathcal{S} = \{S_1, \ldots, S_L\}$ and $\mathcal{V} = \{V_1, \ldots, V_L\}$ be two space-time codes with minimum diversity distances $A(M, L)$ and $A(K, L)$, respectively. The assertion follows by considering the code $\mathcal{S} \oplus \mathcal{V}$ consisting of the matrices $S_i \oplus V_i$, $i = 1, \ldots, L$, where for two matrices A and B we denote by $A \oplus B$ the block diagonal matrix obtained from A and B. This follows from $|\det(A_1 \oplus B_1 - A_2 \oplus B_2)| = |\det(A_1 - A_2) \det(B_1 - B_2)|$ for any $A_1, A_2 \in \mathcal{S}$ and $B_1, B_2 \in \mathcal{V}$.

(3) Follows directly from (2) by setting $K = M$.
(4) Follows from (2) by induction.
(5) Trivial. $\square$

Proposition 2. *We have the following:*

(1) $A(M, 2) = 1$.
(2) $A(M, 3) = \sqrt{3}/2$.

Proof. (1) In view of the previous proposition, we need to show that $A(M, 2) \geq 1$. This is obvious since for $\mathcal{S} := \{I, -I\}$ we have $\zeta(\mathcal{S}) = 1$.

(2) This is slightly more complicated than the previous one. (For a similar argument, see [12].) First, recall the Frobenius norm $||A||$ of a matrix $A = (a_{ij})$ which is defined as $\sum_{i,j} |a_{ij}|^2$. It is well-known that for a unitary $M \times M$-matrix A this is the same as $\sum_i |\rho_i|^2$ where the ρ_i are the eigenvalues of A. As a result, $||A|| = M$ for all unitary $M \times M$-matrices A. This shows that the group of unitary $M \times M$-matrices can be embedded into the sphere $\mathbb{S}^{2M^2-1}$ by mapping the matrix A to $\sum_{i,j} |a_{ij}|^2/M$. Through this mapping the Frobenius norm is mapped to the Euclidean distance on the sphere. By

the arithmetic-geometric-mean inequality we know that $\prod_{i=1}^{M}|\rho_i|^{1/M} \leq \|A\|/M$. This shows that $|\det(A-B)|^{1/M} \leq \|A-B\|/M$, which implies that $|\det(A-B)|$ is upper bounded by the distance of A and B on the sphere. Hence, the maximum minimum diversity distance of a set of three matrices is upper bounded by (half) the maximum minimum distance of a set of three points on the sphere. The latter is $\sqrt{3}/2$. This shows that $A(M,3) \leq \sqrt{3}/2$. The proof of the lower bound $A(M,3) \geq \sqrt{3}/2$ follows from $A(1,3) = \sqrt{3}/2$ and $A(1,3) \leq A(M,3)$ by Proposition 1(4).   □

We do not know the exact value of $A(M,L)$ for other small L. Note that even though optimal spherical codes of small sizes are known for all dimensions, this may not solve our problem, as the embedding of the group $U(M)$ of unitary $M \times M$-matrices into $\mathbb{S}^{2M^2-1}$ is not surjective.

Open Problem 1 *Calculate $A(M,L)$ for all M and other small values of L, e.g., $L \leq$ 10.*

Let us now concentrate on the behavior of $A(M,L)$ for small M and all L. For $M = 1$, the problem becomes that of packing L points on the one-dimensional circle in the two-dimensional plane so that the minimum Euclidean distance between any two points is maximal. This is achieved by putting the points on a regular L-gon. The minimum distance between two points is equal to $2\sin(\pi/L)$, and so the diversity distance of this set equals $\sin(\pi/L)$. Hence, we have

$$A(1,L) = \sin(\pi/L).$$

The case $M = 2$ is more interesting and far less obvious. We start with a parameterization of all unitary 2×2-matrices. It is easily seen that for any such matrix A there exists a point (a_0, a_1, b_0, b_1) on the three dimensional unit sphere $\mathbb{S}^3$, and an angle ϕ mod π such that A equals

$$A = e^{i\phi}\begin{pmatrix} a & b \\ -\overline{b} & \overline{a} \end{pmatrix},$$

where $a = a_0 + ia_1$ and $b = b_0 + ib_1$, and $\overline{z}$ is the complex conjugate of z. Let us concentrate first on the case $\phi = 0$, i.e., $A \in SU(2)$[1] The first application of such matrices to multiple antenna code design was given by Alamouti [1] for the known channel modulation and by Tarokh and Jafarkhani [11] for the unknown channel modulation. The following lemma is from [16].

Lemma 1. *The diversity distance on $SU(2)$ is a metric and $SU(2)$ together with this metric is isomorphic to $\mathbb{S}^3$ with half the standard Euclidean metric. As a result, any packing of $\mathbb{S}^3$ with L points and minimum distance d results in a signal constellation $\mathcal{V}$ in $SU(2)$ with diversity distance $d/2$.*

Proof. First, given the parameterization above, the matrices in $SU(2)$ correspond in a one-to-one manner to the elements of $\mathbb{S}^3$. Let A and B be matrices in $SU(2)$, and let P, Q be points on $\mathbb{S}^3$ corresponding to A and B, respectively. Then one easily verifies that $|\det(A-B)|$ is the square of the Euclidean distance between P and Q. Hence, the first assertion of the lemma follows. The second assertion follows from the first.

---

[1] $U(M)$ and $SU(M)$ denote the group of unitary $M \times M$-matrices and the groups of unitary $M \times M$-matrices of determinant one, respectively.

The lemma shows that $A(2, L) \geq B(2, L)$ where $B(2, L)$ is half the minimum distance of the best spherical code with L points on $\mathbb{S}^3$. This number has been studied quite extensively, and very good upper and lower bounds are known for it [6].

It is not clear whether the best diversity distance for $\mathrm{SU}(2)$ is strictly smaller than the best diversity distance for $\mathrm{U}(2)$. In other words, it is not clear whether the additional parameter ϕ for matrices in $\mathrm{U}(2)$ leads to sets with strictly better diversity distance.

Open Problem 2 *Is $A(2, L) = B(2, L)$? In other words, is the best diversity distance for constellations in $\mathrm{SU}(2)$ the same as the best diversity distance for constellations in $\mathrm{U}(2)$?*

Except the method used in the proof of Proposition 2(2), there is not a whole lot known about upper bounds for the function $A(M, L)$ for $M \geq 3$. Therefore, we will concentrate on lower bounds, i.e., construction of space-time codes $\mathcal{V}$ with large minimum diversity distance. We will accomplish this by using space-time codes that form a group under matrix multiplication. The proper setting to discuss such groups is that of the representation theory of finite groups, which we shall briefly outline in the next section.

4   Fixed-Point-Free Groups

Suppose that the constellation $\mathcal{S}$ of unitary $M \times M$-matrices forms a group under matrix multiplication. Then the diversity distance of $\mathcal{S}$ can be described more easily as

$$\zeta(\mathcal{S}) = \frac{1}{2} \min_{S \neq I} |\det(I - S)|^{1/M}, \tag{5}$$

where I is the $M \times M$-identity matrix. Indeed, $|\det(S - S')| = |\det(S)||\det(I - S^{-1}S')| = |\det(I - S^{-1}S')|$, and since $S^{-1}S'$ belongs to the group, the assertion follows.

The question which finite groups of unitary matrices have a large diversity distance should be preceded by which of these groups have nonzero diversity distance. The latter question is answered using tools from group theory and from the representation theory of finite groups.

A representation of degree M of a finite group G is a homomorphism of G into the group $\mathrm{U}(M)$ of unitary $M \times M$-matrices. Two representations Δ and Δ' are called *equivalent* if there is a unitary matrix T such that $\Delta(g) = T\Delta'(g)T^*$ for all elements g of G. The direct sum of two representations Δ and Δ' is the representation $\Delta \oplus \Delta'$ whose value at a group element g is the block diagonal matrix having the block diagonal entries $\Delta(g)$ and $\Delta'(g)$. A representation is called *irreducible* if it is not equivalent to a direct sum of two representations.

Given (5), a finite group $\mathcal{S}$ of unitary $M \times M$-matrices has nonzero diversity distance if and only if no non-identity element of $\mathcal{S}$ has an eigenvalue 1. Inspired by this, we call a representation Δ of a finite group G *fixed-point-free* (fpf) if the group $\Delta(G)$ has full diversity. A group G is called fpf if it has an irreducible fpf representation.

5 Abelian Groups

Use of space-time codes that form an abelian group was first suggested by Hochwald and Sweldens [8]. It is easy to see that cyclic groups are fpf: any primitive character of the group provides a fpf representation. Moreover, any abelian group that is fpf is necessarily cyclic: if an abelian group is not cyclic, then any character of that group has a nontrivial kernel, hence cannot be fpf.

Suppose we want to construct an fpf-cyclic group whose elements are $M \times M$-matrices. The most general form of such a group is given by

$$
\mathcal{S}(u_2, \ldots, u_M) := \left\{ \begin{pmatrix} \eta^k & 0 & \cdots & 0 \\ 0 & \eta^{u_2} & \cdots & 0 \\ \vdots & \vdots & \ddots & \vdots \\ 0 & 0 & \cdots & \eta^{ku_M} \end{pmatrix} : 0 \le k < L \right\},
$$

where the u_i are pairwise different and co-prime to L, and $\eta = e^{2\pi i/L}$. The obvious question that arises is that of choosing the u_i in such a way that $\zeta(\mathcal{S}(u_2, \ldots, u_M))$ is maximized. This is a very difficult question. The case $M = 2$ was investigated in [15], where it was conjectured that if L is the nth Fibonacci number, then it is best to choose u to be the $(n-1)$st Fibonacci number.

6 Fpf p-Groups

If a group is fpf, then so are all its subgroups. In particular, the p-Sylow subgroups of such a group are fpf as well. However, these can only be of very restricted types, as the following theorem of Burnside [3] shows.

Theorem 1. *Let G be a fpf p-group. If p is odd, then G is a cyclic group. If p is even, then G is either a cyclic group, or a generalized Quaternion group given as $\langle \sigma, \tau \mid \sigma^{2^n} = 1, \tau^2 = \sigma^{2^{n-1}}, \sigma^\tau = \sigma^{-1} \rangle$, where $\sigma^\tau = \tau \sigma \tau^{-1}$. Conversely, all these groups are fpf.*

We remark that the classification of fpf 2-groups was independently rediscovered by Hughes [10] in the context of multiple antenna communication.

The proof of this theorem requires some simple facts from representation theory which we did not elaborate on in Section 4. Readers not familiar with representation theory can skip the proof.

Proof. We only prove the assertion for the case of odd p; the case of even p can be handled similarly. Let $|G| = p^n$. We use induction on n. The case $n = 1$ is obvious, since here G can only be cyclic. Assume now that we have proved the assertion for all groups of size p^{n-1}. Let G be a fpf group of size p^n. By Sylow's theorem, G has a normal subgroup N of size p^{n-1} which is cyclic by the induction hypothesis. Denote by σ a generator of this group, and suppose that τN generates G/N. Then $\tau^p \in \langle \sigma \rangle$, say $\tau^p = \sigma^k$, and $\sigma^\tau = \sigma^\ell$ for some ℓ since N is normal.

Suppose that Δ is a fpf representation of G. Then $\Delta \downarrow N$, the restriction of Δ to N is also fpf. But $\Delta \downarrow N$ is equivalent to the direct sum of characters of N which necessarily have to be primitive. All irreducible representations of G are obtained as inductions/extensions of representations (i.e., characters) of N, and their degree is either 1 (in case of an extension) and p (in case of an induction). This is because the degrees of irreducible representations of G are powers of p and N is cyclic. Suppose G had an irreducible fpf representation Δ of degree p. Δ would be the induction of a primitive character χ of N: $\Delta = \chi \uparrow G$. The value $\chi(\sigma)$ would then be a primitive p^{n-1}st root of unity, which we denote by η. This shows that Δ is equivalent to the representation R given by

$$
R(\sigma) = \begin{pmatrix} \eta & 0 & \cdots & 0 \\ 0 & \eta^{\ell} & \cdots & 0 \\ \vdots & \vdots & \ddots & \vdots \\ 0 & 0 & \cdots & \eta^{\ell^{p-1}} \end{pmatrix}, \qquad R(\tau) = \begin{pmatrix} 0 & 1 & 0 & \cdots & 0 \\ 0 & 0 & 1 & \cdots & 0 \\ \vdots & \vdots & \vdots & \ddots & \vdots \\ 0 & 0 & 0 & \cdots & 1 \\ \eta^k & 0 & 0 & \cdots & 0 \end{pmatrix}.
$$

All elements of G are of the form $\sigma^s \tau^t$ where $0 \le s < p^{n-1}$ and $0 \le t < p$. For $t \not\equiv 0 \bmod p$ the representation R evaluated at such an element is

$$
R(\sigma^s \tau^t) = \left(\begin{array}{cccc|cccc} 0 & 0 & \cdots & 0 & \eta^s & 0 & \cdots & 0 \\ 0 & 0 & \cdots & 0 & 0 & \eta^{s\ell} & \cdots & 0 \\ \vdots & \vdots & \ddots & \vdots & \vdots & \vdots & \ddots & \vdots \\ 0 & 0 & \cdots & 0 & 0 & 0 & \cdots & \eta^{s\ell^{p-t-1}} \\ \hline \eta^{s\ell^{p-t}+k} & 0 & \cdots & 0 & 0 & 0 & \cdots & 0 \\ 0 & \eta^{s\ell^{p-t+1}+k} & \cdots & 0 & 0 & 0 & \cdots & 0 \\ \vdots & \vdots & \ddots & \vdots & \vdots & \vdots & \ddots & \vdots \\ 0 & 0 & \cdots & \eta^{s\ell^{p-1}+k} & 0 & 0 & \cdots & 0 \end{array} \right).
$$

It is not hard to see that $\det(I - R(\sigma^s \tau^t)) = 1 - \eta^{s \sum_{i=0}^{p-1} \ell^i + tk}$. If we can find an integer $t \not\equiv 0 \bmod p$ and another integer s such that the exponent of η is congruent to 0 modulo p^{n-1}, then this proves that $\det(I - R(\sigma^s \tau^t)) = 0$, and shows that G is not fpf. We will show that such a t exists for odd p. To this end, it suffices to prove that if p^r divides $\sum_{i=0}^{p-1} \ell^i$, then p^r also divides k.

First note that $\ell \not\equiv 1 \bmod p^{n-1}$ since we have assumed that G has an irreducible representation of degree p, and hence G is not abelian. Since $\tau^p = \sigma^k$, we have $\sigma^{\tau^p} = \sigma$. But $\sigma^{\tau^p} = \sigma^{\ell^p}$, so $\ell^p \equiv 1 \bmod p^{n-1}$. Since p is odd, the multiplicative group of $\mathbb{Z}/(p^{n-1}\mathbb{Z})$ is cyclic and generated by an element, say ω. This shows that ℓ is in the subgroup generated by $\omega^{p^{n-3}(p-1)}$, which shows that $\ell \equiv 1 \bmod p^{n-2}$. As a result, p divides, and p^2 does not divide $(\ell^p - 1)/(\ell - 1)$.

To finish the proof, it suffices to show that p divides k. Consider $R(\tau^p)$. This is a diagonal matrix with all diagonal entries equal to η^k. On the other hand, $R(\tau^p) = R(\sigma^k)$. The latter is a diagonal matrix with diagonal entries $\eta^k, \eta^{k\ell}, \ldots, \eta^{k\ell^{p-1}}$. So, $\eta^{k\ell} = \eta^k$, which implies $k(\ell - 1) \equiv 0 \bmod p^{n-1}$. Since $\ell \equiv 1 \bmod p^{n-2}$, but $\ell \not\equiv 1 \bmod p^{n-1}$, this implies that p divides k and finishes the proof. $\qquad \square$

For the rest of this section, let us prove that the generalized Quaternion groups are indeed fpf. Let $G := \langle \sigma, \tau \mid \sigma^{2^n} = 1, \tau^2 = \sigma^{2^{n-1}}, \sigma^\tau = \sigma^{-1} \rangle$ be such a group. G has a normal subgroup $H = \langle \sigma \rangle$ of index 2. We induce an fpf representation of H, i.e, a primitive character of H to G. Call the corresponding representation Δ. Then we have (up to equivalence)

$$\Delta(\sigma) = \begin{pmatrix} \eta & 0 \\ 0 & \eta^{-1} \end{pmatrix}, \qquad \Delta(\tau) = \begin{pmatrix} 0 & 1 \\ -1 & 0 \end{pmatrix},$$

where η is a primitive 2^{n-1}st root of unity. It is easy to see that this representation is indeed fpf. Consider $\Delta(\sigma^s \tau^t)$. If $t \equiv 0 \bmod 2$, there is nothing to prove, so let us assume w.l.o.g. that $t = 1$. We have

$$I - \Delta(\sigma^s \tau) = \begin{pmatrix} 1 & \eta^s \\ -\eta^{-s} & 1 \end{pmatrix}.$$

We note in passing that this is (up to scaling) an example of an orthogonal design [1,11], or, (again up to scaling) an element of $SU(2)$ (see Section 3). The determinant of this matrix is $1 + 1 = 2$. This shows that G is fpf.

7   Classification of fpf Groups

The classification of all fpf groups was carried out in large parts by Zassenhaus [20] in the context of classification of finite near-fields. The results were partly rediscovered and partly completed in the context of communication in [16]

One key to the classification of fpf group is the trivial observation that subgroups of fpf groups are themselves fpf. Hence, all p-Sylow subgroups of fpf-groups must be of the forms given in Theorem 1. In particular if G is an fpf-group of odd order, then all p-Sylow subgroups for odd p must be cyclic. It turns out that all such groups have a simple shape. Given a pair of integers (m, r) with r co-prime to m, we implicitly define n to be the order of r modulo m, $r_0 = \gcd(r - 1, m)$, and $t = m/r_0$. We call the pair (m, r) admissible, if $\gcd(n, t) = 1$, and all prime divisors of n divide r_0. Then we have the following theorem [20,16].

Theorem 2. *Suppose that G is a fpf group of odd order. Then there exists an admissible pair (m, r) such that*

$$G \simeq G_{m,r} := \langle \sigma, \tau \mid \sigma^m = 1, \tau^n = \sigma^t, \sigma^\tau = \sigma^r \rangle,$$

where $t = m/r_0$, $r_0 = \gcd(r - 1, m)$, and n is the order of r modulo m. Conversely, all groups $G_{m,r}$ with admissible (m, r) are fpf.

The reader may want to look at [16] for the full-diversity constellations derived from this theorem. The smallest non-cyclic fpf group of odd order is $G_{21,4}$ which has 63 elements. Its corresponding constellation is given by the elements $A^s B^t$, $0 \le s < 21$, $0 \le t < 3$, where

$$A = \begin{pmatrix} \eta & 0 & 0 \\ 0 & \eta^4 & 0 \\ 0 & 0 & \eta^{16} \end{pmatrix}, \qquad B = \begin{pmatrix} 0 & 1 & 0 \\ 0 & 0 & 1 \\ \eta^7 & 0 & 0 \end{pmatrix},$$

where $\eta = e^{2\pi i/21}$. The diversity distance of this constellation is roughly 0.3851.

The classification of fpf groups of even order is a lot more elaborate [20,16] and will not be discussed here for lack of space. It is divided into the simpler part of solvable fpf groups and the more difficult part of non-solvable fpf groups. One of the most interesting groups belonging to the second category is $SL(2, \mathbb{F}_5)$, the group of 2×2-matrices over $\mathbb{F}_5$ with determinant 1. Both irreducible 2-dimensional representations of this group are fpf and give rise to the constellation generated by the two matrices

$$P = \frac{1}{\sqrt{5}} \begin{pmatrix} \eta^2 - \eta^3 & \eta - \eta^4 \\ \eta - \eta^4 & \eta^3 - \eta^2 \end{pmatrix}, \quad Q = \frac{1}{\sqrt{5}} \begin{pmatrix} \eta - \eta^2 & \eta^2 - 1 \\ 1 - \eta^3 & \eta^4 - \eta^3 \end{pmatrix},$$

where $\eta = e^{2\pi i/5}$. The diversity distance of this constellation is $\frac{1}{2}\sqrt{(3 - \sqrt{5})/2} \sim 0.3090$. This is in fact a constellation in $SU(2)$, and hence can be identified with a set of points on $\mathbb{S}^3$. The corresponding polyhedron is a regular polyhedron with 120 vertices. In dimension 4 there are only two such polyhedra. The other one is the dual of the one described here. For more information and historical remarks on this "120-cell," the reader is referred to [17]. Note that if $A(2, L) = B(2, L)$ (see Open Problem 2), then $A(2, 120) = \frac{1}{2}\sqrt{(3 - \sqrt{5})/2}$.

The results of simulations of $SL(2, \mathbb{F}_5)$ as a space-time code can be found in Figure 1.

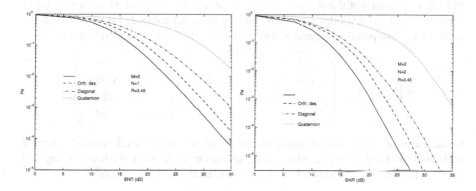

Fig. 1. Comparison of $SL(2, \mathbb{F}_5)$ with orthogonal designs, diagonal codes, and the Quaternion group. The left picture is for one receiver antenna, while the second is for two receiver antennas. The code is designed for $M = 2$ transmit antennas, and has $L = 120$ elements ($R \approx 3.45$).

Even though we now know all fpf finite groups, it may still be that there are full-diversity space-time codes that are not groups themselves, but generate a finite group. This question is open and an answer to that would be very interesting. Here we are not looking for a full classification, as this seems rather difficult.

Open Problem 3 *Find finite sets of unitary matrices with large diversity distance that generate a finite group.*

For some attempts in the direction of answering this question, see [16].

8 Codes from Representations of Compact Lie Groups

Since the classification of fpf finite groups is complete, the next question to ask is that of classification of infinite fpf groups. Here, we require that the group has an irreducible unitary finite-dimensional representation that is fpf. In this direction, Hassibi and Khorrami showed that the only fpf Lie groups are $U(1)$ and $SU(2)$ [7]. But what about other types of groups groups? One such type of groups is given by the group of nonzero elements of a division algebra. In fact, if we can embed a division algebra over $\mathbb{Q}$, say, into $\mathbb{C}^{M \times M}$ for some M such that an infinite subgroup H of the unit group is embedded into $U(M)$, then the image of H will be fpf.

Open Problem 4 *Is it possible to embed a division algebra over $\mathbb{Q}$ into $\mathbb{C}^{M \times M}$ for some M such that an infinite subgroup of the multiplicative group is embedded in $U(M)$?*

An example of such a situation (when $\mathbb{Q}$ is replaced by $\mathbb{R}$) is given by the Hamiltonian Quaternions. Here the elements of norm 1 are mapped into $SU(2)$, and this another reason why $SU(2)$ is fpf.

Even though the theorem of Hassibi and Khorrami rules out the existence of fpf Lie groups other than the obvious ones, the question is open whether there are subsets of Lie groups with high diversity. Compact Lie groups seem to be ideal candidates for such an investigation as all their irreducible representations are unitary and finite dimensional [2]. In [14] we proved that the unique 4-dimensional representation of $SU(2)$ gives rise to very good space-time codes if evaluated at a good spherical code on $\mathbb{S}^3$ in which the angle of any two points is separated by 120 degrees. The representation is given by

$$R \begin{pmatrix} a & b \\ -\overline{b} & \overline{a} \end{pmatrix} = \begin{pmatrix} a^3 & \sqrt{3}a^2 b & \sqrt{3}ab^2 & b^3 \\ -\sqrt{3}a^2\overline{b} & a(|a|^2 - 2|b|^2) & b(2|a|^2 - |b|^2) & \sqrt{3}\overline{a}b^2 \\ \sqrt{3}a\overline{b}^2 & \overline{b}(|b|^2 - 2|a|^2) & \overline{a}(2|b|^2 - |a|^2) & \sqrt{3}\overline{b}\overline{a}^2 \\ -\overline{b}^3 & \sqrt{3}\,\overline{b}^2 a & -\sqrt{3}\,\overline{b}\overline{a}^2 & \overline{a}^3 \end{pmatrix}$$

Good space-time codes in 4 dimensions can be constructed if we take a good spherical code on $\mathbb{S}^3$ and restrict it to a subset of the sphere on which the angle between any two points is bounded away from 120 degrees. In particular, [14] proves that

$$A(4, L) \geq \frac{\sqrt[4]{4\varepsilon(9\varepsilon - 12\varepsilon^2 + 4\varepsilon^3)}}{2}$$

where $\varepsilon = 2A(2, L/0.135)$. The right hand side of the above inequality is, for large L, smaller than $A(2, L)$, which shows that the construction above is worse than the trivial bound. However, this is due to the construction of the spherical code given in [14]. Better constructions will lead to better results.

Open Problem 5 *Are there other ways to use representations of compact Lie groups to derive good space-time codes in other dimensions?*

We will discuss in the next section finite space-time group codes that are not fully diverse. The above construction is an example of an infinite group that is not fully diverse, and in which any non-identity matrix has rank at least 2.

It is relatively easy to compute the eigenvalues of irreducible representations of the groups $SU(M)$, though, for larger M, it is not completely clear how to sample from the corresponding manifold to obtain good space-time codes. As an example, we mention the case of 6-dimensional representation of $SU(3)$. This is obtained by considering the vector space of homogeneous polynomials in three variables of degree 2 as an $SU(3)$-module. A matrix with eigenvalues $\eta, \mu, \eta^{-1}\mu^{-1}$ is mapped to a matrix with eigenvalues $\eta^2, \mu^2, \eta^{-2}\mu^{-2}, \eta\mu, \mu^{-1}, \eta^{-1}$. (This is obtained by considering the action on the maximal torus of $SU(3)$.) As a result, any good space-time code in $SU(3)$ none of whose matrices has eigenvalue ± 1 gives rise to a full-diversity space-time code in $SU(6)$.

9 Finite Groups That Are Not Fully Diverse

A look at the upper bounds on the pairwise probability of error reveals that space-time codes that are not fully diverse could also be used for multiple antenna transmission, as is seen from (4). It is easy to see that for two $M \times M$-matrices A and B the probability of mistaking A for B is given by [13]

$$P(A, B) \leq \frac{1}{2}\left(\frac{\alpha}{\rho}\right)^{N \deg(\tilde{T}(x))} \cdot |\tilde{T}(1)|^{-2N}, \tag{6}$$

where N is the number of receiving antennas and $\tilde{T}(x)$ is the largest factor of the characteristic polynomial of AB^* which is not divisible by $x - 1$. A space-time group code $\mathcal{S}$ contains AB^* for any $A, B \in \mathcal{S}$. Further, the characteristic polynomial AB^* is invariant under conjugation in the group. Hence, the different values of the pairwise

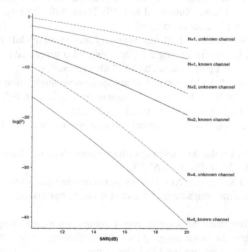

Fig. 2. Block-error rate performance of the group $SL(2, \mathbb{F}_{17})$ versus a good diagonal constellation for $M = 8$ transmitter antennas and $N = 1$ receiver antenna. Both constellations have $L = 4896$ unitary matrices ($R \approx 1.53$).

probability of error can be computed once the characteristic polynomials of the conjugacy classes of S are found. These can be computed from the character table of S, as is shown in [13].

The formula (6) reveals that a non fpf group S performs better if for the rank of $I - A$ is larger for all nontrivial elements of S. An example is given by the group $SL(2, \mathbb{F}_{17})$ of size 4896. Any irreducible 8-dimensional representation of this group has the property that all elements other than those in one conjugacy class have full rank; elements in the distinguished conjugacy class has rank 6. This representation was constructed in [13]. A simulation of this space-time code and a comparison to a diagonal code of the same rate is given in Figure 2.

References

1. S. M. Alamouti. A simple transmitter diversity scheme for wireless communications. *IEEE J. Sel. Area Comm.*, pages 1451–1458, 1998.
2. Th. Bröcker and T. tom Dieck. *Representations of Compact Lie Groups.* Number 98 in Graduate Texts in Mathematics. Springer Verlag, New York, 1985.
3. W. Burnside. On a general property of finite irreducible groups of linear substitutions. *Messenger of Mathematics*, 35:51–55, 1905.
4. K. L. Clarkson, W. Sweldens, and A. Zheng. Fast multiple antenna differential decoding. Technical report, Bell Laboratories, Lucent Technologies, October 1999. Downloadable from http://mars.bell-labs.com.
5. G. J. Foschini. Layered space-time architecture for wireless communication in a fading environment when using multi-element antennas. *Bell Labs. Tech. J.*, 1(2):41–59, 1996.
6. J. Hamkins and K. Zeger. Asymptotically dense spherical codes. *IEEE Trans. Info. Theory*, 43:1774–1798, 1997.
7. B. Hassibi and M. Khorrami. Fully-diverse multiple-antenna signal constellations and fixed-point-free Lie groups. Submitted to IEEE Trans. Info. Theory. Downloadable from http://mars.bell-labs.com, 2000.
8. B. Hochwald and W. Sweldens. Differential unitary space time modulation. Technical report, Bell Laboratories, Lucent Technologies, Mar. 1999. Also submitted to *IEEE Trans. Comm.* Downloadable from http://mars.bell-labs.com.
9. B. M. Hochwald and T. L. Marzetta. Unitary space-time modulation for multiple-antenna communication in Rayleigh flat-fading. *to appear in IEEE Trans. Info. Theory*, Mar. 2000. Downloadable from http://mars.bell-labs.com.
10. B. Hughes. Optimal space-time constellations from groups. *Submitted to IEEE Trans. Info. Theory*, 2000.
11. H. Jafarkhani and V. Tarokh. Multiple transmit antenna differential detection from generalized orthogonal designs. 2000. Preprint, AT&T.
12. Xue-Bin Liang and Xiang-Gen Xia. Unitary signal constellations for differential space-time modulation with two transmit antennas: Parametric codes, optimal designs, and bounds. 2001. Preprint.
13. A. Shokrollahi. Computing the performance of unitary space-time group constellations from their character table. In *Proceedings of ISIT 2001, Washington DC*, 2001. Preprint downloadable from http://mars.bell-labs.com.
14. A. Shokrollahi. Design of unitary space-time codes from representations of $SU(2)$. In *Proceedings of ISIT 2001, Washington DC*, 2001. Preprint downloadable from http://mars.bell-labs.com.

15. A. Shokrollahi. A note on 2-dimensional diagonal space-time codes. In *Proceedings of ISIT 2001, Washington DC*, 2001. Preprint downloadable from http://mars.bell-labs.com.

16. A. Shokrollahi, B. Hassibi, B. Hochwald, and W. Sweldens. Representation theory for high-rate multiple-antenna code design. *IEEE Trans. Inform. Theory*, 2001. To appear. Preprint downloadable from http://mars.bell-labs.com.

17. J. Stilwell. The story of the 120-cell. *Notices of the AMS*, 48:17–24, 2001.

18. V. Tarokh, N. Seshadri, and A. R. Calderbank. Space-time codes for high data rate wireless communication: Performance criterion and code construction. *IEEE Trans. Info. Theory*, 44:744–765, 1998.

19. I. E. Telatar. Capacity of multi-antenna gaussian channels. Technical report, Bell Laboratories, Lucent Technologies, 1995.

20. H. Zassenhaus. Über endliche Fastkörper. *Abh. Math. Sem. Hamburg*, 11:187–220, 1936.

Ideal Error-Correcting Codes: Unifying Algebraic and Number-Theoretic Algorithms

Madhu Sudan [*]

MIT Laboratory for Computer Science,
200 Technology Square,
Cambridge, MA 02139, USA.
madhu@mit.edu,
http://theory.lcs.mit.edu/~madhu

Abstract. Over the past five years a number of algorithms decoding some well-studied error-correcting codes far beyond their "error-correcting radii" have been developed. These algorithms, usually termed as list-decoding algorithms, originated with a list-decoder for Reed-Solomon codes [36,17], and were soon extended to decoders for Algebraic Geometry codes [33,17] and also to some number-theoretic codes [12,6,16]. In addition to their enhanced decoding capability, these algorithms enjoy the benefit of being conceptually simple, fairly general [16], and are capable of exploiting soft-decision information in algebraic decoding [24]. This article surveys these algorithms and highlights some of these features.

1 Introduction

List-decoding was introduced in the late fifties by Elias [7] and Wozencraft [38]. Under this model, a decoder is allowed to output a *list* of possible codewords that a corrupted received word may correspond to. Decoding is considered successful if the transmitted word is included in this list of received words.

While the initial model was introduced to refine the study of probabilistic channels, it has slowly developed into a tool for improving our understanding of error-correction even in adversarial models of error. Strong combinatorial results are known that bound the "list-decoding radius" of an error-correcting code as a function of its rate and its distance (See [5,8,40] for some of the earlier results, and [13,15,21] for some recent progress.) However till the late 90's no non-trivial algorithms were developed to perform efficient list-decoding. In [36], the author gave an algorithm to list-decode Reed-Solomon codes. This was shortly followed up by an algorithm by Shokrollahi and Wasserman [33] to decode algebraic-geometry codes. Subsequently the algorithms have been extended to decode many families of codes. Furthermore the efficiency of the original algorithms has been vastly improved and many applications have been found for this concept. In this paper we describe the basic ideas behind the decoding algorithms. Our

[*] Parts of this work were supported by NSF Grant CCR 9875511, NSF Grant CCR 9912342, and an Alfred P. Sloan Foundation Fellowship.

S. Boztaş and I.E. Shparlinski (Eds.): AAECC-14, LNCS 2227, pp. 36–45, 2001.
© Springer-Verlag Berlin Heidelberg 2001

focus is mostly on the simplicity of these algorithms and not so much on their performance or uses.

2   Reed-Solomon Decoding

Let $\mathbb{F}_q$ denote a field of size q and let $\mathbb{F}_d^k[x]$ denote the vector space of polynomial of degree at most k over $\mathbb{F}_q$. Recall that the Generalized Reed Solomon code of dimension k, is specified by distinct $x_1, \ldots, x_n \in \mathbb{F}_q$ and consists of the evaluations of all polynomials p of degree at most k at the points $x_1, \ldots, x_n$. More formally, letting $\boldsymbol{x} = \langle x_1, \ldots, x_n \rangle$ and letting $p(\boldsymbol{x})$ denote $\langle p(x_1), \ldots, p(x_n) \rangle$, we get that the associated code $\mathrm{RS}_{q,k,\boldsymbol{x}}$ is given by

$$\mathrm{RS}_{q,k,\boldsymbol{x}} = \{p(\boldsymbol{x}) | p \in \mathbb{F}_q^k[x]\}.$$

Viewed from this perspective (as opposed to the dual perspective, where the codewords of the Reed Solomon codes are coefficients of polynomials), the Reed Solomon decoding problem is really a "curve-fitting" problem: Given n-dimensional vectors $\boldsymbol{x}$ and $\boldsymbol{y}$, find all polynomial $p \in \mathbb{F}_q^k[x]$ such that $\Delta(p(\boldsymbol{x}), \boldsymbol{y}) \leq e$. for some error parameter e. (Here and later $\Delta(\cdot, \cdot)$ denotes the Hamming distance.)

Traditional algorithms, starting with those of Peterson [30] attempt to "explain" $\boldsymbol{y}$ as a function of $\boldsymbol{x}$. This part becomes explicit in the work of Welch & Berlekamp [37,3] (see, in particular, the exposition in [35, Appendix A]) where $\boldsymbol{y}$ is interpolated as a rational function of $\boldsymbol{x}$, and this leads to the efficient decoding. (Specifically a rational function $a(x)/b(x)$ can be computed such that for all $i \in \{1, \ldots, n\}$, $a(x_i) = y_i * b(x_i)$.)

Rational functions, however, are limited in their ability to explain data with large amounts of error. To motivate this point, let us consider the following simple (and contrived) channel: The input and output alphabet of the channel are $\mathbb{F}_q$. The channel behavior is as follows: On input a symbol $\alpha \in \mathbb{F}_q$, the channel outputs α with probability $\frac{1}{2}$ and $\omega\alpha$ with probability $\frac{1}{2}$, for some fixed $\omega \in \mathbb{F}_q$. Now this is a channel that makes an error with probability $\frac{1}{2}$, but still the information it outputs is very closely correlated with the input (and its capacity, in the sense of Shannon, is very close to 1). However if we transmit a Reed Solomon codeword on this channel, the typical output vector $\boldsymbol{y}$ does not admit a simple description as a rational function of $\boldsymbol{x}$, and thus traditional decoding algorithms fail.

However it is clear that the output of the channel is explain by some nice algebraic relations: Specifically, there exists a polynomial p (of degree at most k) such that for every i, $y_i = p(x_i)$ or $y_i = \omega \cdot p(x_i)$. The "Or" of two Boolean conditions also has a simple algebraic representation: we simply have that the polynomial $Q(x, y) = (y - p(x)) \cdot (y - \omega \cdot p(x))$ is zero on every given (x_i, y_i). Furthermore such a polynomial $Q(x, y)$ can be found by simple interpolation (which amounts to solving a linear system), and the candidate polynomial $p(x)$ can be determined as a root of the polynomial $Q(x, y)$. (Notice that the factoring will find two polynomials p_1 and p_2 and, if $\omega^2 \neq 1$, the true candidate is p_1 iff it satisfies $p_2 = \omega p_1$.)

The above example illustrates the power of using algebraic curves (over rational functions) in decoding Reed-Solomon codes. The idea of using such functions was proposed by Ar et al. [1] who showed that if, by some fortunate occurrence, the vectors $\boldsymbol{x}$ and $\boldsymbol{y}$ could be explained by some nice algebraic relation, then decomposing the algebraic curve (i.e., factoring) could tell if there exist large subsets of the data that satisfy non-trivial algebraic correlation. However they could not show general conditions under which the vectors $\boldsymbol{x}$ and $\boldsymbol{y}$ could be explained by a nice algebraic curve, and this prevented them from obtaining a general decoding algorithm for Reed Solomon codes.

The complementing result took a few years to emerge, and did so finally in [36], where a simple counting argument is used to show that any pair of vectors $\boldsymbol{x}$ and $\boldsymbol{y}$ has a "nice" algebraic curve explaining it. The x- and y-degree of the curve can be chosen as desired, subject to the condition that the support has at least $n + 1$ coefficients. Putting these two pieces together, and choosing x- and y- degrees appropriately, one obtains the following algorithm and result:

Definition 1. *Let (w_x, w_y)-weighted degree of a monomial $x^i y^j$ be $i \cdot w_x + j \cdot w_y$. The (w_x, w_y)-weighted degree of a polynomial $Q(x, y)$ is the maximum, over all monomials with non-zero coefficient in Q, of their (w_x, w_y)-weighted degree.*

Given $\boldsymbol{x}, \boldsymbol{y} \in \mathbb{F}_q^n$ and k.

1. Compute $Q \neq 0$ with $(1, k)$-weighted degree at most $\lfloor \sqrt{2(k-1)n} \rfloor$ satisfying $Q(x_i, y_i) = 0$ for all $i \in \{1, \ldots, n\}$ (this is simple interpolation).
2. Factor Q and report all polynomials $p \in \mathbb{F}_q^k[x]$ such that $y - p(x)$ is a factor of Q and $p(x_i) = y_i$ for $\lfloor \sqrt{2kn} \rfloor + 1$ values of $i \in \{1, \ldots, n\}$.

Theorem 1 ([36]). *Given vectors $\boldsymbol{x}, \boldsymbol{y} \in \mathbb{F}_q^n$, a list of all polynomials $p \in \mathbb{F}_q^k[x]$ satisfying $p(x_i) = y_i$ for more than $\sqrt{2nk}$ values of $i \in \{1, \ldots, n\}$ can be found in time polynomial in n, provided all pairs (x_i, y_i) are distinct.*

The interesting aspect of the above algorithm is that it takes some very elementary algebraic concepts, such as unique factorization, Bezout's theorem, and interpolation, and makes algorithmic use of these concepts in developing a decoding algorithm for an algebraic code. This may also be a good point to mention some of the significant advances made in the complexity of factoring multivariate polynomials that were made in the 1980's. These algorithms, discovered independently by Grigoriev [14], Kaltofen [22], and Lenstra [25], form the technical foundations of the decoding algorithm above. Modulo these algorithms, the decoding algorithm and its proof rely only on elementary algebraic concepts. Exploiting slightly more sophisticated concepts from commutative algebra, leads to even stronger decoding results that we describe next.

The algorithm of Guruswami and Sudan [17] is best motivated by the following weighted curve fitting question: Suppose in addition to vectors $\boldsymbol{x}$ and $\boldsymbol{y}$, one is also given a vector of positive integers $\boldsymbol{w}$ where w_i determines the "weight" or confidence associated with a given point (x_i, y_i). Specifically we would like to find all polynomials p such that $\sum_{i|p(x_i)=y_i} w_i \geq W$ (for as small a W as possible).

The only prior algorithm (known to this author) that could take such "reliability" consideration into account was the Generalized Minimum Distance (GMD) decoding algorithm of Forney [10]. This algorithm, in combination with Theorem 1, can find such a vector provided $W = \Omega\left(\sqrt{k}\sqrt{\ln n}\sqrt{\sum_{i=1}^{n} w_i^2}\right)$. However, the GMD algorithm is combinatorial, and we would like to look for a more algebraic solution.

How can one interpret the weights in the algebraic setting? A natural way at this stage is to find a "fit" for all the data points that corresponds to the weights: Specifically, find a polynomial $Q(x, y)$ that "passes" through the point (x_i, y_i) at least w_i times. The notion of a curve passing through a point multiple times is a well-studied one. Such points are called singularities. Over fields of characteristic zero, these are algebraically characterized by the fact that the partial derivatives of the curve (all such, upto the $(r-1)$th derivatives, if the point must be visited by the curve r times), vanish at the point. The relevant component of this observation is that insisting that a curve pass through a point r times is placing $\binom{r}{2}$ linear constraints on the coefficients. This fact remains true over finite fields, though the partial derivatives don't yield these linear constraints any more. Using this notion to find curves that fit the points according to the weights, and then factoring the curves, leads to the following algorithm and result.

Given $\boldsymbol{x}, \boldsymbol{y} \in \mathbb{F}_q^n$, $\boldsymbol{w} \in \mathbb{Z}_{\geq 0}^n$, and k.
1. Compute $Q \neq 0$ with $(1, k)$-weighted degree at most $\lfloor\sqrt{k \sum_{i=1}^{n} w_i(w_i+1)}\rfloor$ satisfying $Q(x_i, y_i)$ is a zero of multiplicity w_i, for all $i \in \{1, \ldots, n\}$.
2. Factor Q and report all polynomials $p \in \mathbb{F}_q^k[x]$ such that $y - p(x)$ is a factor of Q and $\sum_{i|p(x_i)=y_i} w_i$ is at least $\lfloor\sqrt{k \sum_{i=1}^{n} w_i(w_i+1)}\rfloor + 1$.

Lemma 1 ([17]). *Given vectors $\boldsymbol{x}, \boldsymbol{y} \in \mathbb{F}_q^n$, a list of all polynomials $p \in \mathbb{F}_q^k[x]$ satisfying $\sum_{i||p(x_i)=y_i} w_i > \lfloor\sqrt{k \sum_{i=1}^{n} w_i(w_i+1)}\rfloor$ can be found in time polynomial in $n, \sum_i w_i$, provided all pairs (x_i, y_i) are distinct.*

At first glance it is not clear if this is better than the GMD bound. The GMD bound is invariant with respect to scaling of the w_i's while the above is not! In fact, it is this aspect that makes the algorithm above intriguing. Fix vectors $\boldsymbol{x}$ and $\boldsymbol{y}$, and consider two possible weight assignments: in the first all weights are 1, and in the second all weights are 2. On the one hand, the weight vectors place the same relative weights on all points, so a "good" solution to the first instance is also a "good" solution to the second instance. On the other hand, a close examination of the bound in Lemma 1 reveals that in the latter case it can find some polynomials that the former can not. The first instance finds all polynomials that agree with the data in $\sqrt{2kn}$ points, while the second finds all polynomials that agree with the data in $\sqrt{\frac{3}{2}kn}$ points. Scaling the weights to larger and larger values, in the limit we find all polynomials that fit the data over more than $\sqrt{kn}$ points. The price we pay is that the running time of the algorithm grows with the scaling factor. However it is easy to see that a finite (polynomial in n) weight suffices to decode up to this bound and this leads to the following theorem:

Theorem 2 ([17]). *Given vectors $x, y \in \mathbb{F}_q^n$, a list of all polynomials $p \in \mathbb{F}_q^k[x]$ satisfying $p(x_i) = y_i$ for more than $\sqrt{nk}$ values of $i \in \{1, \ldots, n\}$ can be found in time polynomial in n, provided all pairs (x_i, y_i) are distinct.*

Note that while the original motivation was to find a better algorithm for the "weighted" decoding problem, the result is a better unweighted decoding algorithm, that uses the weighted version as an intermediate step. Of course, it is also possible to state what the algorithm achieves for a general set of weights. For this part, we will just assume that the weight vector is an artbitrary vector of non-negative reals, and get the following:

Theorem 3 ([17,18]). *Given vectors $x, y \in \mathbb{F}_q^n$, a weight vector $w \in \mathbb{R}_{\geq 0}^n$, and a real number $\epsilon > 0$, a list of all polynomials $p \in \mathbb{F}_q^k[x]$ satisfying $\sum_{i|p(x_i)=y_i} w_i > \sqrt{k(\epsilon + \sum_{i=1}^n w_i^2)}$ can be found in time polynomial in n and $\frac{1}{\epsilon}$, provided the pairs (x_i, y_i) are all distinct.*

This result summarizes the state of knowledge for list-decoding for Reed Solomon codes, subject to the restriction that the decoding algorithm runs in polynomial time. However this criterion, that the decoding algorithm runs in polynomial time, is a very loose one. The practical nature of the problem deserves a closer look at the components involved and efficient strategies to implement these components. This problem has been considered in the literature, with significant success. In particular, it is now known how to implement the interpolation step in $O(n^2)$ time, when the output list size is a constant [29,31]. Similar running times are also known for the root finding problem (which suffices for the second step in the algorithms above) [2,11,28,29,31,39]. Together these algorithms lead to the possibility that a good implementation of list-decoding may actually even be able to compete with the classical Berkelkamp-Massey decoding algorithm in terms of efficiency. A practical implementation of such an algorithm in C++, due to Rasmus Refslund Nielsen, is available from from his homepage (http://www.student.dtu.dk/~p938546/index.html).

3   Ideal Error-Correcting Codes and Decoding

We now move on to other list-decoding algorithms for other algebraic codes. The potential for generalizing the decoding algorithms above to codes other than just the Reed Solomon code, was first shown by Shokrollahi and Wasserman [33]. In their work, they show how to generalize the algorithm above to decode the more general family of algebraic-geometry codes. A full description of this family of codes is out of scope for this article — the reader is encouraged to read the text of Stichtenoth [34] or the article by Høholdt, van Lint, and Pellikaan [20] for a description. However we will attempt to describe the flavor of the results by defining a broad class of codes, that we call "Ideal error-correcting codes".

One way of viewing Reed Solomon codes, is that they are built over a (nice) integral domain $R = \mathbb{F}_q[x]$, [1] The message space $\mathcal{M} = \mathbb{F}_q^k[x]$ is chosen to be a subset of the ring R. Additionally the code is specified by a collection of ideals $I_1, \ldots, I_n$ of R. In the case of Reed Solomon codes, these are the ideals generated by the linear polynomials $x - x_1, \ldots, x - x_n$. The encoding of a message element $p \in R$ is simply its residue modulo n ideals. Thus, in Reed Solomon encoding, $p \mapsto \langle p \bmod (x - x_1), \ldots, p \bmod (x - x_n) \rangle = p(\boldsymbol{x})$ as expected. The following definition summarizes the family of codes obtained this way.

Definition 2 (Ideal error-correcting codes [16]). *An ideal error-correcting code is specified by a triple $(R, \mathcal{M}, \langle I_1, \ldots, I_n \rangle)$, where R is an integral domain, $\mathcal{M} \subseteq R$, and $I_1, \ldots, I_n$ are ideals of R. The code is a subset of $(R/I_1) \times \cdots \times (R/I_n)$, given by the set $\{\langle p \bmod (I_1), \ldots, p \bmod (I_n) \rangle \,|\, p \in \mathcal{M}\}$.*

To quantify the distance properties of such a code, it is useful to impose a notion of size on elements of the ring R. In the case of Reed Solomon codes the size of an element is essentially its degree (though for technical reasons, it is convenient to use $q^{\deg p}$ as a measure of size). The message space usually consists of all elements of small size. To make this space large one needs to know that the ring has sufficiently many small elements. Further the size function is assumed to satisfy some axioms such as $\text{size}(a + b) \leq \text{size}(a) + \text{size}(b)$, $\text{size}(ab) \leq \text{size}(a) \cdot \text{size}(b)$ and so on. Further, if the size of an ideal is defined to be the size of the smallest non-zero element in it, then $\text{size}(J_1 \times J_2)$ should be at least $\text{size}(J_1) \cdot \text{size}(J_2)$. Assuming such, relatively simple axioms it is possible to analyze the minimum distance of an ideal error-correcting code, once the sizes of the ideals I_1 to I_n are known. (We will not cover these definitions formally here - we refer the reader to [16] for a full discussion.) The same axioms guarantee efficient (list-)decoding as well. In fact, the following simple generalization of the algorithm from the previous section gives the algorithm for decoding any ideal error-correcting code. We describe the algorithm informally. Formal specification will involve a careful setting to various parameters.

  Given $I\ y \in R^n$, $w \in \mathbb{Z}_{\geq 0}^n$.
1. Let $J_i = I_i + (y - y_i)$.
2. Compute $Q \in R[y] - \{0\}$ of small degree in y, with small coefficients, satisfying $Q \in \prod_{i=1}^n J_i^{w_i}$.
3. Factor Q and report all elements $p \in \mathcal{M}$ such that $y - p$ is a factor of Q and $y_i \in p + (I_i)$ for sufficiently many i.

---

[1] For the reader that is rusty with the elements of commutative algebra, let us recall that an integral domain is a commutative ring R that has no zero divisors (i.e., $pq = 0$ implies $p = 0$ or $q = 0$). An ideal I in R is a subset that is closed under addition, and $a \in I$ implies $ab \in I$, for all $b \in R$. The quotient of R over I, denoted R/I forms an integral domain and this quotient ring is crucial to many definitions here.

In this setting, the algorithm above may even appear more natural. Note that the ideals J_i above have the following meaning: $y - p$ belongs to the ideal J_i if and only if $y_i \in p + I_i$. Thus we want all elements p such that $y - p$ lies in many of the ideals J_i. To find such an element, we find an element that Q that lies in all of them, and factor it to find any element that lies in many of them.

Why consider this more complicated scheme? Ideal error-correcting codes not only include the class of Reed Solomon codes (as already pointed out), but also all algebraic-geometry codes, and an interesting family of number-theoretic codes termed Redundant Residue Number System (RRNS) codes. As a consequence of the generalization above, one gets a structure for decoding all the above family of codes. Note that we only get a structure, not the algorithm itself. In order to get actual decoding algorithms, one needs to find algorithms to "Compute Q" (the interpolation step) as well as to factor over $R[y]$. Both aspects present their own complexity, as we will illustrate for the RRNS codes. Furthermore, to get the best possible decoding algorithm, one needs to select the parameters, and in particular the weights carefully. We will discuss this more in the next section.

Finally, we point out one important class of codes where the decoding algorithms don't seem to apply. This is the class of Reed-Muller codes where the algorithm of Feng and Rao [9] (see, in particular, the desciption in [20]) decodes up to half the minimum distance. The best known list-decoding algorithm [32] does better than the above algorithm for some choice of parameters, but does not even match up to the above algorithm for other choices of parameters. Extending the list-decoding algorithm given here to apply to the class of Reed-Muller codes seems to require a generalization beyond the class of ideal codes.

4   Redundant Residue Number System Codes

This is the family of ideal error-correcting codes given by $R = \mathbb{Z}$, $\mathcal{M} = \{0, \ldots, K - 1\}$ for some integer K, and $I_i = (p_i)$ for a collection of pairwise prime integers $p_1, \ldots, p_n$. In other words a message is a non-negative integer less than K and its encoding are its residues modulo small integers $p_1, \ldots, p_n$. If we permute the indices so that $p_1 \leq \cdots \leq p_n$, and if $K \leq \prod_{i=1}^{k} p_i$, then this code has minimum distance at least $n - k + 1$, Thus it should be correctible to up to $\frac{n-k}{2}$ errors uniquely, and list-decodable to about $n - \sqrt{nk}$ errors. Turns out Mandelbaum [27] gave a unique decoding algorithm decoding to $\frac{n-k}{2}$ errors. The algorithm runs in polynomial time provided the highest and smallest moduli are relatively close in value. Goldreich et al. [12] gave an algorithm correcting approximately $n - \sqrt{2nk \frac{\log p_n}{\log p_1}}$ errors. Boneh [6] improved this to $n - \sqrt{nk \frac{\log p_n}{\log p_1}}$ errors, and finally Guruswami et al. [16] improved this to correct $n - \sqrt{n(k + \epsilon)}$ errors for arbitrarily small ϵ. They also give a polynomial time unique decoding algorithm correcting up to $\frac{n-k}{2}$ errors.

The algorithms of [12,6,16] illustrate some of technicalities that surface in specializing the algorithm of Section 3. For instance, consider the interpolation step: Even in the simple case when all the w_i's are 1, the case considered in [12], the algorithm for this part is not obvious. We wish to find a polynomial Q with small integer coefficients such that $Q(y_i) = 0 \bmod (\prod_{i=1}^{n} p_i)$. This is a task in Diophantine approximation and no longer a simple linear system. Fortunately, it turns out to be a relatively well-studied problem. The set of polynomials satisfying the condition $Q(y_i) = 0 \bmod (\prod_{i=1}^{n} p_i)$ form a lattice, and finding a polynomial with small coefficients is a "shortest vector problem" in integer lattices and one can use the groundbreaking algorithm of Lenstra, Lenstra, and Lovasz [26] (LLL) to solve this problem near-optimally. In the case of general weights [6,16], the problem remains a short vector problem in a lattice, however it is not simple to express a basis for this lattice explicitly. In the case of uniform, but not unit weights, [6] manages to come up with an explicit description based on some analogies with some problems in cryptography. For the fully general case, [16] do not describe an explicit basis. Instead they give an algorithm that computes this basis from the weights. Thus this step of the process can get quite complicated.in the case of numkber theoretic codes. (In contrast this step remains reasonably simple in the case of algebraic geometry codes.)

Another aspect of the decoding algorithm highlighted by the number-theoretic setting is the choice of weights. Even in the simple case where all the input weights are unit, it is not clear that the best choice of weights is a uniform one. Indeed, the final choice used by [16] gives large weights to the small moduli and smaller weights to the larger modulii. In general, this issue — what is the best choice of weights to the algorithm, and how should they relate to the weights given as input — is far from clear. For example, a recent paper of Kötter and Vardy [24] suggests a completely surprising choice of weights in the case of algebraic geometry codes. This leads to better bounds for decoding these codes that the one given in [17]. A more detailed examination of this question has been carried out by Kötter [23].

Finally, we move on to the second step of the decoding algorithm. In this case the algorithm that is required is an integer root-finding algorithm for integer polynomials. This is again a well-studied problem, with known polynomial time solutions. This step however can get significantly more complicated for other ideals. E.g., in the case of algebraic-geometry codes, the issue becomes that of how the codes are specified. For most well-known families of such codes, the standard specifications do lead to polynomial time solutions [11,28,29,39]. For arbitrary codes, however it is a priori unclear if a natural representation could lead to a polynomial time decoding algorithm. In fact in the absence of a complete characterization of all algebraic geometry codes, it is unclear as to what is a natural representation for all of them. [19] suggest a potential representation that is reasonably succinct (polynomial sized in the generator matrix), that allows this and other necessary tasks to be solved in polynomial time, by using the algorithms of [11] and [29].

Acknowledgments. Thanks to Venkatesan Guruswami for letting me describe many of our joint works here, to Tom Høholdt for enlightening me on many of the developments (both old and new) in the coding theory community, and to Ralf Kötter for clarifying the subtleties in the choice of weights.

References

1. Sigal Ar, Richard J. Lipton, Ronitt Rubinfeld, and Madhu Sudan. Reconstructing algebraic functions from erroneous data. *SIAM Journal on Computing*, 28(2): 487–510, April 1999.
2. Daniel Augot and Lancelot Pecquet. A Hensel lifting to replace factorization in list decoding of algebraic-geometric and Reed-Solomon codes. *IEEE Trans. Info. Theory*, 46(6): 2605-2613, November 2000.
3. Elwyn R. Berlekamp. Bounded distance +1 soft-decision Reed Solomon decoding. *IEEE Transactions on Information Theory*, 42(3):704-720, 1996.
4. Richard E. Blahut. Theory and practice of error control codes. Addison-Wesley Pub. Co., 1983.
5. V. M. Blinovskii. Bounds for codes in the case of list decoding of finite volume. *Problemy Peradachi Informatsii*, 22(1):11–25, January-March 1986.
6. Dan Boneh. Finding smooth integers in short intervals using CRT decoding. (To appear) *Proceedings of the Thirty-Second Annual ACM Symposium on Theory of Computing*, Portland, Oregon, 21-23 May 2000.
7. Peter Elias. List decoding for noisy channels. *WESCON Convention Record*, Part 2, Institute of Radio Engineers (now IEEE), pages 94–104, 1957.
8. Peter Elias. Error-correcting codes for list decoding. *IEEE Transactions on Information Theory*, 37(1):5–12, January 1991.
9. G.-L. Feng and T. R. N. Rao. Decoding algebraic-geometric codes upto the designed minimum distance. *IEEE Transactions on Information Theory*, 39(1):37–45, January 1993.
10. G. David Forney Jr.. *Concatenated Codes*. MIT Press, Cambridge, MA, 1966.
11. S. Gao and M. A. Shokrollahi. Computing roots of polynomials over function fields of curves. *Proceedings of the Annapolis Conference on Number Theory, Coding Theory, and Cryptography*, 1999.
12. Oded Goldreich, Dana Ron, and Madhu Sudan. Chinese remaindering with errors. *IEEE Transactions on Information Theory*. 46(4): 1330–1338, July 2000.
13. Oded Goldreich, Ronitt Rubinfeld, and Madhu Sudan. Learning polynomials with queries: The highly noisy case. *Proceedings of the 36th Annual Symposium on Foundations of Computer Science*, pages 294–303, Milwaukee, Wisconsin, 23-25 October 1995.
14. Dima Grigoriev. Factorization of polynomials over a finite field and the solution of systems of algebraic equations. Translated from *Zapiski Nauchnykh Seminarov Lenningradskogo Otdeleniya Matematicheskogo Instituta im. V. A. Steklova AN SSSR*, 137:20-79, 1984.
15. Venkatesan Guruswami, Johan Håstad, Madhu Sudan, and David Zuckerman. Combinatorial bounds for list decoding. (To appear) *Proceedings of the 38th Annual Allerton Conference on Communication, Control, and Computing*, 2000.
16. Venkatesan Guruswami, Amit Sahai, and Madhu Sudan. "Soft-decision" decoding of Chinese remainder codes. *Proceedings of the 41st Annual Symposium on Foundations of Computer Science*, pages 159-168, Redondo Beach, California, 12-14 November, 2000.
17. Venkatesan Guruswami and Madhu Sudan. Improved decoding of Reed-Solomon codes and algebraic-geometric codes. *IEEE Transactions on Information Theory*, 45(6): 1757–1767, September 1999.
18. Venkatesan Guruswami and Madhu Sudan. List decoding algorithms for certain concatenated codes. Proceedings of the Thirty-Second Annual ACM Symposium on Theory of Computing, pages 181-190, Portland, Oregon, 21-23 May 2000.

19. Venkatesan Guruswami and Madhu Sudan, On representations of algebraic-geometric codes. IEEE Transactions on Information Theory (to appear).
20. T. Høholdt, J. H. van Lint, and R. Pellikaan. Algebraic geometry codes. In *Handbook of Coding Theory*, V. Pless and C. Huffman (Eds.), Elsevier Sciences, 1998.
21. J. Justesen and T. Høholdt. Bounds on list decoding of MDS codes. Manuscript, 1999.
22. Erich Kaltofen. A polynomial-time reduction from bivariate to univariate integral polynomial factorization. *Proceedings of the Fourteenth Annual ACM Symposium on Theory of Computing*, pages 261-266, San Francisco, California, 5-7 May 1982.
23. Ralf Kötter. Personal communication, March 2001.
24. Ralf Kötter and Alexander Vardy. Algebraic soft-decision decoding of Reed-Solomon codes. (To appear) *Proceedings of the 38th Annual Allerton Conference on Communication, Control, and Computing*, 2000.
25. Arjen K. Lenstra. Factoring multivariate polynomials over finite fields. *Journal of Computer and System Sciences*, 30(2):235–248, April 1985.
26. A. K. Lenstra, H. W. Lenstra, and L. Lovasz. Factoring polynomials with rational coefficients. *Mathematische Annalen*, 261:515–534, 1982.
27. D. M. Mandelbaum. On a class of arithmetic codes and a decoding algorithm. *IEEE Transactions on Information Theory*, 21:85–88, 1976.
28. R. Matsumoto. On the second step in the Guruswami-Sudan list decoding algorithm for AG-codes. *Technical Report of IEICE*, pp. 65-70, 1999.
29. R. Refslund Nielsen and Tom Høholdt. Decoding Hermitian codes with Sudan's algorithm. *Proceedings of AAECC-13, LNCS 1719, Springer-Verlag*, 1999, pp. 260-270.
30. W. W. Peterson. Encoding and error-correction procedures for Bose-Chaudhuri codes. *IRE Transactions on Information Theory*, IT-60:459-470, 1960.
31. Ron M. Roth and Gitit Ruckenstein. Efficient decoding of Reed-Solomon codes beyond half the minimum distance. *IEEE Transactions on Information Theory*, 46(1):246-257, January 2000.
32. Madhu Sudan, Luca Trevisan, and Salil Vadhan. Pseudorandom generation without the XOR lemma. *Proceedings of the Thirty-First Annual ACM Symposium on Theory of Computing*, pages 537-546, Atlanta, Georgia, 1-4 May 1999.
33. M. Amin Shokrollahi and Hal Wasserman. List decoding of algebraic-geometric codes. *IEEE Transactions on Information Theory*, 45(2): 432-437, March 1999.
34. Henning Stichtenoth. *Algebraic Function Fields and Codes*. Springer-Verlag, Berlin, 1993.
35. Madhu Sudan. *Efficient Checking of Polynomials and Proofs and the Hardness of Approximations*. ACM Distinguished Theses. Lecture Notes in Computer Science, no. 1001, Springer, 1996.
36. Madhu Sudan. Decoding of Reed Solomon codes beyond the error-correction bound. *Journal of Complexity*, 13(1): 180–193, March 1997.
37. Lloyd Welch and Elwyn R. Berlekamp. Error correction of algebraic block codes. *US Patent* Number 4,633,470, issued December 1986.
38. J. M. Wozencraft. List decoding. Quarterly Progress Report. Research Laboratory of Electronics, MIT, Vol. 48, pp. 90–95, 1958.
39. Xin-Wen Wu and Paul H. Siegel. Efficient list decoding of algebraic geometric codes beyond the error correction bound. *Proc. of International Symposium on Information Theory*, June 2000.
40. V. V. Zyablov and M. S. Pinsker. List cascade decoding. *Problemy Peredachi Informatsii*, 17(4):29–33, October-December 1981.

Self-dual Codes
Using Image Restoration Techniques

A. Baliga[1] and J. Chua[2]

[1] Department of Mathematics, RMIT University, GPO Box 2476V, Melbourne, VIC
3001, Australia.
asha@rmit.edu.au,
[2] School of Computer Science and Software Engineering, PO Box 26, Monash
University, Victoria 3800, Australia.
joselito.chua@csse.monash.edu.au.

Abstract. From past literature it is evident that the search for self-dual codes has been hampered by the computational difficulty of generating the Hadamard matrices required. The use of the cocyclic construction of Hadamard matrices has permitted substantial cut-downs in the search time, but the search space still grows exponentially. Here we look at an adaptation of image-processing techniques for the restoration of damaged images for the purpose of sampling the search space systematically. The performance of this approach is evaluated for Hadamard matrices of small orders, where a full search is possible.

The dihedral cocyclic Hadamard matrices obtained by this technique are used in the search for self-dual codes of length 40, 56 and 72. In addition to the extremal doubly-even [56,28,12] code, and two singly-even [56,28,10] codes, we found a large collection of codes with only one codeword of minimum length.

1 Introduction

In [3], the $[I, A]$ construction was used to obtain doubly-even self-dual codes from $\mathbf{Z}_2^2 \times \mathbf{Z}_t$ - cocyclic Hadamard matrices for t odd. This construction was extended and refined in [1,2] to include the cyclic, the dihedral and the dicyclic groups and the equivalence classes of the codes obtained from groups of order 20 were catalogued. The internal structure of these Hadamard matrices permits substantial cut-downs in the search time for each code found. However the search space for cocyclic Hadamard matrices developed over $\mathbf{D}_{4t}$ grows exponentially with t. Image restoration techniques may provide the answer to this problem, sampling the search space systematically when a full search is computationally infeasible. The performance of this approach is evaluated for $t = 5$ and 7, where a full search was feasible.

The Hadamard matrices thus found were used in the search for all $\mathbf{D}_{4t}$-cocyclic self-dual codes of length 40, 56. In the case of self-dual codes of length 72, this was the only technique used to generate the Hadamard matrices. We catalogue the self-dual codes found in the search, noting the occurrence of self-dual codes with one code word of minimum length.

S. Boztaş and I.E. Shparlinski (Eds.): AAECC-14, LNCS 2227, pp. 46–56, 2001.
© Springer-Verlag Berlin Heidelberg 2001

In Section 2, we outline the structure of dihedral cocyclic Hadamard matrices, detailing the efficiency obtained. In Section 3, the idea of using the search space as an image is explored, and the use of image restoration techniques is discussed, along with a summarised algorithm. Section 4 gives the results we have found so far, including the self-dual codes found using the above techniques.

2 D_{4t} - Cocyclic Hadamard Matrices

In [7], Flannery details the condition for the existence of a Hadamard matrix cocyclic over D_{4t}. Denote by D_{4t} the dihedral group of order $4t, t \geq 1$, given by the presentation

$$< a, b | a^{2t} = b^2 = (ab)^2 = 1 >$$

Cocyclic Hadamard matrices developed over D_{4t} can exist only in the cases $(A, B, K) = (1, 1, 1), (1, -1, 1), (1, -1, -1), (-1, 1, 1)$ for t odd. Here A and B are the inflation variables and K is the transgression variable. We only consider the case $(A, B, K) = (1, -1, -1)$ in this paper since computational results in [7] and [1] suggest that this case contains a large density of cocyclic Hadamard matrices. This case also gives rise to a central extension of $\mathbf{Z}_2$ by D_{4t} called a "group of type Q" [8]. The techniques presented in this paper can be adapted easily for other cases of (A, B, K).

A group developed matrix over the group D_{4t} for the case $(A, B, K) - (1, -1, -1)$ has block form

$$H = \begin{pmatrix} M & N \\ ND & -MD \end{pmatrix} \tag{1}$$

where M and N are $2t \times 2t$ matrices, each of which is the entry wise product of a back circulant and back negacyclic matrix.

D is the matrix obtained by negating every non-initial row of a back circulant $2t \times 2t$ matrix with first row

$$1\ 0\ 0 \cdots 0$$

Following Proposition 6.5 (ii) in [7], we know that H is a cocyclic Hadamard matrix if and only if

$$M^2 + N^2 = 4tI_{2t} \tag{2}$$

Denote the first rows of M and N by $\boldsymbol{m}$ and $\boldsymbol{n}$, respectively. Since the matrices M and N are determined by their first row entries, then $\boldsymbol{m}$ and $\boldsymbol{n}$ can be used to determine whether the corresponding matrices satisfy equation (2) without having to construct the actual matrices. Flannery [7] showed that M and N would satisfy equation (2) for $t \geq 2$ if and only if

$$\boldsymbol{m} \left(\boldsymbol{m} P^i W_i \right)^\top = -\boldsymbol{n} \left(\boldsymbol{n} P^i W_i \right)^\top \quad \text{for } 1 \leq i \leq t - 1 \tag{3}$$

where P is a forward circulant matrix with first row

$$0\ 0\ 0 \cdots 0\ 1$$

and W_i is a $2t \times 2t$ diagonal matrix whose main diagonal is

$$1\ 1 \cdots 1 -1 \cdots -1$$

where the last entry 1 occurs in position $2t - i$.

In the implementation, the matrices P^i and W_i are pre-computed for $i = 1, \ldots, t - 1$ to avoid having to construct them repeatedly for every (m, n) pair. The computational cost of determining whether H is a cocyclic Hadamard matrix is reduced substantially because the calculations in Equation (3) can terminate as soon as the equality fails at an i value.

Flannery [7] also suggests using some symmetries to reduce the search space. For example, if a (m, n) pair satisfies equation (3) then the $(\pm m, \pm n)$ pairs also satisfy the condition. Moreover, a matrix developed from a (m, n) pair satisfying the condition is Hadamard-equivalent to that of $(-m, -n)$. This cuts the search space down by half from 2^{2t} to 2^{2t-1} for each $2t$-tuple.

Similar "paper-folding" symmetries in the case of $(A, B, K) = (1, -1, -1)$ reduce the search space further to $1/8$-th of the entire set of possibilities. It is easy to show that if the matrix which corresponds to (m, n) is Hadamard, then the matrices corresponding to (n, m), $(-n, m)$, $(-m, n)$, $(-m, -n)$, $(-n, -m)$, $(n, -m)$ and $(m, -n)$ will also be Hadamard. These symmetries are illustrated in Figure 1. For the rest of this paper, the discussion is limited to the (m, n) pairs in the shaded region in Figure 1.

Despite the reduction in search space and cut-downs in computational cost, the number of (m, n) pairs we need to consider is still around 2^{4t-3}. In situations where it may not be possible to consider all the 2^{4t-3} pairs, we may need to choose wisely which pairs to consider. This means sampling the search space systematically by making an educated guess of which (m, n) pairs are likely to yield cocyclic Hadamard matrices.

3   The Search Space as an Image

This approach regards the search space as a $2^{2t} \times 2^{2t}$ black and white picture with the black dots representing the (m, n) pairs which yield cocyclic Hadamard matrices. The vectors (m, n) are mapped to a pair of integer coordinates (m, n) by mapping the vector entries $\{1, -1\}$ to $\{1, 0\}$ and interpreting the resulting binary strings as positive integers in base-2 notation.

Figure 2 shows the resulting images for $t = 5$ and $t = 7$ in an octant of the search space. While a cursory glance gives the impression that the dots are scattered uniformly over the search space, a closer examination of the images indicates that the dots are more dense in certain areas, forming distinguishable patterns across the image, thus providing a reason for the use of image processing methods.

The idea here is if the search space is too large to obtain the complete image then it can be sampled uniformly, and the (m, n) coordinates, which yield cocyclic Hadamard matrices, plotted. The resulting sparse plot can now be regarded as a "damaged" version of the image and image-processing techniques

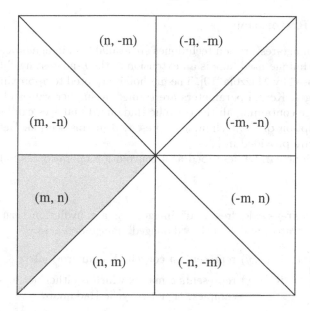

Fig. 1. The (m, n) pairs which yield cocyclic Hadamard matrices in the shaded region determine all the cocyclic Hadamard matrices over the entire search space.

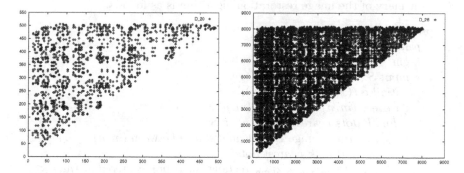

Fig. 2. Sample complete images for $t = 5$ and $t = 7$, respectively. Each image plots the (m, n) coordinates corresponding to all the (m, n) pairs which yield cocyclic Hadamard matrices in an octant of the search space.

applied to "restore" the image. The restored image is therefore an attempt to predict the (m, n) pairs which are likely to yield cocyclic Hadamard matrices. The regions of interest can be identified and the search limited to those regions rather than the entire search space.

3.1   Image Restoration

A number of image-restoration techniques are available in image processing literature. The technique used here is an extension of the k-nearest neighbour-based method proposed by Mazzola [10]. The method is adapted to approximate sparse point-set images. Kernel parameters are trained at smaller values of t, where a complete image containing all the cocyclic Hadamard matrices can be obtained. A brief description of the technique is presented in this section. Details of the development are provided in [4].

The technique can be described as a convolution operation as follows:

$$I = S * \phi \qquad (4)$$

where I is the grey-scale "restored" image, ϕ is a convolution kernel, $*$ is the convolution operator, and S is the "damaged" image defined by:

$$S(x,y) = \begin{cases} 1 & \text{if } (x,y) \text{ represents a cocyclic Hadamard matrix} \\ 0 & \text{if } (x,y) \text{ represents a matrix which is either not among the} \\ & \quad \text{samples, or not cocyclic Hadamard} \end{cases}$$

$$(5)$$

S is a black-and-white image where every black dot (i.e., a pixel value of 1) represents a Hadamard matrix. I is a grey-scale image, where the varying shades of grey represent the relative density of the dots in the corresponding region in S. A summary of the image restoration algorithm is as follows:

```
procedure RestoreImage
  parameter k: integer;
  begin
    input S: black-and-white image;
    initialise I(x, y) := 0   ∀(x, y);
    for every (m, n) such that S(m, n) = 1 do
      find k dots nearest to (m, n) in S;
      r_mn := the average Euclidean distance between (m, n)
              and its k nearest dots;
      R_mn := the 2r_mn × 2r_mn rectangular region centered at (m, n);
      for every (x, y) in R_mn do
        I(x, y) := I(x, y) + φ(x, y);
      end for;
    end for;
    normalise pixel values of I;
    output I: grey-scale image;
  end procedure;
```

The parameter k is the number of neighbouring dots to be considered, and depends on the average density of the dots in S. If S is sparse, then k can only take small values. As more cocyclic Hadamard matrices are found, the average density increases, and k takes on larger values. If N is the number of matrices found so far, then k is estimated as follows:

$$k \approx \sigma_t \cdot N \tag{6}$$

where σ_t is the predicted average density of the dots in S when all cocyclic Hadamard matrices have been found. Known values of t and σ_t predict that σ_t decays exponentially as t increases. An exponential fit estimates an upper bound for σ_t as follows:

$$\sigma_t \approx 0.0000574771 + 92.5021 \cdot e^{-1.81477\,t} \tag{7}$$

The technique described in [5] is used to find k dots nearest to (m, n) in S. These dots are referred to as the k-*nearest neighbours* of (m, n). The size of the convolution window depends on r_{mn}, the average Euclidean distance between the dot at (m, n) and its k-nearest neighbours. The convolution window, R_{mn}, is determined as follows:

$$R_{mn} = \{(x, y) \mid m - r_{mn} < x < m + r_{mn} \text{ and } n - r_{mn} < y < n + r_{mn}\} \tag{8}$$

Given R_{mn}, the convolution operation darkens the corresponding region in I in a manner determined by the convolution kernel, ϕ. Since the aim is to attempt restoration of sparse point-set images, a Poisson kernel with peak response at (m, n) is used:

$$\phi(x, y) = \frac{\lambda^d}{e^\lambda \cdot \Gamma(d + 1)} \tag{9}$$

where $d = \lambda - \sqrt{(x - m)^2 + (y - n)^2}$, $\lambda = \sqrt{2} \cdot r_{mn}$, $(x, y) \in R_{mn}$, and Γ is the Euler gamma function. The values of ϕ are normalised so that the sum of ϕ over R_{mn} is equal to 1. Figure 3 shows ϕ over R_{mn}.

High values of ϕ contribute darker shades of grey in I. The Poisson kernel emphasises the center of R_{mn}, which is at (m, n). The shades become lighter towards the edge of the window. As R_{mn} becomes large, ϕ becomes more spread out, and the peak value at (m, n) becomes smaller. Since R_{mn} in sparse regions of S is large compared to those in dense regions of S, the resulting shades of grey in I correspond to the relative density of the dots in S.

Each pixel in I is the cumulative sum of the convolution operations on the windows determined by the dots around that pixel. Thus, if (x, y) is surrounded by dots in S, then pixel $I(x, y)$ would appear dark even if $S(x, y) = 0$. The pixel values of I are normalised so that the largest and smallest values appear as black and white, respectively, and those in between as degrees of grey.

3.2   Search Method

At the start of the search, S is obtained by sampling the search space uniformly until N is sufficiently large for k to be ≥ 1. The samples are tested using the fast techniques discussed in Section 2. Then, the image restoration technique is

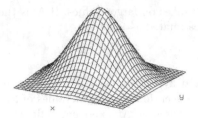

Fig. 3. $\phi(x,y)$ over R_{mn}.

applied on S to obtain the image I. The dark regions of I indicate the areas where large concentrations of cocyclic Hadamard matrices were found. Figure 4 shows the restored image I for $t = 5$ as the search progressed. Note that the light image in Figure 4-a is an indication that N is too small, and the dots too spread out, for the technique to identify areas of particular interest.

The range of the I values are partitioned into ρ intervals, each of length $(I_{\max} - I_{\min})/\rho$. Figure 4-e shows an example of the regions determined by the intervals.

Sample points are selected uniformly (among those not yet known to be cocyclic Hadamard) such that each interval has an equal number of points. Since a Poisson kernel is used over a sparse point-set image, the total area of the regions with high I values can be expected to be much smaller than the total area of the regions with low I values. Thus, regions which correspond to high I values are sampled more densely than those at lower I values. The idea is to put more effort in searching regions around clusters of known cocyclic Hadamard matrices, but without neglecting the bare regions between the clusters.

The search continues by testing the samples using the fast techniques discussed in Section 2. As more cocyclic Hadamard matrices are found from the samples, S is updated and the image I is re-calculated. The search then continues using the new I. σ_t is used to estimate a maximum value for N. The search terminates as soon as N reaches that value. However, frequent re-calculation of I has to be avoided as the computational cost can outweigh the benefits.

The restored image, I, is regarded as the result obtained by using the kernel to evaluate the information provided by the k-nearest neighbours. An obvious way to determine areas of interest using S is to have a rectangular window slide through the search space. As soon as the density of the dots inside the rectangle reaches a threshold, the area is tagged as an area of interest. Image restoration techniques can be thought of as a systematic way of identifying these areas. In the case of the image restoration technique discussed in Section 3.1, the size of the rectangle is adaptive, depending on the local information determined by the k-nearest neighbours. The technique also approximates the likelihood of having values along the gaps between the dots. Rather than having a threshold over the density, the "levels of interest" are determined by the kernel with respect to the relative distances between the points.

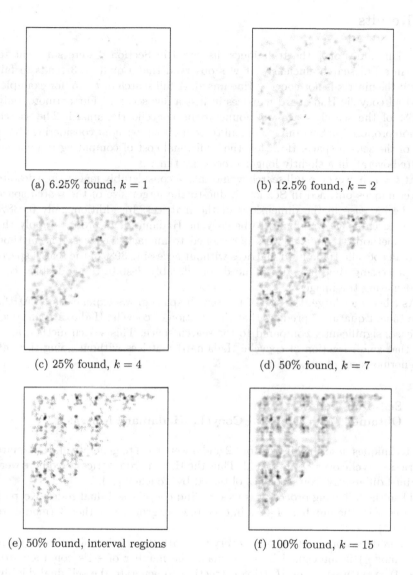

(a) 6.25% found, $k = 1$

(b) 12.5% found, $k = 2$

(c) 25% found, $k = 4$

(d) 50% found, $k = 7$

(e) 50% found, interval regions

(f) 100% found, $k = 15$

Fig. 4. Figures (a) to (d) show the reconstructed image I with $t = 5$ as the search progressed. The parameter k is estimated based on the σ_t. Figure (e) shows the partitioning of the range of I values into $\rho = 5$ intervals. Regions of the same shade of grey, including the black and the white, correspond to points belonging to the same interval. Figure (f) shows the full reconstructed image.

4   Results

For small values of t, the techniques discussed in Section 2 were sufficient to perform a full search efficiently. It was observed that Equation 3 tends to fail early if the matrix is not cocyclic Hadamard. A full search at $t = 5$, for example, found all cocyclic Hadamard matrices in just a few seconds. Furthermore, only 1.066% of the search space was found to be cocyclic Hadamard. The search method proposed in Section 3 found all these matrices without considering about 35% of the search space. However, the additional cost of computing the images resulted overall in a slightly longer processing time.

At $t = 7$, however, a full search would take a considerably longer time despite the techniques outlined in Section 2, due to the larger size of the search space and the increase in the dimensions of the matrices. In addition, only 0.038% of the search space was found to be cocyclic Hadamard. In order to apply the search method, the images were partitioned to manageable sizes. The method found all cocyclic Hadamard matrices without accessing 39% of the search space. The processing time was also reduced considerably despite the additional cost of calculating the images.

As t becomes large, the size of the search space grows exponentially. At the same time, Equation 7 predicts that the fraction of cocyclic Hadamard matrices decreases significantly compared to the search space. This search method aims find that small fraction of cocyclic Hadamard matrices without going through the enormous set of possibilities.

4.1   Self-dual Codes
Obtained from Dihedral Cocyclic Hadamard Matrices

The techniques described in Sections 2 and 3 were used to generate all Hadamard matrices cocyclic over D_{4t} for t odd. Thus the Hadamard matrices used here were obtained differently from the ones obtained by Tonchev [11].

Then the following process was used to find cocyclic self-dual codes: Keep all matrices with the number of +1's in each row congruent to either 3 (mod 4) or 1 (mod 4).

To produce doubly-even codes, every row with the number of +1's congruent to 1 (mod 4) is multiplied by -1 to make the number of +1's congruent to 3 (mod 4). Next we use the $[I, \bar{H}]$ construction to generate the self-dual doubly-even codes. A similar strategy is used to generate the singly-even self-dual codes.

During the search for extremal self-dual codes, we also found codes with only one code word of minimum weight. This interesting case was first encountered in the case $t = 5$, and only among the doubly-even codes in that case.

In the search for self-dual codes for $t = 7$ we found singly-even codes with one codeword of minimum weight, whereas in the case $t = 9$ there are both singly-even and doubly-even codes of this type. Furthermore we found one equivalence class of an extremal doubly-even self-dual [56, 28,12] code and two equivalence classes of singly-even [56,28,10] codes.

The vectors M and N are given in the form of integers. The corresponding vectors are generated by converting the integers to binary, and then replacing all 0's to -1's

In the case of $t = 7$ one equivalence class of a doubly-even extremal code was found. The representative of the class is given below. The Hadamard matrix obtained here is converted into an equivalent form given by Tonchev [11] (see [1] for details) before being used in the $\{I, A\}$ form.

Code	$\{M; N\}$	$|AutC|$
[56,28,12]	2311;6602	$58968 = 2^3 \times 3^4 \times 7 \times 13$

The table below lists the codes found with partial weight enumerators in the form 8:1 meaning 1 codeword of weight 8. The complete weight enumerators can be obtained using Gleason's Theorem [9].

No.	Code	$\{M; N\}$	de or se	Weight Enumerator
1	[40,20,4]	700; 868	de	4:1, 8:309
2	[56,28,8]	430; 1765	se	8:1, 10:248, 12:4116
3	[56,28,8]	2583; 3190	se	8:1, 10:272, 12:4068
4	[56,28,8]	3795; 7632	se	8:1, 10:256, 12:4100
5	[56,28,10]	3487; 7250	se	10:284, 12:4038
6	[56,28,10]	5113; 5908	se	10:268, 12:4070
7	[72,36,8]	11916; 253733	se	8:1, 10:15, 12:556
8	[72,36,8]	132316; 179038	se	8:1, 10:6, 12:722
9	[72,36,8]	70627; 95888	se	8:1, 10:14, 12:536
10	[72,36,8]	616; 94613	de	8:1, 12:1060

5   Conclusion

The fast search techniques discussed in this paper demonstrate two complementing approaches to the problem of finding self-dual codes. One approach is to develop techniques specific to the domain we are searching. The techniques in Section 2, for example, are effective but specific to the structure of D_{4t}-cocyclic Hadamard matrices and the $[I|A]$ construction of self-dual codes.

The second approach is to consider a general framework based on techniques developed in other problem domains. Image restoration techniques have always been concerned with approximating the missing pixel values in a damaged picture. We have adapted that technique to approximate the locations of missing cocyclic Hadamard matrices in the search space. The framework can be applied to any problem domain where the search space can be mapped to a two-dimensional region, and the missing points are unlikely to be uniformly distributed.

6   Further Work

The authors are currently working on implementing the search method in a distributed computing environment, such as the Parallel Parametric Modelling Engine [6] facility at Monash University. Although the search method can be useful in finding the cocyclic Hadamard matrices which can be used in the $[I|A]$ construction of self-dual doubly-even and singly-even codes, identifying the equivalence classes remains rather tedious and time-consuming. We are yet to find a systematic way of doing that task more easily.

References

1. A. Baliga, Cocyclic codes of length 40, *Designs, Codes and Cryptography* (to appear).
2. A. Baliga, "Extremal doubly-even self-dual cocyclic [40, 20] codes", Proceedings of the 2000 IEEE International Symposium on Information Theory, 25-30 June, 2000, pp.114.
3. A. Baliga, New self-dual codes from cocyclic Hadamard matrices, *J. Combin. maths. Combin. Comput.*, 28 (1998) pp. 7-14.
4. J. J. Chua and A. Baliga, An adaptive k-NN technique for image restoration. (In preparation.)
5. J. J. Chua and P. E. Tischer, "Minimal Cost Spanning Trees for Nearest-Neighbour Matching" in *Computational Intelligence for Modelling, Control and Automation: Intelligent Image Processing, Data Analysis and Information Retrieval*, ed M. Mohammadian, IOS Press, 1999, pp. 7–12.
6. http://hathor.cs.monash.edu.au/
7. D.L. Flannery, Cocyclic Hadamard matrices and Hadamard groups are equivalent, *J. Algebra*, 192 (1997), pp 749-779.
8. N. Ito, On Hadamard groups II, *Kyushu J. Math.*, 51(2), (1997) pp. 369-379.
9. F. J. MacWilliams and N. J. A. Sloane, *The Theory of Error-Correcting codes*, North-Holland, Amsterdam, (1977).
10. S. Mazzola, A k-nearest neighbour-based method for restoration of damaged images, *Pattern Recognition*, 23(1/2), (1990) pp. 179-184.
11. V.D. Tonchev, Self-orthogonal designs and extremal doubly-even codes, *J. Combin. Theory, Ser A*, 52,(1989), 197-205.

Low Complexity Tail-Biting Trellises
of Self-dual codes of Length 24, 32 and 40
over $GF(2)$ and $\mathbb{Z}_4$ of Large Minimum Distance

E. Cadic[1], J.C. Carlach[1], G. Olocco[2], A. Otmani[3], and J.P. Tillich[2]

[1] France Telecom R&D, DMR/DDH, CCETT-Rennes, 35512 Cesson-Sévigné, France
[2] LRI, Université Paris-Sud, 91405 Orsay, France.
[3] University of Limoges/LACO, 123 avenue Albert Thomas, 87060 Limoges, France

Abstract. We show in this article how the multi-stage encoding scheme proposed in [3] may be used to construct the $[24, 12, 8]$ Golay code, and two extremal self-dual codes with parameters $[32, 16, 8]$ and $[40, 20, 8]$ by using an extended $[8, 4, 4]$ Hamming base code. An extension of the construction of [3] over $\mathbb{Z}_4$ yields self-dual codes over $\mathbb{Z}_4$ with parameters (for the Lee metric over $\mathbb{Z}_4$) $[24, 12, 12]$ and $[32, 16, 12]$ by using the $[8, 4, 6]$ octacode. Moreover, there is a natural Tanner graph associated to the construction of [3], and it turns out that all our constructions have Tanner graphs that have a cyclic structure which gives tail-biting trellises of low complexity: 16-state tail-biting trellises for the $[24, 12, 8]$, $[32, 16, 8]$, $[40, 20, 8]$ binary codes, and 256-state tail-biting trellises for the $[24, 12, 12]$ and $[32, 16, 12]$ codes over $\mathbb{Z}_4$.

Keywords : self-dual codes, tail-biting trellises, codes over $\mathbb{Z}_4$, Tanner graph.

1 Introduction

We first recall the multi-stage encoding scheme of [3]. It uses a 1/2-rate binary $[2k, k]$ base code and yields a 1/2-rate $[2K, K]$ code for any K which is a multiple of k. To describe the construction we first need to choose a systematic generator matrix $(I|P)$ for the base code. The construction also uses ℓ permutations $\pi_1, \pi_2, \ldots, \pi_\ell$ of $\{1, 2, \ldots, K\}$. The resulting $[2K, K]$ code is obtained as follows. To encode a message $x \overset{\text{def}}{=} (x_1, x_2, \ldots, x_K)$ of length K, x is split into $t \overset{\text{def}}{=} K/k$ blocks of size k namely $(b_1, b_2, \ldots, b_t)$ which are encoded by the base code to yield the redundancy blocks $f(x) \overset{\text{def}}{=} (b_1 P, b_2 P, \ldots, b_t P)$ so as to form a new binary message $y^{(0)}$ of length K. Then we permute the coordinates of $y^{(0)}$ with π_1 and we iterate this process to obtain $y^{(\ell)} \overset{\text{def}}{=} f \pi_\ell f \pi_{\ell-1} \ldots \pi_1 f(x)$. The encoding of x will be the codeword $(x_1, x_2, \ldots, x_K, y_1^{(\ell)}, y_2^{(\ell)}, \ldots, y_K^{(\ell)})$. We say that this is an $(\ell + 1)$-stage code. To avoid cumbersome superscripts we denote by r the message $y^{(\ell)}$.

S. Boztaş and I.E. Shparlinski (Eds.): AAECC-14, LNCS 2227, pp. 57–66, 2001.
© Springer-Verlag Berlin Heidelberg 2001

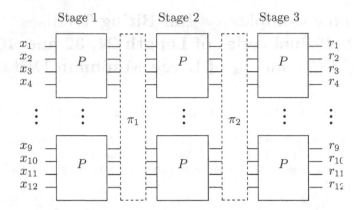

Fig. 1. Architecture of a 3-stage [24, 12] block code

This construction has a few interesting features

- The resulting code is self-dual whenever the base code is itself self-dual [4]. What is more, it is readily checked that if the base code is t-divisible (i.e. all the codeword weights are divisible by t) then the resulting code is also t-divisible if the number of stages is odd. This implies that if the base code is self-dual of type II then all the codes obtained by this construction with an odd number of stages are also self-dual of type II. Making the permutations vary over S_K gives a whole family of self-dual codes for any length which is a multiple of the length of the base code. It seems that many extremal self-dual codes can be obtained in this way when the base code is chosen to be the [8, 4, 4] self-dual extended Hamming code (see [4,5]).
- If the number of stages goes to infinity with the length of the code and if the permutations are chosen at random, then these codes tend to look like random codes. For instance, if the short code is an extended [8, 4, 4] Hamming code and when the number of stages is large enough, then almost all these codes lie on the Gilbert-Varshamov bound and have a binomial distance distribution [11].
- As shown in [10], these codes have a natural iterative decoding algorithm, one round of decoding consists in decoding all small underlying base codes. However this decoding algorithm is highly sensitive to the number of stages, it turns out that the optimal number of stages is 3 for this iterative decoding algorithm [10] to work properly.

This leads us to study what extremal codes can be obtained from this construction by using only 3 stages. Not only do we show in this article how to construct with only 3 stages and with the [8, 4, 4] Hamming code as base code, the [24, 12, 8] Golay code and two extremal [32, 16, 8] and [40, 20, 8] self-dual codes, but it also turns out that the constructions given in this article have an underlying Tanner graph with only a few cycles, which is clearly desirable for iterative decoding. Moreover, by choosing as base code the [8, 4, 6] octacode over

$\mathbb{Z}_4$, we get by using the same structure, two self-dual codes over $\mathbb{Z}_4$ with parameters (for the Lee metric) $[24, 12, 12]$ and $[32, 16, 12]$. By the Gray map these two codes give two non-linear binary codes of rate $1/2$ of length 48 and 64 of distance 12.

What is more, the Tanner graphs associated to our constructions basically yield minimal 16-state tail-biting trellises for the Golay code , the $[32, 16, 8]$ and the $[40, 20, 8]$ codes. For the $[48, 24, 12]$ and $[64, 32, 12]$ non-linear binary codes obtained by the Gray map from the $[24, 12, 12]$ and $[32, 16, 12]$ self-dual codes over $\mathbb{Z}_4$, we also obtain a tail-biting trellis with only very few states, namely 256. This should be compared with the 256-state tail-biting trellis obtained for the $[48, 24, 12]$ quadratic-residue code (see [8]).

2   The Associated Tanner Graph

There is a natural Tanner graph associated to our construction, namely the one which has for variable nodes all the x_j's, the $y_j^{(i)}$'s and the r_j's. We say that a node associated to an x_j or an r_j is a *symbol* node, and the other variable nodes are *state nodes*. This terminology comes from the fact that the symbol nodes form a codeword of the overall code, and the state nodes are only auxiliary nodes which are needed to compute the codeword. Each check node is associated to a P box of the encoding process, and a variable node is associated to a check node if it is either an input or an output of the P box associated to it.

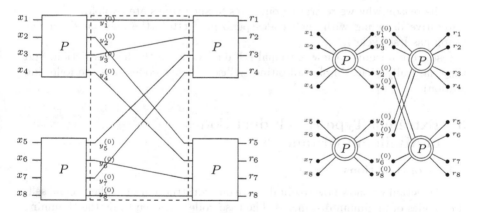

Fig. 2. 2-stage encoding of a $[16,8]$ code with its associated Tanner graph.

In this article we concentrate on codes obtained from our construction using a base code of length 8, with only 3 stages of encoding, and which have an associated Tanner graph with a very simple structure :

if two check nodes have at least one variable node in common, then they have exactly 2 of them.

It is well known how to perform optimal and efficient soft decision decoding on any memoryless channel when the Tanner graph of the code is cycle free [15]. The Tanner graphs of the codes considered in this article have cycles, however by grouping together two variable nodes which are adjacent to a same check node, we obtain a Tanner graph that has less cycles than the original one. We also group the symbol nodes which are associated to the same check node. From now on, we only deal with this "simplified" Tanner graph.

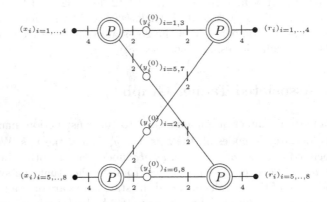

Fig. 3. The simplified Tanner graph.

The reason why we restricted ourselves to such codes are twofold:
- iterative decoding with such codes performs better (the associated Tanner graph has less cycles),
-we still obtain quite a few extremal self-dual codes with this restriction, and they naturally yield simple tail-biting trellises for our codes (see the following section).

3   Extremal Type II Self-dual Codes of Minimum Distance 8

3.1   Constructions

In this section we show how to obtain with our construction several extremal self-dual codes of minimum distance 8. The base code is chosen to be the Hamming code with generator matrix

$$G \stackrel{\text{def}}{=} \begin{bmatrix} 1\,0\,0\,0\,0\,1\,1\,1 \\ 0\,1\,0\,0\,1\,0\,1\,1 \\ 0\,0\,1\,0\,1\,1\,0\,1 \\ 0\,0\,0\,1\,1\,1\,1\,0 \end{bmatrix}$$

In other words $(x_1, x_2, x_3, x_4)P$ is equal to (x_1, x_2, x_3, x_4) if the weight of (x_1, x_2, x_3, x_4) is even and to $(\bar{x}_1, \bar{x}_2, \bar{x}_3, \bar{x}_4)$ otherwise.

An example of a $[24, 12, 8]$ Golay code encoder is shown below in Figure 4 :

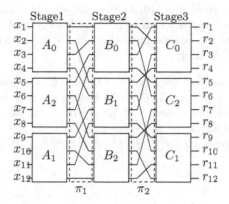

Fig. 4. 3-stage encoding of the $[24, 12, 8]$ Golay code.

π_1 maps $(1, 2, 3, \ldots, 12)$ onto $(1, 5, 2, 6, 3, 9, 4, 10, 7, 11, 8, 12)$, and π_2 maps $(1, 2, 3, \ldots, 12)$ onto $(2, 6, 1, 5, 4, 10, 3, 9, 8, 12, 7, 11)$.

The associated Tanner graph has the following cyclic structure (we identify the check node B_0 at the top with the check node B_0 at the bottom)

Actually, there are many other choices of permutations which lead to the Golay code, the nice feature of this choice is that, by changing the base code to the octacode over $\mathbb{Z}_4$, we obtain with the same choice of permutation a $[24, 12, 12]$

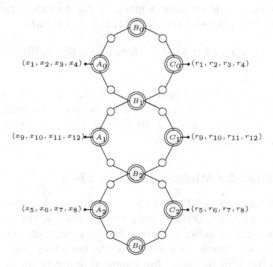

Fig. 5. The Tanner graph of the Golay code

code over $\mathbb{Z}_4$. It is readily checked (and this will be the issue of subsection 5.1) that by merging state nodes which are on the same level we obtain a new Tanner graph which is a cycle and which yields a 16-state tail-biting trellis.

Two extremal $[32, 16, 8]$ and $[40, 20, 8]$ codes are obtained by a similar construction with Tanner graphs which have a similar necklace structure. Since all our Tanner graphs in our article have the same structure, let us define more formally a graph of this kind

Definition 1. *The necklace graph $N(k)$ of order k is a bipartite graph with 9 kinds of nodes : the nodes A_i, B_i, C_i, which are on one side of the bipartition (these nodes are check nodes) and s^i, t^i, u^i, v^i (these are the state nodes), x^i, r^i (these are the symbol nodes) on the other side, for $0 \leq i \leq k - 1$. A_i is adjacent to x^i, s^i, u^i. B_i is adjacent to $u^{(i-1) \bmod k}, v^{(i-1) \bmod k}, s^i, t^i$, and C_i is adjacent to r^i, t^i, v^i.*

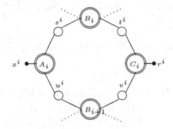

Fig. 6. One link of the necklace

The Tanner graph of the Golay code is nothing else but $N(3)$.

The $[32, 16, 8]$ cp$_{85}$ code encoder is given by the two identical permutations π_1 and π_2 maping the ordered integer set $(1, 2, 3, ..., 16)$ onto

$$(1, 5, 2, 6, 3, 9, 4, 10, 7, 13, 8, 14, 11, 15, 12, 16).$$

Its Tanner graph is $N(4)$.

An example of a $[40, 20, 8]$ type-II self-dual extremal code encoder is given by the two identical permutations π_1 and π_2 maping the ordered integer set $(1, 2, 3, ..., 20)$ onto $(1, 5, 2, 6, 3, 9, 4, 10, 7, 13, 8, 14, 11, 17, 12, 18, 15, 19, 16, 20)$. The associated Tanner graph is $N(5)$.

3.2   Proving That the Minimum Distance Is 8

To obtain the aforementioned bound on the minimum distance, we consider a non-zero codeword and the corresponding values taken at the variable nodes
- the values of the symbol nodes are given by the coordinates of the codeword,
- the values of the state nodes are given by the encoding process explained in the introduction. Note the following fundamental property which holds for every check node of the Tanner graphs of our codes:

(P1) *The binary word of length 8 formed by the the values taken at the symbol or state nodes adjacent to a check node belong to the [8,4,4] extended Hamming code.*

When this binary word of length 8 is not equal to the zero word, we say that the corresponding check node *sees a non-zero Hamming codeword around it*. We also say that a state (or symbol) node *carries weight* α, iff the value taken at the corresponding node has weight α.

The fact that all the codes which are obtained in this way have minimum distance greater than or equal to 8, is a consequence of the fact that all the permutations have been chosen such that the two following properties hold

(P2) *The Tanner graph of our codes is a necklace graph of order $k \geq 3$. State nodes carry values over $\{0,1\}^2$, symbol nodes carry values over $\{0,1\}^4$.*

(P3) *Whenever a check node of degree 4 (that is a check node which is adjacent to exactly 4 nodes which are actually state nodes) sees a non-zero Hamming codeword around it, then at least three of its adjacent nodes carry non-zero values.*

It is straightforward to check that, for a code which has a Tanner graph which satisfies (P1) and (P2), (P3) is necessary to ensure that the minimum distance is greater than 4. It turns out that if the Tanner graph meets all these properties (P1),(P2),(P3), then the resulting code has minimum distance at least 8. This is a straightforward consequence of the two following facts

Fact 1 *To check that the minimum distance is greater than or equal to 8, it is sufficient to check that there is no weight 4 codeword.*

This is a consequence of the more general fact that all the self-dual codes constructed with an odd number of stages with our construction by using the self-dual [8, 4, 4] Hamming code are self-dual of type II, see [4].

Fact 2 *If a symbol node has weight 1, then the 2 other state nodes adjacent to the same check node carry also non-zero values.*

This comes from the fact that if the Hamming weight of a symbol node is 1, then the sum of the weights of the 2 state nodes which are adjacent to the same check node is 3.

We are ready now to prove that the codes which have a Tanner graph satisfying properties (P1),(P2),(P3) have minimum distance at least 8. We distinguish between two different cases.

Case 1 : only one check node of degree 4 (in other words a node of type B, what we denote by B_i) sees a non-zero Hamming codeword around it. There are 2 possibilities

- either the 4 state nodes adjacent to it carry weight 2, this implies that the 4 symbol nodes which are at distance 3 from this check node (i.e $x^i, r^i, x^{(i-1) \bmod k}$, $r^{(i-1) \bmod k}$) have Hamming weight 2. In other words the weight of the codeword is at least $2 + 2 + 2 + 2 = 8$ in this case.

- or at least two of the non-zero state nodes adjacent to this check node carry weight 1, this means that the two check nodes of degree 3 adjacent to these

2 state nodes are adjacent to a symbol node which has weight 3 (since these 2 check nodes are both adjacent to three nodes : one state node of weight 1, another state node of weight 0 by hypothesis, and a symbol node which has to be of weight 3 because the values taken at these 3 nodes should form a codeword of the $[8, 4, 4]$ extended Hamming code). In other words, we have a codeword of weight at least $3 + 3 = 6$ and therefore of weight at least 8 by Fact 1.

Case 2 : at least two check nodes of degree 4 see non-zero Hamming codewords around them. From property (P3) we know that there are at least 4 symbol nodes which are at distance 3 from these two check nodes which are non-zero. Assume that all these 4 symbol nodes have Hamming weight 1 (otherwise we are done by Fact 1), then all are adjacent to a check node for which the two other state nodes adjacent to it are non-zero (Fact 2). Notice now that at least one of these non-zero state nodes is adjacent to a third check node of degree 4 [1] which has to see a non-zero codeword around it too. By using property (P3) again for this check node we obtain a new symbol node which is non-zero. This gives at least 5 non-zero symbol nodes, and yields a codeword of weight at least 5, and therefore of weight at least 8 by Fact 1.

4   Self-dual $\mathbb{Z}_4$ Codes with Minimum Distance 12

The short base code is chosen to be the $[8, 4, 6]$ octacode with generator matrix

$$G \stackrel{\text{def}}{=} \begin{bmatrix} 1 & 0 & 0 & 0 & 2 & 1 & 3 & 1 \\ 0 & 1 & 0 & 0 & 3 & 2 & 1 & 1 \\ 0 & 0 & 1 & 0 & 1 & 3 & 2 & 1 \\ 0 & 0 & 0 & 1 & 1 & 1 & 1 & 2 \end{bmatrix}$$

It turns out that the same choice of permutations which has given the Golay code in Section 3 also works here to give a $[24, 12]$ code with minimum (Lee) distance 12. This code is also self-dual. There are many ways to check this last fact. One may notice that the Tanner graph associated to it is normal [6] (the symbol nodes have degree 1 and the state nodes have degree 2) and all the constraints at the check nodes are self-dual (see [6]), therefore from [6] the associated code is self-dual. Note that this Tanner graph is the necklace graph $N(3)$.

The fact that the minimum distance is 12, follows from a reasoning which is similar to the the one given in the previous section, but it is unfortunately too long to be included here. It will be included in an extended version of this paper. One can also use the method which is sketched in subsection 5.2 to calculate the weight enumerator polynomial to prove this fact.

A $[32, 16, 12]$ self-dual code over $\mathbb{Z}_4$ is also obtained from the octacode with

$$\pi_1(1, 2, 3, ..., 16) = (1, 5, 2, 6, 3, 9, 4, 10, 7, 13, 8, 14, 11, 15, 12, 16)$$
$$\pi_2(1, 2, 3, ..., 16) = (2, 6, 1, 5, 4, 10, 3, 9, 8, 14, 7, 13, 12, 16, 11, 15)$$

---

[1] Note that it is here that we need at least three check nodes of degree 4, so that the order of our necklace graphs has to be at least 3.

5   Conclusion

5.1   Construction of Tail-Biting Trellises of Low Complexity

This construction can be easily put into the form of a tail-biting trellis by merging the nodes of the same type on the same level, i.e A_i is merged with C_i, x^i is merged with r^i, s^i is merged with t^i, and u^i is merged with v^i. The resulting Tanner graph is a cyclic chain : this leads to a tail-biting trellis as can be readily checked from the definition of a tail-biting trellis given for instance in [2]. For the codes of distance 8 that we have constructed in this article, the number of values which can be taken at the new state nodes, that is $4 \times 4 = 16$, gives the state complexity of the tail-biting trellis. From Proposition 6 in [2] we know that the state complexity of any linear tail-biting trellis of a rate $1/2$ binary code with minimum distance 8 is at least 16 : this implies that our tail-biting trellises are minimal with respect to the state complexity.

For the codes of distance 12 that we have constructed, the state complexity is 256. We do not know any bound which would show that this state complexity is minimal, however this state complexity is the same as the state complexity of the tail-biting trellis of the $[48, 24, 12]$ quadratic residue code found by Kötter and Vardy [8].

5.2   Computing the Weight Enumerator Polynomial

It is well known how to compute the weight enumerator of a code which has a Tanner graph which is a tree [1] (this contains as a special case the computation of the weight enumerator of a code from its trellis). In our case we have codes which have Tanner graphs which are cycles (when we merge the nodes as explained in the previous subsection). A slight generalization (see [8] for instance) of this aforementioned method works in our case, and we obtain the well known weight enumerator polynomial $W(Z) = 1 + 759Z^8 + 2576Z^{12} + 759Z^{16} + Z^{24}$ of the Golay code.

5.3   Iterative Decoding

The Tanner graphs that we have constructed also allow to perform sub-optimal iterative decoding by either the min-sum algorithm or the sum-product algorithm [15]. For tail-biting trellises (i.e Tanner graphs which are cycles, which is the case for our Tanner graphs, when we merge the state nodes which are on the same level) such iterative decoding algorithms achieve near-maximum-likelihood performance with only moderate performance [12] and have only very moderate complexity.

Acknowledgement

The authors would like to thank the reviewer for his careful reading of the first draft of this paper and his very helpful comments and suggestions.

References

1. S.M. Aji and R.J. McEliece, "The Generalized Distributive Law", *IEEE Transactions on Information Theory*, vol. 46, no.2, March 2000, pp.325-343.
2. A.R. Calderbank, G.D. Forney, and A. Vardy, "Minimal tail-biting trellises, the Golay code and more", *IEEE Transactions on Information Theory*, vol.45, no.5, July 1999, pp.1435-1455.
3. J.C. Carlach, C. Vervoux, "A New Family of Block Turbo-Codes", AAECC-13, 1999, pp.15-16.
4. J.C. Carlach, A. Otmani, "A Systematic Construction of Self-Dual Codes", submitted to *IEEE Transactions on Information Theory*.
5. J.C. Carlach, A. Otmani and C. Vervoux, "A New Scheme for Building Good Self-Dual Block Codes", ISIT2000, June 2000.
6. G.D. Forney, "Codes on graphs : Normal realizations", *IEEE Transactions on Information Theory*, vol.47, Feb. 2001, pp.520-548.
7. A.R. Hammons, V. Kumar, A.R. Calderbank, N.J.A. Sloane, P. Solé "The $\mathbb{Z}_4$-Linearity of Kerdock, Preparata, Goethals and Related Codes", *IEEE, Transactions on Information Theory*, vol.40, 1996, pp.1072-1092.
8. R. Kötter and A. Vardy, "The theory of tail-biting trellises : bounds and applications", manuscript in preparation, May 1999.
9. R.J. McEliece, "On the BCJR trellis for linear block codes", *IEEE, Transactions on Information Theory*, vol.42, pp.1072-1092, 1996.
10. G. Olocco, J.P. Tillich, "Iterative decoding of a new family of block turbo-codes", Proceedings of 2nd International Symposium on Turbo Codes and Related Topics, Brest Sept. 2000, pp.302-306.
11. G. Olocco, J.P. Tillich, "A family of self-dual codes which behave in many respects like random linear codes of rate $\frac{1}{2}$", submitted to *IEEE Transactions on Information Theory*.
12. A. Reznik, "Iterative Decoding of Codes Defined on Graphs", *S.M. Thesis*, LIDS-TH-2418, Massachusets Institute of Technology, USA, July 1998.
13. D. Tanner, "A Recursive Approach to Low-Complexity Codes", *IEEE, Transactions on Information Theory*, vol.IT-27, Sept. 1981, pp. 533-547.
14. N. Wiberg, H.-A. Loeliger and R. Kötter, "Codes and Iterative Decoding on General Graphs", *European Transactions on Telecommunications*, vol.6, no.5, Sept-Oct 1995, pp.513-525.
15. N. Wiberg, "Codes and Decoding on General Graphs", *Ph.D.Thesis*, Linköping Studies in Science and Technology no.440, Linköping University, Sweden, Apr. 1996.

F_q-Linear Cyclic Codes over F_{q^m}: DFT Characterization

Bikash Kumar Dey and B. Sundar Rajan[*]

Indian Instute of Science, Bangalore 560012, India
bikash@protocol.ece.iisc.ernet.in
bsrajan@ece.iisc.ernet.in

Abstract. Codes over F_{q^m} that form vector spaces over F_q are called F_q-linear codes over F_{q^m}. Among these we consider only cyclic codes and call them F_q-linear cyclic codes (F_qLC codes) over F_{q^m}. This class of codes includes as special cases (i) group cyclic codes over elementary abelian groups ($q = p$, a prime), (ii) subspace subcodes of Reed-Solomon codes and (iii) linear cyclic codes over F_q (m=1). Transform domain characterization of F_qLC codes is obtained using Discrete Fourier Transform (DFT) over an extension field of F_{q^m}. We show how one can use this transform domain structures to estimate a minimum distance bound for the corresponding quasicyclic code by BCH-like argument.

1 Introduction

A code over F_{q^m} (q is a power of a prime) is called linear if it is a vector space over F_{q^m}. We consider F_qLC codes over F_{q^m}, i.e., codes which are cyclic and form vector spaces over F_q. The class of F_qLC codes includes the following classes of codes as special cases:

1. **Group cyclic codes over elementary abelian groups:** When $q = p$ the class of F_pLC codes becomes group cyclic codes over an elementary abelian group C_p^m (a direct product of m cyclic groups of order p). A length n group code over a group G is a subgroup of G^n under componentwise operation. Group codes constitute an important ingredient in the construction of geometrically uniform codes [4]. Hamming distance properties of group codes over abelian groups is closely connected to the Hamming distance properties of codes over subgroups that are elementary abelian [5]. Group cyclic codes over C_p^m have been studied and applied to block coded modulation schemes with phase shift keying [8]. It is known [13],[19] that group cyclic codes over C_p^m contain MDS codes that are not linear over F_{p^m}.

2. **SSRS codes:** With $n = q^m - 1$, the class of F_qLC codes includes the subspace subcodes of Reed-Solomon (SSRS) codes [7], which contain codes with larger number of codewords than any previously known code for some lengths and minimum distances.

---

[*] This work was partly supported by CSIR, India, through Research Grant (22(0298)/99/EMR-II) to B. S. Rajan

S. Boztaş and I.E. Shparlinski (Eds.): AAECC-14, LNCS 2227, pp. 67–76, 2001.

3. **Linear cyclic codes over finite fields:** Obviously, with $m = 1$, the $F_q LC$ codes are the extensively studied class of linear cyclic codes.

A code is m-quasicyclic if cyclic shift of components of every codeword by m positions gives another codeword [11]. If $\{\beta_0, \beta_1, \cdots, \beta_{m-1}\}$ is a F_q-basis of F_{q^m}, then any vector $(a_0, a_1, \cdots, a_{n-1}) \in F_{q^m}^n$ can be seen with respect to this basis as $(a_{0,0}, a_{0,1}, \cdots, a_{0,m-1}, \cdots, a_{n-1,0}, a_{n-1,1}, \cdots, a_{n-1,m-1}) \in F_q^{mn}$, where $a_i = \sum_{j=0}^{m-1} a_{i,j} \beta_j$. This gives a 1-1 correspondence between the class of $F_q LC$ codes of length n over F_{q^m} and the class of m-quasicyclic codes of length mn over F_q. Unlike in [3], which considers $(nm, q) = 1$, $F_q LC$ codes gives rise to m-quasicyclic codes of length mn with $(n, q) = 1$.

It is well known [1], [14] that cyclic codes over F_q and over the residue class integer rings Z_m are characterizable in the transform domain using Discrete Fourier Transform (DFT) over appropriate Galois fields and Galois rings [12] respectively and so are the wider class of abelian codes over F_q and Z_m using a generalized DFT [15],[16]. The transform domain description of codes is useful for encoding and decoding [1],[17]. DFT approach for cyclic codes of arbitrary length is discussed in [6]. In this correspondence, we obtain DFT domain characterization of $F_q LC$ codes over F_{q^m} using the notions of certain invariant subspaces of extension fields of F_{q^m}, two different kinds of cyclotomic cosets and linearized polynomials.

The proofs of all the theorems and lemmas are omitted due to space limitations.

2 Preliminaries

Suppose $\mathbf{a} = (a_0, a_1, \cdots, a_{n-1}) \in F_{q^m}^n$, where $(n, q) = 1$. From now on, r will denote the smallest positive integer such that $n | (q^{mr} - 1)$ and $\alpha \in F_{q^{mr}}$ an element of multiplicative order n. The set $\{0, 1, \cdots, n-1\}$ will be denoted by I_n. The Discrete Fourier Transform (DFT) of $\mathbf{a}$ is defined to be $\mathbf{A} = (A_0, A_1, \cdots, A_{n-1}) \in F_{q^{mr}}^n$, where $A_j = \sum_{i=0}^{n-1} \alpha^{ij} a_i$, $\quad j \in I_n$. A_j is called the j-th DFT coefficient or the j-th transform component of $\mathbf{a}$. The vectors $\mathbf{a}$ and $\mathbf{A}$ will be referred as time-domain vector and the corresponding transform vector respectively.

For any $j \in I_n$, the **q-cyclotomic coset modulo n** of j is defined as $[j]_n^q = \{i \in I_n | j \equiv iq^t \bmod n \text{ for some } t \geq 0\}$, and the q^m**-cyclotomic coset modulo n** of j is defined as $[j]_n^{q^m} = \{i \in I_n | j \equiv iq^{mt} \bmod n \text{ for some } t \geq 0\}$. We'll denote the cardinalities of $[j]_n^q$ and $[j]_n^{q^m}$ as e_j and r_j respectively.

Example 1. Table 1 shows $[j]_{15}^2$, $[j]_{15}^{2^2}$, $[j]_{15}^{2^3}$ and $[j]_{15}^{2^4}$ for $j \in I_{15}$.

Mostly we'll have n for the modulus. So we'll drop the modulus when not necessary. Clearly, a q-cyclotomic coset is a disjoint union of some q^m-cyclotomic cosets. If $J \subseteq I_n$, we write $[J]_n^q = \cup_{j \in J} [j]_n^q$ and $[J]_n^{q^m} = \cup_{j \in J} [j]_n^{q^m}$.

If $\mathbf{b}$ is the cyclically shifted version of $\mathbf{a}$, then $B_j = \alpha^j A_j$ for $j \in I_n$. This is the **cyclic shift property** of DFT. The DFT components satisfy **conjugacy**

Table 1. Cyclotomic cosets modulo 15

$2/2^3$-cycl. cosets	{0}	{1,2,4,8}		{3,6,9,12}		{5,10}		{7,13,11,14}	
cardinality	1	4		4		2		4	
2^2-cycl. cosets	{0}	{1,4}	{2,8}	{3,12}	{6,9}	{5}	{10}	{7,13}	{14,11}
cardinality	1	2	2	2	2	2		2	2
2^4-cycl. cosets	{0}	{1}	{2}	{4}	{8}	{3}	{6}	{9}	{12}	{5}	{10}	{7}	{13}	{11}	{14}
cardinality	1	1	1	1	1	1	1	1	1	1	1	1	1	1	1

constraint[1], given by $A_{(q^m j) \bmod n} = A_j^{q^m}$. So, conjugacy constraint relates the transform components in same q^m-cyclotomic coset.

Let $I_1, I_2, \cdots, I_l$ be some disjoint subsets of I_n and suppose $R_{I_j} = \{(A_i)_{i \in I_j} \mid \mathbf{a} \in \mathcal{C}\}$ for $j = 1, 2, \cdots, l$. The sets of transform components $\{A_i | i \in I_j\}$; $1 \le j \le l$ are called **unrelated** for $\mathcal{C}$ if $\{((A_i)_{i \in I_1}, (A_i)_{i \in I_2}, \cdots, (A_i)_{i \in I_l}) \mid \mathbf{a} \in \mathcal{C}\} = R_{I_1} \times R_{I_2} \times \cdots \times R_{I_l}$.

For a code $\mathcal{C}$, we say, A_j takes values from $\{\sum_{i=0}^{n-1} \alpha^{ij} a_i \mid \mathbf{a} \in \mathcal{C}\} \subseteq F_{q^{mr}}$. For linear cyclic codes, A_j takes values from $\{0\}$ or $F_{q^{mr_j}}$ and transform components in different q^m-cyclotomiccosets are unrelated.

For any element $s \in F_{q^l}$, the set $[s]^q = \{s, s^q, s^{q^2}, \cdots, s^{q^{e-1}}\}$, where e is the smallest positive integer such that $s^{q^e} = s$, is called the q-conjugacy class of s. Note that, if $\alpha \in F_{q^l}$ is of order n and $s = \alpha^j$, then there is an 1-1 correspondence between $[j]_n^q$ and $[s]^q$, namely $jq^t \mapsto s^{q^t}$. So, $|[s]^q| = |[j]_n^q| = e_j$.

For any element $s \in F_{q^l}$, an F_q-subspace U of F_{q^l} is called **s-invariant** (or $[s, q]$-subspace in short) if $sU = U$. An $[s, q]$-subspace of F_{q^l} is called minimal if it contains no proper $[s, q]$-subspace. If U and V are two $[s, q]$-subspaces of F_{q^l}, then so are $U \cap V$ and $U + V$. If e is the exponent of $[s]^q$, then $Span_{F_q}\{s^i | i \ge 0\} \simeq F_{q^e}$. So, for any $g \in F_{q^l} \setminus \{0\}$, the minimal $[s, q]$-subspace containing g is gF_{q^e}. Clearly, if $s' \in [s]^q$, then $[s, q]$-subspaces and $[s', q]$-subspaces are same.

Example 2. The minimal $[\alpha^5, 2]$ and $[\alpha^{10}, 2]$-subspaces of F_{2^4} are $V_1 = F_4 = \{0, 1, \alpha^5, \alpha^{10}\}$, $V_2 = \alpha F_4$, $V_3 = \alpha^2 F_4$, $V_4 = \alpha^3 F_4$, $V_5 = \alpha^4 F_4$. The $[\alpha^k, 2]$-subspaces, for $k \ne 0, 5, 10$ are $\{0\}$ and F_{16}. Every subset $\{0, x \in F_{16}^*\}$ is a minimal $[\alpha^0, 2]$-subspace.

3   Transform Domain Characterization of *FqLC* Codes

By the cyclic shift property, in an $F_q LC$ code $\mathcal{C}$, the values of A_j constitute an $[\alpha^j, q]$-subspace of $F_{q^{mr}}$. However, this is not sufficient for $\mathcal{C}$ to be an $F_q LC$ code.

Example 3. Consider length 15, F_2-linear codes over $F_{16} = \{0, 1, \alpha, \alpha^2, \cdots, \alpha^{14}\}$. We have $q = 2, m = 4$ and $r = 1$. In Table 2, the code $\mathcal{C}_3$ is not cyclic, though each transform component takes values from appropriate invariant subspaces. Other five codes in the same table are $F_2 LC$ codes. As DFT kernel, we have taken a primitive element $\alpha \in F_{16}$ with minimal polynomial $X^4 + X + 1$.

The characterization of $F_q LC$ codes is in terms of certain decompositions of the codes. In the following subsection, we discuss the decomposition of $F_q LC$ codes and in Subsection 3.2 present the characterization.

3.1 Decomposition of $F_q LC$ Codes

We start from the following notion of minimal generating set of subcodes for F_q-linear codes.

A set of F_q-linear subcodes $\{C_\lambda | \lambda \in \Lambda\}$ of a F_q-linear code C is said to be a generating set of subcodes if $C = \Sigma_{\lambda \in \Lambda} C_\lambda$. A generating set of subcodes $\{C_\lambda | \lambda \in \Lambda\}$ of C is called a **minimal generating set of subcodes (MGSS)** if $\Sigma_{\lambda \neq \lambda'} C_\lambda \neq C$ for all $\lambda' \in \Lambda$. MGSS of an F_q-linear code is not unique. For example, consider the length 3 F_2-linear code over F_{2^2}, $C = \{c_1 = (00,00,00), c_2 = (01,01,01), c_3 = (10,10,10), c_4 = (11,11,11)\}$. The sets of subcodes $\{\{c_1,c_2\}, \{c_1,c_3\}\}$ and $\{\{c_1,c_2\}, \{c_1,c_4\}\}$ are both MGSS for C.

Suppose A_j takes values from $V \subset F_{q^{mr}}$, $V \neq \{0\}$ for an F_q-linear code C. Let V_1 be an F_q-subspace of $F_{q^{mr}}$.We call $C' = \{a | a \in C, A_j \in V_1\}$ as the F_q-linear subcode obtained by restricting A_j in V_1. For example, the subcode C_1 of Table 2 can be obtained from C_4 by restricting A_5 to $\{0\}$. Clearly, if C is cyclic and V_1 is an $[\alpha^j, q]$-subspace, then C' is also cyclic. If $S \subseteq I_n$, then the subcode obtained by restricting the transform components A_j; $j \notin S$ to 0 is called the S-subcode of C and is denoted as C_S.

Lemma 1. *Suppose in an F_q-linear code C, A_j takes values from a subspace $V \in F_{q^{mr}}$. Let $V_1, V_2 \subseteq V$ be two subspaces of V such that $V = V_1 + V_2$. (i) If C_1 and C_2 are the subcodes of C, obtained by restricting A_j in V_1 and V_2 respectively, then $C = C_1 + C_2$. (ii) If V_1 and V_2 are $[\alpha^j, q]$-subspaces, then C is cyclic if and only if C_1 and C_2 are cyclic.*

Suppose for an F_q-linear code C, A_j takes values from a nonzero F_q-subspace V of $F_{q^{mr}}$, and V intersects with more than one minimal $[\alpha^j, q]$-subspace. Then, we have two nonzero $[\alpha^j, q]$-subspaces V_1 and V_2 such that $V \subseteq V_1 \oplus V_2$ and $V \cap V_1 \neq \phi$ and $V \cap V_2 \neq \phi$. Then, we can decompose the code as the sum of two smaller codes C_1 and C_2 obtained by restricting A_j to V_1 and V_2 respectively, i.e., $C = C_1 + C_2$. So by successively doing this for each j, we can decompose C into a generating set of subcodes, in each of which, for any $j \in I_n$, transform component A_j takes values from a F_q-subspace of a minimal $[\alpha^j, q]$-subspace. In particular, if the original code was an $F_q LC$ code, all the subcodes obtained this way will have A_j from minimal $[\alpha^j, q]$-subspaces. The following are immediate consequences of this observation and Lemma 1.

1. In a minimal $F_q LC$ code, any nonzero transform component A_j takes values from a minimal $[\alpha^j, q]$-subspace of $F_{q^{mr}}$. For example, for the codes C_1 and C_2 in Table 2, A_5 and A_{10} take values from minimal $[\alpha^5, 2]$-subspaces.
2. A code is $F_q LC$ if and only if all the subcodes obtained by restricting any nonzero transform component A_j in minimal $[\alpha^j, q]$-subspaces of $F_{q^{mr}}$ are $F_q LC$. The statement is also true without the word 'minimal'.

Suppose in an F_q-linear code $\mathcal{C}$, transform components A_j, $j \in I_n$ take values from F_q-subspaces V_j of $F_{q^{mr}}$. A set of transform components $\{A_l | l \in L \subseteq I_n\}$ is called a **maximal set of unrelated components (MSUC)** if they are unrelated for $\mathcal{C}$ and any other transform component A_k, $k \notin L$ can be expressed as $A_k = \sum_{l \in L} \sigma_{kl} A_l$ such that σ_{kl} is an F_q-homomorphism of V_l into V_k.

If some disjoint sets of transform components are unrelated in two codes $\mathcal{C}'$ and $\mathcal{C}''$, then so is true for the code $\mathcal{C}' + \mathcal{C}''$. However, the converse is not true. For instance, for the codes $\mathcal{C}_0$ and $\mathcal{C}_1$ in Table 2, A_5 and A_{10} are related but they are unrelated for the sum $\mathcal{C}_4 = \mathcal{C}_0 + \mathcal{C}_1$.

Theorem 1. *If $\mathcal{C}$ is an $F_q LC$ code over F_{q^m} where any nonzero transform component A_j takes values from a minimal $[\alpha^j, q]$-subspace V_j of $F_{q^{mr}}$, then there is an MSUC $\{A_l | l \in L \subset I_n\}$ for $\mathcal{C}$.*

Clearly, for a code as described in Theorem 1, if $l \in L$, the code $\mathcal{C}_l = \{\mathbf{a} \in \mathcal{C} | A_j = 0 \text{ for } j \in L \setminus \{l\}\}$ is a minimal $F_q LC$ code. So $\mathcal{C}$ can be decomposed into an MGSS as $\mathcal{C} = \oplus_{l \in L} \mathcal{C}_l$. Since any code can be decomposed into a minimal generating set of subcodes with nonzero transform components taking values from minimal invariant subspaces by restricting the components to minimal invariant subspaces, a minimal generating set of minimal $F_q LC$ subcodes can be obtained by further decomposing each of the subcodes as above. So, we have,

Theorem 2. *Any $F_q LC$ code can be decomposed as direct sum of minimal $F_q LC$ codes.*

Suppose, in an $F_q LC$ code, A_j and A_k take values from the $[\alpha^j, q]$-subspace V_1 and $[\alpha^k, q]$-subspace V_2 respectively. Suppose A_k is related to A_j by an F_q homomorphism $\sigma : V_1 \mapsto V_2$ i.e. $A_k = \sigma(A_j)$. Then, since the code is cyclic,

$$\sigma(\alpha^j v) = \alpha^k \sigma(v) \qquad \forall \quad v \in V_1. \qquad (1)$$

Clearly, for such a homomorphism, $Ker(\sigma)$ is an $[\alpha^j, q]$-subspace.

Lemma 2. *Let $\mathcal{C}$ be an $F_q LC$ code over F_{q^m} where each nonzero transform component A_j takes values from a minimal $[\alpha^j, q]$-subspace of $F_{q^{mr}}$. If $A_k = \sum_{i=1}^t \sigma_{j_i} A_{j_i}$, where A_{j_i}, $i = 1, 2, \cdots, t$ take values freely from some respective minimal invariant subspaces, then σ_{j_i}, $i = 1, 2, \cdots, t$ are all F_q-isomorphisms.*

3.2 Transform Characterization

The following theorem characterizes $F_q LC$ codes in the DFT domain.

Theorem 3. *Let $\mathcal{C} \subset F_{q^m}^n$ be an n-length F_q-linear code over F_{q^m} Then, $\mathcal{C}$ is $F_q LC$ if and only if all the subcodes of an MGSS obtained by restricting the transform components to minimal invariant subspaces satisfy the conditions:*
1. For all $j \in I_n$, the set of jth transform components is α^j-invariant.
2. There is an MSUC $\{A_j | j \in J\}$ where A_j takes values from a minimal $[\alpha^j, q]$-subspace V_j and $A_k = \sum_{j \in J} \sigma_{kj} A_j$ for all $k \notin J$, where σ_{kj} is an F_q-isomorphism of V_j onto V_k satisfying

$$\sigma_{kj}(\alpha^j v) = \alpha^k \sigma_{kj}(v) \ \forall v \in V_j. \qquad (2)$$

Example 4. In Table 2, the codes obtained by restricting A_{10} to V_5 and V_1 for the code C_5 are respectively C_0 and C_2. In both C_0 and C_2, the nonzero transform components A_5 and A_{10} take values from minimal $[\alpha^5, 2]$ invariant subspaces and sum of C_0 and C_2 is C_5. So, $\{C_0, C_2\}$ is an MGSS of C_5. In both C_0 and C_2, A_5 and A_{10} are related by isomorphisms. It can be checked that the isomorphisms satisfy the condition (2).

Since for an $F_q LC$ code, transform components can be related by homomorphisms satisfying (1), we characterize such homomorphisms in Section 4. We also show that for $F_q LC$ codes, A_j and A_k can be related iff $k \in [j]_n^q$.

4   Connecting Homomorphisms for $F_q LC$ Codes

Throughout the section an endomorphism will mean an F_q-endomorphism.

A polynomial of the form $f(X) = \sum_{i=0}^{t} c_i X^{q^i} \in F_{q^l}[X]$ is called a *q*-**polynomial or a linearized polynomial** [10] over F_{q^l}. Each *q*-polynomial of degree less than q^l induces a distinct F_q-linear map of F_{q^l}. So, considering the identical cardinalities, we have $End_{F_q}(F_{q^l}) = \{\sigma_f : x \mapsto f(x) | f(X) = \sum_{i=0}^{l-1} c_i X^{q^i} \in F_{q^l}[X]\}$

For any $y \in F_{q^l} \setminus \{0\}$, the automorphism induced by $f(X) = yX$ will be denoted by σ_y. The subset $\{\sigma_y | y \in F_{q^l} \setminus \{0\}\}$ forms a cyclic subgroup of $Aut_{F_q}(F_{q^l})$, generated by $\sigma_{\beta_{q^l}}$, where $\beta_{q^l} \in F_{q^l}$ is a primitive element of F_{q^l}. In this subgroup, $\sigma_y^i = \sigma_{y^i}$. We shall denote this subgroup as $S_{q,l}$ and $S_{q,l} \cup \{0\}$ as $\mathbf{S}_{q,l}$, where 0 denotes the zero map. Clearly, $\mathbf{S}_{q,l}$ forms a field isomorphic to F_{q^l}.

We shall denote the map $\sigma_{X^q} : y \mapsto y^q$ of F_{q^l} onto F_{q^l}, induced by the polynomial $f(X) = X^q$, as $\theta_{q,l}$. Clearly, $\theta_{q,l}\sigma_x = \sigma_x^q \theta_{q,l}$ i.e., $\theta_{q,l}\sigma_x \theta_{q,l}^{-1} = \sigma_x^q$ for all $x \in F_{q^l}$. The map induced by the polynomial $f(X) = X^{q^i}$ is $\theta_{q,l}^i$. So, for any $f(X) = \sum_{i=0}^{l-1} c_i X^{q^i}$, $\sigma_f = \sum_{i=0}^{l-1} \sigma_{c_i} \theta_{q,l}^i$. Thus we have $End_{F_q}(F_{q^l}) = \oplus_{i=0}^{l-1} \mathbf{S}_{q,l} \theta_{q,l}^i$ i.e., any endomorphism $\sigma \in End_{F_q}(F_{q^l})$ can be decomposed uniquely as $\sigma = \sum_{i=0}^{l-1} \sigma_{(i)}$ where $\sigma_{(i)} \in \mathbf{S}_{q,l} \theta_{q,l}^i$. We shall call this decomposition as canonical decomposition of σ.

Theorem 4. *Suppose $x_1, x_2 \in F_{q^l}$. Then, $[x_1]^q = [x_2]^q \Leftrightarrow \exists \sigma \in Aut_{F_q}(F_{q^l})$ such that $\sigma(x_1 x) = x_2 \sigma(x) \; \forall x \in F_{q^l}$.*

Lemma 3. *Let $V_1 \subseteq F_{q^l}$ be a minimal $[x_1, q]$-subspace and $\sigma : V_1 \longrightarrow F_{q^l}$ be a nonzero homomorphism of V_1 into F_{q^l}, satisfying $\sigma(x_1 v) = x_2 \sigma(v) \; \forall \; v \in V_1$. Then $[x_1]^q = [x_2]^q$.*

Theorem 5. *Suppose $x_1, x_2 \in F_{q^l}$. Let $V_1 \subset F_{q^l}$ be a $[x_1, q]$-subspace and σ is as in Lemma 3. Then (i) $[x_1]^q = [x_2]^q$ and (ii) $\sigma(V_2)$ is a $[x_1, q]$-subspace for any $[x_1, q]$-subspace $V_2 \subset V_1$.*

Theorem 6. *In an F_qLC code, the transform components of different q- cyclotomic cosets are mutually unrelated.*

Corollary 1. *Any minimal F_qLC code takes nonzero values only in one q-cyclotomic coset in transform domain and any minimal F_qLC code which has nonzero transform components in $[j]_n^q$ has size q^{e_j}.*

So, if $J_1, J_2, \cdots, J_t$ are the distinct q-cyclotomic cosets of I_n, then any F_qLC code $\mathcal{C}$ can be decomposed as $\mathcal{C} = \oplus_{i=1}^t \mathcal{C}_{J_i}$. Corresponding m-quasi-cyclic codes are called primary components [9] or irreducible components [2]. If $\mathbf{a} \in F_{q^m}^n$, then the intersection of all the F_qLC codes containing $\mathbf{a}$ is called the F_qLC code generated by $\mathbf{a}$. We call such F_qLC codes as one-generator F_qLC codes. Clearly, For a one-generator F_qLC code $\mathcal{C}$, each component $\mathcal{C}_{J_i}$ is minimal.

Suppose V_1 and V_2 are two subspaces of F_{q^l}. Suppose $y \in F_{q^l}$ such that V_1 is y-invariant and i is a nonnegative integer. Then, we define $Hom_{F_q}(V_1, V_2, y, i) = \left\{\sigma \in Hom_{F_q}(V_1, V_2) | \sigma y x = y^{q^i} \sigma x \ , \ \forall x \in V_1 \right\}$. Clearly, $Hom_{F_q}(V_1, V_2, y, i)$ is a subspace of $Hom_{F_q}(V_1, V_2)$. Since $y^{q^{e_y+i}} = y^{q^i}$, we shall always assume $i < e_y$. We are interested in $Hom_{F_q}(V_1, V_2, y, i)$ since, if for an F_qLC code, $A_j \in V_1$ and $A_{jq^i} \in V_2$, then A_j and A_{jq^i} can be related by a homomorphism $\sigma : V_1 \to V_2$ if and only if $\sigma \in Hom_{F_q}(V_1, V_2, \alpha^j, i)$.

Theorem 7. *Any $\sigma \in Hom_{F_q}(x_1 F_{q^{e_y}}, x_2 F_{q^{e_y}}, y, l)$ is induced by a polynomial $f(X) = cX^{q^i}$ for some unique constant $c \in x_2 x_1^{-1} F_{q^{e_y}}$.*

For $y = \alpha^j$, this theorem specifies all possible homomorphisms by which A_{jq^l} can be related to A_j for an F_qLC code when A_j takes values from a minimal $[\alpha^j, q]$-subspace.

Example 5. Clearly, in the codes $\mathcal{C}_0$ and $\mathcal{C}_2$ in Table 2, A_5 is related to A_{10} by homomorphisms. Suppose $A_5 = \sigma_f(A_{10})$ where $f(X)$ is a q-polynomial over F_{q^l}. For $\mathcal{C}_0$, $f(X) = \alpha^8 X^2$ and for $\mathcal{C}_2$, $f(X) = \alpha X^2$.

The following theorem specifies the possible relating homomorphisms when A_j takes values from a nonminimal $[\alpha^j, q]$-subspace.

Theorem 8. *Suppose $V \subseteq F_{q^l}$ is a $[y, q]$-subspace and $V = \oplus_{j=0}^{t-1} V_j$ where V_j are minimal $[y, q]$-subspaces. Then, for any $\sigma \in Hom_{F_q}(V, F_{q^l}, y, i)$, there is a unique polynomial of the form $f(X) = \sum_{j=0}^{t-1} a_j X^{q^{j e_y + i}}$, $a_j \in F_{q^l}$ such that $\sigma = \sigma_f$. So, $Hom_{F_q}(V, F_{q^l}, y, i) = \{\sigma_f | f(X) = \sum_{j=0}^{t-1} a_j X^{q^{j e_y + i}}, a_j \in F_{q^l}\}$*

So, if $j_1, \cdots, j_w \in [k]_n^q$ and A_k is related to $A_{j_1}, \cdots, A_{j_w}$ by homomorphisms i.e., if $A_k = \sigma_1(A_{j_1}) + \cdots + \sigma_w(A_{j_w})$, where $\sigma_1, \cdots, \sigma_w$ are homomorphisms, then the relation can be expressed as $A_k = \sum_{h_1=0}^{l_1-1} c_{1,h_1} A_{j_1}^{q^{h_1 e_k + t_1}} + \cdots + \sum_{h_w=0}^{l_w-1} c_{w,h_w} A_{j_w}^{q^{h_w e_k + t_w}}$, where $k \equiv j_i^{q^{t_i}} \mod n$ for $i = 1, \cdots, w$.

Example 6. In the code $\mathcal{C}_5$ in Table 2, A_5 is related to A_{10} by a homomorphism induced by the polynomial $f(X) = \alpha^{14} X^2 + \alpha^8 X^8$.

5 Parity Check Matrix and Minimum Distance of Quasicyclic Codes

For linear codes, Tanner used BCH like argument [18] to estimate minimum distance bounds from the parity check equations over an extension field.

With respect to any basis of F_{q^m}, there is a 1-1 correspondence between n-length $F_q LC$ codes and m-quasi-cyclic codes of length nm over F_q. Here we describe how in some cases one can directly get a set of parity check equations of a quasi-cyclic code from the transform domain structure of the corresponding $F_q LC$ code. We first give a theorem from [3] for the distance bound.

Theorem 9. *[3] Suppose, the components of the vector* $\mathbf{v} \in F_{q^r}^n$ *are nonzero and distinct. If for each* $k = k_0, k_1, \cdots, k_{\delta-2}$, *the vectors* $\mathbf{v}^k$ *are in the span of a set of parity check equations over* F_{q^r}, *then the minimum distance of the code is at least that of the cyclic code of length* $q^r - 1$ *with roots* β^k, $k = k_0, k_1, \cdots, k_{\delta-2}$ *where* β *is a primitive element of* F_{q^r}.

So, If $k_i = k_0 + i$, BCH bound gives $d_{min} \geq \delta$.

Let us fix a basis $\{\beta_0, \beta_1, \cdots, \beta_{m-1}\}$ of F_{q^m} over F_q. By our characterization of $F_q LC$ codes in DFT domain, we know that for any $j \in [0, n-1]$, A_j can take values from any $[\alpha^j, q]$-subspace of $F_{q^{rm_j}}$. In particular, A_j can take values from subspaces of the form $c^{-1} F_{q^l}$ where $e_j | l$ and $l | m r_j$. Then,

$$
(cA_j)^{q^l} = cA_j \Leftrightarrow \left(c \sum_{i=0}^{n-1} \alpha^{ij} a_i \right)^{q^l} = c \sum_{i=0}^{n-1} \alpha^{ij} a_i
$$

$$
\Leftrightarrow \left(c \sum_{i=0}^{n-1} \alpha^{ij} \sum_{x=0}^{m-1} a_{ix} \beta_x \right)^{q^l} = c \sum_{i=0}^{n-1} \alpha^{ij} \sum_{x=0}^{m-1} a_{ix} \beta_x.
$$

This gives a parity check vector $\mathbf{h} = (h_{0,0}, h_{0,1}, \cdots, h_{0,m-1}, \cdots, h_{n-1,0}, \cdots, h_{n-1,m-1})$ with $h_{i,x} = \left(c^{q^l} \alpha^{ijq^l} \beta_x^{q^l} - c\alpha^{ij} \beta_x \right)$. If $A_j = 0$, it gives a parity check vector $\mathbf{h}$ with $h_{i,x} = \beta_x$.

Now, for $F_q LC$ code, A_k can be related to several other transform components $A_{j_1}, A_{j_2}, \cdots, A_{j_w}$ by homomorphisms, where $j_1, \cdots, j_w \in [k]_n^q$. Then, $A_k = \sum_{h_1=0}^{l_1-1} c_{1,h_1} A_{j_1}^{q^{h_1 e_k + t_1}} + \cdots + \sum_{h_w=0}^{l_w-1} c_{w,h_w} A_{j_w}^{q^{h_w e_k + t_w}}$ for some constants $c_{i,h_i} \in F_{q^{mr}}$. It can be checked in the same way that, this gives a parity check vector $\mathbf{h}$ with $h_{i,x} = \beta_x \alpha^{ik} - \sum_{h_1=0}^{l_1-1} c_{1,h_1} \beta_x^{q^{h_1 e_k + t_1}} \alpha^{ij_1 q^{h_1 e_k + t_1}} - \cdots - \sum_{h_w=0}^{l_w-1} c_{w,h_w} \beta_x^{q^{h_w e_k + t_w}} \alpha^{ij_w q^{h_w e_k + t_w}}$.

The component wise conjugate vectors of the parity check vectors obtained in these ways and the vectors in their span are also parity check vectors of the code. However, in general for any $F_q LC$ code, the components may not be related simply by homomorphisms or components may not take values from the subspaces of the form $c^{-1} F_{q^l}$. In those cases, the parity check vectors obtained in the above ways may not specify the code completely. But still those equations can be used for estimating a minimum distance bound by Theorem 9.

Since the DFT components in different q-cyclotomic cosets modulo n are unrelated, the set of parity check equations over $F_{q^{mr}}$ are union of the check equations corresponding to each q-cyclotomic coset modulo n. Clearly, for any one generator code, a set of parity check vectors completely specifying the code can be obtained in this way. There are however other codes for which complete set of parity check vectors can be derived. In fact, codes can be constructed by imposing simple transform domain restrictions and thus allowing derivations of a complete set of parity check equations over $F_{q^{mr}}$. We illustrate this with the following example. If β is a primitive element of $F_{q^{mr}}$, then we use $\alpha = \beta^{\frac{q^{mr}-1}{n}}$ as the DFT kernel and we take the basis $\{1, \beta, \beta^2, \cdots, \beta^{m-1}\}$.

Example 7. We consider the F_2LC code of length $n = 3$ over F_{2^4} given by the transform domain restrictions $A_0 = 0$ and $A_2 = \beta^4 A_1^2 + \beta^{10} A_1^8$. With the chosen basis, these two restrictions give the parity check vectors of the underlying 4-quasi-cyclic code $\mathbf{h}_{(1)} = (1, \beta, \beta^2, \beta^3, 1, \beta, \beta^2, \beta^3, 1, \beta, \beta^2, \beta^3)$ and $\mathbf{h}_{(2)} = (\beta^8, \beta^5, \beta^{12}, \beta^6, \beta^3, 1, \beta^7, \beta, \beta^{13}, \beta^{10}, \beta^2, \beta^{11})$ respectively. Component-wise conjugates of these vectors are also parity check vectors. Moreover, $\mathbf{h}_{(2)}^3 = (\beta^9, 1, \beta^6, \beta^3, \beta^9, 1, \beta^6, \beta^3, \beta^9, 1, \beta^6, \beta^3) = \beta \mathbf{h}_{(1)} + \beta^8 \mathbf{h}_{(1)}^2 + \beta^6 \mathbf{h}_{(1)}^4 + \mathbf{h}_{(1)}^8$ and $\mathbf{h}_{(2)}^0 = (1,1,1,1,1,1,1,1,1,1,1,1) = \beta^{11}\mathbf{h}_{(1)} + \beta^7\mathbf{h}_{(1)}^2 + \beta^{15}\mathbf{h}_{(1)}^4 + \beta^{13}\mathbf{h}_{(1)}^8$. So, the underlying quasi-cyclic code is a $[12, 4, 6]$ code. This code is actually same as the $[12, 4, 6]$ code discussed in [18].

Table 2. Few Length 15 F_2-Linear Codes over F_{16}

[Only nonzero transform components are shown. The elements of F_{16}^* are represented by the corresponding power of the primitive element and 0 is represented by -1.]

a_0	a_1	a_2	a_3	a_4	a_5	a_6	a_7	a_8	a_9	a_{10}	a_{11}	a_{12}	a_{13}	a_{14}	A_5	A_{10}	a_0	a_1	a_2	a_3	a_4	a_5	a_6	a_7	a_8	a_9	a_{10}	a_{11}	a_{12}	a_{13}	a_{14}	A_5	A_{10}
C_0																	C_2																
-1	-1	-1	-1	-1	-1	-1	-1	-1	-1	-1	-1	-1	-1	-1	-1	-1	-1	-1	-1	-1	-1	-1	-1	-1	-1	-1	-1	-1	-1	-1	-1	-1	-1
0	2	8	0	2	8	0	2	8	0	2	8	0	2	8	1	4	4	3	7	4	3	7	4	3	7	4	3	7	4	3	7	1	0
8	0	2	8	0	2	8	0	2	8	0	2	8	0	2	6	14	7	4	3	7	4	3	7	4	3	7	4	3	7	4	3	6	10
2	8	0	2	8	0	2	8	0	2	8	0	2	8	0	11	9	3	7	4	3	7	4	3	7	4	3	7	4	3	7	4	11	5
C_1																	C_3																
-1	-1	-1	-1	-1	-1	-1	-1	-1	-1	-1	-1	-1	-1	-1	-1	-1	-1	-1	-1	-1	-1	-1	-1	-1	-1	-1	-1	-1	-1	-1	-1	-1	-1
4	9	14	4	9	14	4	9	14	4	9	14	4	9	14	-1	4	0	2	8	0	2	8	0	2	8	0	2	8	0	2	8	1	4
14	4	9	14	4	9	14	4	9	14	4	9	14	4	9	-1	14	5	7	13	5	7	13	5	7	13	5	7	13	5	7	13	6	9
9	14	4	9	14	4	9	14	4	9	14	4	9	14	4	-1	9	10	12	3	10	12	3	10	12	3	10	12	3	10	12	3	11	14
$C_4 = C_0 + C_1$																	$C_5 = C_0 + C_2$																
-1	-1	-1	-1	-1	-1	-1	-1	-1	-1	-1	-1	-1	-1	-1	-1	-1	-1	-1	-1	-1	-1	-1	-1	-1	-1	-1	-1	-1	-1	-1	-1	-1	-1
4	9	14	4	9	14	4	9	14	4	9	14	4	9	14	-1	4	4	3	7	4	3	7	4	3	7	4	3	7	4	3	7	1	0
14	4	9	14	4	9	14	4	9	14	4	9	14	4	9	-1	14	7	4	3	7	4	3	7	4	3	7	4	3	7	4	3	6	10
9	14	4	9	14	4	9	14	4	9	14	4	9	14	4	-1	9	3	7	4	3	7	4	3	7	4	3	7	4	3	7	4	11	5
0	2	8	0	2	8	0	2	8	0	2	8	0	2	8	1	4	0	2	8	0	2	8	0	2	8	0	2	8	0	2	8	1	4
1	11	6	1	11	6	1	11	6	1	11	6	1	11	6	1	-1	1	6	11	1	6	11	1	6	11	1	6	11	1	6	11	-1	1
3	10	12	3	10	12	3	10	12	3	10	12	3	10	12	1	9	9	10	13	9	10	13	9	10	13	9	10	13	9	10	13	11	2
7	13	5	7	13	5	7	13	5	7	13	5	7	13	5	1	14	14	12	5	14	12	5	14	12	5	14	12	5	14	12	5	6	8
8	0	2	8	0	2	8	0	2	8	0	2	8	0	2	6	14	8	0	2	8	0	2	8	0	2	8	0	2	8	0	2	6	14
5	7	13	5	7	13	5	7	13	5	7	13	5	7	13	6	9	5	14	12	5	14	12	5	14	12	5	14	12	5	14	12	11	3
6	1	11	6	1	11	6	1	11	6	1	11	6	1	11	6	-1	11	1	6	11	1	6	11	1	6	11	1	6	11	1	6	-1	11
12	3	10	12	3	10	12	3	10	12	3	10	12	3	10	6	4	13	9	10	13	9	10	13	9	10	13	9	10	13	9	10	11	12
2	8	0	2	8	0	2	8	0	2	8	0	2	8	0	11	9	2	8	0	2	8	0	2	8	0	2	8	0	2	8	0	11	9
10	12	3	10	12	3	10	12	3	10	12	3	10	12	3	11	14	10	13	9	10	13	9	10	13	9	10	13	9	10	13	9	6	7
13	5	7	13	5	7	13	5	7	13	5	7	13	5	7	11	4	12	5	14	12	5	14	12	5	14	12	5	14	12	5	14	1	13
11	6	1	11	6	1	11	6	1	11	6	1	11	6	1	11	-1	6	11	1	6	11	1	6	11	1	6	11	1	6	11	1	-1	6

Acknowledgement

We thank P. Fitzpatrick and R. M. Tanner for sending us preprint/unpublished versions of their work.

References

1. R. E. Blahut, *Theory and Practice of Error Control Codes*, Addison Wesley, 1983.
2. J. Conan and G. Seguin, "Structural Properties and Enumeration of Quasi Cyclic Codes", *Applicable Algebra in Engineering Communication and Computing*, pp. 25-39, Springer-Verlag 1993.
3. B. K. Dey and B. Sundar Rajan, "DFT Domain Characterization of Quasi-Cyclic Codes", Submitted to IEEE Trans. Inform. Theory.
4. G. D. Forney Jr., *Geometrically Uniform Codes*, IEEE Trans. Inform. Theory,IT-37 (1991), pp. 1241-1260.
5. G. D. Forney Jr., *On the Hamming Distance Properties of Group Codes*, IEEE Trans. Inform. Theory, IT-38 (1992), pp. 1797-1801.
6. G. Gunther, *A Finite Field Fourier Transform for Vectors of Arbitrary Length*, Communications and Cryptography: Two Sides of One Tapestry, R. E. Blahut, D. J. Costello, U. Maurer, T. Mittelholzer (Eds), Kluwer Academic Pub., 1994.
7. M. Hattori, R. J. McEliece and G. Solomon, *Subspace Subcodes of Reed-Solomon Codes*, IEEE Trans. Inform. Theory, IT-44 (1998), pp. 1861-1880.
8. M. Isaksson and L. H. Zetterberg, *Block-Coded M-PSK Modulation over $GF(M)$*, IEEE Trans. Inform. Theory, IT-39 (1993), pp. 337-346.
9. K. Lally and P. Fitzpatrick, "Algebraic Structure of Quasicyclic Codes", to appear in *Discrete Applied Mathematics*.
10. R. Lidl and H. Niederreiter, *Finite Fields*, Encyclopedia of Mathematics and Its Applications, vol. 20, Cambridge University Press.
11. F. J. MacWilliams and N. J. A. Sloane, *The Theory of Error-Correcting Codes*, North-Holland, 1988.
12. McDonald B. R., *Finite rings with identity*, Marcel Dekker, New York, 1974.
13. M. Ran and J. Snyders, *A Cyclic [6,3,4] group code and the hexacode over $GF(4)$*", IEEE Trans. Inform. Theory, IT-42 (1996), pp. 1250-1253.
14. B. Sundar Rajan and M. U. Siddiqi, *Transform Domain Characterization of Cyclic Codes over Z_m*, Applicable Algebra in Engineering, Communication and Computing, Vol. 5, No. 5, pp. 261-276, 1994.
15. B. Sundar Rajan and M. U. Siddiqi, *A Generalized DFT for Abelian Codes overZ_m*, IEEE Trans. Inform. Theory, IT-40 (1994), pp. 2082-2090.
16. B. Sundar Rajan and M. U. Siddiqi, *Transform Domain Characterization of Abelian Codes*, IEEE Trans. Inform. Theory, IT-38 (1992), pp. 1817-1821.
17. B. Sundar Rajan and M. U. Siddiqi, *Transform Decoding of BCH Codes over Z_m*, International J. of Electronics, Vol 75, No. 6, pp. 1043-1054, 1993.
18. R. M. Tanner, "A Transform Theory for a Class of Group-Invariant Codes", *IEEE Trans. Inform. Theory*, vol. 34, pp. 752-775, July 1988.
19. A. A. Zain and B. Sundar Rajan, *Algebraic Characterization of MDS Group Codes over Cyclic Groups*, IEEE Trans. Inform. Theory, IT-41 (1995), pp. 2052-2056.

Cyclic Projective Reed-Muller Codes

Thierry P. Berger and Louis de Maximy

LACO, University of Limoges,
123, av. A. Thomas, Limoges 87060, FRANCE
thierry.berger@unilim.fr, http://www.unilim.fr/laco/

Abstract. The Projective Reed-Muller codes (PRM codes) were introduced by G. Lachaud [4] in 1988. A change in the choice of the set of representatives of the projective space gives two PRM codes that are not equivalent by permutation. In this paper, we present some criteria in the choice of the set of representatives to construct cyclic or quasi-cyclic PRM codes.

Key words: projective Reed-Muller codes, quasi-cyclic codes, projective linear group.

1 Introduction

The Projective Reed-Muller codes are obtained by evaluating the homogeneous polynomials of degree ν on a set of representatives of the projective space P_m. They are studied in many papers [4,5,3,6], in particular their parameters are known.

In [4], G. Lachaud constructs the set of representatives of P_m using the lexicographic order. He explains that a change of the set of representatives gives two codes equivalent by automorphism (in fact, by scalars multiplications on each component), but these codes are not permutation-equivalent. In [3], A. Sorensen constructs some cyclic PRM codes.

In this paper we give a general criteria on the set of representatives to construct a PRM code invariant under a permutation chosen in the Projective Linear Group. Then we apply this result to give some conditions to be cyclic or quasi-cyclic.

2 Preliminaries

Let $\mathbb{F}_q$ be the finite field with q elements, where $q = p^r$ is a power of the prime p. Let $\mathbb{F}_{q^m}$ be an extension of degree m of $\mathbb{F}_q$.

Let V_m be the vector space of dimension m over $\mathbb{F}_q$ and P_m be the associate projective space (the set of all linear lines over V_m). Set $n = \frac{q^m - 1}{q - 1} = |P_m|$.

Let $\mathbb{F}_q[X_1, \cdots, X_m]$ the set of polynomials with m indeterminates over $\mathbb{F}_q$, A_ν be the subset of $\mathbb{F}_q[X_1, \cdots, X_m]$ generated by homogeneous polynomials of degree ν.

S. Boztaş and I.E. Shparlinski (Eds.): AAECC-14, LNCS 2227, pp. 77–81, 2001.

Definition 1. *Let* $S = (S_1, \ldots, S_n)$, $S_i \in V_m^*$ *be a system of representatives of the projective space* P_m. *The projective Reed-Muller code* PRM_ν *of order* ν, *of length* n, *associated to* S *is the vector space*

$$\{(f(S_1), \ldots, f(S_n)) \mid f \in A_\nu\}$$

The parameters of PRM codes are known (cf. [4,3]):
the length of PRM_ν is $n = (q^m - 1)/(q - 1)$,
the dimension is $k_\nu = \sum_{t \equiv \nu \bmod (q-1), 0 < t \leq \nu} \left(\sum_{j=0}^m \binom{m}{j} \binom{t-jq+m-1}{t-jq} \right)$
and the minimum distance is $d_\nu = (q-1)(q-s)q^{m-r-2}/(q-1)$, where $\nu - 1 = r(q-1) + s$, $0 \leq s < \nu$.

For $\nu > (m-1)(q-1)$, the PRM codes are trivial. We suppose $\nu \leq (m-1)(q-1)$.

3   Main Result

Let σ be an element of the linear group $GL(m,q)$ considered as a linear permutation on V_m^*. Such an element induces a permutation $\tilde{\sigma}$ on the set of representatives S of P_m as follows: if S_i is an element of S, set $U = \sigma(S_i) \in V_m$ and S_j the representative of U, i.e. the element of S such that there exists a $\lambda \in \mathbb{F}_q$ satisfying $U = \lambda S_j$. Then $\tilde{\sigma}(S_i) = S_j$. The projective linear group $PGL(m,q)$ is the set of permutations $\tilde{\sigma}$, $\sigma \in GL(m,q)$.

Definition 2. *Let* S *be a system of representatives of the projective space and* σ *be an element of* $GL(m,q)$. S *is compatible with* σ *if* $\sigma(S) = S$, *i.e.* $\sigma(S_i) = \tilde{\sigma}(S_i)$ *for all* i.

Theorem 1. *Let* S *be a system of representatives compatible with an element* σ *in* $GL(m,q)$. *The projective Reed-Muller code* PRM_ν *is then invariant under the permutation* $\tilde{\sigma}$.

Proof. Suppose that S is compatible with σ. Let $c = (f(S_1), \ldots, f(S_n), f \in A_\nu)$ be an element of PRM_ν. So $\tilde{\sigma}(c) = (f(\sigma(S_1)), \ldots, f(\sigma(S_n)))$.
Set $S_k = (S_{k,1}, \ldots, S_{k,m})$. Since σ is a linear transformation, there exist some scalars $\sigma_{i,j}$ such that $\sigma(S_k) = (\sum_{j=1}^m S_{k,j}\sigma_{j,1}, \ldots, \sum_{j=1}^m S_{k,j}\sigma_{j,m})$.
Therefore $f(\sigma(S_k)) = (\sum_{i_1+\ldots+i_m=d}(\sum_{j=1}^m S_{k,j}\sigma_{j,1})^{i_1} \ldots (\sum_{j=1}^m S_{k,j}\sigma_{j,m})^{i_m})$.
The polynomial
$f^\sigma(X_1, \ldots, X_m) = \sum_{i_1+\ldots+i_m=d}(\sum_{j=1}^m \sigma_{j,1}X_j)^{i_1} \ldots (\sum_{j=1}^m \sigma_{j,m}X_m)^{i_m}$ is clearly homogeneous of degree ν.
Then we have $\tilde{\sigma}((f(S_1), \ldots, f(S_n)) = (f^\sigma(S_1), \ldots, f^\sigma(S_n)))$.
Thus $\tilde{\sigma}((f(S_1), \ldots, f(S_n)))$ is an element of PRM_ν, and PRM_ν is invariant under σ. $\square$

4   Cyclicity and Quasi-cyclicity

4.1   Cyclic Case

In order to construct cyclic PRM codes, we need some sets of representatives that are an orbit of length n under a permutation $\sigma \in GL(m, q)$. Using a basis of $\mathbb{F}_{q^m}$ over $\mathbb{F}_q$ we identify V_m and $\mathbb{F}_{q^m}$.

Lemma 1. *Suppose that n and $q - 1$ are coprime. Let α be a primitive root of $\mathbb{F}_{q^m}$. Set $\beta = \alpha^{q-1}$. The set $(\beta, \beta^2, \ldots, \beta^n)$ is a set of representatives of the projective space P_m.*

Proof. Suppose that $\frac{\beta^i}{\beta^j}$ is in $\mathbb{F}_q$, $0 < j \leq i \leq n$. Then $\alpha^{(q-1)(i-j)}$ is in $\mathbb{F}_q$. That implies n divides $(q-1)(i-j)$. Since n and $q-1$ are coprime, n divides $i - j < n$. We deduce $i = j$ and $(\beta, \beta^2, \ldots, \beta^n)$ is a set of representatives of P_m.   □

Let $\sigma_\beta \in GL(m, q)$ be the linear permutation of $V_m = \mathbb{F}_{q^m}$ defined by $\sigma_\beta(g) = \beta g$. Clearly, $(\beta, \beta^2, \ldots, \beta^n)$ is compatible with σ_β. We can deduce the following Theorem, which is a part of Theorem 3 in [3].

Theorem 2. *If n and $q - 1$ are coprime, then the code PRM_ν is cyclic for all ν, $0 \leq \nu \leq (m - 1)(q - 1)$.*

Remark 1. In [3], Sorensen gave another examples of cyclic PRM codes. The cyclicity of these codes does not depend on the set of representatives, but only of the value on ν.

Example 1:   Set $q = 3$ and $m = 3$. Then the length of PRM codes is $n = 13$ which is coprime to $q - 1 = 2$. We obtain cyclic codes over $GF(3)$ with the following parameters:

ν	1	2	3	4
k	3	6	10	12
d	9	6	3	2

Where k and d are respectively the dimension and the minimum distance of the PRM_ν code.

Example 2:   Set $q = 4$ and $m = 4$. The length of PRM codes is $n = 85$. We obtain cyclic codes over $GF(4)$ with the following parameters:

ν	1	2	3	4	5	6	7	8	9
k	4	10	20	35	50	64	75	81	84
d	64	48	32	16	12	8	4	3	2

4.2   Quasi-cyclic Case

In this section, we generalize the results of Theorem 2 to the case $(n, q - 1) \neq 1$. This leads to quasi-cyclic codes.

Lemma 2. *Let α be a primitive root of $\mathbb{F}_{q^m}$. Let $c = (n, q-1)$ be the gcd of n and $q-1$. Set $\gamma = \alpha^{c(q-1)}$. If $q-1$ and n/c are coprime, then $S = (\gamma, \ldots, \gamma^{n/c}, \alpha\gamma, \ldots, \alpha\gamma^{n/c}, \ldots, \alpha^{c-1}\gamma, \ldots, \alpha^{c-1}\gamma^{n/c})$ is a set of representatives of the projective space.*

Proof. Assume that $\frac{\alpha^i\gamma^j}{\alpha^l\gamma^r}$ is in $\mathbb{F}_q$. Since $\gamma = \alpha^{c(q-1)}$, we deduce $\alpha^{(i-l)+c(q-1)(j-r)} \in \mathbb{F}_q$. As previously, n must divide $(i-l)+c(q-1)(j-r)$.

This implies c divides $(i-l)+c(q-1)(j-r)$, and then c divides $(i-l)$. Since $|i-l| < c$, we have $i = l$.

Our hypothesis is reduced to $\alpha^{c(q-1)(j-r)} \in \mathbb{F}_q$. We know that n divides $c(q-1)(j-r)$, which means that n/c divides $(q-1)(j-r)$.

Since n/c and $q-1$ are coprime, n/c divides $(j-r)$. But $|j-r|$ is less than n/c and so $j = r$.

For $\alpha^i\gamma^j \neq \alpha^l\gamma^r$, $\frac{\alpha^i\gamma^j}{\alpha^l\gamma^r}$ is not in $\mathbb{F}_q$ and S is a set of representatives of the projective space. □

Theorem 3. *Let α be a primitive root of $\mathbb{F}_{q^m}$. Let $c = (n, q-1)$ be the gcd of n and $q-1$. Suppose that $q-1$ and n/c are coprime. For all ν, the code PRM_ν associated to the set of representatives S defined in Lemma 2 is quasi-cyclic of index c and order n/c.*

Proof. Let σ be the element of $GL(m, q)$ such that $\sigma(g) = \gamma g$ for every $g \in \mathbb{F}_{q^m}$.

The set S is invariant under σ and σ induces a quasi-cyclic permutation over PRM_ν with cycles of length n/c. □

Example 3: Set $q = 4$ and $m = 3$. The length of PRM codes is $n = 21$. We obtain $c = (21, 3) = 3$ and $(n/c, c) = (7, 3) = 1$. This leads to quasi-cyclic codes over $GF(4)$ of index 3 and order 7 with the following parameters:

ν	1	2	3	4	5	6
k	3	6	10	15	18	20
d	16	12	8	4	3	2

If n/c and $q-1$ are not coprime, it is possible to extend the results of Theorem 3 as follows:

Set $c_1 = (n, q-1)$, $c_2 = (n/c_1, q-1)$, $c_3 = (n/(c_1c_2), q-1)$ until $c_s = 1$. Let c be $c = \prod_{i=1}^{s-1} c_i$. Then c is the least divisor of n such that n/c and $q-1$ are co-prime.

Lemma 3. *Let α be a primitive root of $\mathbb{F}_{q^m}$ and $\delta = \alpha^c$ where c is defined as previously. Then*

$$S = (\delta, \ldots, \delta^{n/c}, \alpha\delta, \ldots, \alpha\delta^{n/c}, \ldots, \alpha^{c-1}\delta, \ldots, \alpha^{c-1}\delta^{n/c})$$

is a set of representatives of the projective space.

Proof. It is the direct generalization of those of Lemma 2. □

Theorem 4. *For the set of representatives S defined in Lemma 3, the code PRM_ν is quasi-cyclic of index c and order n/c for all ν.*

Proof. The proof is the same as for Theorem 3.   □

Example 4:   Set $q = 3$ and $m = 4$. Then the length of PRM codes is $n = 40$. We obtain $c_1 = (40, 2) = 2$, $c_2 = (20, 2) = 2$, $c_3 = (10, 2) = 2$ and $c_4 = (10/2, 2) = 1$. Then $c = 8$, we obtain quasi-cyclic codes over $GF(3)$ of index 8 and order $40/8 = 5$ with the following parameters:

ν	1	2	3	4
k	3	6	10	12
d	9	6	3	2

Remark 2. It is possible that n may be totally decomposed using Lemma 3. In that case, the code is not quasi-cyclic, since its order is 1. However, we do not find such an example under the condition $n = (q^m - 1)/(q - 1)$, $q = p^r$ a power of a prime p.

References

1. T.P. Berger: Automorphism Groups of Homogeneous and Projective Reed-Muller Codes, submitted.
2. T.P. Berger: Groupes d'automorphismes des codes de Reed-Muller homogènes et projectifs, C.R. Acad. Sci. Paris, **331** série I, 935–938, 2000.
3. A.B. Sorensen: Projective Reed-Muller Codes, IEEE Trans. Inform . Theory, vol.IT-37, **6**, 1567–1576, 1991.
4. G. Lachaud: Projective Reed-Muller Codes, Lect. Notes in Comp. Sci., **311**, Berlin, Springer,1988.
5. G. Lachaud: The parameters of projective Reed-Muller codes, Discret Math., **81**, 217–221, 1990.
6. G. Lachaud, I. Lucien, D.J. Mercier, R. Rolland: Group Structure on projective Spaces and Cyclic Codes over Finite Fields, to appear in Finite Fields and Their Applications.

Codes Identifying Sets of Vertices

Tero Laihonen[1] and Sanna Ranto[2]

[1] Department of Mathematics and Turku Centre for Computer Science TUCS,
University of Turku, FIN-20014 Turku, Finland
terolai@utu.fi
[2] Turku Centre for Computer Science TUCS,
University of Turku, FIN-20014 Turku, Finland
samano@utu.fi

Abstract. We consider identifying and strongly identifying codes. Finding faulty processors in a multiprocessor system gives the motivation for these codes. Constructions and lower bounds on these codes are given. We provide two infinite families of optimal $(1, \leq 2)$-identifying codes, which can find malfunctioning processors in a binary hypercube F_2^n. Also two infinite families of optimal codes are given in the corresponding case of strong identification. Some results on more general graphs are as well provided.

1 Introduction

Let F_2^n be the Cartesian product of n copies of the binary field F_2. The Hamming distance $d(x, y)$ between vectors (words) x and y of F_2^n is the number of coordinates in which they differ; the Hamming weight $w(x)$ of x is defined as $d(x, 0)$. A nonempty subset of F_2^n is called a *code*.

In the seminal paper [12] by Karpovsky, Chakrabarty and Levitin, the problem of locating malfunctioning processors in a multiprocessor system was introduced.

Assume that 2^n processors are labelled by the distinct binary vectors of F_2^n and the processors are connected (with a communication link) if and only if the Hamming distance of the corresponding labels equals one. Any processor can check the processors within Hamming distance t. It reports "NO" if problems are detected in its neighbourhood and "YES" otherwise. Assuming that there are at most l malfunctioning processors, we want to choose a subset of processors (i.e., a code $C \subseteq F_2^n$) in such a way that based on their reports we know where the faulty processors are. Of course, the smaller the subset the better.

Let us be more precise. We denote by $|X|$ the cardinality of a set X and the Hamming sphere by $B_t(x) = \{y \in F_2^n \mid d(x, y) \leq t\}$. Let $C \subseteq F_2^n$. For any $X \subseteq F_2^n$ we define

$$I_t(X) = I_t(C; X) = \left(\bigcup_{x \in X} B_t(x) \right) \cap C.$$

S. Boztaş and I.E. Shparlinski (Eds.): AAECC-14, LNCS 2227, pp. 82–91, 2001.

Definition 1. *Let t and l be non-negative integers. A code $C \subseteq F_2^n$ is called $(t, \leq l)$-identifying, if for all $X, Y \subseteq F_2^n$, $X \neq Y$, with $|X| \leq l$ and $|Y| \leq l$, we have $I_t(X) \neq I_t(Y)$.*

When we receive $I_t(C; X)$, we immediately know the set of faulty processors X if C is $(t, \leq l)$-identifying and $|X| \leq l$.

When maintaining multiprocessor systems with the model above, we expect to receive correct reports also from the malfunctioning processors that are in the code. If, however, faulty processors may either send the wrong report (i.e., be silent) or the correct report, then we need the following concept of strong identification to handle the situation.

In order to find the malfunctioning processors in this case we require that C satisfies the following. Let for any different subsets X and Y of F_2^n ($|X|, |Y| \leq l$) the sets $I_t(X) \setminus S$ and $I_t(Y) \setminus T$ be distinct for all $S \subseteq X \cap C$ and $T \subseteq Y \cap C$. Then obviously we can always distinguish between X and Y.

Definition 2. *[10] Let $C \subseteq F_2^n$. Let further t and l be non-negative integers. Define*

$$\mathcal{I}_t(X) = \{U \mid I_t(X) \setminus (X \cap C) \subseteq U \subseteq I_t(X)\} \tag{1}$$

for every $X \subseteq F_2^n$.

If for all $X, Y \subseteq F_2^n$, where $X \neq Y$ and $|X|, |Y| \leq l$, we have $\mathcal{I}_t(X) \cap \mathcal{I}_t(Y) = \emptyset$, then we say that C is a strongly $(t, \leq l)$-identifying code.

Let us denote $I_1(X) = I(X)$, $I_t(\{x_1, \ldots, x_s\}) = I_t(x_1, \ldots, x_s)$ and $I_t'(y) = I_t(y) \setminus \{y\}$. The smallest cardinality of a $(t, \leq l)$-identifying code and a strongly $(t, \leq l)$-identifying code of length n is denoted by $M_t^{(\leq l)}(n)$ and $M_t^{(\leq l)SID}(n)$, respectively. A code attaining the smallest cardinality is called *optimal*. We say that x t-*covers* y, if $d(x, y) \leq t$, and we omit t, if $t = 1$.

In this paper we focus on $(1, \leq 2)$-identifying and strongly $(1, \leq 2)$-identifying codes in binary Hamming spaces. We will provide infinite sequences of optimal codes in both cases. Results on $l = 1$ can be found in the case of identifying codes in [1,2,12,7,6] and in the case of strong identification in [10,11]. For results on $l \geq 3$, consult [13,14].

In the last section, we will discuss about some properties of identification also in other graphs (results in the case $l = 1$ can be found, for instance, from [12,3]).

2   Lower Bounds

Let us first give a lemma, which is needed throughout the paper.

Lemma 1. *For $a, b \in F_2^n$ we have*

$$|B_1(a) \cap B_1(b)| = \begin{cases} n + 1 & \text{if } a = b \\ 2 & \text{if } d(a, b) = 1 \text{ or } 2 \\ 0 & \text{otherwise.} \end{cases} \tag{2}$$

Theorem 1. *[13,15] Let $l \geq 2$. Then*

$$M_1^{(\leq l)}(n) \geq \left\lceil (2l - 1)\frac{2^n}{n+1} \right\rceil$$

and

$$M_1^{(\leq l)SID}(n) \geq \left\lceil (2l - 1)\frac{2^n}{n} \right\rceil.$$

Proof. Let us prove the latter inequality. Let C be a strongly $(1, \leq l)$-identifying code. If $x \notin C$, then $|I(x)| \geq 2l - 1$. Indeed, otherwise if $I(x) = \{c_1, \ldots, c_{2l-2}\}$ and x_i $(i = 1, \ldots, l - 1)$ is the unique word $(x_i \neq x)$ at distance one from both c_{2i-1} and c_{2i}, we have $I(x_1, \ldots, x_{l-1}) = I(x_1, \ldots, x_{l-1}, x)$, which is a contradiction. Obviously less than $2l - 2$ codewords in $I(x)$ is also impossible.

Assume then that $x \in C$. Let $I(x) = \{c_1, \ldots, c_{2l-2}, x\}$ and define x_i as above for all $i = 1, \ldots, l - 1$. Now $I(x_1, \ldots, x_{l-1}) = I(x_1, \ldots, x_{l-1}, x) \setminus \{x\}$ which is impossible and hence $|I(x)| \geq 2l$.

Thus we obtain $|C|(n + 1) \geq 2l|C| + (2l - 1)(2^n - |C|)$ which gives the claim. $\square$

As we shall see, these estimates can often be attained.

3   Optimal Codes for $(1, \leq 2)$-Identification

In the sequel we need two initial codes in order to provide the two infinite sequences of optimal codes. These initial codes are given next and can be found from [9,16].

Theorem 2. *The following code is $(1, \leq 2)$-identifying*

$$\{00100, 00010, 00001, 11000, 10100, 10010, 01100, 01001,$$
$$00011, 11010, 11001, 10101, 01110, 10111, 01111, 11111\}.$$

Corollary 1. $M_1^{(\leq 2)}(5) = 16$.

Denote by $\mathcal{H}_3$ the Hamming code of length seven with the parity check matrix

$$H = \begin{pmatrix} 0 & 0 & 0 & 1 & 1 & 1 & 1 \\ 0 & 1 & 1 & 0 & 0 & 1 & 1 \\ 1 & 0 & 1 & 0 & 1 & 0 & 1 \end{pmatrix}.$$

Let $C_1 = \mathcal{H}_3 + 1011001$ and $C_2 = \mathcal{H}_3 + 0000100$ be two cosets of $\mathcal{H}_3$. Let further P_1 and P_3 be the codes obtained by permuting C_1 using the permutations $(7, 3)(4, 2)$ and $(6, 3)(4, 1)$, respectively. By P_2 we denote the code obtained from C_2 using the permutation $(1, 2)(3, 5)$. It is easy to check (with computer) that $U = P_1 \cup P_2 \cup P_3$ is $(1, \leq 2)$-identifying. The following result now follows from Theorem 1.

Theorem 3. $M_1^{(\leq 2)}(7) = 48$.

Next we describe a construction which together with the initial codes gives the two infinite sequences of optimal $(1, \leq 2)$-identifying codes.

Theorem 4. *[16] If $C \subseteq F_2^n$ is a $(1, \leq 2)$-identifying code, then also*

$$C' = \{(\pi(u), u, u + v) \mid u \in F_2^n, v \in C\},$$

where $\pi(u)$ denotes a parity check bit on u, is a $(1, \leq 2)$-identifying code of length $2n + 1$.

Proof. (Sketch) By Theorem [16, Theorem 2] the code C covers each word at least three times, and thus by [4, Theorems 3.4.3 and 14.4.3] the code C' also does. Since C' covers each word at least three times, and the intersection of three different spheres of radius one contains at most one element, C' is $(1, \leq 1)$-identifying. Moreover, all single words and pairs of words are distinguishable. Indeed, consider the sets $\{x\}$ and $\{y, z\}$, $y \neq z$. Then without loss of generality $y \neq x$, and because $|B_1(y) \cap B_1(x)| \leq 2$, we know that $B_1(y) \cap C$ contains at least one codeword which is not in $B_1(x)$.

Thus we only need to check that all pairs are identified from one another.

Let us divide the words of F_2^{2n+1} into two classes by their first bit and consider the codewords which cover a word in each class. Let $x = (a, u, u + v) \in F_2^{2n+1}$.

I If $a = \pi(u)$ then $I(x) = \{(\pi(u), u, u + c) \mid c \in C, d(c, v) \leq 1\}$.
II If $a \neq \pi(u)$ then $I(x) = A \cup \{(a, u', u + v) \mid d(u', u) = 1, \exists c \in C : u + v = u' + c\}$. Here $A = \{(\pi(u), u, u + v)\}$ if $v \in C$, and $A = \emptyset$ if $v \notin C$.

So in both classes we are interested in codewords $c \in C$ such that $d(c, v) \leq 1$. Namely in the class II the properties $d(u', u) \leq 1$ and $u + v = u' + c$ imply that also $d(v, c) \leq 1$. If $I(x) = \{(b_i, s_i, t_i) \mid i = 1, 2, \ldots, k\}$, then in both cases $I(C; v) = \{s_i + t_i \mid i = 1, 2, \ldots, k\}$.

Suppose there were words x, y, z and w in F_2^{2n+1} such that

$$I(x, y) = I(z, w) \text{ and } \{x, y\} \neq \{z, w\}, x \neq y, z \neq w . \tag{3}$$

If v_1, v_2, v_3 and v_4 are v's of x, y, z and w respectively, then in F_2^n by the previous discussion $I(C; v_1, v_2) = I(C; v_3, v_4)$. Since C is a $(1, \leq 2)$-identifying code we must have $\{v_1, v_2\} = \{v_3, v_4\}$. We will show that (3) cannot hold. Assume to the contrary that (3) holds.

Because $|I(x)| \geq 3$, we know by (3) that at most one of the sets $I(x) \cap I(z)$ and $I(x) \cap I(y)$ has cardinality one or less. A similar remark applies to $I(y)$, $I(z)$ and $I(w)$. Hence we can without loss of generality assume that $|I(x) \cap I(z)| \geq 2$ and $|I(y) \cap I(w)| \geq 2$. Depending on which class x belongs to, also z belongs to the same class. Similarly y and w belong to the same class.

If $x = (\pi(u_1), u_1, u_1 + v_1)$ and $y = (\pi(u_2), u_2, u_2 + v_2)$ are words in the class I, then also $z = (\pi(u_3), u_3, u_3 + v_3)$ and $w = (\pi(u_4), u_4, u_4 + v_4)$ are. Since in $I(x)$ and $I(z)$ the codewords begin with the same $n + 1$ bits, we get $u_1 = u_3$. Similarly

$u_2 = u_4$. We can assume that $|I(z) \cap I(y)| \geq 1$ (or that $|I(x) \cap I(w)| \geq 1$, which is a symmetric case): otherwise $I(z) = I(x)$ and $I(w) = I(y)$ and hence $z = x$ and $w = y$. Since z and y are both words in the class I, we get $u_2 = u_1$. The fact that $\{v_1, v_2\} = \{v_3, v_4\}$ now implies that $\{x, y\} = \{z, w\}$.

Assume $x = (\pi(u_1) + 1, u_1, u_1 + v_1)$ and $y = (\pi(u_2) + 1, u_2, u_2 + v_2)$ and so also $z = (\pi(u_3) + 1, u_3, u_3 + v_3)$ and $w = (\pi(u_4) + 1, u_4, u_4 + v_4)$ are words in the class II. Since in $I(x)$ and $I(z)$ the codewords end with the same n bits as x and z we get $u_1 + v_1 = u_3 + v_3$, and similarly $u_2 + v_2 = u_4 + v_4$. If now $v_1 = v_3$ and $v_2 = v_4$ we are done, since then $x = z$ and $y = w$. Suppose therefore that $v_2 = v_3$ and $v_1 = v_4$. As in the previous case, we can assume that $|I(z) \cap I(y)| \geq 1$. Now the last n bits must be the same in $I(z)$ and $I(y)$, and thus $u_3 + v_3 = u_2 + v_2$ and we get $u_3 = u_2$, i.e., $y = z$. The word z cannot cover the whole $I(x)$, otherwise $z = x$, since $|I(x)| \geq 3$ and the intersection of three different spheres of radius one has cardinality at most one. This would imply that $x = z = y$. So w must cover at least one word from $I(x)$ which implies $u_1 + v_1 = u_4 + v_4$ and now $u_1 = u_4$, i.e., $x = w$. Therefore $\{x, y\} = \{z, w\}$.

The proof of the final case where x and y belong to different classes can be found from [16].   □

Corollary 2. *[16]* $M_1^{(\leq 2)}(2n + 1) \leq 2^n M_1^{(\leq 2)}(n)$.

Corollary 3. *[16]*

$$\text{For } k \geq 1: \quad M_1^{(\leq 2)}(3 \cdot 2^k - 1) = 2^{3 \cdot 2^k - k - 1}.$$
$$\text{For } k \geq 3: \quad M_1^{(\leq 2)}(2^k - 1) = 3 \cdot 2^{2^k - k - 1}.$$

Proof. By Corollary 1 we know that $M_1^{(\leq 2)}(5) = 16$. Using Corollary 2 recursively and the lower bound from Theorem 1 we get the first equation. Similarly by Theorem 3 we get the second claim.   □

4   Optimal Codes for Strong Identification

In this section we provide optimal codes for strong $(1, \leq 2)$-identification. We denote the direct sum [4, p. 63] of the codes A and B by $A \oplus B$.

Lemma 2. *[15] Let $C \subseteq F_2^n$ be $(1, \leq 2)$-identifying and $a, b \in F_2^n$, $a \neq b$. Then*

$$|I(a, b) \setminus \{a, b\}| \geq \begin{cases} 2 \text{ if } d(a, b) = 1 \\ 3 \text{ if } d(a, b) = 2 \\ 4 \text{ if } d(a, b) \geq 3. \end{cases}$$

Theorem 5. *[15] If C is a $(1, \leq 2)$-identifying code, then $D = C \oplus F_2$ is strongly $(1, \leq 2)$-identifying.*

Proof. (Sketch) By [9, Theorem 7] we know that D is $(1, \leq 2)$-identifying and by [10, Theorem 3] we know that hence D is strongly $(1, \leq 1)$-identifying. Thus to prove the claim it suffices to check the following two sets of inequalities for all x, y, z and w in F_2^{n+1} and for all sets J where $I'(a) \subseteq J(a) \subseteq I(a)$ and $I(a, b) \setminus \{a, b\} \subseteq J(a, b) \subseteq I(a, b)$: the first set is

$$J(x) \neq J(z, w) \tag{4}$$

where $z \neq w$ and $J(x) \neq I(x)$ or $J(z, w) \neq I(z, w)$, and the second set is

$$J(x, y) \neq J(z, w) \tag{5}$$

where $\{x, y\} \neq \{z, w\}$ and $J(x, y) \neq I(x, y)$ or $J(z, w) \neq I(z, w)$.

By [9, Theorem 3] we have $|I(C; x)| \geq 3$ for all $x \in F_2^n$ and therefore also $|I(y)| \geq 3$ for all $y \in F_2^{n+1}$ and, moreover, $|I(y)| \geq 4$ if $y \in D$. Thus $|I'(y)| \geq 3$ for all $y \in F_2^{n+1}$.

Step 1: Let us first look at the inequalities (4). Either $x \neq z$ or $x \neq w$, say $x \neq z$. By (2), $|B_1(x) \cap B_1(z)| \leq 2$. If $d(x, z) \neq 2$, then there are at least two codewords in $I'(z)$ which are not in $I(x)$ and only one of them can be removed from $J(z, w)$. If $d(x, z) = 2$, then there can be only one such codeword and it can be removed from $I(z, w)$ if it is w. However, then the words in $I'(w)$ cannot be in $I(x)$, since $d(x, w) = 3$. This shows that (4) is satisfied.

Step 2: Consider next the inequalities (5). We denote by z' the word obtained by puncturing the last coordinate of $z \in F_2^{n+1}$. In the sequel we will often use the fact that $I(a) \cap (F_2^n \oplus \{1\})$, where $a = a'0 \in F_2^{n+1}$, contains the unique word $a'1$ if $a \in D$ and otherwise it is empty. Denote $L_0 := F_2^n \oplus \{0\}$ and $L_1 := F_2^n \oplus \{1\}$.

Case 1: Let $x, y, z, w \in L_0$. In the inequalities (5) we may assume that x is removed from $I(x, y)$. Thus $x \in D$. Consequently, by the fact above $x \in \{z, w\}$, say $x = z$. If also $y \in D$, then $\{x, y\} = \{z, w\}$. Similarly, we can assume that $w \notin D$. Let then $y \notin D$ and thus y cannot be removed from $I(x, y)$. Hence it suffices to verify that $I(x, y) \setminus \{x\} \neq I(z, w)$ and $I(x, y) \setminus \{x\} \neq I(z, w) \setminus \{z\}$. The first is immediately clear, since $x \in I(z, w)$. The second follows, because D is $(1, \leq 2)$-identifying and hence we have $I(x, y) \neq I(z, w)$. This proves (5) in this case.

Case 2: Let $x, y \in L_0$ and $z, w \in L_1$. Evidently, $|I(z, w) \cap L_0| \leq 2$. Therefore, by Lemma 2, we only need to examine the case, where $d(x, y) = 1$, both x and y are removed from $I(x, y)$ and $|(I'(x) \cap L_0) \setminus I(y)| = |(I'(y) \cap L_0) \setminus I(x)| = 1$. By symmetry, we can assume that the analogous premises hold for z and w as well, and thus only the inequality $I(x, y) \setminus \{x, y\} \neq I(z, w) \setminus \{z, w\}$ is left to be verified. Let $(I(x, y) \cap L_0) \setminus \{x, y\} = \{c_1, c_2\}$ for some $c_1, c_2 \in D$. If the inequality fails, we must have $z = c_1'1$ and $w = c_2'1$. Similarly, $(I(z, w) \cap L_1) \setminus \{z, w\} = \{x'1, y'1\}$. But this is a contradiction, since now $I(C; \{x', y'\}) = I(C; \{z', w'\})$ although $\{x', y'\} \neq \{z', w'\}$.

The complete proof of the other (two) cases is given in [15].   □

Example 1. Let us look at the other direction. Shortening (on any coordinate) and with respect to 0 (1 in the case of the fifth coordinate) of an optimal strongly

$(1, \leq 2)$-identifying code consisting of the words

$$
\begin{array}{l}
10100 \ 01000 \ 00100 \ 00010 \ 00001 \ 11000 \\
10010 \ 01100 \ 01010 \ 01001 \ 00011 \ 00101 \\
11010 \ 10110 \ 11001 \ 10101 \ 00111 \ 01110 \\
11110 \ 11101 \ 10111 \ 11111
\end{array}
$$

is *not* $(1, \leq 2)$-identifying code of length four.

Corollary 4. *[15]* $M_1^{(\leq 2)SID}(n) \leq 2M_1^{(\leq 2)}(n-1)$.

We are now in the position to give the two infinite families of optimal codes.

Corollary 5. *[15]*

$$
\text{For } k \geq 1: \quad M_1^{(\leq 2)SID}(3 \cdot 2^k) = 2^{3 \cdot 2^k - k}.
$$
$$
\text{For } k \geq 3: \quad M_1^{(\leq 2)SID}(2^k) = 3 \cdot 2^{2^k - k}.
$$

Proof. From Corollary 3 we know that $M_1^{(\leq 2)}(n) = 2^{3 \cdot 2^k - k - 1}$, if $n = 3 \cdot 2^k - 1$ $(k \geq 1)$, and $M_1^{(\leq 2)}(n) = 3 \cdot 2^{2^k - k - 1}$, if $n = 2^k - 1$ $(k \geq 3)$. Combining this with Corollary 4 and the lower bound from Theorem 1 we obtain the equations. $\square$

No infinite family of optimal regularly or strongly $(1, \leq 1)$-identifying codes is known.

Theorem 6. *[15] Let C be a strongly $(1, \leq 2)$-identifying code of length n. The code $C' = \{(\pi(u), u, u + c) \mid u \in F_2^n, c \in C\}$ is a strongly $(1, \leq 2)$-identifying code of length $2n + 1$.*

5   On General Graphs

Let $G = (V, E)$ be a connected undirected graph where V is the set of vertices and E is the set of edges. A nonempty subset $C \subseteq V$ is called a *code* and its elements are *codewords* of C. Let $d(u, v)$ denote the number of edges in any shortest path between u and v. Define $B_t(x) = \{y \in V \mid d(x, y) \leq t\}$. Denote $(A \subseteq V)$

$$
I_t(A) = \left(\bigcup_{a \in A} B_t(a) \right) \cap C.
$$

A code $C \subseteq V$ is called $(t, \leq l)$-*identifying* if for any two sets $X, Y \subseteq V$ $(|X|, |Y| \leq l)$ we have $I_t(X) \neq I_t(Y)$. If moreover for all distinct sets $X, Y \subseteq V$ of cardinality at most l we get $I_t(X) \setminus S \neq I_t(Y) \setminus T$ for every $S \subseteq X \cap C$ and $T \subseteq Y \cap C$, then C is called *strongly* $(t, \leq l)$-*identifying*.

Table 1. Bounds on regular and strong $(1, \leq 2)$-identification (see [9,16,15]).

n	$M_1^{(\leq 2)}(n)$	$M_1^{(\leq 2)SID}(n)$
4	11	-
5	16	22
6	30–32	32
7	48	55–64
8	90–96	96
9	154–176	171–192
10	289–352	308–352
11	512	559–704
12	972-1024	1024
13	1756–2048	1891–2048
14	3356–4096	3511–4096
15	6144	6554–8192
16	11566–12288	12288

Theorem 7. *Let* $G = (V, E)$ $(|V| < \infty)$ *be as above. If* $C \subset V$ *is strongly* $(t, \leq l)$-*identifying, then*

$$2^{|C|} \geq \sum_{i=0}^{l} \sum_{j=0}^{i} 2^j \binom{|C|}{j} \binom{|V| - |C|}{i - j}.$$

Proof. Assume that C is strongly $(t, \leq l)$-identifying. Let $X \subseteq V$, where $|X| = i$ $(0 \leq i \leq l)$, consist of exactly j $(0 \leq j \leq i)$ codewords of C and $i - j$ noncodewords. Then we have $|\{U \mid I_t(X) \setminus (X \cap C) \subseteq U \subseteq I_t(X)\}| = 2^j$, i.e., X implies the existence of 2^j distinct subsets of C. There are

$$\binom{|C|}{j} \binom{|V| - |C|}{i - j}$$

such X's where it has exactly j codewords in it.

The total number of different subsets of C is therefore at least

$$\sum_{i=0}^{l} \sum_{j=0}^{i} 2^j \binom{|C|}{j} \binom{|V| - |C|}{i - j}$$

and there are $2^{|C|}$ subsets of C all in all. This gives the claim. □

The set of vertices adjacent to a vertex $x \in V$ is denoted by $\Gamma(x)$. The *degree* of x is $d(x) = |\Gamma(x)|$. The *minimal degree* of the vertices of G is denoted by $\delta = \delta(G)$.

Theorem 8. *Let* $G = (V, E)$ *be again as above with* $|V| \geq 3$. *If there is a* $(1, \leq l)$-*identifying code* $C \subseteq V$, *then* $l \leq \delta(G)$.

Proof. Let $x \in V$ be a vertex giving the minimal degree, i.e., $d(x) = \delta(G) \geq 1$. Assume that C is $(1, \leq l)$-identifying. Denote $\Gamma(x) = \{v_1, v_2, \ldots, v_\delta\}$. If $l > \delta(G)$, then $I_1(v_1, v_2, \ldots, v_\delta) = I_1(v_1, \ldots, v_\delta, x)$ which is impossible.   □

Example 2. Consider any path P_ℓ of length $2 < \ell < \infty$. Then the code C consisting of all the vertices of P_ℓ is $(1, \leq 1)$-identifying, but according to the previous theorem is not $(1, \leq 2)$-identifying. This example shows that we can have $(1, \leq \delta)$-identifying codes.

For specific graphs we can of course say more, for example, there does not exist a $(1, \leq l)$-identifying code in F_2^n if $l \geq n/2 + 2$ and n is even or if $l \geq \lceil n/2 \rceil + 1$ and n is odd.

Acknowledgement

The first author would like to thank the Academy of Finland for financial support (grant 46186).

References

1. U. Blass, I. Honkala, S. Litsyn: Bounds on identifying codes. Discrete Math., to appear
2. U. Blass, I. Honkala, S. Litsyn: On binary codes for identification. J. Combin. Des., **8** (2000) 151-156
3. I. Charon, I. Honkala, O. Hudry, A. Lobstein: General bounds for identifying codes in some infinite regular graphs. Electronic Journal of Combinatorics, submitted
4. G. Cohen, I. Honkala, S. Litsyn, A. Lobstein: Covering Codes. Elsevier, Amsterdam, the Netherlands (1997)
5. G. Cohen, I. Honkala, A. Lobstein, G. Zémor: On identifying codes. In: Barg, A., Litsyn, S. (eds.): Codes and Association Schemes. DIMACS Series in Discrete Mathematics and Theoretical Computer Science **56** (2001) 97–109
6. G. Exoo: Computational results on identifying t-codes. preprint
7. I. Honkala: On the identifying radius of codes. Proceedings of the Seventh Nordic Combinatorial Conference, Turku, (1999) 39–43
8. I. Honkala: Triple systems for identifying quadruples. Australasian J. Combinatorics, to appear
9. I. Honkala, T. Laihonen, S. Ranto: On codes identifying sets of vertices in Hamming spaces. Des. Codes Cryptogr., to appear
10. I. Honkala, T. Laihonen, S. Ranto: On strongly identifying codes. Discrete Math., to appear
11. I. Honkala, T. Laihonen, S. Ranto: Codes for strong identification. Electronic Notes in Discrete Mathematics, to appear
12. M.G. Karpovsky, K. Chakrabarty, L. B. Levitin, On a new class of codes for identifying vertices in graphs. IEEE Trans. Inform. Theory, **44** (1998) 599–611
13. T. Laihonen: Sequences of optimal identifying codes. IEEE Trans. Inform. Theory, submitted
14. T. Laihonen: Optimal codes for strong identification, European J. Combinatorics, submitted

15. T. Laihonen, S. Ranto: Families of optimal codes for strong identification. Discrete Appl. Math., to appear
16. S. Ranto, I. Honkala, T. Laihonen: Two families of optimal identifying codes in binary Hamming spaces. IEEE Trans. Inform. Theory, submitted

Duality and Greedy Weights
of Linear Codes and Projective Multisets

Hans Georg Schaathun

Departement of Informatics, University of Bergen, N-5020 Bergen, Norway
georg@ii.uib.no
http://www.ii.uib.no/~georg/

Abstract. A projective multiset is a collection of projective points, which are not necessarily distinct. A linear code can be represented as a projective multiset, by taking the columns of a generator matrix as projective points. Projective multisets have proved very powerful in the study of generalised Hamming weights. In this paper we study relations between a code and its dual.

1 Background

A linear code is a normed space and the weights (or norms) of codewords are crucial for the code's performance. One of the most important parameters of a code is the minimum distance or minimum weight of a codeword.

The concept of weights can be generalised to subcodes or even arbitrary subsets of the code. (This is often called support weights or support sizes.) One of the key papers is [16], where Wei defined the rth generalised Hamming weight to be the least weight of a r-dimensional subcode. After Wei's work, we have seen many attempts to determine the generalised Hamming weights of different classes of codes.

Weights are alpha and omega for codes. Yet we know very little about the weight structure of most useful codes. The generalised Hamming weights give some information, and several practical applications are known, including finding bounds on the trellis complexity [8,7]. Still they do not fully answer our questions.

Several other parameters describing weights of subcodes have been introduced, and they can perhaps contribute to understanding the structure of linear codes. The support weight distribution appeared as early as 1977 in [9]. The chain condition from [17] have received a lot of attention. Chen and Kløve [4,3] introduded the greedy weights, inspired by a set of parameters from [5].

It is well known that a code and its dual are closely related. Kløve [11] has generalised the MacWilliams identities to give a relation for the support weight distributions. Wei [16] found a simple relation between the weight hierarchies of a code and its dual. We will find a similar result for the greedy weights.

We consider a linear $[n, k]$ code C. We usually define a linear code by giving the generator matrix G. The rows of G make a basis for C, and as such they are much studied. Many works consider the columns instead. This gives rise

S. Boztaş and I.E. Shparlinski (Eds.): AAECC-14, LNCS 2227, pp. 92–101, 2001.
© Springer-Verlag Berlin Heidelberg 2001

to the *projective multisets* [6]. The weight hierarchy is easily recognised in this representation [10,15]. Other terms for projective multisets include projective systems [15] and value assignments [2].

There are at least two ways to develop the correspondence between codes and multisets. Most coding theorists will probably just take the columns of some generator matrix (e.g. [10,2]). Some mathematicians (e.g. [6,15]) develop the projective multisets abstractly. They take the elements to be the coordinate forms on C, and get a multiset on the dual space of C (this is *not* the dual code). Hence their argument does not depend on the (non-unique) generator matrix of C.

We will need the abstract approach for our results, but we will try to carefully explain the connections between the two approaches, in the hope to reach more readers. For the interested reader, we refer to a more thorough report [14], where we use the present techniques to address some other problems, including support weight distributions, in addition to the present results.

2 Definitions and Notation

2.1 Vectors, Codes, and Multisets

A multiset is a collection of elements, which are not necessarily distinct. More formally, we define a multiset γ on a set S as a map $\gamma : S \to \{0, 1, 2, \ldots\}$. The number $\gamma(s)$ is the number of occurences of s in the collection γ. The map γ is always extended to the power set of S,

$$\gamma(S') = \sum_{s \in S'} \gamma(s), \quad \forall S' \subset S.$$

The number $\gamma(s)$, where $s \in S$ or $s \subset S$, is called the value of s. The size of γ is the value $\gamma(S)$.

We will be concerned with multisets of vectors. We will always keep the informal view of γ as a collection in mind.

We consider a fixed finite field $\mathbf{F}$ with q elements. A message word is a k-tuple over $\mathbf{F}$, while a codeword is an n-tuple over $\mathbf{F}$. Let $\mathbf{M}$ be a vector space of dimension k (the message space), and $\mathbf{V}$ a vector space of dimension n (the channel space). The generator matrix G gives a linear, injective transformation $G : \mathbf{M} \to \mathbf{V}$, and the code C is simply the image under G.

As vector spaces, $\mathbf{M}$ and C are clearly isomorphic. For every message word $\mathbf{m}$, there is a unique codeword $\mathbf{c} = \mathbf{m}G$.

A codeword $(c_1, c_2, \ldots, c_n) = \mathbf{m}G$ is given by the value c_i in each coordinate position i. If we know $\mathbf{m}$, we obtain this value as the inner product of $\mathbf{m}$ and the ith column $\mathbf{g}_i$ of G, i.e.

$$c_i = \mathbf{g}_i \cdot \mathbf{m} = \sum_{j=1}^{k} m_j g_{i,j}, \quad \text{where} \quad \mathbf{g}_i = (g_{i,1}, g_{i,2}, \ldots, g_{i,k}), \tag{1}$$

$$\text{and} \quad \mathbf{m} = (m_1, m_2, \ldots, m_k).$$

The columns $\mathbf{g}_i$ are elements of $\mathbf{M}$. These vectors are not necessarily distinct, so they make a multiset

$$\gamma_C : \mathbf{M} \to \{0, 1, 2, \ldots\}.$$

If we reorder the columns of G, we get an equivalent code. Hence γ_C defines C up to equivalence. If we replace a column with a proportional vector, we also get an equivalent code. Therefore many papers consider γ_C as a multiset on the projective space $\mathbf{P}(\mathbf{M})$, and a projective multiset will also define the code up to equivalence.

We say that two multisets γ and γ' on $\mathbf{M}$ are equivalent if $\gamma' = \gamma \circ \phi$ for some automorphism ϕ on $\mathbf{M}$. Such an automorphism is given by $\phi : \mathbf{g} \mapsto \mathbf{g}A$ where A is a square matrix of full rank. Replacing all the $\mathbf{g}_i$ by $\mathbf{g}_i A$ in (1) is equivalent to replacing $\mathbf{m}$ by $A\mathbf{m}$. In other words, equivalent multisets give different encoding, but they give the same code. This is an important observation, because it implies that the coordinate system on $\mathbf{M}$ is not essential.

Now we seek a way to represent the elements of γ_C as vectors of $\mathbf{V}$.

Let $\mathbf{b}_i$ be the ith coordinate vector of $\mathbf{V}$, that is the vector with 1 in position i and 0 in all other positions. The set of all coordinate vectors is denoted by

$$\mathcal{B} := \{\mathbf{b}_1, \mathbf{b}_2, \ldots, \mathbf{b}_n\}.$$

If we know the codeword $\mathbf{c}$ corresponding to $\mathbf{m}$, the ith coordinate position c_i is given as the inner product of $\mathbf{b}_i$ and $\mathbf{c}$.

$$c_i = \mathbf{b}_i \cdot \mathbf{c} = \sum_{j=1}^{n} c_j b_{i,j}, \quad \text{where} \quad \mathbf{b}_i = (b_{i,1}, b_{i,2}, \ldots, b_{i,k}), \tag{2}$$

$$\text{and} \quad \mathbf{c} = (c_1, c_2, \ldots, c_k).$$

We note that $\mathbf{b}_i$ takes the role of $\mathbf{g}_i$, and $\mathbf{c}$ takes the role of $\mathbf{m}$ from (1).

However, $\mathbf{b}_i$ is not the only vector of $\mathbf{V}$ with this property. In fact, for any vector $\mathbf{c}' \in C^\perp$, we have $(\mathbf{b}_i + \mathbf{c}') \cdot \mathbf{c} = c_i$. Therefore, we can consider the vector $\mathbf{b}_i$ as the coset $\mathbf{b}_i + C^\perp$ of $C^\perp$. The set of such cosets is usually denoted $\mathbf{V}/C^\perp$, and it is a vector space of dimension

$$\dim \mathbf{V}/C^\perp = \dim \mathbf{V} - \dim C^\perp = n - (n - k) = k = \dim \mathbf{M}.$$

Hence $\mathbf{M} \cong \mathbf{V}/C^\perp$ as vector spaces. Obviously $\mathbf{b}_i + C^\perp$ corresponds to $\mathbf{g}_i$.

We let $\mu : \mathbf{V} \to \mathbf{V}/C^\perp$ be the natural endomorphism, i.e. $\mu : \mathbf{g} \mapsto \mathbf{g} + C^\perp$. This map is not injective, so if $S \subset \mathbf{V}$, it is reasonable to view the image $\mu(S)$ as a multiset. Our analysis gives this lemma.

Lemma 1. *A code $C \subset \mathbf{V}$ is given by the vector multiset $\gamma_C := \mu(\mathcal{B})$ on $\mathbf{V}/C^\perp \cong \mathbf{M}$.*

Given a collection $\{s_1, s_2, \ldots, s_m\}$ of vectors and/or subsets of a vector space $\mathbf{V}$, we write $\langle s_1, s_2, \ldots, s_m \rangle$ for its span. In other words $\langle s_1, s_2, \ldots, s_m \rangle$ is the intersection of all subspaces containing $s_1, s_2, \ldots, s_m$.

2.2   Weights

We define the support $\chi(\mathbf{c})$ of $\mathbf{c} \in C$ to be the set of coordinate positions not equal to zero, that is

$$\chi(\mathbf{c}) := \{i \mid c_i \neq 0\}, \quad \text{where } \mathbf{c} = (c_1, c_2, \ldots, c_n).$$

The support of a subset $S \subset C$ is

$$\chi(S) = \bigcup_{\mathbf{c} \in S} \chi(\mathbf{c}).$$

The weight (or support size) $w(S)$ is the cardinality of $\chi(S)$. The ith minimum support weight $d_i(C)$ is the smallest weight of an i-dimensional subcode $D_i \subset C$. The subcode D_i will be called a minimum i-subcode. The weight hierarchy of C is $(d_1(C), d_2(C), \ldots, d_k(C))$. The following Lemma was proved in [10], and the remark is a simple consequence of the proof.

Lemma 2. *There is a one-to-one correspondence between subcodes $D \subset C$ of dimension r and subspaces $U \subset \mathbf{M}$ of codimension r, such that $\gamma_C(U) = n - w(D)$.*

Remark 1. Consider two subcodes D_1 and D_2, and the corresponding subspaces U_1 and U_2. We have that $D_1 \subset D_2$ is equivalent with $U_2 \subset U_1$.

We define $d_{k-r}(\gamma_C)$ such that $n - d_{k-r}(\gamma_C)$ is the largest value of an r-space $\Pi_r \subset \mathsf{PG}(k-1, q)$. From Lemma 2 we get this corollary.

Corollary 1. *If C is a linear code and γ_C is the corresponding multiset, then $d_i(\gamma_C) = d_i(C)$.*

Definition 1. *We say that a code is* chained *if there is a chain $0 = D_0 \subset D_1 \subset \ldots \subset D_k = C$, where each D_i is a minimum i-subcode of C.*

In terms of vector systems, the chain of subcodes corresponds to a chain of maximum value subspaces by remark 1. The difference sequence $(\delta_0, \delta_1, \ldots, \delta_{k-1})$ is defined by $\delta_i = d_{k-i} - d_{k-1-i}$, and is occasionally more convenient than the weight hierarchy.

2.3   Submultisets

Viewing the multiset γ as a collection, we probably have an intuitive notion of a submultiset. A submultiset $\gamma' \subset \gamma$ is a multiset with the property that $\gamma'(x) \leq \gamma(x)$ for all x.

If γ is a multiset on some vector space $\mathbf{V}$, we define a special kind of submultiset, namely the cross-sections. If $U \subset \mathbf{V}$ is a subspace, then the cross-section $\gamma|_U$ is the multiset defined by $\gamma|_U(x) = \gamma(x)$ for $x \in U$, and $\gamma|_U(x) = 0$ otherwise.

If U has dimension r, we call $\gamma|_U$ an r-dimensional cross-section. In some cases it is easier to deal with cross-sections and their sizes, than with subspaces and their values. In Lemma 2, we can consider the cross-section $\gamma_C|_U$ rather than the subspace U. In particular, we have that $n - d_{k-r}(\gamma_C)$ is the size of the largest r-dimensional cross-section of γ_C.

2.4 Duality

Consider a code $C \subset \mathbf{V}$ and its orthogonal code $C^\perp \subset \mathbf{V}$. Write $(d_1, \ldots, d_k)$ for the weight hierarchy of C, and $(d_1^\perp, \ldots, d_{n-k}^\perp)$ for the weight hierarchy of $C^\perp$. Let $\mathcal{B}$ be the set of coordinate vectors for $\mathbf{V}$, and let μ be the natural endomorphism,

$$\mu : \mathbf{V} \to \mathbf{V}/C^\perp,$$

$$\mu : \mathbf{v} \mapsto \mathbf{v} + C^\perp.$$

According to Lemma 1, the vector multiset corresponding to C, is $\gamma_C := \mu(\mathcal{B})$.

Let $B \subset \mathcal{B}$. Then $\mu(B)$ is a submultiset of γ_C. Every submultiset of γ_C is obtained this way. Obviously $\dim\langle B \rangle = \#B$. Let $D := \langle B \rangle \cap C^\perp$ be the largest subcode of $C^\perp$ contained in $\langle B \rangle$. Then D is the kernel of $\mu|_{\langle B \rangle}$, the restriction of μ to $\langle B \rangle$. Hence

$$\dim\langle \mu(B) \rangle = \dim\langle B \rangle - \dim D. \tag{3}$$

Clearly $\#B \geq w(D)$.

With regard to the problem of support weights, we are not interested in arbitrary submultisets of γ_C. We are only interested in cross-sections. Therefore, we ask when $\mu(B)$ is a cross-section of $\mu(\mathcal{B})$. This is of course the case if and only if $\mu(B)$ equals the cross-section $\mu(\mathcal{B})|_{\langle \mu(B) \rangle}$.

Let $U \subset \mathbf{V}/C^\perp$ be a subspace. We have $\mu(\mathcal{B})|_U = \mu(B)$, where $B = \{\mathbf{b} \in \mathcal{B} \mid \mu(\mathbf{b}) \in U\}$. Hence we have $\mu(B) = \mu(\mathcal{B})|_{\langle \mu(B) \rangle}$ if and only if there exists no point $\mathbf{b} \in \mathcal{B} \backslash B$ such that $\mu(\mathbf{b}) \in \langle \mu(B) \rangle$.

It follows from (3) that a large cross-section $\mu(B)$ of a given dimension, must be such that $\langle B \rangle$ contains a large subcode of $C^\perp$ of sufficiently small weight.

Define for any subcode $D \subset C^\perp$,

$$\beta(D) := \{\mathbf{b}_x \mid x \in \chi(D)\} \subset \mathcal{B}.$$

Obviously $\beta(D)$ is the smallest subset of $\mathcal{B}$ such that D is contained in its span. It follows from the above argument that if D is a minimum subcode and $\mu(\beta(D))$ is a cross-section, then $\mu(\beta(D))$ is a maximum cross-section for C. Thus we are lead to the following two lemmas.

Lemma 3. *If $n - d_r = d_i^\perp$, $B \subset \mathcal{B}$, and $\#B = n - d_r$, then $\mu(B)$ is a cross-section of maximum size and codimension r if and only $B = \beta(D_i)$ for some minimum i-subcode $D_i \subset C^\perp$.*

Lemma 4. *Let r be an arbitrary number, $0 < r \leq n - k$. Let i be such that $d_i^\perp \leq n - d_r < d_{i+1}^\perp$, and let $D_i \subset C^\perp$ be a minimum i-subcode. Then $\mu(\langle B \rangle)$ is a maximum r-subspace for any $B \subset \mathcal{B}$ such that $D_i \subset \langle B \rangle$ and $\#B = n - d_r$.*

As an example of our technique, we include two old results from [16,17], with new proofs based on the argument above.

Proposition 1 (Wei 1991). *The weight sets*

$$\{d_1, d_2, \ldots, d_k\} \quad and \quad \{n+1-d_1^\perp, n+1-d_2^\perp, \ldots, n+1-d_{n-k}^\perp\}$$

are disjoint, and their union is $\{1, 2, \ldots, n\}$.

Proof. Suppose for a contradiction that $d_i = n - s$ and $d_j^\perp = s + 1$ for some i, j, and s. Let $D_j \subset C^\perp$ be a minimum j-subcode. Let $B_i \subset \mathcal{B}$ such that $\mu(B_i)$ is a maximum cross-section of codimension i. We have $\#\beta(D_j) = \#B_i + 1$ and thus $\dim\langle B_i\rangle \cap C^\perp < j$. Hence $\dim\mu(B_i) \geq \dim\mu(\beta(D_j))$. Thus $\mu(B_i)$ cannot be maximum cross-section, contrary to assumption.

Proposition 2 (Wei and Yang 1993). *If a C is a chained code, then so is $C^\perp$, and vice versa.*

Proof. Suppose $C^\perp$ is a chained code. Let

$$\{0\} = D_0 \subset D_1 \subset \ldots \subset D_k = C^\perp$$

be a chain of subcodes of minimum weight. Choose a coordinate ordering, such that

$$\chi(D_i) = \{1, 2, \ldots, d_i^\perp\}, \quad \forall i.$$

For each $r = 1, 2, \ldots, n$, let $B_r \subset \mathcal{B}$ be the set of the r first coordinate vectors. By our argument, $\mu(B_r)$ is a cross-section of maximum size except if $d_i^\perp = r + 1$ for some i; in which case there is no cross-section of maximum size and r elements. Obviously $\mu(B_r) \subset \mu(B_{r+1})$ for all r.

3 Greedy Weights

3.1 Definitions

Definition 2 (Greedy r-subcode). *A (bottom-up) greedy 1-subcode is a minimum 1-subcode. A (bottom-up) greedy r-subcode, $r \geq 2$, is any r-dimensional subcode containing a (bottom-up) greedy $(r-1)$-subcode, such that no other such code has lower weight.*

Definition 3 (Greedy subspace). *Given a vector multiset γ, a (bottom-up) greedy hyperplane is a hyperplane of maximum value. A (bottom-up) greedy space of codimension r, $r \geq 1$, is a subspace of codimension r contained in a (bottom-up) greedy space of codimension $r - 1$, such that no other such subspace has higher value.*

A greedy r-subcode corresponds to a greedy subspace of codimension r, and the r-th greedy weight may be defined from either, as follows.

Definition 4 (Greedy weights). *The rth (bottom-up) greedy weight e_r is the weight of a (bottom-up) greedy r-subcode. For a vector multiset, $n - e_r$ is the value of a (bottom-up) greedy space of codimension r.*

Remark 2. We have obviously that $d_1 = e_1$ and $d_k = e_k$, for any k-dimensional code. For most codes $e_2 > d_2$ [5]. The chain condition is satisfied if and only if $e_r = d_r$ for all r.

We introduce a new set of parameters, the top-down greedy weights. It is in a sense the dual of the greedy weights, and we will see later on that top-down greedy weights can be computed from the greedy weights of the orthogonal code, and vice versa.

Definition 5 (Top-Down Greedy Subspace). *A top-down greedy 0-space of a vector multiset is* $\{0\}$. *A top-down greedy r-space is an r-space containing a top-down greedy $(r-1)$-subspace such that no other such subspace has higher value.*

Definition 6 (Top-Down Greedy Weights). *The r-th top-down greedy weight $\tilde{e}_r$ is $n - \gamma_C(\Pi)$, where Π is a top-down greedy subspace of codimension r.*

Remark 3. The top-down greedy weights share many properties with the (bottom-up) greedy weights. For all codes $\tilde{e}_r \geq d_r$. The chain condition holds if and only if $\tilde{e}_r = d_r$ for all r. In general, $\tilde{e}_r$ may be equal to, greater than, or less than e_r.

We will occasionally speak of (top-down) greedy cross-sections, which is just $\gamma_C|_U$ for some (top-down) greedy space U.

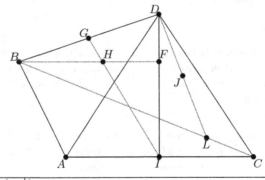

$\gamma(p) =$	for
0	$p \in \langle A, B, C \rangle \backslash \{A, D\}$, $p \in \{F, H, I, J\}$
1	$p \in \langle B, F \rangle \backslash \{B, F, H\}$, $p \in \langle G, I \rangle \backslash \{G, H, I\}$, $p = D$
3	$p = C$
2	otherwise

Fig. 1. Case B, Construction 1 from [1].

Example 1. We take an example of a code from [1] (Case B). The projective multiset is presented in Fig. 1. A chain of greedy subspaces is

$$\emptyset \subset \langle A \rangle \subset \langle A, L \rangle \subset \langle A, B, C \rangle \subset \mathsf{PG}(4, q),$$

and a chain of top-down greedy subspaces is

$$\emptyset \subset \langle C \rangle \subset \langle C, D \rangle \subset \langle A, C, D \rangle \subset \mathsf{PG}(4, q).$$

In the binary case, we get greedy weights $(4, 6, 9, 12)$, and top-down greedy weights $(3, 6, 10, 12)$. The weight hierarchy is $(3, 6, 9, 12)$.

3.2   Basic Properties

Theorem 1 (Monotonicity). *If $(e_1, e_2, \ldots, e_k)$ are greedy weights for some code C, then $0 = e_0 < e_1 < e_2 < \ldots < e_k$. Similarily, if $(\tilde{e}_1, \tilde{e}_2, \ldots, \tilde{e}_k)$ are top-down greedy weights for some code C, then $0 = \tilde{e}_0 < \tilde{e}_1 < \tilde{e}_2 < \ldots < \tilde{e}_k$.*

Proof. Let

$$\{0\} = \Pi_0 \subset \Pi_1 \subset \ldots \subset \Pi_k = \mathbf{M},$$

be a chain of greedy subspaces. We are going to show that $\gamma_C|_{\Pi_i}$ contains more points than $\gamma_C|_{\Pi_{i-1}}$ for all i. It is sufficient to show that $\gamma_C|_{\Pi_i}$ contains a set of points spanning Π_i.

Since γ_C is non-degenerate, it contains a set of points spanning Π_k. Suppose that $\gamma_C|_{\Pi_r}$ contains a set of points spanning Π_r. Consider Π_{r-1}. Suppose $\dim \langle \gamma_C|_{\Pi_{r-1}} \rangle < r - 1$. Obviously there is a point $x \in \gamma_C|_{\Pi_r} - \gamma_C|_{\Pi_{r-1}}$. Hence we can replace Π_{r-1} by $\langle \gamma_C|_{\Pi_{r-1}}, x \rangle$ and get a subspace $\Pi'_{r-1} \subset \Pi_r$ with larger value. This contradicts the assumption that Π_{r-1} is a greedy subspace.

We can replace the Π_i with a chain of top-down greedy subspaces, and repeat the proof to prove the second statement of the lemma.

Monotonicity also holds for the weight hierarchy by a similar argument [16].

3.3   Duality

Lemma 5. *Suppose $\tilde{e}_{i+1} > \tilde{e}_i + 1$ where $0 \leq i \leq k$, and define $s := n - \tilde{e}_i + i - k$. Then U is a top-down greedy cross-section of codimension i if and only if $U = \mu(\beta(D_s))$ for some greedy s-subcode $D_s \subset C$.*

Proof. Let $\bar{i}$ be the largest value of $i \leq k-1$ such that $\tilde{e}_{i+1} > \tilde{e}_i + 1$. Then $\delta_j = 1$ for $0 \leq j \leq k - 1 - (\bar{i} + 1)$. It follows that any subset B_j of $j \leq k - 1 - \bar{i}$ elements, gives rise to a top-down greedy cross-section $\mu(B_j)$ of dimension j (and size j). The codimension of such a $\mu(B_j)$ is $k - j \geq \bar{i} + 1$.

Hence $\mu(B_{k-\bar{i}})$ is a top-down greedy cross-section of codimension $\bar{i}$, if and only if it is a maximum value cross-section of codimension $\bar{i}$. Hence, for $i = \bar{i}$, the lemma follows from Lemma 3.

Suppose $\tilde{e}_{m+1} > \tilde{e}_m + 1$, and assume the lemma holds for all i, $\bar{i} \geq i > m$. We will prove the lemma by induction. Define

$$j := \max\{j > m \mid \tilde{e}_j - \tilde{e}_{m+1} = j - (m + 1)\}.$$

Clearly, $\tilde{e}_{j+1} - \tilde{e}_j > 1$.

Now consider a top-down greedy subspace $\mu(B)$ of codimension m, where $B \subset \mathcal{B}$. Clearly there is $B' \subset B$ such that $\mu(B')$ is a top-down greedy subspace of codimension j. By the induction hypothesis, $B' = \beta(D_r)$ for some greedy r-subcode $D_r \subset C^\perp$ where $r = n - k - \tilde{e}_j + j$. Also,

$$\#B' = w(D_r) = e_r^\perp = n - \tilde{e}_j.$$

Note that we can make top-down greedy cross-sections of codimension x for $m < x \le j$ by adding $j - x$ random elements $\mathbf{b}_y$ to B'. This implies also that there cannot be a subcode D_{r+1} of dimension $r + 1$ such that $D_r \subset D_{r+1} \subset C$ and $w(D_{r+1}) \le w(D_r) + 1 + j - x$. Hence

$$e_{r+1}^\perp \ge n - \tilde{e}_j + 1 + j - m. \tag{4}$$

Let $B'' = B_{k-(m+1)} \subset \mathcal{B}$ be such that $\mu(B'')$ is a top-down greedy cross-section of codimension $m + 1$ with $B' \subset B'' \subset B$. Note that $D_r = \langle B'' \rangle \cap C^\perp$.
  Let

$$z := \#B - \#B'' = (n - \tilde{e}_m) - (n - \tilde{e}_{m+1}) = \tilde{e}_{m+1} - \tilde{e}_m.$$

Write $D := \langle B \rangle \cap C^\perp$. Since $\dim \mu(B) - \dim \mu(B'') = 1$, we must have $B = \beta(D)$, and there must be a chain of z subcodes

$$D_r \subset D_{r+1} \subset D_{r+2} \subset \ldots \subset D_{r+z-1} = D$$

where D_i has dimension i for $r \le i < r + z$ and $w(D_i) = w(D_{i+1}) - 1$ for $r < i \le r + z - 2$. By the bound (4), we get

$$w(D_i) = n - \tilde{e}_j + 1 + j - m + i - r - 1 = e_i^\perp.$$

And in particular

$$w(D) = w(D_{r+z-1}) = n - \tilde{e}_j + j - m + z - 1 = e_{r+z-1}^\perp.$$

It remains to show that $s = r + z - 1$ (where s is given in the lemma). Consider

$$r + z - 1 - s = (n - k - \tilde{e}_j + j) + (\tilde{e}_{m+1} - \tilde{e}_m) - 1 - (n - k - \tilde{e}_m + m)$$
$$= j - \tilde{e}_j + \tilde{e}_{m+1} - (m + 1) = 0,$$

by the definition of j.

Corollary 2. *If i and s are as given in Lemma 5, then $e_s^\perp = n - \tilde{e}_i$.*

Theorem 2 (Duality). *Let $(e_1, \ldots e_k)$ be the greedy weight hierarchy of a code C, and $(\tilde{e}_1^\perp, \ldots, \tilde{e}_{n-k}^\perp)$ the top-down greedy weight hierarchies for $C^\perp$. Then*

$$\{\tilde{e}_1, \tilde{e}_2, \ldots, \tilde{e}_k\} \quad and \quad \{n + 1 - e_1^\perp, n + 1 - e_2^\perp, \ldots, n + 1 - e_{n-k}^\perp\}$$

are disjoint sets whose union is $\{1, \ldots n\}$.

Proof. Let $i_1 < i_2 < \ldots$ be the values of i for which $\tilde{e}_i > \tilde{e}_{i-1}$. Going to the proof of Lemma 5, with $m = i_x$, we get $j = i_{x+1}$. The proof showed that $n - \tilde{e}_y + 1 \ne e_s^\perp$ for all s, for all y, $i_x \le y < i_{x+1}$. This holds for all x, hence the theorem.

References

1. Wende Chen and Torleiv Kløve. The weight hierarchies of q-ary codes of dimension 4. *IEEE Trans. Inform. Theory*, 42(6):2265–2272, November 1996.
2. Wende Chen and Torleiv Kløve. Bounds on the weight hierarchies of extremal non-chain codes of dimension 4. *Applicable Algebra in Engineering, Communication and Computing*, 8:379–386, 1997.
3. Wende Chen and Torleiv Kløve. On the second greedy weight for binary linear codes. In M. Fossorier et al., editor, *Applied Algebra, Algebraic Algorithms and Error-Correcting Codes*, volume 1719 of *Springer Lecture Notes in Computer Science*, pages 131–141. Springer-Verlag, 1999.
4. Wende Chen and Torleiv Kløve. On the second greedy weight for linear codes of dimension 3. *Discrete Math.*, 2001. To appear.
5. Gérard D. Cohen, Sylvia B. Encheva, and Gilles Zémor. Antichain codes. *Des. Codes Cryptogr.*, 18(1-3):71–80, 1999.
6. S. Dodunekov and J. Simonis. Codes and projective multisets. *Electron. J. Combin.*, 5(1), 1998. Research Paper 37.
7. G. David Forney, Jr. Dimension/length profiles and trellis complexity of linear block codes. *IEEE Trans. Inform. Theory*, 40(6):1741–1752, 1994.
8. T Fujiwara, T Kasami, S Lin, and T Takata. On the optimum bit orders with respect to the state complexity of trellis diagrams for binary linear codes. *IEEE Trans. Inform. Theory*, 39(1):242–245, 1993.
9. Tor Helleseth, Torleiv Kløve, and Johannes Mykkeltveit. The weight distribution of irreducible cyclic codes with block lengths $n_1((q^l − 1)/n)$. *Discrete Math.*, 18:179–211, 1977.
10. Tor Helleseth, Torleiv Kløve, and Øyvind Ytrehus. Generalized Hamming weights of linear codes. *IEEE Trans. Inform. Theory*, 38(3):1133–1140, 1992.
11. Torleiv Kløve. Support weight distribution of linear codes. *Discrete Math.*, 106/107:311–316, 1992.
12. Hans Georg Schaathun. Upper bounds on weight hierarchies of extremal non-chain codes. *Discrete Math.*, 2000. To appear in the Tverberg Anniversary Volume.
13. Hans Georg Schaathun. The weight hierarchy of product codes. *IEEE Trans. Inform. Theory*, 46(7):2648–2651, November 2000.
14. Hans Georg Schaathun. Duality and weights for linear codes and projective multisets. Technical report, Department of Informatics, University of Bergen, 2001. Also available at http://www.ii.uib.no/$\sim$georg/sci/inf/coding/public/.
15. Michael A. Tsfasman and Serge G. Vlăduţ. Geometric approach to higher weights. *IEEE Trans. Inform. Theory*, 41(6, part 1):1564–1588, 1995. Special issue on algebraic geometry codes.
16. Victor K. Wei. Generalized Hamming weights for linear codes. *IEEE Trans. Inform. Theory*, 37(5):1412–1418, 1991.
17. Victor K. Wei and Kyeongcheol Yang. On the generalized Hamming weights of product codes. *IEEE Trans. Inform. Theory*, 39(5):1709–1713, 1993.

Type II Codes over $\mathbb{F}_{2^r}$

Koichi Betsumiya[1], Masaaki Harada[2], and Akihiro Munemasa[3]

[1] Nagoya University, Graduate School of Mathematics
Nagoya 464–8602, Japan
koichi@math.nagoya-u.ac.jp
[2] Yamagata University, Department of Mathematical Sciences
Yamagata 990–8560, Japan
mharada@sci.kj.yamagata-u.ac.jp
[3] Kyushu University, Department of Mathematics
Fukuoka 812–8581, Japan
munemasa@math.kyushu-u.ac.jp

Abstract. Motivated by the work of Pasquier and Wolfmann in 1980's, we define Type II codes over $\mathbb{F}_{2^r}$ as self-dual codes with the property that their binary images with respect to a trace-orthogonal basis are doubly-even. We give a classification of Type II codes of length 8 over $\mathbb{F}_8, \mathbb{F}_{16}$ and $\mathbb{F}_{32}$. We also characterize all Type II codes whose binary images are the extended Golay $[24, 12, 8]$ code.

1 Introduction

In 1980's, Pasquier and Wolfmann studied self-dual codes over the finite field $\mathbb{F}_{2^r}$ whose binary images with respect to a trace-orthogonal basis (TOB) are Type II (that is, doubly-even self-dual) codes including the extended Hamming code and the extended Golay code, noting that the binary images of self-dual codes over $\mathbb{F}_{2^r}$ with respect to a TOB are self-dual (see [6], [7], [8], [10], [11] and [12]). In their papers, extended Reed–Solomon codes and H-codes were widely investigated. More precisely, extended Reed–Solomon codes and H-codes whose binary images are extremal Type II codes of lengths $32, 40$ and 64 were constructed (see [8]), and two classes of H-codes whose binary images with respect to a TOB are Type II were found [12]. Recently Type II codes over $\mathbb{F}_4$ have been introduced in [3] as self-dual codes whose binary images with respect to the unique TOB are Type II. In this paper, we define Type II codes over $\mathbb{F}_{2^r}$ as codes whose binary images with respect to a TOB are Type II.

The organization of this paper is as follows. In Sect. 2, we introduce Type II codes over $\mathbb{F}_{2^r}$ along with giving basic properties. We also investigate Lee weight enumerators of Type II codes. In Sects. 3, 4 and 5, we investigate Type II codes over $\mathbb{F}_8, \mathbb{F}_{16}$ and $\mathbb{F}_{32}$, respectively. The classification of Type II codes over $\mathbb{F}_8, \mathbb{F}_{16}$ and $\mathbb{F}_{32}$ of length 8 is given. In Sect. 4, it is shown that there is no extremal Type II $\mathbb{F}_{16}$-code of length 12. In the final section, we characterize all Type II codes over $\mathbb{F}_{2^r}$ whose binary images are the extended Golay $[24, 12, 8]$ code, and we show that the extended Reed–Solomon codes $RS_{32}(a)$ are not extremal for any primitive element a of $\mathbb{F}_{32}$.

S. Boztaş and I.E. Shparlinski (Eds.): AAECC-14, LNCS 2227, pp. 102–111, 2001.

2   Basic Results

In this section, we introduce Type II codes over $\mathbb{F}_{2^r}$ and give properties of these codes.

A basis $B = \{b_1, b_2, \ldots, b_r\}$ of $\mathbb{F}_{2^r}$ over $\mathbb{F}_2$ viewed as an $\mathbb{F}_2$-vector space, is called a trace-orthogonal basis (TOB) of $\mathbb{F}_{2^r}$ if $\mathrm{Tr}(b_i b_j) = \delta_{ij}$, where Tr denotes the trace of $\mathbb{F}_{2^r}$ over $\mathbb{F}_2$ and δ_{ij} is the Kronecker symbol.

Definition 1. *Let $B = \{b_1, b_2, \ldots, b_r\}$ be a TOB of $\mathbb{F}_{2^r}$. Let $x = \sum_{j=1}^{r} a_j b_j$ be an element of $\mathbb{F}_{2^r}$ where $a_j \in \mathbb{F}_2$. Then the* Lee weight *of x with respect to B is defined as the number of i's with $a_i = 1$. The* Lee weight *$wt_B(c)$ of a vector c with respect to B is the sum of the Lee weights of its components. A* Type II *code over $\mathbb{F}_{2^r}$ with respect to B is a self-dual code with the property that the Lee weights of the codewords with respect to B are divisible by four.*

If $r = 2$, then there is a unique TOB $B = \{\omega, \omega^2\}$ of $\mathbb{F}_4 = \mathbb{F}_2[\omega]/(\omega^2+\omega+1)$, and Type II codes over $\mathbb{F}_4$ with respect to B have been investigated in [2,3].

We remark that our definition of Type II codes depends on the choice of a TOB. However, the definition of Type II codes with respect to a TOB was recently shown to be independent of the choice of a TOB, by the first author [1].

For $x = (x_1, x_2, \ldots, x_n) \in \mathbb{F}_{2^r}^n$, the binary image $\phi_B(x)$ of x with respect to a basis B is obtained by replacing each component x_i by $(x_i^1, x_i^2, \ldots, x_i^r)$ where $x_i = \sum_{j=1}^{r} x_i^j b_j$. Note that the definition of $\phi_B(x)$ depends on the ordering of the elements of B. However, the resulting binary image $\phi_B(C)$ of a code C is determined up to permutation-equivalence. If C is a Type II $\mathbb{F}_{2^r}$-code of length n and minimum Lee weight d_B with respect to a TOB B, then the binary image $\phi_B(C)$ is a binary Type II $[rn, rn/2, d_B]$ code. In particular, rn is divisible by 8, and $d_B \leq 4 \left[\frac{rn}{24} \right] + 4$ holds (see [9]). A Type II code meeting this bound is called *extremal* (with respect to the TOB B). Note that $d_B \leq n$ since a self-dual code contains the all-one vector.

We are interested in the classification of the binary images of Type II codes, as well as the classification of Type II codes themselves. Since the binary image of a Type II code depends on the choice of a TOB, we will first classify TOB's and Type II codes, then determine the binary images of Type II codes with respect to each of the TOB's.

We now give some results on Type II codes.

Proposition 2. *Let C be a linear $[n, n/2]$ code over $\mathbb{F}_{2^r}$ and let B be a TOB. If every codeword of C has Lee weight divisible by four, then C is Type II.*

Proof. $\phi_B(C)$ is a binary doubly-even code. Thus it is self-orthogonal. For $x, y \in C$ and $\alpha \in \mathbb{F}_{2^r}$,

$$\mathrm{Tr}(\alpha(x \cdot y)) = \mathrm{Tr}(\alpha x \cdot y) = \phi_B(\alpha x) \cdot \phi_B(y) = 0$$

since $\alpha x, y \in C$. Then we have that $x \cdot y = 0$. Thus C is self-orthogonal. Since C is a linear $[n, n/2]$ code, C is self-dual.   □

Theorem 3 (Munemasa [5]). *The total number of Type II* $\mathbb{F}_{2^r}$*-codes of length* n *with respect to a fixed TOB is given by*

$$N_{II,r}(n) = \prod_{i=0}^{n/2-2} (2^{ri} + 1) , \tag{1}$$

if $rn \equiv 0 \pmod 8$ *and* $n \equiv 0 \pmod 4$*, and 0 otherwise.*

The formula (1) is called the mass formula, and will be used to check that the classifications in Sections 3, 4 and 5 are complete.

The *Lee weight enumerator* of a Type II $\mathbb{F}_{2^r}$-code C of length n with respect to a TOB B is defined as $W_B(C) = \sum_{c \in C} x^{rn - wt_B(c)} y^{wt_B(c)}$. It is easy to see that the Lee weight enumerator of C is the same as the Hamming weight enumerator of the binary image $\phi_B(C)$ which is a binary Type II code. By Gleason's theorem, $W_B(C) \in \mathbb{C}[W_{e_8}, W_{g_{24}}]$, where W_{e_8} and $W_{g_{24}}$ are the Hamming weight enumerators of the binary extended Hamming [8, 4, 4] code e_8 and the extended Golay [24, 12, 8] code g_{24}, respectively. Let $\mathbb{C}[W_{e_8}, W_{g_{24}}]_f$ denote the f-th homogeneous part of $\mathbb{C}[W_{e_8}, W_{g_{24}}]$. The Lee weight enumerator of a Type II code over $\mathbb{F}_{2^r}$ belongs to $\oplus_{f=0}^{\infty} \mathbb{C}[W_{e_8}, W_{g_{24}}]_{8rf}$ or $\oplus_{f=0}^{\infty} \mathbb{C}[W_{e_8}, W_{g_{24}}]_{4rf}$, if r is odd or even, respectively. In particular, for $r = 3$, we have

$$\bigoplus_{f=0}^{\infty} \mathbb{C}[W_{e_8}, W_{g_{24}}]_{24f} = \mathbb{C}[W_{e_8}^3, W_{g_{24}}] = \mathbb{C}[\varphi_{3,1}, \varphi_{3,2}] \tag{2}$$

where we define $\varphi_{3,1} = W_{g_{24}}$, $\varphi_{3,2} = \frac{1}{42}(W_{e_8}^3 - W_{g_{24}})$ for later use. We will show that the Lee weight enumerators of Type II codes over $\mathbb{F}_8$ generate the ring (2).

Similarly for $r = 4$, we have

$$\bigoplus_{f=0}^{\infty} \mathbb{C}[W_{e_8}, W_{g_{24}}]_{16f} = \mathbb{C}[W_{e_8}^2, W_{e_8} W_{g_{24}}, W_{g_{24}}^2] = \mathbb{C}[\varphi_{4,1}, \varphi_{4,2}, \varphi_{4,3}] \tag{3}$$

where we define $\varphi_{4,1} = W_{e_8}^2$, $\varphi_{4,2} = W_{e_8}^4 - 56 W_{e_8} \varphi_{3,2}$, $\varphi_{4,3} = W_{e_8}^6 - 84 W_{e_8}^3 \varphi_{3,2} + 246 \varphi_{3,2}^2$ for later use. Note that $\varphi_{3,1}, \varphi_{4,1}, \varphi_{4,2}$ and $\varphi_{4,3}$ are the weight enumerators of extremal binary Type II codes of lengths 24, 16, 32 and 48, respectively. We will show that the Lee weight enumerators of Type II codes over $\mathbb{F}_{16}$ generate the ring (3).

There is a class of self-dual codes known as H-codes (see [8]). In particular for length 8, such a code has a generator matrix of the form (I , A) where the rows of A are (a_1, a_2, a_3, a_4), (a_2, a_1, a_4, a_3), (a_3, a_4, a_1, a_2), (a_4, a_3, a_2, a_1) with $a_1 + a_2 + a_3 + a_4 = 1$. In this paper, we call a code with generator matrix (I , A) with the above A simply an H-code with generator (a_1, a_2, a_3, a_4). It is obvious that an H-code has a fixed-point-free involutive automorphism. We will see that any Type II code of length 8 over $\mathbb{F}_8$, $\mathbb{F}_{16}$ or $\mathbb{F}_{32}$ whose permutation automorphism group has even order is permutation-equivalent to an H-code.

3   Type II Codes over $\mathbb{F}_8$

In this section, we investigate Type II codes over $\mathbb{F}_8 = \mathbb{F}_2[\alpha]/(\alpha^3 + \alpha + 1)$. The unique TOB of $\mathbb{F}_8$ is $B_8 = \{\alpha^3, \alpha^5, \alpha^6\}$.

We first give a classification of Type II codes of length 8 with respect to the unique TOB B_8. It turns out that all Type II codes over $\mathbb{F}_8$ belong to the class of H-codes. Let $C_{8,1}, C_{8,2}$ and $C_{8,3}$ be the H-codes with generators $(0, 1, 1, 1)$, $(\alpha^3, \alpha^3, \alpha^4, \alpha^5)$ and $(1, \alpha^2, \alpha^4, \alpha)$, respectively. These three codes are all Type II, and we give the orders of the permutation automorphism groups and the Lee weight enumerators of these codes in Table 1. The fourth column gives the binary image using the notation in [9, Table VII]. The minimum Hamming weight d_H of $C_{8,i}$ and the minimum Lee weight d_B, that is, the minimum weight of the binary image are also listed in the table. Note that $C_{8,3}$ is an MDS code. We find $\sum_{C \in \mathcal{C}_8} \frac{8!}{|\mathrm{PAut}(C)|} = 1170 = N_{II,3}(8)$ where $\mathcal{C}_8 = \{C_{8,1}, C_{8,2}, C_{8,3}\}$.

Theorem 3 shows that there are no other Type II codes over $\mathbb{F}_8$.

Table 1. All Type II $\mathbb{F}_8$-codes of length 8

Code	|PAut|	Lee weight enumerator	Binary image	d_H	d_B
$C_{8,1}$	1344	$\varphi_{3,1} + 42\varphi_{3,2}$	C24	4	4
$C_{8,2}$	96	$\varphi_{3,1} + 6\varphi_{3,2}$	C27	4	4
$C_{8,3}$	56	$\varphi_{3,1}$	C28 (g_{24})	5	8

Theorem 4. *There are three Type II $\mathbb{F}_8$-codes of length 8, up to permutation-equivalence, one of which is extremal.*

Corollary 5. *The ring $\mathbb{C}[\varphi_{3,1}, \varphi_{3,2}]$ given in (2) is generated by the Lee weight enumerators of Type II codes over $\mathbb{F}_8$.*

The code $C_{8,1}$ is the extended binary Hamming code e_8 regarded as a code over $\mathbb{F}_8$. The code $C_{8,2}$ is given in [8, Theorem 6.8]. It follows from Pasquier's result [6] that $C_{8,3}$ is permutation-equivalent to the extended Reed–Solomon code RS_8 (see Sect. 6.2). The code $C_{8,3}$ has the Hamming weight enumerator $1 + 392y^5 + 588y^6 + 1736y^7 + 1379y^8$.

We have restricted ourselves so far to binary images with respect to a TOB, but one could consider the binary images with respect to arbitrary bases of $\mathbb{F}_8$ over $\mathbb{F}_2$. However, the results for the above three codes are uninteresting. Since $C_{8,1}$ has a binary generator matrix, its binary image is independent of the choice of a basis. As for $C_{8,2}$ and $C_{8,3}$, the binary image with respect to a basis B is self-dual if and only if B is a scalar multiple of the TOB B_8, and in this case, the binary image with respect to B is the same as the binary image with respect to B_8.

By Theorem 3, the number of distinct Type II codes of length 16 is given by $N_{II,3}(16) = \prod_{i=0}^{6}(8^i + 1)$. The order of the permutation automorphism group of each code is at most 16!. Since $N_{II,3}(16)/16! > 1009611$, there are at least 1009612 Type II codes of length 16 up to permutation-equivalence. Therefore it seems infeasible to classify all Type II codes of this length.

4   Type II Codes over $\mathbb{F}_{16}$

In this section, we study Type II codes over $\mathbb{F}_{16} = \mathbb{F}_2[\alpha]/(\alpha^4 + \alpha + 1)$. Let σ be the Frobenius automorphism defined by $\alpha^\sigma = \alpha^2$.

Lemma 6. *There are two TOB's of* $\mathbb{F}_{16}$, *namely* $B_{16,1} = \{\alpha^3, \alpha^7, \alpha^{12}, \alpha^{13}\}$ *and* $B_{16,2} = B_{16,1}^\sigma = \{\alpha^6, \alpha^{14}, \alpha^9, \alpha^{11}\}$.

There is a unique Type II code of length 4 [5]. In this section, we classify Type II $\mathbb{F}_{16}$-codes of length 8 with respect to the TOB $B_{16,1}$. According to a result of the first author [1], Type II codes with respect to $B_{16,1}$ are also Type II codes with respect to $B_{16,2}$, hence our result automatically implies the classification of Type II codes with respect to $B_{16,2}$. However, since Theorem 3 is independent of the choice of a TOB, and we can check that all codes we found are also Type II codes with respect to $B_{16,2}$ by direct computation, our classification is in fact independent of [1].

Every Type II code C of length 8 is permutation-equivalent to a code with generator matrix of the form (I , A) where A is a 4×4 matrix over $\mathbb{F}_{16}$. Thus we only need to consider the set of 4×4 matrices A, rather than the set of generator matrices. The set of matrices A was constructed, row by row, using a back-tracking algorithm. Permuting the rows of A gives rise to a generator matrix of a code which is permutation-equivalent to the code generated by (I , A). We considered only those matrices A which are smallest among all matrices obtained from A by permuting its rows, where the ordering is defined by regarding an $\mathbb{F}_{16}$-vector as an integral vector by some fixed ordering. Our computer search was done by constructing all matrices A such that the code generated by (I , A) is Type II. It is obvious that any Type II code is permutation-equivalent to one of the codes constructed.

We list all Type II codes over $\mathbb{F}_{16}$ up to permutation-equivalence in Table 2. The second column gives the orders of the permutation automorphism groups, the third column gives the Lee weight enumerators, and the fourth column gives the binary images using the notation in [9, Tables III and IV]. The minimum Hamming weight d_H and the minimum Lee weight d_B, that is, the minimum weight of the binary image are also listed in the table.

Note that the binary images of the codes $C_{16,1}, \ldots, C_{16,5}$ and $C_{16,5}^\sigma$ turned out to be independent of the choice of a TOB. This is obvious for the codes $C_{16,1}, \ldots, C_{16,4}$. Indeed, the orders of permutation automorphism groups tell us that these codes are permutation-equivalent to its Frobenius image. Since $\phi_B(C) = \phi_{B^\sigma}(C^\sigma)$ for a code C and a TOB B, we see that for each of these four codes, the binary image with respect to $B_{16,1}$ is permutation-equivalent to the binary image with respect to $B_{16,1}^\sigma = B_{16,2}$. For $C_{16,5}$, we have verified directly that the binary image with respect to $B_{16,1}$ is permutation-equivalent to the binary image with respect to $B_{16,2}$. This implies that the same is true for $C_{16,5}^\sigma$. It follows that the Lee weight enumerator of each of the Type II codes of length 8 over $\mathbb{F}_{16}$ is independent of the choice of a TOB. We will see in Sect. 5 that there exists a Type II code over $\mathbb{F}_{32}$ whose binary image depends on the choice of a TOB.

In the same way as in the case of $\mathbb{F}_8$, one can consider the binary images with respect to arbitrary bases. Since $C_{16,2}$ has a binary generator matrix, its binary image is independent of the choice of a basis. The binary images with respect to a scalar multiple of a TOB B are the same as the binary images with respect to B. If B is not a scalar multiple of a TOB and $C \neq C_{16,2}$ is one of the codes listed in Table 2, then the binary image of C with respect to B is self-dual if and only if $C = C_{16,1}$ and B is a scalar multiple of $\{\alpha, \alpha^2, \alpha^6, \alpha^7\}^{\sigma^i}$ for some integer i, and in this case the binary image is C24.

Theorem 7. *There are six Type II $\mathbb{F}_{16}$-codes of length 8, up to permutation-equivalence, one of which is extremal.*

As a check, we verify the mass formula: $\sum_{C \in \mathcal{C}_{16}} \frac{8!}{|\mathrm{PAut}(C)|} = 8738 = N_{II,4}(8)$ where $\mathcal{C}_{16} = \{C_{16,1}, C_{16,2}, C_{16,3}, C_{16,4}, C_{16,5}, C_{16,5}^\sigma\}$.

Table 2. All Type II $\mathbb{F}_{16}$-codes of length 8

Code	|PAut|	Lee weight enumerator	Binary image	d_H	d_B
$C_{16,1}$	288	$\varphi_{4,1}^2$	C24	3	4
$C_{16,2}$	1344	$\varphi_{4,1}^2$	C24	4	4
$C_{16,3}$	15	$\varphi_{4,2} - 3(\varphi_{4,2} - \varphi_{4,1}^2)/28$	C60	3	4
$C_{16,4}$	8	$\varphi_{4,2}$	C85	4	8
$C_{16,5}, C_{16,5}^\sigma$	96	$\varphi_{4,2} - (\varphi_{4,2} - \varphi_{4,1}^2)/7$	C67	4	4

The code $C_{16,1}$ is the H-code with generator $(0, 0, \alpha^5, \alpha^{10})$ which is the direct sum of the unique Type II code of length 4. The code $C_{16,2}$ is the extended binary Hamming code e_8 regarded as a code over $\mathbb{F}_{16}$. The code $C_{16,4}$ is the H-code with generator $(1, \alpha^{10}, \alpha^{11}, \alpha^{14})$ given in [8, Theorem 6.9]. The code $C_{16,5}$ is the H-code with generator $(\alpha^{11}, \alpha, \alpha, \alpha^{12})$. The code $C_{16,3}$ has generator matrix $(I, A_{16,3})$ where the rows of $A_{16,3}$ are $(\alpha, \alpha^{12}, \alpha^{11}, \alpha)$, $(\alpha^7, \alpha^{13}, 1, \alpha^5)$, $(\alpha^3, \alpha^4, \alpha^{14}, \alpha^4)$, $(0, 0, \alpha^{10}, \alpha^5)$.

By Theorem 3, the number of distinct Type II codes of length 12 is given by $N_{II,4}(12) = \prod_{i=0}^{4}(16^i + 1)$. The order of the permutation automorphism group of each code is at most 12!. Since $N_{II,4}(12)/12! > 4898$, there are at least 4899 Type II codes of length 12 up to permutation-equivalence. This is a weak bound, thus it seems infeasible to classify all Type II codes of this length.

Let C be a $[6+t, t]$ code with $t \geq 1$ such that the minimum Lee weight is at least 12. By shortening C, we obtain a $[6+t-1, t-1]$ code such that the minimum Lee weight is at least 12. This means that we can construct all $[6+t, t]$ codes such that the minimum Lee weight is at least 12 and that all the

Lee weights are divisible by 4, by computer search starting from $t = 0, 1, 2, \ldots$. There is such a $[10, 4]$ code, however, we have found no such $[11, 5]$ code. Thus, we obtain the following result.

Proposition 8. *There is no extremal Type II $\mathbb{F}_{16}$-code of length* 12.

We have found a Type II code of length 12 with the Lee weight enumerator

$$\frac{93}{82}\varphi_{4,1} - \frac{279}{82}\varphi_{4,2} + \frac{134}{41}\varphi_{4,3} = x^{48} + 558x^{40}y^8 + 12832x^{36}y^{12} + 550719x^{32}y^{16} + \cdots .$$

Hence we have the following:

Corollary 9. *The ring* $\mathbb{C}[\varphi_{4,1}, \varphi_{4,2}, \varphi_{4,3}]$ *given in (3) is generated by the Lee weight enumerators of Type II codes over* $\mathbb{F}_{16}$.

It is known that there is an extremal Type II $\mathbb{F}_{16}$-code of length 16 (see [7]).

5   Type II Codes over $\mathbb{F}_{32}$

In this section, we give the classification of Type II codes of length 8 over $\mathbb{F}_{32} = \mathbb{F}_2[\alpha]/(\alpha^5 + \alpha^2 + 1)$. The method used is similar to the one given in the previous section, so we list the results only. Again we denote by σ the Frobenius automorphism defined by $\alpha^\sigma = \alpha^2$.

Lemma 10. *There are six TOB's of* $\mathbb{F}_{32}$, *namely* $B_{32,1} = \{\alpha^5, \alpha^9, \alpha^{10}, \alpha^{18}, \alpha^{20}\}$, $B_{32,2} = \{\alpha^3, \alpha^{12}, \alpha^{13}, \alpha^{20}, \alpha^{26}\}$, $B_{32,2}^\sigma$, $B_{32,2}^{\sigma^2}$, $B_{32,2}^{\sigma^3}$ *and* $B_{32,2}^{\sigma^4}$.

According to a result of the first author [1], Type II codes with respect to $B_{32,1}$ are also Type II codes with respect to any other TOB, hence the classification is independent of the choice of the TOB. By the same reason we indicated in Sect. 4, this fact can be directly checked in our case.

We list all Type II codes over $\mathbb{F}_{32}$ up to permutation-equivalence in Table 3. The second and fifth columns give the order of the permutation automorphism group of the code, the third and sixth columns give the minimum Hamming weight d_H.

Theorem 11. *There are* 12 *Type II $\mathbb{F}_{32}$-codes of length 8, up to permutation-equivalence.*

As a check, we verify the mass formula: $\sum_{C \in \mathcal{C}_{32}} \frac{8!}{|\mathrm{PAut}(C)|} = 67650 = N_{II,5}(8)$ where $\mathcal{C}_{32} = \{C_{32,1}, C_{32,2}, C_{32,3}^{\sigma^j}, C_{32,4}^{\sigma^j} \mid j = 0, 1, \ldots, 4\}$.

The code $C_{32,1}$ is the extended binary Hamming code e_8 regarded as a code over $\mathbb{F}_{32}$. The code $C_{32,4}$ is the H-code with generator $(1, \alpha^5, \alpha^{14}, \alpha^{21})$ given in [8, Theorem 6.10]. The code $C_{32,3}$ is the H-code with generator $(\alpha^{22}, \alpha^7, \alpha^{17}, \alpha^{17})$. The code $C_{32,2}$ has generator matrix $(\ I\ ,\ A_{32,2}\ )$ where the rows of $A_{32,2}$ are $(\alpha^5, \alpha^5, \alpha^{12}, \alpha^{23})$, $(\alpha^{29}, \alpha^{21}, \alpha^7, \alpha^2)$, $(\alpha^{27}, \alpha^{14}, \alpha^{11}, \alpha^4)$, $(\alpha^{19}, 1, \alpha^{13}, \alpha^9)$. The codes $C_{32,2}$ and $C_{32,4}^{\sigma^j}$ $(j = 0, 1, \ldots, 4)$ are MDS codes.

Table 3. All Type II $\mathbb{F}_{32}$-codes of length 8

Code	|PAut|	d_H	Code	|PAut|	d_H
$C_{32,1}$	1344	4	$C_{32,2}$	1	5
$C_{32,3}, C^\sigma_{32,3}, C^{\sigma^2}_{32,3}, C^{\sigma^3}_{32,3}, C^{\sigma^4}_{32,3}$	96	4	$C_{32,4}, C^\sigma_{32,4}, C^{\sigma^2}_{32,4}, C^{\sigma^3}_{32,4}, C^{\sigma^4}_{32,4}$	8	5

Table 4. Binary images of Type II codes

	$B_{32,1}$	$B_{32,2}$	$B^\sigma_{32,2}$	$B^{\sigma^2}_{32,2}$	$B^{\sigma^3}_{32,2}$	$B^{\sigma^4}_{32,2}$		$B_{32,1}$	$B_{32,2}$	$B^\sigma_{32,2}$	$B^{\sigma^2}_{32,2}$	$B^{\sigma^3}_{32,2}$	$B^{\sigma^4}_{32,2}$
$C_{32,1}$	$D_{40,1}$	$D_{40,1}$	$D_{40,1}$	$D_{40,1}$	$D_{40,1}$	$D_{40,1}$	$C_{32,2}$	$D_{40,2}$	$D_{40,3}$	$D_{40,3}$	$D_{40,3}$	$D_{40,3}$
$C_{32,3}$	$D_{40,4}$	$D_{40,4}$	$D_{40,4}$	$D_{40,5}$	$D_{40,4}$	$D_{40,4}$	$C_{32,4}$	$D_{40,6}$	$D_{40,9}$	$D_{40,6}$	$D_{40,7}$	$D_{40,8}$	$D_{40,6}$

The occurrence of self-dual codes with trivial automorphism group has been investigated (see [4, Sect. 7.5]). In particular, it is a problem to determine the smallest length for which there is a code with a trivial automorphism group for each class of self-dual codes. Note that $C_{32,2}$ is the first example of a Type II $\mathbb{F}_{2^r}$-code with a trivial permutation automorphism group for $r \geq 2$.

We now consider the binary images of the Type II codes in $\mathcal{C}_{32}$ with respect to the six TOB's. The binary images are listed in Table 4, and descriptions of the entries of Table 4 are given in Table 5. We do not list the binary images of the codes $C^{\sigma^j}_{32,3}$, $C^{\sigma^j}_{32,4}$ ($j = 1, \ldots, 4$), since they can be derived from those of the codes $C_{32,3}$, $C_{32,4}$, respectively.

Theorem 12. *The six Type II $\mathbb{F}_{32}$-codes $C_{32,2}$, $C^{\sigma^j}_{32,4}$ ($j = 0, \ldots, 4$) are extremal with respect to each of the six TOB's.*

By Theorem 3, the number of distinct Type II codes of length 16 is given by $N_{II,5}(16) = \prod_{i=0}^{6}(32^i + 1)$. The order of the permutation automorphism group of each code is at most 16!. Since $N_{II,5}(16)/16! \geq 4 \cdot 10^{18}$, there are great many Type II codes of length 16 up to permutation-equivalence. Therefore it seems infeasible to classify all Type II codes of this length.

An H-code over $\mathbb{F}_{32}$ of length 16 is given in [8, Theorem 6.13]. It was verified in [8] that the code is Type II with respect to $B_{32,1}$ and $B^\sigma_{32,2}$. We have verified that the minimum Lee weights of the code with respect to $B_{32,1}$ and $B^{\sigma^j}_{32,2}$ ($j = 0, \ldots, 4$) are 12, that is, non-extremal.

6   Other Codes

6.1   Reconstruction of the Extended Golay Code

One reason of interest in Type II codes over $\mathbb{F}_{2^r}$ comes from the existence of the Type II $\mathbb{F}_8$-code of length 8 which gives a new construction of the binary extended Golay code g_{24} [6,10]. In this subsection, we classify all Type II codes over $\mathbb{F}_{2^r}$ whose binary images are g_{24}.

Table 5. Equivalences of the binary images

Binary code	$|\mathrm{Aut}(D_{40,i})|$	Minimum weight	Weight enumerator
$D_{40,1}$	526232406856826880	4	$W_{e_8}^5$
$D_{40,2}$	240	8	$W_{e_8}^5 - 70W_{e_8}^2\varphi_{3,2}$
$D_{40,3}$	6	8	$W_{e_8}^5 - 70W_{e_8}^2\varphi_{3,2}$
$D_{40,4}$	12079595520	4	$W_{e_8}^5 - 60W_{e_8}^2\varphi_{3,2}$
$D_{40,5}$	45298483200	4	$W_{e_8}^5 - 60W_{e_8}^2\varphi_{3,2}$
$D_{40,6}$	128	8	$W_{e_8}^5 - 70W_{e_8}^2\varphi_{3,2}$
$D_{40,7}$	64	8	$W_{e_8}^5 - 70W_{e_8}^2\varphi_{3,2}$
$D_{40,8}$	6144	8	$W_{e_8}^5 - 70W_{e_8}^2\varphi_{3,2}$
$D_{40,9}$	4608	8	$W_{e_8}^5 - 70W_{e_8}^2\varphi_{3,2}$

By Theorem 3, it is sufficient to consider extremal Type II codes over $\mathbb{F}_{2^r}$ of length n with $(r,n) = (2,12)$, $(3,8)$ and $(6,4)$. It was shown in [2] that there is a unique extremal Type II $\mathbb{F}_4$-code of length 12, up to permutation-equivalence. In Theorem 4, we showed that there is a unique extremal Type II $\mathbb{F}_8$-code of length 8, up to permutation-equivalence, with respect to the unique TOB. Munemasa [5] shows that there is a unique Type II $\mathbb{F}_{2^r}$-code of length 4 with respect to a fixed TOB, for even r. The unique code contains a codeword of Lee weight 4 for any TOB. Therefore we have the following:

Theorem 13. *The only Type II codes over $\mathbb{F}_{2^r}$ whose binary images are the binary extended Golay code are the unique extremal Type II $\mathbb{F}_4$-code of length 12 with respect to the unique TOB and the unique extremal Type II $\mathbb{F}_8$-code of length 8 with respect to the unique TOB.*

6.2   Reed–Solomon Codes

Let $RS_{2^r}(a)$ be the extended Reed–Solomon $\mathbb{F}_{2^r}$-code of length 2^r which is the extended cyclic code generated by the polynomial $\prod_{i=1}^{2^{r-1}-1}(x - a^i)$, where a is a primitive element of $\mathbb{F}_{2^r}$. In this section, we discuss the Type II property and the extremality of all the extended Reed–Solomon codes of length 8, 16 and 32 with respect to all TOB's. To do this, it is enough to consider representatives of equivalence classes under permutations and the Frobenius automorphisms. We can easily verify that $RS_{2^r}(a^{-1})$ is permutation-equivalent to $RS_{2^r}(a)$, and that $RS_{2^r}(a^2)$ is the Frobenius image of $RS_{2^r}(a)$. Hence for $r = 3$ and 4, without loss of generality we may take the element a to be a primitive element α satisfying $\alpha^3 + \alpha + 1 = 0$ and $\alpha^4 + \alpha + 1 = 0$, respectively. Let RS_8 (resp. RS_{16}) denote the extended Reed–Solomon code $RS_8(\alpha)$ (resp. $RS_{16}(\alpha)$) for this α.

Proposition 14 (Pasquier [8]). *The code RS_8 is an extremal Type II $\mathbb{F}_8$-code of length 8 with respect to the TOB B_8, and the code RS_{16} is an extremal Type II $\mathbb{F}_{16}$-code of length 16 with respect to the two TOB's $B_{16,1}$ and $B_{16,2}$.*

For $r = 5$, we take α to be a primitive element satisfying $\alpha^5 + \alpha^2 + 1 = 0$ as in Sect. 5. Then it suffices to consider $RS_{32}(\alpha)$, $RS_{32}(\alpha^3)$ and $RS_{32}(\alpha^5)$. It was shown in [8] that $RS_{32}(\alpha)$ is a Type II code with respect to the two TOB's given in [8] with minimum Lee weight $d_L \in \{20, 24, 28\}$. Note that the generator polynomial of the code was incorrectly reported. The correct polynomial is

$$g(x) = x^{15} + \alpha^7 x^{14} + \alpha^{17} x^{13} + \alpha^5 x^{12} + \alpha^{22} x^{11} + \alpha^{13} x^{10} + \alpha^{27} x^9 + \alpha^{28} x^8$$
$$+ \alpha^5 x^7 + \alpha^{20} x^6 + \alpha^{22} x^5 + \alpha^{16} x^4 + \alpha^{15} x^3 + \alpha^{12} x^2 + \alpha^{18} x + \alpha^{27} \ .$$

Proposition 15. *The codes $RS_{32}(\alpha)$, $RS_{32}(\alpha^3)$ and $RS_{32}(\alpha^5)$ are non-extremal Type II $\mathbb{F}_{32}$-codes with respect to each of the six TOB's.*

Proof. It is easy to check that $RS_{32}(\alpha)$, $RS_{32}(\alpha^3)$, $RS_{32}(\alpha^5)$ are Type II with respect to each of the TOB's. Moreover, we have found a codeword of Lee weight 24 for each of these codes. □

References

1. Betsumiya, K.: Type II property for self-dual codes over finite fields of characteristic two. (submitted).
2. Betsumiya, K., Gulliver, T.A., Harada, M. and Munemasa, A.: On Type II codes over $\mathbb{F}_4$. IEEE Trans. Inform. Theory. (to appear).
3. Gaborit, P., Pless, V., Solé, P. and Atkin, O.: Type II codes over $\mathbb{F}_4$. (preprint).
4. Huffman, W.C.: "Codes and groups," in Handbook of Coding Theory, V.S. Pless and W.C. Huffman (Eds.), Elsevier, Amsterdam 1998, pp. 1345–1440
5. Munemasa, A.: Type II codes and quadratic forms over a finite field of characteristic two. The Ohio State Univ. Math. Research Institute Monograph Ser. (to appear).
6. Pasquier, G.: The binary Golay code obtained from an extended cyclic code over $\mathbb{F}_8$. Eurp. J. Combin. **1** (1980) 369–370
7. Pasquier, G.: Binary images of some self-dual codes over $GF(2^m)$ with respect to trace-orthogonal basis. Discrete Math. **37** (1981) 127–129
8. Pasquier, G.: Binary self-dual codes construction from self-dual codes over a Galois field $\mathbb{F}_{2^m}$. Ann. Discrete Math. **17** (1983) 519–526
9. Rains, E., Sloane, N.J.A.: "Self-dual codes," in Handbook of Coding Theory, V.S. Pless and W.C. Huffman (Eds.), Elsevier, Amsterdam 1998, pp. 177–294
10. Wolfmann, J.: A new construction of the binary Golay code $(24, 12, 8)$ using a group algebra over a finite field. Discrete Math. **31** (1980) 337–338.
11. Wolfmann, J.: A class of doubly even self dual binary codes. Discrete Math. **56** (1985) 299–303.
12. Wolfmann, J.: A group algebra construction of binary even self dual codes. Discrete Math. **65** (1987) 81–89.

On Senary Simplex Codes

Manish K. Gupta[1], David G. Glynn[1], and T. Aaron Gulliver[2]

[1] Department of Mathematics, University of Canterbury,
Private Bag 4800, Christchurch, New Zealand
m.k.gupta@ieee.org
d.glynn@math.canterbury.ac.nz
[2] Department of Electrical and Computer Engineering, University of Victoria,
P.O. Box 3055, STN CSC, Victoria, B.C., Canada V8W 3P6
agullive@engr.uvic.ca

Abstract. This paper studies senary simplex codes of type α and two punctured versions of these codes (type β and γ). Self-orthogonality, torsion codes, weight distribution and weight hierarchy properties are studied. We give a new construction of senary codes via their binary and ternary counterparts, and show that type α and β simplex codes can be constructed by this method.

1 Introduction

There has been much interest in codes over finite rings in recent years, especially the rings $\mathbb{Z}_{2k}$ where $\mathbb{Z}_{2k}$ denotes the ring of integers modulo $2k$. In particular codes over $\mathbb{Z}_4$ have been widely studied [1], [5],[6],[7],[8], [9],[10],[11], [12]. More recently $\mathbb{Z}_4$-simplex codes (and their Gray images), have been investigated by Bhandari, Gupta and Lal in [2]. Good binary linear and non-linear codes can be obtained from codes over $\mathbb{Z}_4$ via the Gray map. Thus it is natural to investigate simplex codes over the ring $\mathbb{Z}_{2k}$. In particular, one can construct mixed binary/ternary codes via senary codes by applying the Chinese Gray map (see Example 1). Motivated by this (apart from practical applications such as PSK modulation [4]), in this paper we consider senary simplex codes, and investigate their fundamental properties. We also study their Chinese product type construction.

A *linear code* $\mathcal{C}$, of length n, over $\mathbb{Z}_6$ is an additive subgroup of $\mathbb{Z}_6^n$. An element of $\mathcal{C}$ is called a *codeword of* $\mathcal{C}$ and a *generator matrix* of $\mathcal{C}$ is a matrix whose rows generate $\mathcal{C}$. The *Hamming weight* $w_H(x)$ of a vector x in $\mathbb{Z}_6^n$ is the number of non-zero components. The *Lee weight* $w_L(x)$ of a vector $x = (x_1, x_2, \ldots, x_n)$ is $\sum_{i=1}^{n} \min\{|x_i|, |6 - x_i|\}$. The *Euclidean weight* $w_E(x)$ of a vector x is $\sum_{i=1}^{n} \min\{x_i^2, (6-x_i)^2\}$. The Euclidean weight is useful in connection with lattice constructions. The *Chinese Euclidean weight* $w_{CE}(x)$ of a vector $x \in \mathbb{Z}_m^n$ is $\sum_{i=1}^{n} \left\{ 2 - 2\cos\left(\frac{2\pi x_i}{m}\right) \right\}$. This is useful for $m-PSK$ coding [4]. The Hamming, Lee and Euclidean distances $d_H(x,y)$, $d_L(x,y)$ and $d_E(x,y)$ between two vectors x and y are $w_H(x-y)$, $w_L(x-y)$ and $w_E(x-y)$, respectively. The minimum Hamming, Lee and Euclidean weights, d_H, d_L and d_E, of $\mathcal{C}$ are the

S. Boztaş and I.E. Shparlinski (Eds.): AAECC-14, LNCS 2227, pp. 112–121, 2001.
© Springer-Verlag Berlin Heidelberg 2001

smallest Hamming, Lee and Euclidean weights among all non-zero codewords of $\mathcal{C}$, respectively.

The *Chinese Gray map* $\phi : \mathbb{Z}_6^n \to \mathbb{Z}_2^n \mathbb{Z}_3^n$ is the coordinate-wise extension of the function from $\mathbb{Z}_6$ to $\mathbb{Z}_2 \mathbb{Z}_3$ defined by $0 \to (0,0), 1 \to (1,1), 2 \to (0,2), 3 \to (1,0), 4 \to (0,1)$ and $5 \to (1,2)$. The inverse map ϕ^{-1} is a ring isomorphism and so is $\phi[6]$. The image $\phi(\mathcal{C})$, of a linear code $\mathcal{C}$ over $\mathbb{Z}_6$ of length n by the Chinese Gray map, is a mixed binary/ternary code of length $2n$.

The *dual code* $\mathcal{C}^\perp$ of $\mathcal{C}$ is defined as $\{x \in \mathbb{Z}_6^n \mid x \cdot y = 0 \text{ for all } y \in \mathcal{C}\}$ where $x \cdot y$ is the standard inner product of x and y. $\mathcal{C}$ is *self-orthogonal* if $\mathcal{C} \subseteq \mathcal{C}^\perp$ and $\mathcal{C}$ is *self-dual* if $\mathcal{C} = \mathcal{C}^\perp$.

Two codes are said to be *equivalent* if one can be obtained from the other by permuting the coordinates and (if necessary) changing the signs of certain coordinates. Codes differing by only a permutation of coordinates are called *permutation-equivalent*.

In this paper we define $\mathbb{Z}_6$-simplex codes of type α, β and γ namely, S_k^α, S_k^β and S_k^γ, and determine some of their fundamental parameters. Section 2 contains some preliminaries and notations. Definitions and basic parameters of $\mathbb{Z}_6$-simplex codes of type α, β and γ are given in Section 3. Section 4 investigates their Chinese product type construction.

2   Preliminaries and Notations

Any linear code $\mathcal{C}$ over $\mathbb{Z}_6$ is permutation-equivalent to a code with generator matrix G (the rows of G generate $\mathcal{C}$) of the form

$$G = \begin{bmatrix} I_{k_1} & A_{1,2} & A_{1,3} & A_{1,4} \\ 0 & 2I_{k_2} & 2A_{2,3} & 2A_{2,4} \\ 0 & 0 & 3I_{k_3} & 3A_{3,4} \end{bmatrix}, \tag{1}$$

where the $A_{i,j}$ are matrices with entries 0 or 1 for $i > 1$, and I_k is the identity matrix of order k. Such a code is said to have rank $\{1^{k_1}, 2^{k_2}, 3^{k_3}\}$ or simply rank $\{k_1, k_2, k_3\}$ and $|\mathcal{C}| = 6^{k_1} 3^{k_2} 2^{k_3}$ [1]. If $k_2 = k_3 = 0$ then the rank of $\mathcal{C}$ is $\{k_1, 0, 0\}$ or simply $k_1 = k$.

To each code $\mathcal{C}$ one can associate two *residue codes* viz $\mathcal{C}_2$ and $\mathcal{C}_3$ defined as

$$\mathcal{C}_2 = \{v \mid v \equiv w \pmod{2}, \ w \in \mathcal{C}\},$$

and

$$\mathcal{C}_3 = \{v \mid v \equiv w \pmod{3}, \ w \in \mathcal{C}\}.$$

Code $\mathcal{C}_2$ is permutation-equivalent to a code with generator matrix of the form

$$\begin{pmatrix} I_{k_1} & A_{1,2} & A_{1,3} & A_{1,4} \\ 0 & 0 & 3I_{k_3} & 3A_{3,4} \end{pmatrix}, \tag{2}$$

where $A_{i,j}$ are binary matrices for $i > 1$. Note that $\mathcal{C}_2$ has dimension $k_1 + k_3$. The ternary code $\mathcal{C}_3$ is permutation-equivalent to a code with generator matrix of the form

$$\begin{pmatrix} I_{k_1} & A_{1,2} & A_{1,3} & A_{1,4} \\ 0 & 2I_{k_2} & 2A_{2,3} & 2A_{2,4} \end{pmatrix}, \tag{3}$$

where $A_{i,j}$ are binary matrices for $i > 1$. Note that C_3 has dimension $k_1 + k_2$.

One can also associate two *torsion codes* with C viz $C_2^{\ast}$ and $C_3^{\ast}$ defined as

$$C_2^{\star} = \left\{ \frac{c}{3} \mid c = (c_1, \ldots, c_n) \in C \text{ and } c_i \equiv 0 \pmod{3} \text{ for } 1 \le i \le n \right\}$$

and

$$C_3^{\star} = \left\{ \frac{c}{2} \mid c = (c_1, \ldots, c_n) \in C \text{ and } c_i \equiv 0 \pmod{2} \text{ for } 1 \le i \le n \right\}.$$

If $k_2 = k_3 = 0$ then $C_i = C_i^{\star}$ for $i = 2, 3$.

A linear code C over $\mathbb{Z}_6$ of length n and rank $\{k_1, k_2, k_3\}$ is called an $[n; k_1, k_2, k_3]$ code. If $k_2 = k_3 = 0$, C is called an $[n, k]$ code. In the case of simplex codes we indeed have $k_2 = k_3 = 0$.

Let $C : [n; k_1, k_2, k_3]$ be a code over $\mathbb{Z}_6$. For $r_1 \le k_1, r_2 \le k_2, r_1 + r_2 + r_3 \le k_1 + k_2 + k_3$, the *Generalized Hamming Weight* of C is defined by

$$d_{r_1, r_2, r_3} = \min \left\{ w_S(\mathcal{D}) \mid \mathcal{D} \text{ is an } [n; r_1, r_2, r_3] \text{ subcode of } C \right\},$$

where $w_S(\mathcal{D})$, called *support size* of $\mathcal{D}$, is the number of coordinates in which some codeword of $\mathcal{D}$ has a nonzero entry. The set $\{d_{r_1, r_2, r_3}\}$ is called the *weight hierarchy* of C.

We have the following Lemma connecting the support weight and the Chinese Euclidean weight.

Lemma 1. *Let $\mathcal{D} : [n; r_1, r_2, r_3]$ be a senary linear code then*

$$\sum_{c \in \mathcal{D}} w_{CE}(c) = 2^{r_1 + r_3 + 1} \cdot 3^{r_1 + r_2} \cdot w_S(\mathcal{D}).$$

Proof. Consider the $(r \times n)$ array of all the codewords in $\mathcal{D}$ (where $r = 6^{r_1} 3^{r_2} 2^{r_3}$). It is easy to see that each column consists of either

1. only zeros
2. 0 and 3 equally often
3. 0, 2 and 4 equally often
4. 0, 1, 2, 3, 4, 5 equally often.

Let $n_i, i = 0, 1, 2, 3$ be the number of columns of each type. Then $n_1 + n_2 + n_3 = w_S(\mathcal{D})$. Now applying the standard arguments to evaluate the sum yields the result.

Thus for any linear code C over $\mathbb{Z}_6$, d_{r_1, r_2, r_3} may also be defined by

$$d_{r_1, r_2, r_3} = \frac{1}{2^{r_1 + r_3 + 1} \cdot 3^{r_1 + r_2}} \min \left\{ \sum_{c \in \mathcal{D}} w_{CE}(c) \mid \mathcal{D} \text{ is an } [n; r_1, r_2, r_3] \text{ subcode of } C \right\}.$$

3   Senary Simplex Codes of Type α, β and γ

Let G_k^α be a $k \times 2^k 3^k$ matrix over $\mathbb{Z}_6$ consisting of all possible distinct columns. Inductively, G_k^α may be written as

$$G_k^\alpha = \begin{bmatrix} 00\cdots0 & 11\cdots1 & 22\cdots2 & 33\cdots3 & 44\cdots4 & 55\cdots5 \\ \hline G_{k-1}^\alpha & G_{k-1}^\alpha & G_{k-1}^\alpha & G_{k-1}^\alpha & G_{k-1}^\alpha & G_{k-1}^\alpha \end{bmatrix}_{k\times 6^k}$$

with $G_1^\alpha = [012345]$. The code S_k^α generated by G_k^α has length 6^k and the rank of S_k^α is $\{k, 0, 0\}$.

The following observations are useful to obtain the weight distribution of S_k^α.

Remark 1. If A_{k-1} denotes the $(6^{k-1} \times 6^{k-1})$ array consisting of all codewords in S_{k-1}^α, and if J is the matrix of all $1's$ then the $(6^k \times 6^k)$ array of codewords of S_k^α is given by

$$\begin{bmatrix} A_{k-1} & A_{k-1} & A_{k-1} & A_{k-1} & A_{k-1} & A_{k-1} \\ A_{k-1} & J+A_{k-1} & 2J+A_{k-1} & 3J+A_{k-1} & 4J+A_{k-1} & 5J+A_{k-1} \\ A_{k-1} & 2J+A_{k-1} & 4J+A_{k-1} & A_{k-1} & 2J+A_{k-1} & 4J+A_{k-1} \\ A_{k-1} & 3J+A_{k-1} & A_{k-1} & 3J+A_{k-1} & A_{k-1} & 3J+A_{k-1} \\ A_{k-1} & 4J+A_{k-1} & 2J+A_{k-1} & A_{k-1} & 4J+A_{k-1} & 2J+A_{k-1} \\ A_{k-1} & 5J+A_{k-1} & 4J+A_{k-1} & 3J+A_{k-1} & 2J+A_{k-1} & 1J+A_{k-1} \end{bmatrix}.$$

Remark 2. If $R_1, R_2, ..., R_k$ denote the rows of the matrix G_k^α then $w_H(R_i) = 5 \cdot 6^{k-1}, w_L(R_i) = 9 \cdot 6^{k-1}, w_E(R_i) = 19 \cdot 6^{k-1}$ and $w_{CE}(R_i) = 2 \cdot 6^k$.

It may be observed that each element of $\mathbb{Z}_6$ occurs equally often in every row of G_k^α. Let $\mathbf{c} = (c_1, c_2, \ldots, c_n) \in \mathcal{C}$. For each $j \in \mathbb{Z}_6$ let $\omega_j(\mathbf{c}) = |\{i \mid c_i = j\}|$. We have the following lemma.

Lemma 2. *Let $\mathbf{c} \in S_k^\alpha, \mathbf{c} \neq 0$.*

1. *If for at least one i, c_i is a unit $(1$ or $5)$ then $\forall j \in \mathbb{Z}_6$ $\omega_j = 2^{k-1} \cdot 3^{k-1}$ in $\mathbf{c}$.*
2. *If $\forall i$, $c_i \in \{0, \pm2\}$ then $\forall j \in \{0, \pm2\}$ $\omega_j = 2^k \cdot 3^{k-1}$ in $\mathbf{c}$.*
3. *If $\forall i$, $c_i \in \{0, 3\}$ then $\forall j \in \{0, 3\}$ $\omega_j = 2^{k-1} \cdot 3^k$ in $\mathbf{c}$.*

Proof. By Remark 1, any $x \in S_{k-1}^\alpha$ gives rise to six codewords of S_k^α:

$$\begin{aligned}
y_1 &= \left(x|x|x|x|x|x \right), \\
y_2 &= \left(x|1+x|2+x|3+x|4+x|5+x \right), \\
y_3 &= \left(x|2+x|4+x|x|2+x|4+x \right), \\
y_4 &= \left(x|3+x|x|3+x|x|3+x \right), \\
y_5 &= \left(x|4+x|2+x|x|4+x|2+x \right), \\
&\quad \text{and} \\
y_6 &= \left(x|5+x|4+x|3+x|2+x|1+x \right), \text{ where } \mathbf{i} = (iii...i).
\end{aligned}$$

Now the result can be easily proved by induction on k.

Now we recall some known facts about binary and ternary simplex codes. Let $G(S_k)$ (columns consisting of all non-zero binary k-tuples) be a generator matrix for an $[n, k]$ binary simplex code S_k. Then the extended binary simplex code (also known as a type α binary simplex code), $\hat{S}_k$ is generated by the matrix $G(\hat{S}_k) = [\mathbf{0} \; G(S_k)]$. Inductively,

$$G(\hat{S}_k) = \left[\begin{array}{c|c} 00\ldots0 & 11\ldots1 \\ \hline G(\hat{S_{k-1}}) & G(\hat{S_{k-1}}) \end{array} \right] \quad \text{with } G(\hat{S}_1) = [01]. \tag{4}$$

The ternary simplex code of type α is defined inductively by

$$T_k^\alpha = \left[\begin{array}{c|c|c} 00\cdots0 & 11\cdots1 & 22\cdots2 \\ \hline T_{k-1}^\alpha & T_{k-1}^\alpha & T_{k-1}^\alpha \end{array} \right] \quad \text{with } T_1^\alpha = [012], \tag{5}$$

and the ternary simplex code is defined by the usual generator matrix as

$$T_k^\beta = \left[\begin{array}{c|c} 11\cdots1 & 00\cdots0 \\ \hline T_{k-1}^\alpha & T_{k-1}^\beta \end{array} \right] \quad \text{with } T_2^\beta = \left[\begin{array}{c|c} 111 & 0 \\ 012 & 1 \end{array} \right].$$

Now we determine the torsion codes of the senary simplex code of type α.

Lemma 3. *The binary (ternary) torsion code of S_k^α is equivalent to 3^k copies of the binary type α simplex code $\left(2^k \text{ copies of the ternary type } \alpha \text{ simplex code}\right)$.*

Proof. We will prove the binary case by induction on k. The proof of ternary case is similar and so is omitted. Observe that the binary torsion code of S_k^α is the set of codewords obtained by replacing 3 by 1 in all 2-linear combinations of the rows of the matrix

$$3G_k^\alpha = \left[\begin{array}{c|c|c|c|c|c} 00\cdots0 & 33\cdots3 & 00\cdots0 & 33\cdots3 & 00\cdots0 & 33\cdots3 \\ \hline 3G_{k-1}^\alpha & 3G_{k-1}^\alpha & 3G_{k-1}^\alpha & 3G_{k-1}^\alpha & 3G_{k-1}^\alpha & 3G_{k-1}^\alpha \end{array} \right]. \tag{6}$$

Clearly the result holds for $k = 2$. Assuming that the binary torsion code is equivalent to the 3^{k-1} copies of the extended binary simplex code, we have $\left[3G(\hat{S_{k-1}}) | \cdots | 3G(\hat{S_{k-1}}) \right]$ in place of $3G_{k-1}^\alpha$ in the above matrix. Now regrouping the columns in the above matrix according to (4) yields the desired result.

As a consequence of Lemmas 2 and 3, one gets the weight distribution of S_k^α.

Theorem 1. *The Hamming, Lee, Euclidean and C-Euclidean weight distributions of S_k^α are*

1. $A_H(0) = 1, A_H(3 \cdot 6^{k-1}) = (2^k - 1), A_H(4 \cdot 6^{k-1}) = (3^k - 1),$
   $A_H(5 \cdot 6^{k-1}) = (2^k - 1)(3^k - 1).$
2. $A_L(0) = 1, A_L(8 \cdot 6^{k-1}) = (3^k - 1), A_L(9 \cdot 6^{k-1}) = 3^k(2^k - 1) - 1.$
3. $A_E(0) = 1, A_E(27 \cdot 6^{k-1}) = (2^k - 1), A_E(16 \cdot 6^{k-1}) = (3^k - 1),$
   $A_E(19 \cdot 6^{k-1}) = (2^k - 1)(3^k - 1).$
4. $A_{CE}(0) = 1, A_{CE}(2 \cdot 6^k) = 3^k \cdot 2^k - 1,$
   where $A_H(i)$ $(A_L(i))$ denotes the number of vectors of Hamming (Lee) weight i in S_k^α, and similarly for the Euclidean weights of both types.

Proof. By Lemma 2, each non-zero codeword of S_k^α has Hamming weight either $3 \cdot 6^{k-1}$, $4 \cdot 6^{k-1}$, or $5 \cdot 6^{k-1}$ and Lee weight either $8 \cdot 6^{k-1}$ or $9 \cdot 6^{k-1}$. Since the dimension of the binary torsion code is k, there will be $2^k - 1$ codewords of the Hamming weight $3 \cdot 6^{k-1}$, and the dimension of the ternary torsion code is k, so there will be $3^k - 1$ codewords of the Hamming weight $4 \cdot 6^{k-1}$. Hence the number of codewords having Hamming weight $5 \cdot 6^{k-1}$ will be $6^k - (3^k + 2^k - 1)$. Similar arguments hold for the other weights.

The *symmetrized weight enumerator (swe)* of a senary code $\mathcal{C}$ is defined as

$$swe_\mathcal{C}(a,b,c,d) := \sum_{x \in \mathcal{C}} a^{n_0(x)} b^{n_1(x)} c^{n_2(x)} d^{n_3(x)},$$

where $n_i(x)$ denotes the number of j such that $x_j = \pm i$. Let $\bar{S}_k^\alpha$ be the punctured code of S_k^α obtained by deleting the zero coordinate. Then the swe of $\bar{S}_k^\alpha$ is

$$swe_{\bar{S}_k^\alpha}(a,b,c,d) = 1 + (2^k - 1)d(ad)^{3 \cdot 6^{k-1}-1} + $$
$$(3^k - 1)a^{2 \cdot 6^{k-1}-1} c^{4 \cdot 6^{k-1}} + (2^k - 1)(3^k - 1)d(ad)^{6^{k-1}-1}(bc)^{2 \cdot 6^{k-1}}.$$

Remark 3.   1. $\bar{S}_k^\alpha$ is an equidistant code with respect to Chinese Euclidean distance whereas the binary (quaternary i.e, over $\mathbb{Z}_4$) simplex code is equidistant with respect to Hamming (Lee) distance.
2. The minimum weights of $\bar{S}_k^\alpha$ are: $d_H = 3 \cdot 6^{k-1}, d_L = 8 \cdot 6^{k-1}, d_E = 16 \cdot 6^{k-1}, d_{CE} = 2 \cdot 6^k$.

Example 1. Consider the $6^4 = 1296$ codewords of the senary code generated by the following generator matrix

$$11111111$$
$$22220000$$
$$22002200$$
$$20202020\,.$$
$$33334444$$
$$33443344$$
$$34343434$$

Using the Chinese Gray map results in a mixed code with 8 binary and 8 ternary coordinates, which gives $N(8,8,4) \geq 1296$, while the ternary code of length 8, dimension 4 and distance 4 is optimal [3].

Let Λ_k be the $k \times 3^k \cdot (2^k - 1)$ matrix defined inductively by $\Lambda_1 = [135]$ and

$$\Lambda_k = \left[\begin{array}{c|c|c|c|c|c} 0 \dots 0 & 1 \cdots 1 & 2 \cdots 2 & 3 \cdots 3 & 4 \cdots 4 & 5 \cdots 5 \\ \hline \Lambda_{k-1} & G_{k-1}^\alpha & \Lambda_{k-1} & G_{k-1}^\alpha & \Lambda_{k-1} & G_{k-1}^\alpha \end{array} \right],$$

for $k \geq 2$; and let μ_k be the $k \times 2^{k-1} \cdot (3^k - 1)$ matrix defined inductively by $\mu_1 = [12]$ and

$$\mu_k = \left[\begin{array}{c|c|c|c} 0 \cdots 0 & 1 \cdots 1 & 2 \cdots 2 & 3 \cdots 3 \\ \hline \mu_{k-1} & G_{k-1}^\alpha & G_{k-1}^\alpha & \mu_{k-1} \end{array} \right],$$

for $k \geq 2$, where G^α_{k-1} is the generator matrix of S^α_{k-1}.

Now let G^β_k be the $k \times \frac{(2^k-1)(3^k-1)}{2}$ matrix defined inductively by

$$G^\beta_2 = \left[\begin{array}{c|c|c|c}111111 & 0 & 222 & 33 \\ \hline 012345 & 1 & 135 & 12\end{array}\right],$$

and for $k > 2$

$$G^\beta_k = \left[\begin{array}{c|c|c|c}11\cdots1 & 00\cdots0 & 22\cdots2 & 33\cdots3 \\ \hline G^\alpha_{k-1} & G^\beta_{k-1} & \Lambda_{k-1} & \mu_{k-1}\end{array}\right],$$

where G^α_{k-1} is the generator matrix of S^α_{k-1}. Note that G^β_k is obtained from G^α_k by deleting $\frac{(2^k+1)(3^k-1)+2^{k+1}}{2}$ columns. By induction it is easy to verify that no two columns of G^β_k are multiples of each other. Let S^β_k be the code generated by G^β_k. Note that S^β_k is a $\left[\frac{(2^k-1)(3^k-1)}{2}, k\right]$ code. To determine the weight distributions of S^β_k we first make some observations.

The proof of the following proposition is similar to that of Proposition 2.

Proposition 1. *Each row of G^β_k contains 6^{k-1} units and*
$$w_2 + w_4 = 3^{k-1}(2^{k-1} - 1), w_3 = 2^{k-2}(3^{k-1} - 1), w_0 = \frac{(2^{k-1}-1)(3^{k-1}-1)}{2}.$$

Remark 4. Each row of G^β_k has Hamming weight $\left(3^{k-1} \cdot (2^k - 1) + 2^{k-2} \cdot (3^{k-1} - 1)\right)$, Lee weight $\left(2 \cdot 3^{k-1}(3 \cdot 2^{k-2} - 1) + 3 \cdot 2^{k-2}(3^{k-1} - 1)\right)$, Euclidean weight $\left(3^{k-1}(19 \cdot 2^{k-2} - 4) - 9 \cdot 2^{k-2}\right)$, and Chinese Euclidean weight $6^k - 2^k - 3^k$.

The proof of the following lemma is similar to the proof of Lemma 2.

Lemma 4. *Let $c \in S^\beta_k, c \neq 0$.*

1. *If for at least one i, c_i is a unit then $\forall j \in \mathbb{Z}_6$ $w_1 + w_5 = 6^{k-1}$,*
   $w_2 + w_4 = 3^{k-1}(2^{k-1} - 1), w_3 = 2^{k-2}(3^{k-1} - 1), w_0 = \frac{(2^{k-1}-1)(3^{k-1}-1)}{2}$ *in c.*
2. *If $\forall i, c_i \in \{0, \pm2\}$ then $\forall j \in \{0, \pm2\}$ $w_2 + w_4 = 3^{k-1}(2^k - 1), w_0 = \frac{(2^k-1)(3^{k-1}-1)}{2}$ in c.*
3. *If $\forall i, c_i \in \{0, 3\}$ then $\forall j \in \{0, 3\}$ $w_3 = 2^{k-2}(3^k - 1), w_0 = \frac{(2^{k-1}-1)(3^k-1)}{2}$ in c.*

The proof of the following lemma is similar to that of Lemma 3 and is omitted.

Lemma 5. *The binary (ternary) torsion code of S^β_k is equivalent to $\frac{(3^k-1)}{2}$ copies of the binary simplex code $\left((2^k - 1)$ copies of the ternary simplex code$\right)$.*

The proof of the following theorem is similar to that of Theorem 1 and is omitted.

Theorem 2. *The Hamming, Lee weight, Euclidean and C-Euclidean weight distributions of S^β_k are:*

1. $A_H(0) = 1, A_H\left(2^{k-2} \cdot (3^k - 1)\right) = (2^k - 1), A_H\left(3^{k-1} \cdot (2^k - 1)\right) = (3^k - 1)$,
   $A_H\left(3^{k-1} \cdot (2^k - 1) + 2^{k-2} \cdot (3^{k-1} - 1)\right) = (2^k - 1)(3^k - 1)$.

2. $A_L(0) = 1, A_L\left(3 \cdot 2^{k-2}(3^k - 1)\right) = (2^k - 1), A_L\left(2 \cdot 3^{k-1}(2^k - 1)\right) = (3^k - 1),$
   $A_L\left(2 \cdot 3^{k-1}(3 \cdot 2^{k-2} - 1) + 3 \cdot 2^{k-2}(3^{k-1} - 1)\right) = (2^k - 1)(3^k - 1).$

3. $A_E(0) = 1, A_E\left(9 \cdot 2^{k-2}(3^k - 1)\right) = (2^k - 1), A_E\left(4 \cdot 3^{k-1}(2^k - 1)\right) = (3^k - 1), A_E\left(3^{k-1}(19 \cdot 2^{k-2} - 4) - 9 \cdot 2^{k-2}\right) = (2^k - 1)(3^k - 1).$

4. $A_{CE}(0) = 1, A_{CE}(6^k - 2^k) = (2^k - 1), A_{CE}(6^k - 3^k) = (3^k - 1),$
   $A_{CE}(6^k - 2^k - 3^k) = (2^k - 1)(3^k - 1),$

   where $A_H(i)$ $(A_L(i))$ denotes the number of vectors of Hamming (Lee) weight i in S_k^α, and similarly for the Euclidean weights of both types.

Remark 5. 1. The swe of S_k^β is given as

$$swe(a,b,c,d) = 1 + 3^{-k}p(k)a^{n(k-1)+p(k-1)}d^{2^{-1}q(k)} + $$
$$2^{-k+1}q(k)a^{n(k-1)}\left\{a^{q(k-1)}c^{3^{-1}p(k)} + 3^{-k}p(k)b^{6^{k-1}}c^{p(k-1)}d^{q(k-1)}\right\}.$$

where $n(k) = \frac{(2^k - 1)(3^k - 1)}{2}$, $p(k) = 3^k(2^k - 1)$ and $q(k) = 2^{k-1}(3^k - 1)$.

2. The minimum weights of S_k^β are: $d_H = 2^{k-2}(3^k - 1), d_L = 2 \cdot 3^{k-1}(2^k - 1), d_E = 4 \cdot 3^{k-1}(2^k - 1), d_{CE} = 6^k - 2^k - 3^k.$

Let G_k^γ be the $k \times 2^{k-1}(3^k - 2^k)$ matrix defined inductively by

$$G_2^\gamma = \begin{bmatrix} 111111 & 0 & 2 & 3 & 4 \\ 012345 & 1 & 1 & 1 & 1 \end{bmatrix},$$

and for $k > 2$

$$G_k^\gamma = \begin{bmatrix} 11 \cdots 1 & 00 \cdots 0 & 22 \cdots 2 & 33 \cdots 3 & 44 \cdots 4 \\ G_{k-1}^\alpha & G_{k-1}^\gamma & G_{k-1}^\gamma & G_{k-1}^\gamma & G_{k-1}^\gamma \end{bmatrix},$$

where G_{k-1}^α is the generator matrix of S_{k-1}^α. Note that G_k^γ is obtained from G_k^α by deleting $2^{k-1}(2^k + 3^k)$ columns. By induction it is easy to verify that no two columns of G_k^γ are multiples of each other. Let S_k^γ be the code generated by G_k^γ. Note that S_k^γ is a $\left[2^{k-1}(3^k - 2^k), k\right]$ code.

Proposition 2. *Each row of G_k^γ contains $6^{(k-1)}$ units and*
$\omega_0 = \omega_2 = \omega_3 = \omega_4 = 2^{k-2}(3^{k-1} - 2^{k-1}).$

Proof. Clearly the assertion holds for the first row. Assume that the result holds for each row of G_{k-1}^γ. Then the number of units in each row of G_{k-1}^γ is $6^{(k-2)}$. By Lemma 2, the number of units in any row of G_{k-1}^α is $2^{k-1} \cdot 3^{k-2}$. Hence the total number of units in any row of G_k^γ will be $2^{k-1} \cdot 3^{k-2} + 4 \cdot 2^{k-2} \cdot 3^{k-2} = 2^{k-1} \cdot 3^{k-1}$. A similar argument holds for the number of 0's, 2's, 3's and 4's.

Remark 6. Each row of G_k^γ has Hamming weight $3 \cdot 2^{k-2}\left[5 \cdot 3^{k-2} - 2^{k-1}\right]$, Lee weight $2^{k-2}\left[3^{k+1} - 7 \cdot 2^{k-1}\right]$, Euclidean weight $2^{k-2}\left[19 \cdot 3^{k-1} - 17 \cdot 2^{k-1}\right]$, and Chinese Euclidean weight $6^k - 5 \cdot 4^{k-1}$.

The various weight distributions of S_k^γ can be obtained using arguments similar to other simplex codes. To save the space we omit them.

The weight hierarchy of S_k^α is given by the following theorem.

Theorem 3. *The weight hierarchy of S_k^α is given by*

$$d_{r_1,r_2,r_3}(S_k^\alpha) = 6^k - 3^{k-r_1-r_2} \cdot 2^{k-r_1-r_3}.$$

Proof. By Remark 3 and the definition of d_{r_1,r_2,r_3} after the Lemma 1.

4   Chinese Product Type Construction

The Chinese remainder theorem (CRT) plays an important role in the study of codes over $\mathbb{Z}_{2^k}$ [4,6]. In particular, given binary and ternary linear codes of length n and dimension k, one can construct a senary code (over $\mathbb{Z}_6$) of length n using CRT. The following theorem is from [4,6].

Theorem 4. *[4,6] If B and T are linear codes of length n over $GF(2)$ and $GF(3)$, respectively, then the set $CRT(B,T) = \{\phi^{-1}(\mathbf{c}_b, \mathbf{c}_t) \mid \mathbf{c}_b \in B, \mathbf{c}_t \in T\}$ is a linear code of length n over $\mathbb{Z}_6$. Moreover if B and T are self-orthogonal then $CRT(B,T)$ is also self-orthogonal.*

If generator matrices of B, T and $CRT(B,T)$ are $G(B), G(T)$ and $G(CRT(B,T))$, respectively, then we have $\phi(G(CRT(B,T))) = [G(B)|G(T)]$, where ϕ is the Chinese Gray map. If the codes B and T are of different lengths, say, n_1 and n_2 then it seems that no non-trivial method is known to construct a code over $\mathbb{Z}_6$ from these codes. In the trivial case of course one can add extra zero columns to the generator matrix of the code of shorter length and then use Theorem 4. Here we present a new construction of a generator matrix of senary code from codes of different lengths.

Let $G(B) = [x_1 x_2 \ldots x_{n_1}]$ and $G(T) = [y_1 y_2 \ldots y_{n_2}]$ where x_i, y_i are the corresponding columns. Now form the matrix $G(B) \star G(T)$ consisting of the $n_1 n_2$ pairs of total $2n_1 n_2$ columns $\{x_i y_1 x_i y_2 \ldots x_i y_{n_2}\}_{i=1}^{n_1}$. These pairs of columns give a generator matrix of length $n_1 n_2$ (the product of the lengths of the binary and ternary codes) over $\mathbb{Z}_6$ using the inverse Chinese Gray map. In particular, if $n_1 = n_2 = n$ then we get a code of length n^2. Note that if we use the Theorem 4 to construct a generator matrix for the case of $n_1 = n_2 = n$, we obtain a code of length n with generator matrix $[x_1 y_1 x_2 y_2 \ldots x_n y_n]$. In this case, the resulting code will be self orthogonal if the corresponding binary and ternary codes are self orthogonal [6]. Similarly it is easy to see that

Lemma 6. *The senary codes constructed by $G(B) * G(T)$ will be self orthogonal if the corresponding codes B and T are self orthogonal.*

The next two results show that self-orthogonal simplex codes of type α and β can be obtained from the construction $G(B) * G(T)$.

Theorem 5. *The codes S_k^α and S_k^β can be obtained via the construction $G(B) * G(T)$.*

Proof. We will only prove the result for S_k^α, since the other case is similar. If we apply the Chinese Gray map to the generator matrix G_k^α, we see that it is equivalent to the matrix $G(\hat{S}_k) \star T_k^\alpha$, where T_k^α is defined in (5).

Theorem 6. *The codes S_k^α ($k \geq 3$) and S_k^β ($k \geq 2$) are self orthogonal.*

Proof. The result follows from Lemma 6 and Theorem 5. It can also be proved by induction on k since the rows of the generator matrices are pairwise orthogonal and each of the rows has Euclidean weight a multiple of 12 [1].

Remark 7. The code S_k^γ is not self-orthogonal as the Euclidean weights of the rows of G_k^γ are not a multiple of 12.

Acknowledgement

The authors would like to thank Patrick Solé for providing copies of [6] and [12], and also Patric R.J. Östergård for providing a copy of [13].

References

1. Bannai E., Dougherty S.T., Harada M. and Oura M., *Type II codes, even unimodular lattices and invariant rings.* IEEE Trans. Inform. Theory **45** (1999), 1194–1205.
2. Bhandari M. C., Gupta M. K. and Lal, A. K. *On $\mathbb{Z}_4$ simplex codes and their gray images* Applied Algebra, Algebraic Algorithms and Error-Correcting Codes, AAECC-13, Lecture Notes in Computer Science **1719** (1999), 170–180.
3. Brouwer, A.E., Hamalainen, H.O., Östergård, P.R.J. and Sloane, N.J.A. *Bounds on mixed binary/ternary codes.* IEEE Trans. Inform. Theory **44** Jan. 1998, 140 –161.
4. Chen, C. J., Chen T. Y. and Loeliger, H. A. *Construction of linear ring codes for 6 PSK.* IEEE Trans. Inform. Theory **40** (1994), 563–566.
5. Dougherty S.T., Gulliver T.A. and Harada M., *Type II codes over finite rings and even unimodular lattices.* J. Alg. Combin., **9** (1999), 233–250.
6. Dougherty S.T., Harada M. and Solé P., *Self-dual codes over rings and the Chinese Remainder Theorem.* Hokkaido Math. J., **28** (1999), 253–283.
7. Gulliver T.A. and Harada M., *Double circulant self dual codes over $\mathbb{Z}_{2k}$.* **44** (1998), 3105–3123.
8. Gulliver T.A. and Harada M., *Orthogonal frames in the Leech Lattice and a Type II code over $\mathbb{Z}_{22}$.* J. Combin. Theory Ser. A (to appear).
9. A. R. Hammons, P. V. Kumar, A. R. Calderbank, N. J. A. Sloane, and P. Solé. *The $\mathbb{Z}_4$-linearity of kerdock, preparata, goethals, and related codes.* IEEE Trans. Inform. Theory, **40** (1994), 301–319.
10. Harada M., *On the existence of extremal Type II codes over $\mathbb{Z}_6$.* Des. Math. **223** (2000), 373–378.
11. Harada M., *Extremal Type II codes over $\mathbb{Z}_6$ and their lattices.* (submitted).
12. Ling, S. and Solé P. *Duadic Codes over $\mathbb{Z}_{2k}$.* IEEE Trans. Inform. Theory, **47** (2001), 1581–1589.
13. Östergård, P. R. J., *Classification of binary/ternary one-error-correcting codes* Disc. Math. **223** (2000) 253–262.
14. F.W.Sun and H.Leib, *Multiple-Phase Codes for Detection Without Carrier Phase Reference* IEEE Trans. Inform. Theory, **vol 44** No. 4, 1477-1491 (1998).

Optimal Double Circulant $\mathbb{Z}_4$-Codes

T. Aaron Gulliver[1] and Masaaki Harada[2]

[1] Department of Electrical and Computer Engineering
University of Victoria
P.O. Box 3055, STN CSC, Victoria, B.C.
Canada V8W 3P6
agullive@ece.uvic.ca
[2] Department of Mathematical Sciences, Yamagata University
Yamagata 990–8560, Japan
mharada@sci.kj.yamagata-u.ac.jp

Abstract. Recently, an optimal formally self-dual $\mathbb{Z}_4$-code of length 14 and minimum Lee weight 6 has been found using the double circulant construction by Duursma, Greferath and Schmidt. In this paper, we classify all optimal double circulant $\mathbb{Z}_4$-codes up to length 32. In addition, double circulant codes with the largest minimum Lee weights for this class of codes are presented for lengths up to 32.

1 Introduction

Some of the best known non-linear binary codes which are better than any comparable linear codes are the Nordstorm-Robinson, Kerdock and Preparata codes. The Nordstorm-Robinson and Preparata codes are twice as large as the best linear codes for the same parameters. The Nordstorm-Robinson, Kerdock and Preparata codes are the Gray map images $\phi(C)$ of some extended cyclic linear codes C over $\mathbb{Z}_4$ [6]. In particular, the Nordstorm-Robinson code is the Gray map image $\phi(O_8)$ of the octacode O_8, which is the unique self-dual $\mathbb{Z}_4$-code ($C = C^\perp$) of length 8 and minimum Lee weight 6.

The $\mathbb{Z}_4$-dual of $\phi(C)$ is defined as $\phi(C^\perp)$ [6]. $\phi(C)$ is called formally self-dual if $\phi(C)$ and $\phi(C^\perp)$ have identical weight enumerators. The Gray map image of a self-dual code over $\mathbb{Z}_4$ is formally self-dual [2]. A $\mathbb{Z}_4$-code with the same symmetrized weight enumerator as its dual code is called formally self-dual. The Gray map image of a formally self-dual $\mathbb{Z}_4$-code is also formally self-dual. Moreover, there exist formally self-dual $\mathbb{Z}_4$-codes which have a better minimum Lee weight than any self-dual code of that length. Recently, an optimal formally self-dual $\mathbb{Z}_4$-code of length 14 and minimum Lee weight 8 has been found using the double circulant construction [4]. It is natural to consider formally self-dual $\mathbb{Z}_4$-codes in order to construct better binary non-linear formally self-dual codes. This motivates our investigation of optimal double circulant $\mathbb{Z}_4$-codes with respect to Lee weights. Double circulant codes are a class of isodual codes and formally self-dual codes [1]. Optimal double circulant codes over $\mathbb{Z}_4$ with respect to Euclidean weights were found in [1] to construct dense isodual lattices.

S. Boztaş and I.E. Shparlinski (Eds.): AAECC-14, LNCS 2227, pp. 122–128, 2001.
© Springer-Verlag Berlin Heidelberg 2001

This paper is organized as follows. In Sect. 2, we give basic definitions and known results for codes over $\mathbb{Z}_4$. Sect. 3 presents a unique double circulant code of length 14. In Sect. 4, we give a classification of optimal double circulant codes of lengths 6 and 10. In Sect. 5, double circulant codes with the largest minimum Lee weight for this class of codes are constructed for lengths up to 32. It is also demonstrated that there are no optimal double circulant codes for lengths $n = 8, 12$ and $16 \leq n \leq 32$, which completes the classification of all optimal double circulant codes up to length 32. Table 1 gives the largest minimum Lee weights $d(\text{DCC})$ of double circulant codes up to length 32. If the minimum Lee weight $d(\text{DCC})$ is optimal (see Sect. 2 for the definition), it is marked by $*$. For comparison purposes, the largest minimum Lee weights $d(\text{SDC})$, of self-dual codes are also given. The values of $d(\text{SDC})$ are from [8] and [9].

Table 1. Largest minimum Lee weights

Length	$d(\text{DCC})$	$d(\text{SDC})$	Length	$d(\text{DCC})$	$d(\text{SDC})$
2	2^*	2	18	8	8
4	4^*	4	20	9	8
6	4^*	4	22	10	8
8	4	6	24	10	12
10	6^*	4	26	10	≤ 10
12	6	4	28	12	≤ 12
14	8^*	6	30	12	≤ 12
16	8	8	32	12	≤ 16

2 Codes over $\mathbb{Z}_4$

A code C of length n over $\mathbb{Z}_4$ (or a $\mathbb{Z}_4$-code of length n) is a $\mathbb{Z}_4$-submodule of $\mathbb{Z}_4^n$ where $\mathbb{Z}_4$ is the ring of integers modulo 4. A generator matrix of C is a matrix whose rows generate C. An element of C is called a codeword of C. The symmetrized weight enumerator of C is given by

$$swe_C(a, b, c) = \sum_{x \in C} a^{n_0(x)} b^{n_1(x)+n_3(x)} c^{n_2(x)},$$

where $n_i(x)$ is the number of components of $x \in C$ that are equal to i. The *Lee weight* $wt_L(x)$ of a vector x is defined as $n_1(x) + 2n_2(x) + n_3(x)$. The minimum Lee weight d_L of C is the smallest Lee weight among all non-zero codewords of C. We say that two codes are *equivalent* if one can be obtained from the other by permuting the coordinates and (if necessary) changing the signs of certain coordinates. For $x = (x_1, \ldots, x_n)$ and $y = (y_1, \ldots, y_n)$, we define the inner product of x and y in $\mathbb{Z}_4^n$ by $x \cdot y = x_1 y_1 + \cdots + x_n y_n$. The dual code $C^\perp$ of C is defined as $C^\perp = \{x \in \mathbb{Z}_4^n \mid x \cdot y = 0 \text{ for all } y \in C\}$. C is *self-dual* if $C = C^\perp$, and C is *isodual* if C and $C^\perp$ are equivalent [1]. We say that C is *formally self-dual* if $swe_C(a, b, c) = swe_{C^\perp}(a, b, c)$. Of course, a self-dual code is also isodual and an isodual code is also formally self-dual. The symmetrized weight enumerators of isodual codes over $\mathbb{Z}_4$ were investigated in [1].

The *Gray map* ϕ is defined as a map from $\mathbb{Z}_4^n$ to $\mathbb{F}_2^{2n}$ mapping $(x_1, \ldots, x_n)$ to $(\varphi(x_1), \ldots, \varphi(x_n))$, where $\varphi(0) = (0, 0)$, $\varphi(1) = (0, 1)$, $\varphi(2) = (1, 1)$ and $\varphi(3) = (1, 0)$. This is an isometry from $(\mathbb{Z}_4^n$, Lee distance) to $(\mathbb{F}_2^{2n}$, Hamming distance). The Gray map image $\phi(C)$ of a $\mathbb{Z}_4$-code C need not be $\mathbb{F}_2$-linear and the dual code may not even be defined. The $\mathbb{Z}_4$-dual of $\phi(C)$ is defined as $\phi(C^\perp)$ [6]. Note that the weight distributions of $\phi(C)$ and $\phi(C^\perp)$ are MacWilliams transforms of one another. In addition, we say that $\phi(C)$ is formally self-dual if the weight distributions of $\phi(C)$ and $\phi(C^\perp)$ are the same [2]. Of course, the Gray map image of a self-dual code over $\mathbb{Z}_4$ is formally self-dual [2]. The Gray map image of a formally self-dual code keeps this property.

A (pure) *double circulant* code (DCC) has a generator matrix of the form

$$(I_n \; R),\tag{1}$$

where I_n is the identity matrix of order n and R is an n by n circulant matrix. The matrix $(1, 1)$ generates the unique (trivial) DCC of length 2. For length 4, it is easy to see that a DCC is equivalent to one of the three codes $D_{4,1}$, $D_{4,2}$ and $D_{4,3}$ with R in the generator matrices (1) given by the following matrices

$$\begin{pmatrix} 1 & 0 \\ 0 & 1 \end{pmatrix}, \begin{pmatrix} 1 & 1 \\ 1 & 1 \end{pmatrix} \text{ and } \begin{pmatrix} 2 & 1 \\ 1 & 2 \end{pmatrix},$$

respectively. Since the symmetrized weight enumerators of the three codes are distinct, the codes are inequivalent. $D_{4,3}$ is equivalent to the self-dual code with $d_L = 4$. $D_{4,3}$ is the unique double circulant code with $d_L = 4$.

We say that a $\mathbb{Z}_4$-code C of length n and minimum Lee weight d_L is *optimal* if there is no binary $(2n, |C|, d_L + 1)$ code (including non-linear) by the sphere-packing bound. For example, since there are no binary codes with parameters $(4, 2^2, 3)$ and $(8, 2^4, 5)$ by the sphere-packing bound, the above double circulant codes of length 2 (resp. 4) and $d_L = 2$ (resp. 4) are optimal. We will demonstrate that there are optimal double circulant codes for lengths 2, 4, 6, 10 and 14. A double circulant code is called *DCC-optimal* if it has the largest minimum Lee weight among all double circulant codes of that length.

3   A Unique Optimal Double Circulant Code of Length 14

Recently Duursma, Greferath and Schmidt [4] have found an optimal formally self-dual $\mathbb{Z}_4$-code C_{14} of length 14 (that is, it has minimum Lee weight 8). This code was discovered by an exhaustive search under some condition on the binary reduction. This motivates our interest in the classification of all double circulant codes with $d_L = 8$.

By exhaustive search, we have found all distinct double circulant codes with $d_L = 8$. The codes have generator matrices (1) with the following first rows of R:

$$(3133211), (1311233), (2331311) \text{ and } (2113133).$$

Let C be a double circulant code with generator matrix (1). It is obvious that a cyclic shift of the first row of R defines an equivalent double circulant code. In addition, the codes with the following generator matrices

$$\left(I_n \ -R \right), \left(I_n \ R^T \right) \text{ and } \left(I_n \ -R^T \right),$$

are double circulant codes which are equivalent to C, where R^T denotes the transpose of R. This can reduce the number of codes which must be checked further for equivalence to complete the classification. For the above codes, this is sufficient to determine that the four codes are equivalent.

Theorem 1. *There is a unique optimal double circulant $\mathbb{Z}_4$-code of length 14 and minimum Lee weight 8, up to equivalence.*

The unique code has the following symmetrized weight enumerator:

$$a^{14} + 21a^{10}c^4 + 168a^8b^4c^2 + 336a^7b^4c^3 + 64a^7c^7 + 112a^6b^8 + 896a^6b^7c$$
$$+504a^6b^4c^4 + 35a^6c^8 + 336a^5b^8c + 672a^5b^4c^5 + 1008a^4b^8c^2 + 4480a^4b^7c^3$$
$$+504a^4b^4c^6 + 1568a^3b^8c^3 + 336a^3b^4c^7 + 224a^2b^{12} + 1008a^2b^8c^4 + 2688a^2b^7c^5$$
$$+168a^2b^4c^8 + 7a^2c^{12} + 448ab^{12}c + 336ab^8c^5 + 224b^{12}c^2 + 112b^8c^6 + 128b^7c^7.$$

4 Optimal Double Circulant Codes of Lengths 6, 8 and 10

– **Length 6:** The largest minimum Lee weight among all double circulant codes of length 6 is 4. Any formally self-dual code of this length with $d_L = 4$ is optimal by considering the sphere-packing bound on the Gray map image. A double circulant code with $d_L = 4$ is equivalent to one of the codes $D_{6,1}$, $D_{6,2}$, $D_{6,3}$ and $D_{6,4}$ with first rows of R given by

$$(210), (311), (221) \text{ and } (321),$$

respectively. $D_{6,3}$ is equivalent to $\mathcal{D}_6^\oplus$ in [3] which is the unique self-dual code with $d_L = 4$. The symmetrized weight enumerators of $D_{6,1}$, $D_{6,2}$ and $D_{6,4}$ are as follows:

$$swe_{D_{6,1}} = a^6 + 3a^4c^2 + 12a^3b^2c + 3a^2c^4 + 24ab^4c + 12ab^2c^3 + 8b^6 + c^6$$
$$swe_{D_{6,2}} = a^6 + 3a^4c^2 + 6a^3b^2c + 6a^2b^4 + 12a^2b^2c^2 + 3a^2c^4 + 12ab^4c$$
$$\qquad +6ab^2c^3 + 8b^6 + 6b^4c^2 + c^6$$
$$swe_{D_{6,4}} = a^6 + 4a^3c^3 + 6a^2b^4 + 24a^2b^3c + 3a^2c^4 + 12ab^4c + 6b^4c^2 + 8b^3c^3.$$

We now consider the binary Gray map images $\phi(D_{6,1})$, $\phi(D_{6,2})$, $\phi(D_{6,3})$ and $\phi(D_{6,4})$ of the above four optimal double circulant codes. We have verified that $\phi(D_{6,1})$ and $\phi(D_{6,3})$ are linear while $\phi(D_{6,2})$ and $\phi(D_{6,4})$ are non-linear. Since the Hamming weight distributions of $\phi(D_{6,2})$ and $\phi(D_{6,4})$ are distinct, the two codes are inequivalent. Our computer search shows that $\phi(D_{6,3})$ is equivalent to the unique self-dual $[12, 6, 4]$ code; and $\phi(D_{6,1})$ is a formally self-dual $[12, 6, 4]$ code which is not self-dual and is equivalent to $C_{12,2}$ in [5]. Therefore the Gray map images of the four codes are inequivalent.

Proposition 1. *There are exactly four inequivalent optimal double circulant codes of length 6 and minimum Lee weight 4. The Gray map images of the four codes are also inequivalent.*

- **Length 8:** We have verified that there is no DCC of length 8 and minimum Lee weight 6. In fact, the largest minimum Lee weight of a DCC of length 8 is 4. The octacode $\mathcal{O}_8$ is the unique self-dual code of this length and minimum Lee weight 6. It is well known that $\mathcal{O}_8$ is optimal and its Gray map image is the Nordstorm-Robinson code.
- **Length 10:** By exhaustive search, we have found all distinct double circulant codes of length 10 and minimum Lee weight 6. All distinct codes $D_{10,1}, \ldots, D_{10,6}$ and $D_{10,7}$ which must be checked further for equivalence to complete the classification have the following first rows of R:

$$(21110), (12110), (32110), (23110), (31210), (32211) \text{ and } (32121),$$

respectively. For the cases $(i, j) = (1, 2), (4, 5)$ and $(6, 7)$, we have verified that $D_{10,i}$ and $D_{10,j}$ are permutation-equivalent. We have also verified that $D_{10,i}$ and $D_{10,j}$ are not permutation-equivalent for $(i, j) = (1, 3), (1, 4)$ and $(4, 6)$. However, our computer search shows that the codes $D_{10,1}$ and $D_{10,3}$ are equivalent, and the codes $D_{10,4}$ and $D_{10,6}$ are equivalent.

Table 2. Codes $D_{10,1}$ and $D_{10,4}$

Code	$M_6(5)$	$m_6(5)$	$M_6(4)$	$m_6(4)$	$M_6(3)$	$m_6(3)$	$M_6(2)$	$m_6(2)$	$M_6(1)$	$m_6(1)$
$D_{10,1}$	9	0	17	7	32	20	59	53	99	99
$D_{10,4}$	10	0	17	7	32	20	59	53	99	99

Code	$M_8(5)$	$m_8(5)$	$M_8(4)$	$m_8(4)$	$M_8(3)$	$m_8(3)$	$M_8(2)$	$m_8(2)$	$M_8(1)$	$m_8(1)$
$D_{10,1}$	70	60	104	96	142	138	188	186	240	240
$D_{10,4}$	70	62	104	96	142	138	188	186	240	240

We now use the following method given in [7] to check the inequivalence of codes $D_{10,1}$ and $D_{10,4}$. Let C be a code of length $2n$. Let $M_t = (m_{ij})$ be the $A_t \times 2n$ matrix with rows composed of the codewords of Hamming weight t in C, where A_i denotes the number of codewords of Hamming weight i in C. For an integer k ($1 \le k \le 2n$), let $n_t(j_1, \ldots, j_k)$ be the number of r ($1 \le r \le A_t$) such that $m_{rj_1} \cdots m_{rj_k} \ne 0$ over $\mathbb{Z}$ for $1 \le j_1 < \cdots < j_k \le 2n$. We consider the set

$$S_t(k) = \{n_t(j_1, \ldots, j_k) | \text{ for any } k \text{ distinct columns } j_1, \ldots, j_k\}.$$

Let $M_t(k)$ and $m_t(k)$ be the maximum and minimum numbers in $S_t(k)$, respectively. The values of $M_t(k)$ and $m_t(k)$ ($t = 6, 8$, $k = 1, \ldots, 5$) for the two codes are listed in Table 2, and establish that $D_{10,1}$ and $D_{10,4}$ are inequivalent.

Codes $D_{10,1}$ and $D_{10,4}$ have the following symmetrized weight enumerator:

$$a^{10} + 15a^6c^4 + 60a^5b^4c + 30a^4b^6 + 120a^4b^4c^2 + 15a^4c^6$$
$$+120a^3b^6c + 120a^3b^4c^3 + 180a^2b^6c^2 + 120a^2b^4c^4 + 120ab^6c^3$$
$$+60ab^4c^5 + 32b^{10} + 30b^6c^4 + c^{10}.$$

The Gray map images $\phi(D_{10,1})$ and $\phi(D_{10,4})$ are both non-linear. We have determined via computer that the codewords of weight 6 in both codes form 1-$(20, 6, 27)$ designs. Moreover, Magma was used to show that the two 1-designs are non-isomorphic. Hence the two codes $\phi(D_{10,1})$ and $\phi(D_{10,4})$ are inequivalent. Therefore we have the following:

Proposition 2. *There are exactly two inequivalent optimal double circulant codes of length 10 and minimum Lee weight 6. The binary Gray map images are also inequivalent.*

5   DCC-Optimal Double Circulant Codes

By exhaustive search, we have found all distinct double circulant codes up to length 32. Table 3 lists some DCC-optimal codes of length 12. The coefficients of a^8c^4 and a^6b^6 are also given in the table. These coefficients show that the codes have distinct symmetrized weight enumerators, and so are inequivalent.

Table 3. DCC-optimal double circulant codes of length 12

Code	First row	Coeff. of a^8c^4	Coeff. of a^6b^6
$D_{12,1}$	211100	15	12
$D_{12,2}$	121100	15	24
$D_{12,3}$	321100	15	40
$D_{12,4}$	221010	6	0
$D_{12,5}$	322010	6	28
$D_{12,6}$	211110	6	30
$D_{12,7}$	311110	15	48
$D_{12,8}$	121110	6	32
$D_{12,9}$	131110	15	64
$D_{12,10}$	222110	6	2
$D_{12,11}$	203210	6	40
$D_{12,12}$	321111	15	16
$D_{12,13}$	312111	15	0

Table 4. DCC-optimal double circulant codes

Code	First row	d_L
D_{16}	32121100	8
D_{18}	121210000	8
D_{20}	2321011000	9
D_{22}	12312110000	10
D_{24}	123312100000	10
D_{26}	3100302220100	10
D_{28}	21331312010000	12
D_{30}	212313201000000	12
D_{32}	2311221010000000	12

Table 4 gives DCC-optimal double circulant codes $D_{16}, \ldots, D_{30}$ for lengths $16, \ldots, 32$. These codes complete Table 1. We have determined that there are no optimal double circulant $\mathbb{Z}_4$-codes of lengths $n = 12$ and $16 \leq n \leq 32$ (see also Table 1). Hence the classification of optimal double circulant codes up to length 32 is complete.

References

1. C. Bachoc, T.A. Gulliver and M. Harada, Isodual codes over $\mathbb{Z}_{2k}$ and isodual lattices, *J. Alg. Combin.* **12** (2000) pp. 223–240.
2. A. Bonnecaze, A.R. Calderbank and P. Solé, Quaternary quadratic residue codes and unimodular lattices, *IEEE Trans. Inform. Theory* **41** (1995) pp. 366–377.
3. J.H. Conway and N.J.A. Sloane, Self-dual codes over the integers modulo 4, *J. Combin. Theory Ser. A* **62** (1993) pp. 30–45.
4. I.M. Duursma, M. Greferath and S.E. Schmidt, On the optimal $\mathbb{Z}_4$ codes of Type II and length 16, *J. Combin. Theory Ser. A* **92** (2000) pp. 77–82.
5. T.A. Gulliver and M. Harada, Classification of extremal double circulant formally self-dual even codes, *Designs, Codes and Cryptogr.* **11** (1997) pp. 25–35.
6. A.R. Hammons, Jr., P.V. Kumar, A.R. Calderbank, N.J.A. Sloane and P. Solé, The $\mathbb{Z}_4$-linearity of Kerdock, Preparata, Goethals and related codes, *IEEE Trans. Inform. Theory* **40** (1994) pp. 301–319.
7. M. Harada, New extremal ternary self-dual codes, *Austral. J. Combin.* **17** (1998) pp. 133–145.
8. E. Rains, Bounds for self-dual codes over $\mathbb{Z}_4$, *Finite Fields and Their Appl.* **6** (2000) pp. 146–163.
9. E. Rains and N.J.A. Sloane, Self-dual codes, in Handbook of Coding Theory, V.S. Pless and W.C. Huffman, eds., (Elsevier, Amsterdam, 1998) 177–294.

Constructions of Codes from Number Fields

Venkatesan Guruswami

MIT Laboratory for Computer Science
200 Technology Square, Cambridge, MA 01239.
venkat@theory.lcs.mit.edu

Abstract. We define number-theoretic error-correcting codes based on algebraic number fields, thereby providing a generalization of Chinese Remainder Codes akin to the generalization of Reed-Solomon codes to Algebraic-geometric codes. Our construction is very similar to (and in fact less general than) the one given by Lenstra [8], but the parallel with the function field case is more apparent, since we only use the non-archimedean places for the encoding. We prove that over an alphabet size as small as 19, there even exist *asymptotically good* number field codes of the type we consider. This result is based on the existence of certain number fields that have an infinite class field tower in which some primes of small norm split completely.

1   Introduction

Algebraic Error-correcting Codes. For a finite field $\mathbb{F}_q$, an $[n, k, d]_q$-code $\mathcal{C}$ is a subset of $\mathbb{F}_q^n$ of size q^k such that if $c_1 \neq c_2 \in \mathcal{C}$ are two distinct codewords then they differ in at least d of the n positions. If $\mathcal{C}$ is actually a *subspace* of dimension k of the vector space $\mathbb{F}_q^n$ (over $\mathbb{F}_q$) then it is called a *linear code*. The parameters n, k, d are referred to as the blocklength, dimension, and minimum distance (or simply distance) of the code $\mathcal{C}$. For non-linear codes, the "dimension" need not be an integer – we just use it to refer to the quantity $\log_q |\mathcal{C}|$. The *rate* of the code, denoted by $R(\mathcal{C})$, is the quantity $\frac{\log_q |\mathcal{C}|}{n}$.

A broad and very useful class of error-correcting codes are *algebraic-geometric* codes (henceforth AG-codes), where the message is interpreted as specifying an element of some "function field" and it is encoded by its evaluations at a certain fixed set of "points" on an underlying well-behaved algebraic curve. A simple example is the widely used class of Reed-Solomon codes where messages are low degree polynomials and the codewords correspond to evaluations of such a polynomial at a fixed set of points in a finite field. The distance of the code follows from the fact that a low-degree polynomial cannot have too many zeroes in any field, and this is generalized for the case of algebraic-geometric codes using the fact that any "regular function" on an algebraic curve cannot have more zeroes than poles.

The class of algebraic-geometric codes are a broad class of very useful codes that include codes which beat the Gilbert-Varshamov bound for alphabet sizes $q \geq 49$ (see for example [17,2]). In addition to achieving such good performance, they possess a nice algebraic structure which has enabled design of efficient decoding algorithms to decode even in the presence of a large number of errors [14,5].

Motivation behind our work. Another family of algebraic codes that have received some study are number-theoretic redundant residue codes called the "Chinese Remainder

S. Boztaş and I.E. Shparlinski (Eds.): AAECC-14, LNCS 2227, pp. 129–140, 2001.
© Springer-Verlag Berlin Heidelberg 2001

codes" (henceforth called CRT codes). Here the messages are identified with integers with absolute value at most K (for some parameter K that governs the rate) and a message m is encoded by its residues modulo n primes $p_1 < p_2 < \cdots < p_n$. If $K = p_1 \cdot p_2 \cdots p_k$ and $n > k$, this gives a redundant encoding of m and the resulting "code" (which is different from usual codes in that symbols in different codeword positions are over different alphabets) has distance $n - k + 1$.

In light of the progress in decoding algorithms for Reed-Solomon and algebraic-geometric codes, there has also been progress on decoding CRT codes [3,1,6] in the presence of very high noise, and the performance of the best known algorithm matches the number of errors correctable for Reed-Solomon codes [5]. Since Reed-Solomon codes are a specific example of the more general family of AG-codes, it is natural to ask if CRT codes can also be realized as certain kind of AG-codes, and further whether there is a natural generalization of CRT codes akin to the generalization of Reed-Solomon codes to algebraic-geometric codes. It is this question which is addressed in this work.

Our Results. For those familiar with the algebraic-geometric notion of *schemes*, it is not hard to see that the CRT code can be captured by a geometric framework using the idea of "one-dimensional schemes" and can thus be cast as a geometric code via an appropriately defined non-singular curve (namely $\mathrm{Spec}(\mathbb{Z})$ which is space of all prime ideals of $\mathbb{Z}$) and viewing integers (which are the messages) as regular functions on that curve. More generally, using this idea we are able to define error-correcting codes based on any number field (a finite field extension of the field $\mathbb{Q}$ of rational numbers) – we call such codes *number field codes* (or NF-codes).

We prove that over a large enough alphabet ($\mathrm{GF}(19)$ suffices), there in fact exist *asymptotically good* number field codes. A code family $\{\mathcal{C}_i\}$ of $[n_i, k_i, d_i]_q$ codes of increasing blocklength $n_i \to \infty$ is called *asymptotically good* if $\liminf \frac{k_i}{n_i} > 0$ and $\liminf \frac{d_i}{n_i} > 0$. Explicit constructions of asymptotically good codes is a central problem in coding theory and several constructions are known, the best ones (for large enough q) being certain families of algebraic-geometric codes. Our construction of asymptotically good number fields uses concepts from class field theory and in particular is based on the existence of certain number fields that have an infinite Hilbert class field tower in which several primes of small norm split completely all the way up the tower. Obtaining such a construction over as small an alphabet size as possible is one of the primary focuses of this paper.

Comparison with [8]. It is our pleasure to acknowledge here that Lenstra [8] (see also the account in [15]) had long back already considered the construction of codes from algebraic number fields, and we are therefore by no means the first to consider this question. Unfortunately, we were unaware of his work when we came up with our constructions. The main point of difference between his constructions and ours is the following. In our constructions, messages are taken to be an appropriate subset of elements of the ring of integers in a number field and they are encoded by their residues modulo certain *non-archimedean* (also referred to as *finite*) places. This corresponds exactly to the "ideallic" view of codes (see [6]) since we have an underlying ring and messages are encoded by their residues modulo a few prime ideals. The construction in [8] is actually more general, and also allows *archimedean* (also referred to as *infinite*) places to be used for encoding. (We stress that it is not *necessary* to use the archimedean places for the constructions in [8], but doing so enabled [8] to prove the existence of asymptotically good codes more easily.)

The use of archimedean places as in [8] is extremely insightful and cute, and also makes it easier to get good code parameters. But, not using them maintains the parallel with the function field situation, and gives a base case of NF-code constructions which is most amenable to encoding/decoding, assuming algorithms for these will eventually be studied.

Finally, the results in [8] are of an asymptotic flavor, i.e. focus on what can be achieved in the limit for large alphabet size q, and do not imply asymptotically good codes exist for some reasonably small q. Also, no unconditional result (i.e. without assuming the Generalized Riemann Hypothesis (GRH)) guaranteeing the existence of asymptotically good codes can be directly inferred from [8] **if** one modifies the constructions therein to include *only* codeword positions corresponding to the finite places. We are able to prove, using some results on the existence of infinite class field towers, that asymptotically good codes of the kind we construct exist for reasonable values of q (for example, $q = 19$ suffices).

2   Algebraic Codes: Construction Philosophy

We now revisit, at a high level, the basic principle that underlies the construction of all algebraic error-correcting codes, including Reed-Solomon codes, Algebraic-geometric codes, and the Chinese Remainder code. A similar discussion can also be found in [6].

An algebraic error-correcting code is defined based on an underlying ring R (assume it is an integral domain) whose elements r come equipped with some notion of "size", denoted size(r). For example, for Reed-Solomon codes, the ring is polynomial ring $F[X]$ over a (large enough) finite field F, and the "size" of $f \in F[X]$ is simply its degree as a polynomial in X. Similarly, for the CRT code, the ring is $\mathbb{Z}$, and the "size" is the usual absolute value.

The messages of the code are the elements of the ring R whose size is at most a parameter Λ (this parameter governs the rate of the code). The encoding of a message $m \in R$ is given by

$$m \mapsto \mathsf{Enc}(m) = (m/I_1, m/I_2, \cdots, m/I_n)$$

where I_j, $1 \leq j \leq n$ are n (distinct) prime ideals of R. (For instance, in the case of Reed-Solomon codes, we have $R = F[X]$ and $I_j = (X - \alpha_j)$ – the ideal generated by the polynomial $(X - \alpha_j)$ – for $1 \leq j \leq n$, where $\alpha_1, \ldots, \alpha_n$ are *distinct* elements of F. This ideal-based view was also at the heart of the decoding algorithm for CRT codes presented in [6].)

There are two properties of a code that of primary concern in its design, namely (a) its rate, and (b) its minimum distance. The rate property of the code constructed by the above scheme follows from an estimate of the number of elements m of R that have size$(m) \leq \Lambda$. The distance of the code follows by using further properties of the size$(\cdot)$ function which we mention informally below.

1. For elements $a, b \in R$, size$(a - b)$ is "small" whenever size(a) and size(b) are both "small".

2. If $f \neq 0$ belongs to "many" ideals among $I_1, I_2, \ldots, I_n$, then $\mathrm{size}(f)$ cannot be "too small".

It is not difficult to see that, together these two properties imply that if $m_1 \neq m_2$ are distinct messages, then their encodings $\mathrm{Enc}(m_1)$ and $\mathrm{Enc}(m_2)$ cannot agree in too many places, and this gives the distance property of the code.

3 Constructing Codes from Number Fields

The previous section described how to construct codes from rings provided an appropriate notion of size can be defined on it. We now focus on the specific problem of constructing codes based on the ring of integers of number fields.

An algebraic number field (or number field for short) is a finite (algebraic) extension of the field $\mathbb{Q}$ of rational numbers. Given some algebraic number field $K/\mathbb{Q}$ of degree $[K : \mathbb{Q}] = m$ (i.e., $K = \mathbb{Q}(\alpha)$ where α satisfies an irreducible polynomial of degree m over $\mathbb{Q}$), the code will comprise of a subset of elements from its ring of integers, denoted $\mathcal{O}_K$. (Recall that the ring of integers of a number field K is the integral closure of $\mathbb{Z}$ in K, i.e., it consists of elements of K that satisfy some monic polynomial over $\mathbb{Z}$.) It is well known that $\mathcal{O}_K$ is a *Dedekind domain* with several nice properties. For reasons of space, we refer the reader to any standard algebraic number theory text (for example [11,10]) for the necessary background on number fields.

3.1 Norms of Ideals and Elements in a Ring of Integers

"Norms" of Ideals: For every non-zero ideal I of a ring of integers R, R/I is finite. One can thus define a norm function on ideals as:

Definition 1 [Norm of Ideals]: *The norm of a non-zero ideal $I \subset R$, denoted $\|I\|$, is defined as $\|I\| = |R/I|$. Note that for a prime ideal $\mathfrak{p}$, $\|\mathfrak{p}\| = p^{f(\mathfrak{p}|p)}$ if $\mathfrak{p}$ lies above $p \in \mathbb{Z}$ and $f(\mathfrak{p}|p)$ is the inertia degree of $\mathfrak{p}$ over p.*

Definition 2 [Norm of Elements in $\mathcal{O}_K$]: *The norm of an element $x \in R$, also denoted $\|x\|$ by abuse of notation, is defined as $\|(x)\|$, i.e., the norm of the ideal generated by x. (Define $\|0\| = 0$.)*

The following fact will be very useful for us later:

Fact 3 *For a number field K, If $x \in I$ for some ideal $I \subset \mathcal{O}_K$, then $\|I\|$ divides $\|x\|$.*

3.2 Defining Size of an Element

By Fact 3, it is tempting to define the size of an element f as $\mathrm{size}(f) = \|f\|$. In fact, this satisfies one of the properties we required of size, namely that if $m \neq 0$ has small size, then it cannot belong to several ideals I_i. Unfortunately, the other property we would like our size function to satisfy, namely $\mathrm{size}(a - b)$ is "small" whenever $\mathrm{size}(a)$ and $\mathrm{size}(b)$ are both small, is not satisfied in general for all number fields by the definition size$(f) = \|f\|$.[1] We thus need a different notion of size of an element. To this end, we will appeal to the valuation-theoretic point of view of the theory of number fields. We refer the reader to the book by Neukirch [11] for an excellent exposition of the valuation-theoretic approach to algebraic number theory; we rapdily review the most basic definitions and facts about valuations.

A **valuation** of a field K is a function $|\ |: K \to \mathbb{R}$ with the properties:[2] (i) $|x| \geq 0$, and $|x| = 0 \iff x = 0$, (ii) $|xy| = |x||y|$, and (iii) There exists a constant $c \geq 1$ such that for all $x, y \in K$, $|x + y| \leq c \cdot \max\{|x|, |y|\}$ ("triangle inequality").[3]

Two valuations $|\ |_1$ and $|\ |_2$ on K are said to be *equivalent* iff there exists a real number $s > 0$ such that $|x|_1 = |x|_2^s$ for all $x \in K$. A **place** of K is an equivalence class of valuations of K. Whenever we refer to a valuation from now on, we implicitly mean any member of its associated place. A valuation (place) $|\ |$ is called **non-archimedean** (or **ultrametric**) if $|n|$ stays bounded for all $n \in \mathbb{N}$. Otherwise it is called **archimedean**. Alternatively, a valuation $|\ |$ is non-archimedean if and only if it satisfies the triangle inequality of Condition (3) above with $c = 1$, i.e., if $|x + y| \leq \max\{|x|, |y|\}$ for all $x, y \in K$.

One can define a non-archimedean valuation of the fraction field of any domain R based on any non-zero prime ideal of R (the trivial valuation $|x| = 1$ for all $x \neq 0$ corresponds to the zero ideal), similar to the valuation $|\ |_p$ defined on $\mathbb{Q}$ above. In fact the non-zero prime ideals of the ring of integers $\mathcal{O}_K$ of a number field K correspond precisely to the non-archimedean places of K, and are called the **finite places** of K. In addition, the archimedean valuations of K correspond to the **infinite places** of K. The infinite places are important objects in the study of number fields and we review them next.

Infinite places and a notion of "size": Let $K/\mathbb{Q}$ be a field extension of degree $[K : \mathbb{Q}] = M$. Then there are M distinct field homomorphisms (called embeddings) $\tau_i : K \to \mathbb{C}$ of the field K into $\mathbb{C}$ which leave $\mathbb{Q}$ fixed. Out of these let r of the embeddings be into the reals, say $\tau_1, \ldots, \tau_r : K \to \mathbb{R}$, and let the remaining $2s = M - r$ embeddings be complex. We refer to this pair (r, s) as the *signature* of K. These $2s$ embeddings come in s pairs of complex conjugate non-real embeddings, say $\sigma_j, \bar{\sigma}_j : K \to \mathbb{C}$ for $1 \leq j \leq s$. The following fundamental result shows the correspondence between the archimedean valuations and the embeddings of K into $\mathbb{C}$.

Fact 4 *There are precisely* $(r + s)$ *infinite places (which we denote by* $\mathfrak{q}_1, \mathfrak{q}_2, \ldots, \mathfrak{q}_{r+s}$ *throughout) of a number field* K *that has signature* (r, s) *(with* $r + 2s = [K : \mathbb{Q}]$*). The* r *infinite places* $\mathfrak{q}_1, \mathfrak{q}_2, \ldots, \mathfrak{q}_r$ *corresponding to the* r *real embeddings* $\tau_1, \ldots, \tau_r$ *are given by the (archimedean) valuations* $|x|_{\mathfrak{q}_i} \overset{\text{def}}{=} |\tau_i(x)|$ *for* $1 \leq i \leq r$ *and the* s *infinite places* $\mathfrak{q}_{r+1}, \ldots, \mathfrak{q}_{r+s}$ *corresponding to the* s *pairs of complex conjugate embeddings* σ_j *are given by* $|x|_{\mathfrak{q}_{r+j}} = |\sigma_j(x)|^2$. □

We now come to our definition of the size of an element x in a number field K.

---

[1] For example, let $K = \mathbb{Q}(\alpha)$ where α is a root of $x^2 + Dx + 1 = 0$. Then one can easily see that $\|\alpha\| = 1$ (for example using Proposition 6) and of course $\|1\| = 1$, but $\|\alpha - 1\| = D + 2$, and thus $\|x - y\|$ can be arbitrarily larger than both $\|x\|$ and $\|y\|$ even for quadratic extensions.

[2] Several textbooks call a "valuation" with these properties as an absolute value.

[3] When this condition is met with any $c \leq 2$, then it is an easy exercise to show that in fact $|x + y| \leq |x| + |y|$ for all x, y, which is the "familiar" triangle inequality.

Definition 5 [Size]: *Let K be a number field with signature (r, s). Let Let $| \ |_{q_1}, | \ |_{q_2}, \ldots, | \ |_{q_r}$ be the archimedean valuations of x corresponding to the r real embeddings of K, and let $| \ |_{q_{r+1}}, \ldots, | \ |_{q_{r+s}}$ be the archimedean valuations of x corresponding to the s complex conjugate embeddings of K. The size of an element $x \in K$ is defined as*

$$\mathsf{size}(x) \overset{\text{def}}{=} \sum_{i=1}^{r} |x|_{q_1} + \sum_{i=1}^{s} 2\sqrt{|x|_{q_{r+s}}} \, . \tag{1}$$

The following shows an important property of the above definition of $\mathsf{size}(\cdot)$ (which was lacking in the attempted definition $\mathsf{size}(x) = \|x\|$):

Lemma 1. *Let K be a number field with signature (r, s) and let $a, b \in \mathcal{O}_K$. Then $\mathsf{size}(a - b) \leq \mathsf{size}(a) + \mathsf{size}(b)$.*

Proof: The proof follows from the definition of $\mathsf{size}(x)$ and the (easy to check) facts that $|x - y|_q \leq |x|_q + |y|_q$ for the real infinite places q of K, and $\sqrt{|x - y|_q} \leq \sqrt{|x|_q} + \sqrt{|y|_q}$ for the complex (infinite) places of K.   □

The following central and important result (see any textbook, eg. [11], for a proof), relates the norm of an element (recall Definition 2) to its size, and is crucial for lower bounding the distance of the our codes.

Proposition 6 *For a number field K and for any element $x \in \mathcal{O}_K$ in its ring of integers, we have*

$$\|x\| = |x|_{q_1} \cdot |x|_{q_2} \cdots |x|_{q_\ell},$$

where $q_1, \ldots, q_\ell$ are the archimedean (infinite) places of K.

Using the above, we get the following useful upper bound on $\|x\|$ in terms of $\mathsf{size}(x)$.

Lemma 2. *For a number field K with $[K : \mathbb{Q}] = M$ and any $x \in \mathcal{O}_K$, we have $\|x\| \leq \left(\frac{\mathsf{size}(x)}{M}\right)^M$.*

Proof: Let (r, s) be the signature of K. The claimed result follows using Equation (1), Proposition 6 and an application of the Arithmetic Mean-Geometric Mean inequality to the M numbers $(|x|_{q_1}, \ldots, |x|_{q_r}, \sqrt{|x|_{q_{r+1}}}, \sqrt{|x|_{q_{r+1}}}, \cdots, \sqrt{|x|_{q_{r+s}}}, \sqrt{|x|_{q_{r+s}}})$. The arithmetic mean of these numbers equals $\frac{\mathsf{size}(x)}{M}$ and their geometric mean equals $\|x\|^{1/M}$.
□

Corollary 7 *Let K be a number field of degree $[K : \mathbb{Q}] = M$. If $a, b \in \mathcal{O}_K$ are such that $\mathsf{size}(a) \leq B$ and $\mathsf{size}(b) \leq B$, then $\|a - b\| \leq \left(\frac{2B}{M}\right)^M$.*

Proof: By Lemma 1, we have $\mathsf{size}(a - b) \leq 2B$, and using Lemma 2, we get $\|a - b\| \leq \left(\frac{2B}{M}\right)^M$.   □.

3.3   The Code Construction

For the rest of this section, let K be a number field of degree $[K : \mathbb{Q}] = M$ and signature (r, s). A number field code (NF-code for short) $\mathcal{C} = \mathcal{C}_K$, based on a number field K, has parameters $(n, \mathfrak{p}_1, \mathfrak{p}_2, \ldots, \mathfrak{p}_n; B)$ where n is the blocklength of the code, $\mathfrak{p}_1, \mathfrak{p}_2, \ldots, \mathfrak{p}_n$ are distinct (non-zero) prime ideals of $\mathcal{O}_K$, and B is a positive real. The ith position of the code $\mathcal{C}$ is defined over an alphabet of size $\|\mathfrak{p}_i\|$ for $1 \le i \le n$; let us assume w.l.o.g that $\|\mathfrak{p}_1\| \le \|\mathfrak{p}_2\| \le \cdots \le \|\mathfrak{p}_n\|$.

We are now almost in a position to define our code $\mathcal{C} = \mathcal{C}_K$, but for a technical reason that will become clear in Section 3.5, we will need to define the code with one extra "shift" parameter $\mathbf{z} \in \mathbb{R}^r \times \mathbb{C}^s$. Given such a $\mathbf{z} = \langle z_1, z_2, \ldots, z_{r+s} \rangle$ with $z_i \in \mathbb{R}$ for $1 \le i \le r$ and $z_j \in \mathbb{C}$ for $r < j \le r+s$, the $\mathbf{z}$-shifted size $\mathsf{size}_{\mathbf{z}}(x)$ of $x \in \mathcal{O}_K$ is defined as follows. Let $\tau_1, \ldots, \tau_r$ be the embeddings $K \to \mathbb{R}$, and let $\sigma_j, 1 \le j \le s$, be the non-conjugate complex embeddings $K \to \mathbb{C}$. For $i = 1, 2, \ldots, r$, define $a_i^{(x)} = |\tau_i(x) - z_i|$, and for $1 \ge j \le s$, define $b_j^{(x)} = |\sigma_j(x) - z_{r+j}|^2$. (Thus the archimedean valuations are just "shifted" with respect to $\mathbf{z}$.) Now define

$$\mathsf{size}_{\mathbf{z}}(x) = \sum_{i=1}^{r} a_i^{(x)} + \sum_{j=1}^{s} 2\sqrt{b_j^{(x)}}. \tag{2}$$

Lemma 3. *Let K be a number field of signature (r, s), with $[K : \mathbb{Q}] = r + 2s = M$, and $\mathbf{z} \in \mathbb{R}^r \times \mathbb{C}^s$. If $a, b \in \mathcal{O}_K$ are such that $\mathsf{size}_{\mathbf{z}}(a) \le B$ and $\mathsf{size}_{\mathbf{z}}(b) \le B$, then $\|a - b\| \le \left(\frac{2B}{M}\right)^M$.*

Proof: One can show, similarly to Lemma 1, that if $\mathsf{size}_{\mathbf{z}}(a) \le B$ and $\mathsf{size}_{\mathbf{z}}(b) \le B$, then $\mathsf{size}(a - b) \le 2B$. The proof then follows using Lemma 2. $\qquad\square$

We now formally specify our code construction with the "shift parameter" added in.

Definition 8 *The code $\mathcal{C} = \mathcal{C}_K$ based on a number field K with parameters $(n, \mathfrak{p}_1, \mathfrak{p}_2, \ldots, \mathfrak{p}_n; B; \mathbf{z})$ is defined as follows. The message set of $\mathcal{C}$ is $\{m \in \mathcal{O}_K : \mathsf{size}_{\mathbf{z}}(m) \le B\}$. The encoding function is $\mathsf{Enc}_{\mathcal{C}}(m) = \langle m/\mathfrak{p}_1, m/\mathfrak{p}_2, \ldots, m/\mathfrak{p}_n \rangle$.*

3.4 Distance of the Code

We now estimate the distance of the code. If $\mathsf{Enc}_{\mathcal{C}}(m_1)$ and $\mathsf{Enc}_{\mathcal{C}}(m_2)$ agree in t places, say $1 \le i_1 < i_2 < \cdots < i_t \le n$, then $m_1 - m_2 \in \mathfrak{p}_{i_1} \cdots \mathfrak{p}_{i_t}$. By Fact 3 and the ordering of the $\mathfrak{p}_i$'s in increasing order of norm, we get $\|m_1 - m_2\| \ge \prod_{i=1}^{t} \|\mathfrak{p}_i\|$. On the other hand, by Lemma 3, we have $\|m_1 - m_2\| \le (2B/M)^M$. Thus if $\prod_{i=1}^{t} \|\mathfrak{p}_i\| > (2B/M)^M$, then we must have $m_1 = m_2$, and thus two distinct codewords can agree in at most $(t - 1)$ places. We have thus shown the following:

Lemma 4. *For a number field code $\mathcal{C} = \mathcal{C}_K$ based on a field K with $[K : \mathbb{Q}] = M$ with parameters $(n, \mathfrak{p}_1, \mathfrak{p}_2, \ldots, \mathfrak{p}_n; B; \mathbf{z})$, if t $(1 \le t \le n)$ is such that $\|\mathfrak{p}_1\| \times \|\mathfrak{p}_2\| \cdots \times \|\mathfrak{p}_t\| > (2B/M)^M$, then the distance $d(\mathcal{C})$ of $\mathcal{C}$ is at least $(n - t + 1)$. In particular,*

$$d(\mathcal{C}) > n - \frac{M \log(2B/M)}{\log \|\mathfrak{p}_1\|}. \qquad\square$$

3.5   The Rate of the Code

To estimate the rate of the code we need a lower bound (or good estimate) of the number of elements x of $\mathcal{O}_K$ with $\text{size}_{\mathbf{z}}(x) \leq B$. The key quantity in such a lower bound is the *discriminant* of a number field.

Given a number field K of degree M with M embeddings $\zeta_1, \ldots, \zeta_M : K \to \mathbb{C}$, the *discriminant* of any M-tuple of elements $\alpha_1, \ldots, \alpha_M \in K$, denoted $\text{disc}(\alpha_1, \ldots, \alpha_M)$, is defined as the square of the determinant of the $M \times M$ matrix having $\zeta_i(\alpha_j)$ as its $(i, j)^{\text{th}}$ entry. $\text{disc}(\alpha_1, \ldots, \alpha_M) \in \mathbb{Q}$ and if $\alpha_i \in \mathcal{O}_K$, then $\text{disc}(\alpha_1, \ldots, \alpha_M) \in \mathbb{Z}$.

The **discriminant** of K, denoted D_K, is defined as $\text{disc}(\beta_1, \ldots, \beta_M)$ where $\beta_1, \ldots, \beta_M$ is *any* integral basis of $\mathcal{O}_K$ over $\mathbb{Z}$ (it can be shown to be independent of the choice of the basis). Lastly, the root discriminant of K, denoted rd_K, is defined as $|D_K|^{1/M}$.

The following proposition gives a lower bound on the number of elements of bounded size in the ring of integers of a number field. It uses the geometry of the Minkowski lattice and is based on an averaging argument similar to the one used by Lenstra [8].

Proposition 9 ([8]) *For any number field K with signature (r, s) and discriminant D_K, and any $B \in \mathbb{R}_+$, there exists a $\mathbf{z} \in \mathbb{R}^r \times \mathbb{C}^s$, such that*

$$\left| \{ x \in \mathcal{O}_K : \text{size}_{\mathbf{z}}(x) \leq B \} \right| \geq \frac{2^r \pi^s}{\sqrt{|D_K|}} \frac{B^M}{M!} . \tag{3}$$

The following proposition records the quantitative parameters (rate and distance) of the NF-code construction we gave in Section 3.3. (All logarithms are to the base 2.)

Proposition 10 *Let K be a number field of degree $[K : \mathbb{Q}] = M$ and signature (r, s). Let $\mathcal{C} = \mathcal{C}_K$ be a number field code defined with parameters $(n, \mathfrak{p}_1, \mathfrak{p}_2, \ldots, \mathfrak{p}_n; B; \mathbf{z})$ with $\|\mathfrak{p}_1\| \leq \cdots \leq \|\mathfrak{p}_n\|$. Then there exists a choice of the "shift" $\mathbf{z}$ for which the rate $R(\mathcal{C})$ of $\mathcal{C}$ is at least $\frac{1}{n \log \|\mathfrak{p}_n\|} \cdot \left(\log \left(2^r \pi^s B^M \right) - \log M! - \log \sqrt{|D_K|} \right)$, and the distance $d(\mathcal{C})$ of $\mathcal{C}$ is greater than $n - \frac{M \log (2B/M)}{\log \|\mathfrak{p}_1\|}$. In particular we have*

$$R(\mathcal{C}) > \frac{(n - d(\mathcal{C})) \log \|\mathfrak{p}_1\| + s \log(\pi/4) + M \log e - M \log \sqrt{\text{rd}_K} - \log 3M}{n \log \|\mathfrak{p}_n\|} .$$

Proof: The proof follows easily from Lemma 4 and Proposition 9, and using Stirling's approximation that $M! \simeq \sqrt{2\pi M} \left(\frac{M}{e} \right)^M \leq 3M \left(\frac{M}{e} \right)^M$ for all $M \geq 1$. $\square$

4   Constructing an Asymptotically Good Code

By Proposition 10, in order to have good rate, one would like to define codes based on number fields K with small root discriminant rd_K. In addition, in order to define a code of blocklength n over a alphabet of size q, we need $\mathcal{O}_K$ to have n prime ideals of norm at most q. In particular, if one hopes to construct a family of asymptotically good codes over an alphabet of size q based on this approach, then one needs a family of number fields $\{K_n\}$ such that (a) K_n has small root discriminant (the best one can hope for is of the form c for some constant c by existing lower bounds on the discriminant, see the survey [12]), and (b) it has $\Omega([K_n : \mathbb{Q}])$ prime ideals of norm at most q. Constructions of sequences of number fields with bounded root discriminant are obtained in the literature using the existence of infinite Hilbert class field towers. For our application we will need such towers with the added restriction that certain primes of small norm split completely all the way up the tower. We next review the main definition from class field theory that will be necessary for our number field constructions.

4.1 Class Fields with Splitting Conditions: Definitions

We will quickly review the basic notation: a finite extension K/k of number fields is (i) *unramified* if no place (including the infinite ones: i.e. real places stay real) of k is ramified in K (this implies $\mathrm{disc}(K/k) = (1)$); (ii) *abelian* if K/k is Galois with abelian Galois group; (iii) a p-extension if K/k is Galois with $[K : k]$ a power of p.

Definition 11 [T-decomposing p-Class Field]: *For any number field k and a set of primes T (of $\mathcal{O}_k$), the maximal unramified abelian p-extension of k in which every prime in T splits completely, denoted k_p^T, is called the T-decomposing p-class field of k.*

Definition 12 [T-decomposing p-Class Field Tower]: *For any number field k and a set of primes T, the T-decomposing p-class field tower of k is obtained by repeatedly taking T-decomposing p-class fields: It is the sequence of fields $k_0 = k$, $k_1 = k_p^T$ and for $i \geq 2$, $k_i = (k_{i-1})_p^{T_{i-1}}$, where T_i is the set of primes in k_i lying above T. We say that k has an infinite T-decomposing p-class field tower if this tower does not terminate for any finite i.*

4.2 The Construction Approach

The basic approach behind constructing number fields K with infinite class field towers is the Golod-Šafarevič theory [4] (cf. [13]). For our purposes, it suffices to use the following result which gives a specific sufficient condition for quadratic extensions to have infinite 2-class field towers with certain added splitting constraints. This result appears as Corollary 6.2 in [16] and is proved using techniques which also appear in related works like [9,7].

Proposition 13 ([16]) *Let $P = \{p_1, \ldots, p_s\}$ and $Q = \{q_1, \ldots, q_r\}$ be disjoint sets of primes. Consider a imaginary quadratic extension $K/\mathbb{Q}$ that is ramified exactly at those primes in Q. Let T be the set of primes ideals of $\mathcal{O}_K$ that lie above the primes in P, and let $|T| = t$. Suppose further that*

$$r \geq 3 + t - s + 2\sqrt{2 + t}. \tag{4}$$

Then K has an infinite T-decomposing 2-class field tower.

The above is a very useful proposition and we believe plugging in specific values into it will lead to many asymptotically good number field code constructions. In the next two subsections, our aim is to present concrete examples of code constructions based on the above proposition and we therefore focus on a specific setting of parameters which will lead to an asymptotically good code over a reasonably small alphabet.

4.3 Specific Constructions

We now apply Proposition 13 to get a specific construction of a number field with an infinite 2-class field tower.

Lemma 5. *Let* $d = 3 \cdot 5 \cdot 7 \cdot 11 \cdot 13 \cdot 17 \cdot 19 = 4849845$, *and let* $K = \mathbb{Q}(\sqrt{-d})$. *Then:*

(i) $\mathrm{rd}_K = \sqrt{4d} \simeq 4404.4727$

(ii) $\mathcal{O}_K$ *has a set* T *of two prime ideals of norm* 29.

(iii) K *has an infinite* T-*decomposing* 2-*class field tower.*

Proof: The first two parts follow from standard properties of imaginary quadratic extensions (cf. [10]). To prove Part (iii), we apply Proposition 13 to $K/\mathbb{Q}$ with $Q = \{2, 3, 5, 7, 11, 13, 17, 19\}$ and $P = \{29\}$. The prime 29 splits into a set T of two primes in $K/\mathbb{Q}$, we thus have $r = 8$, $s = 1$ and $t = 2$ in Proposition 13. Since these values satisfy Condition (4), we conclude that K has an infinite T-decomposing 2-class field tower. $\qquad\square$

4.4 Obtaining an Asymptotically Good Number Field Code

Let $K_0 = K$ be the number field from the previous section. Let $K_0 \subset K_1 \subset K_2 \subset \cdots$ be the (infinite) T-decomposing 2-class field tower of K_0. We construct a family of codes $\mathcal{C}_n$ based on the number fields K_n below.

Fix an n and let $[K_n : \mathbb{Q}] = M$ (note that M will be a power of 2 but this will not be important for us). Since K_n is totally complex, the signature of K_n is $(0, M/2)$. By Lemma 5, the prime 29 *splits completely* in the extension $K_n/\mathbb{Q}$, and thus $\mathcal{O}_{K_n}$ has M prime ideals, say $\mathfrak{p}_1, \mathfrak{p}_2, \ldots, \mathfrak{p}_M$, each of norm 29.

Now consider the code $\mathcal{C}_n$ (defined as in Section 3.3) based on K_n with parameters $(M, \mathfrak{p}_1, \ldots, \mathfrak{p}_M; B; \mathbf{z})$ where $B = c_0 M$ for some some constant $c_0 > 0$ to be specified later, and $\mathbf{z} \in \mathbb{C}^{M/2}$ is a "shift parameter" as guaranteed by Proposition 9. Now let us analyze the parameters of this code family $\{\mathcal{C}_n\}_{n \geq 0}$. Define the *designed distance* of the code $\mathcal{C}_n$ to be

$$d'(\mathcal{C}_n) = M - \frac{M \log(2c_0)}{\log 29}. \tag{5}$$

Then, by Proposition 10, the distance of the code $d(\mathcal{C}_n)$ is at least $d'(\mathcal{C}_n)$, and the rate of the code $R(\mathcal{C}_n)$ is greater than

$$\frac{1}{M} \cdot \left(M - d'(\mathcal{C}_n) - \frac{M}{\log 29} \cdot \log\left(\frac{2}{e}\sqrt{\frac{\mathrm{rd}_{K_n}}{\pi}}\right) - \frac{\log 3M}{\log 29} \right). \tag{6}$$

Combining Equations (5) and (6) above and using $\mathrm{rd}_{K_n} = \mathrm{rd}_K = 4404.4727$, we obtain, in the limit of large $M \to \infty$,

$$R(\mathcal{C}_n) \geq 1 - \frac{d'(\mathcal{C}_n)}{M} - \frac{\log 27.55}{\log 29} > 0.015 - \frac{d'(\mathcal{C}_n)}{M}. \tag{7}$$

Thus if $\frac{d'(\mathcal{C}_n)}{M} \leq 0.015$, we can get asymptotically good codes. By Equation (5) this will be the case if $c_0 > 29^{0.985}/2$, or if $c_0 \geq 13.79$. Also we must have $c_0 < 29/2 = 14.5$ in order to have $d'(\mathcal{C}_n)/M > 0$. By varying c_0 in this range ($13.79 \leq c_0 < 14.5$), we can achieve asymptotically good codes over an alphabet of size 29 for any value of relative distance δ in the range $0 < \delta \leq 0.015$. We have thus proved the following:

Theorem 14. *There exist asymptotically good families of number field codes. In particular, such codes exist over* GF(29). □

4.5 Asymptotically Good Codes Over a Smaller Alphabet

We now present a different construction that achieves even smaller alphabet size. The codes are defined by using prime ideals of different norms (since NF-codes are anyway non-linear, this presents no problems). We sketch this construction below. The proof of the following lemma is similar to that of Lemma 5.

Lemma 6. *Let* $d' = 3 \cdot 5 \cdot 7 \cdot 11 \cdot 13 \cdot 23 \cdot 29 \cdot 37 \cdot 41$, *and let* $K' = \mathbb{Q}(\sqrt{-d'})$. *Then:*

(i) $\mathrm{rd}_{K'} = \sqrt{4d'} \simeq 246515.72$
(ii) $\mathcal{O}_{K'}$ *has a set* T' *of four prime ideals, two of which lie above* 17 *(and have norm* 17*) and two of which lie above* 19 *(and have norm* 19*).*
(iii) K' *has an infinite* T'*-decomposing 2-class field tower.*

One can now construct a family of codes from the infinite T'-decomposing 2-class field tower $K_0' = K' \subset K_1' \subset K_2' \subset \ldots$ of K', similar to the construction in Section 4.4. The code $\mathcal{C}_n'$ based on K_n' will have parameters $(N, \mathfrak{p}_1, \ldots, \mathfrak{p}_M, \mathfrak{q}_1, \ldots, \mathfrak{q}_M; B; \mathbf{z})$. Here the $\mathfrak{p}_i$'s (resp. $\mathfrak{q}_i$'s) are the M primes in K_n' that lie above the prime integer 17 (resp. 19), $N = 2M$ is the blocklength of the code, $B = c_0'M$ for some some constant $c_0' > 0$, and $\mathbf{z} \in \mathbb{C}^{M/2}$ is an appropriate "shift parameter". Now using Proposition 10 and arguments similar to those in Section 4.4, we obtain in the limit of large $M \to \infty$,

$$R(\mathcal{C}_n') \geq \left(1 - \frac{d(\mathcal{C}_n')}{N}\right)\frac{\log 17}{\log 19} - \frac{1}{2}\frac{\log\left(\frac{2}{e}\sqrt{\frac{\mathrm{rd}_{K'}}{\pi}}\right)}{\log 19} > \frac{\log 17}{\log 19}\left(1 - \frac{d(\mathcal{C}_n')}{N} - \frac{\log 14.5}{\log 17}\right). \tag{8}$$

Thus we can get asymptotically good codes for any value of relative minimum distance that is at most $(1 - \frac{\log 14.5}{\log 17}) \simeq 0.056$. We therefore conclude the following strengthening of Theorem 14.

Theorem 15. *There exist asymptotically good number field codes over an alphabet of size* 19. □

5 Concluding Remarks

We conclude with some specific questions: Can one prove unconditionally, without assuming the GRH, that there exist codes that beat the Gilbert-Varshamov bound for a *not too large* alphabet size? If so, what is the smallest alphabet size one can achieve for such a result, and what is the best asymptotic performance one can achieve in the limit of large (but constant) alphabet size?

Acknowledgments

I am extremely grateful to Farshid Hajir for several useful discussions on these topics early on during this work. I thank Amin Shokrollahi for crucial pointers to prior work on number field codes. My sincere thanks to Hendrik Lenstra for bringing to my attention the necessity to use the averaging argument in the proof of Proposition 9, for numerous other useful discussions, and for sending me a copy of his paper [8]. I am grateful to Michael Tsfasman and Serge Vlădut for pointers to [15,16] and for sending me a copy of their paper [16] (which was crucial in simplifying some of my asymptotically good code constructions). I would like to thank Andrew Odlyzko and Madhu Sudan for useful discussions.

References

1. D. Boneh. Finding Smooth integers in short intervals using CRT decoding. *Proceedings of the 32nd Annual ACM Symposium on Theory of Computing*, Portland, Oregon, May 2000, pp. 265-272.
2. A. Garcia and H. Stichtenoth. A tower of Artin-Schreier extensions of function fields attaining the Drinfeld-Vladut bound. *Invent. Math.*, 121 (1995), pp. 211-222.
3. O. Goldreich, D. Ron and M. Sudan. Chinese Remaindering with errors. *IEEE Trans. on Information Theory*, to appear. Preliminary version appeared in *Proc. of the 31st Annual ACM Symposium on Theory of Computing*, Atlanta, Georgia, May 1999, pp. 225-234.
4. E. S. Golod and I. R. Šafarevič. On class field towers. (Russian), Izv. Akad Nauk SSSR, 28 (1964), 261-272; (English translation) Amer. Math. Soc. Transl. Ser 2, 48 (1965), pp. 91-102.
5. V. Guruswami and M. Sudan. Improved decoding of Reed-Solomon and Algebraic-geometric codes. *IEEE Trans. on Information Theory*, 45 (1999), pp. 1757-1767.
6. V. Guruswami, A. Sahai and M. Sudan. "Soft-decision" decoding of Chinese Remainder codes. *Proceedings of the 41st IEEE Symposium on Foundations of Computer Science (FOCS)*, Redondo Beach, California, November 12-14,2000.
7. Y. Ihara. How many primes decompose completely in an infinite unramified Galois extension of a global field? *J. Math. Soc. Japan*, 35 (1983), pp. 693-709.
8. H. W. Lenstra. Codes from algebraic number fields. In: M. Hazewinkel, J.K. Lenstra, L.G.L.T. Meertens (eds), *Mathematics and computer science II, Fundamental contributions in the Netherlands since 1945*, CWI Monograph 4, pp. 95-104, North-Holland, Amsterdam, 1986.
9. C. Maire. Finitude de tours et p-pours T-ramifiées moderées, S-décomposées. *J. de Théorie des Nombres de Bordeaux*, **8** (1996), pp. 47-73.
10. B. A. Marcus. *Number Fields*. Universitext, Springer-Verlag.
11. J. Neukirch. *Algebraic Number Theory*. Grundlehren der mathematischen Wissenschaften, Volume 322, Springer-Verlag, 1999.
12. A. M. Odlyzko. Bounds for discriminants and related estimates for class numbers, regulators and zeros of zeta functions: a survey of recent results. *Séminaire de Théorie des Nombres, Bordeaux*, pp. 1-15, 1989.
13. P. Roquette. On class field towers. In: Algebraic Number Theory (Eds. J.W.S. Cassels, A. Frölich). Acad. Press, 1967, pp. 231-249.
14. M. A. Shokrollahi and H. Wasserman. List decoding of algebraic-geometric codes. *IEEE Trans. on Information Theory*, Vol. 45, No. 2, March 1999, pp. 432-437.
15. M. A. Tsfasman. Global fields, codes and sphere packings. *Journees Arithmetiques 1989, - "Asterisque"*, 1992, v.198-199-200, pp.373-396.
16. M. A. Tsfasman and S. G. Vlădut. *Asymptotic properties of global fields and generalized Brauer-Siegel Theorem.* Preprint IML-98-35 (Marseille), 1998
17. M. A. Tsfasman, S. G. Vlădut and T. Zink. Modular curves, Shimura curves, and codes better than the Varshamov-Gilbert bound. *Math. Nachrichten*, 109:21-28, 1982.

On Generalized Hamming Weights
for Codes over Finite Chain Rings

Hiroshi Horimoto[1] and Keisuke Shiromoto[2*]

[1] Kumamoto National College of Technology
2659-2, Suya, Nishigoshi, Kikuchi, Kumamoto, 861-1102, Japan
[2] Department of Mathematics, Kumamoto University
2-39-1, Kurokami, Kumamoto, 860-8555, Japan
keisuke@math.sci.kumamoto-u.ac.jp

Abstract. In this paper, we introduce the generalized Hamming weights with respect to rank (GHWR), from a module theoretical point of view, for linear codes over finite chain rings. We consider some basic properties of GHWR.

1 Introduction

For an $[n, k]$ code C over a finite field $\mathbb{F}_q$ and $1 \leq r \leq k$, the rth *generalized Hamming weight* (GHW) $d_r(C)$ of C is defined by Wei ([10]) as follows:

$$d_r(C) := \min\{|\mathrm{Supp}(D)| : D \text{ is a } [n, r] \text{ subcode of } C\},$$

where $\mathrm{Supp}(D) := \cup_{\boldsymbol{x} \in D} \mathrm{supp}(\boldsymbol{x})$ and $\mathrm{supp}(\boldsymbol{x}) := \{i \mid x_i \neq 0\}$ for $\boldsymbol{x} = (x_1, \ldots, x_n) \in \mathbb{F}_q^n$. A lot of papers dealing with GHW for codes over finite fields have been published (see [9] etc.).

On the other hand, in the last few years, linear codes over finite rings have been in the focus of the coding research (see [3], [5], [6], [7] and [11], etc.). In particular, Ashikhmin, Yang, Helleseth *et al.* ([1], [12], [13] and [4]) introduced the rth *generalized Hamming weight with respect to order* (GHWO) $d_r(C)$ for a linear code C of length n over $\mathbb{Z}_4$ and $1 \leq r \leq \log_4 |C|$ as follows:

$$d_r(C) := \min\{|\mathrm{Supp}(D)| : D \text{ is a submodule of } C \text{ with } \log_4 |D| = r\}.$$

And they exactly determined $d_r(C)$ of Preparata, Kerdock, Goethals codes *et al.* over $\mathbb{Z}_4$ for some r.

In this paper, we shall introduce a concept of *rank* for linear codes over finite chain rings and consider some fundamental properties of a generalized Hamming weight with respect to rank for these codes.

In this paper, all rings are assumed to be finite and associative with $1 \neq 0$. In any module, 1 is assumed to act as the identity.

---

* Research Fellow of the Japan Society for the Promotion of Science.

S. Boztaş and I.E. Shparlinski (Eds.): AAECC-14, LNCS 2227, pp. 141–150, 2001.
© Springer-Verlag Berlin Heidelberg 2001

2   Codes over Finite Chain Rings

A finite ring R with Jacobson radical $J(R) \neq 0$ is called a *chain ring* if the principal left ideals of R form a chain (see [8] and [5]). We remark that a finite chain ring R can be viewed as a local ring with $J(R) = R\theta$ for any $\theta \in J(R) \backslash J(R)^2$. For example, the ring $\mathbb{Z}/q\mathbb{Z}$ of integers module q, where q is a prime power, the Galois ring $GR(q, m)$ of characteristic q with q^m elements and $\mathbb{F}_2 + u\mathbb{F}_2$ $(u^2 = 0)$ are chain rings. On the other hand, $\mathbb{Z}/k\mathbb{Z}$, where k is not a prime power, and $\mathbb{F}_2 + v\mathbb{F}_2$ $(v^2 = v)$ are not chain rings. Let m be the index of nilpotency of $J(R)$ and let R^* be the group of units of R. In addition, since R is a local ring, we denote by a prime power q the cardinality of the finite field $R/J(R)$, that is, $R/J(R) \cong \mathbb{F}_q$ and $|R| = q^m$. Let R^n be the free R-module of rank n consisting of all n-tuples of elements of R. With respect to component-wise addition and right/left multiplication, R^n has the structure of an (R, R)-bimodule. A *right* (resp., *left*) *linear code* C of length n over R is a right (resp., left) R-submodule of R^n. If C is a free R-submodule of R^n, then we shall call C a *free code*. For a right (left) linear code C over R, we define the *rank* of C, denoted by rank(C), as the minimum number of generators of C and define the *free rank* of C, denoted by frank(C), as the maximum rank of the free R-submodules of C. In this case, C is isomorphic, as an R-module, to a direct sum:

$$C \cong \bigoplus_{i=0}^{m-1} (R/R\theta^{m-i})^{k_i},$$

where $R\theta^i := \{r\theta^i \mid r \in R\} = \{x \in R \mid x\theta^{m-i} = 0\}$, for each $i \in \{0, 1, \ldots, m-1\}$. We note that rank($C$) $= \sum_{i=0}^{m-1} k_i$ and frank(C) $= k_0$, and define the *type* of C, denoted by type(C), as the sequence $(k_0, k_1, \ldots, k_{m-1})$.

For an R-module M, the *socle* of M, that is, the sum of all simple submodules of M, is denoted by Soc(M). For a right (resp., left) linear code C over R, we note that

$$\text{Soc}(C) = \{\boldsymbol{x} \in C \mid \boldsymbol{x}\theta = \boldsymbol{0}\}$$
$$(\text{resp., Soc}(C) = \{\boldsymbol{x} \in C \mid \theta\boldsymbol{x} = \boldsymbol{0}\}).$$

For a right (left) linear code C over R, we define $I(C)$ as a minimal free R-submodule of R^n which contains C and define $F(C)$ as a maximal free R-submodule of C. If C is a right (resp., left) linear code of length n over R, then $I(C)$ is a right (resp., left) free code of length n with rank($I(C)$) $=$ rank(C) and $F(C)$ is a right (resp., left) free code of length n with rank($F(C)$) $=$ frank(C) (cf. [7]).

For a vector $\boldsymbol{x} = (x_1, \ldots, x_n) \in R^n$, the *support* of $\boldsymbol{x}$ is defined by

$$\text{supp}(\boldsymbol{x}) := \{i \mid x_i \neq 0\}$$

and the *Hamming weight* wt($\boldsymbol{x}$) of $\boldsymbol{x}$ is defined to be the order of the support of $\boldsymbol{x}$. The *minimum Hamming weight* of a linear code C of length n over R is

$$d(C) := \min\{\text{wt}(\boldsymbol{x}) \mid (\boldsymbol{0} \neq)\boldsymbol{x} \in C\}.$$

If $\mathrm{Soc}(R) \cong R/J(R)$ as right R-modules and as left R-modules, then R is called as a *Frobenius ring* ([8], [7] and [11]). Since a chain ring R is a Frobenius ring, we have an R-isomorphism $\phi : \mathrm{Soc}(R) \cong R/J(R)$. In this case, ϕ induces the following R-isomorphism:

$$\phi^n : \mathrm{Soc}(R)^n \cong (R/J(R))^n$$
$$: \boldsymbol{x} = (x_1, \ldots, x_n) \mapsto \phi^n(\boldsymbol{x}) = (\phi(x_1), \ldots, \phi(x_n)),$$

(cf. [8] and [7]). We have the following proposition.

Proposition 2.1 ([7]). *If C is a right (left) linear code of length n over R, then $\phi^n(\mathrm{Soc}(C))$ is a linear $[n, \mathrm{rank}(C), d(C)]$ code over the finite field $R/J(R)$.*

For two vectors $\boldsymbol{x} = (x_1, \ldots, x_n) \in R^n$ and $\boldsymbol{y} = (y_1, \ldots, y_n) \in R^n$, we define the *inner product*

$$\langle \boldsymbol{x}, \boldsymbol{y} \rangle := x_1 y_1 + \cdots + x_n y_n.$$

For a subset $C \subseteq R^n$, we define the *right dual code* $C^\perp$ and the *left dual code* $^\perp C$ of C as follows:

$$C^\perp := \{ \boldsymbol{y} \in R^n \mid \langle \boldsymbol{x}, \boldsymbol{y} \rangle = 0, \forall \boldsymbol{x} \in C \}$$
$$^\perp C := \{ \boldsymbol{y} \in R^n \mid \langle \boldsymbol{y}, \boldsymbol{x} \rangle = 0, \forall \boldsymbol{x} \in C \}.$$

If C is a right (resp., left) linear code of length n over R, then

$$\mathrm{rank}(C) + \mathrm{frank}(^\perp C) = n$$
$$(\text{resp.,}\ \mathrm{rank}(C) + \mathrm{frank}(C^\perp) = n)$$

and $(^\perp C)^\perp = C$ (resp., $^\perp(C^\perp) = C$) (cf. [5] and [7]).

A generator matrix of a right (resp., left) linear code C of length n over R is a $\mathrm{rank}(C) \times n$ matrix over R whose rows form a minimal set of generators of C. Similarly, a *parity check matrix* of C is an $n \times (n - \mathrm{frank}(C))$ matrix over R whose columns form a minimal set of generators of $^\perp C$ (resp., $C^\perp$).

In the remaining part of this paper, we shall concentrate on right linear codes because all results and proofs for left linear codes always go through as well as those for right linear codes.

3   Generalized Hamming Weights

For a subset $C \subseteq R^n$, we define the *support* of C by

$$\mathrm{Supp}(C) := \bigcup_{\boldsymbol{x} \in C} \mathrm{supp}(\boldsymbol{x}).$$

Evidently we note that if C_1 and C_2 are subsets of R^n such that $C_1 \subseteq C_2$, then $|\mathrm{Supp}(C_1)| \leq |\mathrm{Supp}(C_2)|$.

Definition 3.1. For a right linear code C of length n over R and $1 \leq r \leq \mathrm{rank}(C)$, the rth *generalized Hamming weight with respect to rank* (GHWR) of C is defined by

$$d_r(C) := \min\{ |\mathrm{Supp}(D)| : D \text{ is an } R\text{-submodule of } C \text{ with } \mathrm{rank}(D) = r \}.$$

The *weight hierarchy* of C is the set of integers $\{ d_r(C) : 1 \leq r \leq \mathrm{rank}(C) \}$

Example 3.2. Let C be a linear code over $\mathbb{Z}_4$ with generator matrix

$$G = \begin{pmatrix} 1\,0\,0\,1\,1\,2\,3 \\ 0\,2\,0\,2\,2\,2\,0 \\ 0\,0\,2\,2\,0\,2\,2 \end{pmatrix}.$$

Then $d_1(C) = 4, d_2(C) = 6$ and $d_3(C) = 7$.

The following lemma is essential.

Lemma 3.3. *If C is a right (left) linear code of length n over R, then*

$$\mathrm{Soc}(C) = \mathrm{Soc}(I(C)).$$

Proof. Evidently, $\mathrm{Soc}(C) \subseteq \mathrm{Soc}(I(C))$. From Proposition 2.1, both of them have the same order. The lemma follows.   □

The following result is a generalization of Proposition 2.1 with respect to GHWR.

Theorem 3.4. *Let C be a right linear code C of length n over R. Then*

$$d_r(C) = d_r(\mathrm{Soc}(C)) = d_r(I(C)),$$

for any r, $1 \le r \le \mathrm{rank}(C)$.

Proof. For any r, $1 \le r \le \mathrm{rank}(C)$, let D_r be a R-submodule of C with $\mathrm{rank}(D_r) = r$ and $|\mathrm{Supp}(D_r)| = d_r(C)$. Since $\mathrm{Soc}(D_r)$ is also an R-submodule of D_r and C, and $\mathrm{rank}(\mathrm{Soc}(D_r)) = r$, we have

$$d_r(C) \le |\mathrm{Supp}(\mathrm{Soc}(D_r))| \le |\mathrm{Supp}(D_r)| = d_r(C).$$

By Lemma 3.3, the second equality in the theorem follows from the first one.   □

Remark 3.5. The above theorem also claims that all free R-submodules of R^n which contain C and have the same rank as C have the same weight hierarchy as C.

Example 3.6. Let $\mathcal{P}_m$ be the *Preparata* code of length 2^m over $\mathbb{Z}_4$ with parity check matrix

$$H = \begin{pmatrix} 1\,1\,1\ \ 1 & \cdots & 1 \\ 0\,1\,\beta\,\beta^2 & \cdots & \beta^{2^m-2} \end{pmatrix},$$

where β is a unit of order $2^m - 1$ in the Galois ring $GR(4, m)$ of characteristic 4 with 4^m elements (cf. [3], [13], etc.). Then it is well-known that $\phi^{2^m}(\mathrm{Soc}(\mathcal{P}_m))$ is the extended binary Hamming code $\mathcal{H}_m$ ([3]). And the weight hierarchy of $\mathcal{H}_m$ is found in [10]. Thus we have the following:

$$\begin{aligned} \{d_r(\mathcal{P}_m) : \ &1 \le r \le 2^m - m - 1\} \\ = \{d_r(\mathcal{H}_m) : \ &1 \le r \le 2^m - m - 1\} \\ = \{2, 3, \ldots, 2^m\} &\setminus \{2^s + 1 : \ 0 \le s \le m - 1\}. \end{aligned}$$

Using the above theorem, we have the following results from Theorem 1 and Corollary 1 in [10].

Corollary 3.7. *For a right linear code C of length n over R with* $\mathrm{rank}(C) = k > 0$,

$$1 \leq d_1(C) < d_2(C) < \cdots < d_k(C) \leq n.$$

Corollary 3.8. *For a right linear code C of length n over R and any r, $1 \leq r \leq \mathrm{rank}(C)$,*

$$d_r(C) \leq n - \mathrm{rank}(C) + r.$$

If C meets the above bound, i.e., $d_r(C) = n - \mathrm{rank}(C) + r$, then C is called an rth *MDS code* over R. In [2] and [7], the first MDS codes over the finite rings (simply called MDR or MDS codes in these papers) are studied. In particular, the code considered in Example 3.2 is a second MDS code and so is a third MDS code over $\mathbb{Z}_4$.

For a right linear code C of length n and $M \subseteq N := \{1, 2, \ldots, n\}$, we set

$$R^n(M) := \{\boldsymbol{x} \in R^n \mid \mathrm{supp}(\boldsymbol{x}) \subseteq M\}$$
$$C(M) := C \cap R^n(M) = \{\boldsymbol{x} \in C \mid \mathrm{supp}(\boldsymbol{x}) \subseteq M\}.$$

Clearly, $R^n(M)$ is a free R-module of rank $|M|$ and $C(M)$ is also a right linear code of length n over R. And for right linear codes C and D over R and a linear map $\psi : C \to D$, we define

$$C^* := \mathrm{Hom}_R(C, R)$$
$$\psi^* \ : \ D^* \to C^*$$
$$: \ g \mapsto g\psi.$$

Moreover, there is the following isomorphism as left R-modules:

$$f : R^n \to (R^n)^*$$
$$: \boldsymbol{x} \mapsto (f(\boldsymbol{x}) : \boldsymbol{y} \mapsto \langle \boldsymbol{x}, \boldsymbol{y} \rangle).$$

Then the following proposition is essential.

Proposition 3.9 ([6]). *Let C be a right linear code of length n over R. Then the sequence*

$$0 \longrightarrow {}^\perp C(N - M) \xrightarrow{\mathrm{inc}} R^n(N - M) \xrightarrow{f} C^* \xrightarrow{\mathrm{res}} C(M)^* \longrightarrow 0$$

is exact as left R-modules for any $M \subseteq N$, where the maps inc, res denote the inclusion map, the restriction map, respectively.

In [6], they proved the Singleton type bound for codes over finite quasi-Frobenius rings by using this proposition. In this paper, we prove a duality for GHWR of codes over finite chain rings using this proposition.

Lemma 3.10. *Let C be a right linear code of length n over R. Then*

$$d_r(C) = \min\{|M| : \ \mathrm{rank}(C(M)) \geq r, \ M \subseteq N\},$$

for all r, $1 \leq r \leq \mathrm{rank}(C)$.

Proof. For any R-submodule D of C, we note that

$$|\text{Supp}(D)| = \min\{|M| : \ D \subseteq C(M), \ M \subseteq N\}.$$

The lemma follows from the above equality, immediately.   □

Lemma 3.11. *Let $\mathcal{A}$, $\mathcal{B}$, $\mathcal{C}$ and $\mathcal{D}$ be left R-modules and assume that the sequence*

$$0 \to \mathcal{A} \xrightarrow{\alpha} \mathcal{B} \xrightarrow{\beta} \mathcal{C} \xrightarrow{\gamma} \mathcal{D} \to 0$$

is exact as left R-modules. Then we have an isomorphism:

$$(\mathcal{B}/\alpha(\mathcal{A}))^* \simeq \mathcal{C}^*/\gamma^*(\mathcal{D}^*).$$

Moreover if $\mathcal{B}$ and $\mathcal{C}$ are free, then

$$\text{rank}(\mathcal{A}) + \text{rank}(\mathcal{C}) = \text{rank}(\mathcal{B}) + \text{rank}(\mathcal{D}). \tag{1}$$

Proof. By the assumption we have the short exact sequence :

$$0 \to \mathcal{B}/\alpha(\mathcal{A}) \xrightarrow{\overline{\beta}} \mathcal{C} \xrightarrow{\gamma} \mathcal{D} \to 0$$

where $\overline{\beta} : b + \alpha(\mathcal{A}) \mapsto \beta(b)$. From the injectivity of $_R R$, we have the dual sequence :

$$0 \to \mathcal{D}^* \xrightarrow{\gamma^*} \mathcal{C}^* \xrightarrow{\overline{\beta}^*} (\mathcal{B}/\alpha(\mathcal{A}))^* \to 0.$$

Thus we have the following isomorphism:

$$(\mathcal{B}/\alpha(\mathcal{A}))^* \simeq \mathcal{C}^*/\gamma^*(\mathcal{D}^*).$$

We suppose that $\mathcal{B}$ and $\mathcal{C}$ are free. Since R is a chain ring, the types of quotient modules $\mathcal{B}/\alpha(\mathcal{A})$ and $\mathcal{C}^*/\gamma^*(\mathcal{D}^*)$ only depend on the types of their submodules $\alpha(\mathcal{A})$ and $\gamma^*(\mathcal{D}^*)$ respectively. Therefore we have the equation (1).   □

A duality for GHW of codes over finite fields is proved in [10] and similarly, a duality for GHWO of codes over Galois rings is proved in [1]. As in these case, we have a similar duality relation for GHWR of codes over finite chain rings as follows:

Theorem 3.12. *Let C be a right linear code of length n over R with $\text{rank}(C) = k$. Then*

$$\{d_r(C) : \ 1 \leq r \leq k\} = \{1, 2, \ldots, n\} \backslash \{n + 1 - d_r(F(^{\perp}C)) : \ 1 \leq r \leq n - k\}.$$

Proof. Since $^{\perp}I(C)$ is a left free R-submodule of $^{\perp}C$ and $\text{rank}(^{\perp}I(C)) = n - k = \text{frank}(^{\perp}C)$, we can take $F(^{\perp}C) = {}^{\perp}I(C)$. Conversely, if we take any $F(^{\perp}C)$, then we can take $I(C)$ such that $I(C) = F(^{\perp}C)^{\perp}$. It is sufficient to prove $d_r(C) \neq n + 1 - d_{r'}(F(^{\perp}C))$ for any $1 \leq r \leq k$ and $1 \leq r' \leq n - k$.

For any r, $1 \leq r \leq n - k$, we set $t = k + r - d_r(F(^{\perp}C))$. First, we prove that $d_t(C) < n + 1 - d_r(F(^{\perp}C))$. From Lemma 3.10, let M be a subset of N with

$|N - M| = d_r(F(^{\perp}C))$ and $\mathrm{rank}(F(^{\perp}C)(N - M)) \geq r$. Combining Proposition 3.9 for $I(C)$ with Lemma 3.11, we have

$$\mathrm{rank}(I(C)(M)) = \mathrm{rank}(I(C)) + \mathrm{rank}(F(^{\perp}C)(N - M)) - \mathrm{rank}(R^n(N - M))$$
$$\geq k + r - d_r(F(^{\perp}C)).$$

From Lemma 3.10, the following inequality follows:

$$d_t(C) = d_t(I(C)) \leq |M| = n - d_r(F(^{\perp}C)) < n + 1 - d_r(F(^{\perp}C)).$$

By using Corollary 3.7, so we have

$$d_1(C) < \cdots < d_{t-1}(C) < d_t(C) < n + 1 - d_r(F(^{\perp}C)).$$

Next, we show that $d_{t+\Delta}(C) \neq n+1-d_r(F(^{\perp}C))$ for any $\Delta \geq 1$. We assume that $d_{t+\Delta}(C) = n + 1 - d_r(F(^{\perp}C))$ for some $\Delta \geq 1$. Let M be a subset of N with $|M| = d_{t+\Delta}(I(C))$ and $\mathrm{rank}(I(C)(M)) \geq t+\Delta$. Then we have the following equation by using Proposition 3.9 for $I(C)$ and Lemma 3.11:

$$\mathrm{rank}(F(^{\perp}C)(N - M)) = \mathrm{rank}(I(C)(M)) + |N - M| - \mathrm{rank}(I(C))$$
$$\geq (t + \Delta) + (n - d_{t+\Delta}(I(C))) - k$$
$$= r + \Delta - 1$$

From Lemma 3.10, the following inequality follows:

$$d_s(F(^{\perp}C)) \leq |N - M| = n - d_{t+\Delta}(I(C)) = d_r(F(^{\perp}C)) - 1,$$

where $s = r + \Delta - 1$, contradicting Corollary 3.7. We complete the proof.   □

Example 3.13. Let C be the linear code defined in Example 3.2. Then the dual code $C^{\perp}$ has a generator matrix (cf. [3]):

$$G^{\perp} = \begin{pmatrix} 3\,1\,1\,1\,0\,0\,0 \\ 3\,1\,0\,0\,1\,0\,0 \\ 2\,1\,1\,0\,0\,1\,0 \\ 1\,0\,1\,0\,0\,0\,1 \\ 0\,2\,0\,0\,0\,0\,0 \\ 0\,0\,2\,0\,0\,0\,0 \end{pmatrix}.$$

So $F(C^{\perp})$ is a linear code having a generator matrix:

$$G_F^{\perp} = \begin{pmatrix} 3\,1\,1\,1\,0\,0\,0 \\ 3\,1\,0\,0\,1\,0\,0 \\ 2\,1\,1\,0\,0\,1\,0 \\ 1\,0\,1\,0\,0\,0\,1 \end{pmatrix}.$$

By calculating, we have $\{d_r(F(C^{\perp})) : r = 1, 2, 3, 4\} = \{3, 5, 6, 7\}$ and $\{d_r(C) : r = 1, 2, 3\} \cup \{8 - d_r(F(C^{\perp})) : r = 1, 2, 3, 4\} = \{1, 2, \ldots, 7\}$.

Though we have many possibilities of taking $F(C)$ for a right linear code C, the following result follows from the above theorem.

Corollary 3.14. *If C is a right linear code of length n over R with $\mathrm{frank}(C) = k_0$, then all right free R-submodules of C with rank k_0 have the same weight hierarchy determined by that of $^\perp C$.*

Now we introduce a weight for a vector in R^n which is a generalization of the Lee weight for a vector in $\mathbb{Z}_4^n$. For an element $(0 \neq) x \in R$, we define the *socle weight* $s(x)$ of $(0 \neq)x \in R$ as follows:

$$s(x) = \begin{cases} q - 1 & (x \notin \mathrm{Soc}(R)) \\ q & (x \in \mathrm{Soc}(R)) \end{cases},$$

and set $s(0) = 0$. For example, if $R = \mathbb{Z}_{27} = \{0, 1, 2, \ldots, 26\}$, then $s(x) = 2$ for $x \neq 0, 9, 18$ and $s(x) = 3$ for $x = 9, 18$. For a vector $\boldsymbol{x} = (x_1, \ldots, x_n) \in R^n$, the *socle weight* $w_S(\boldsymbol{x})$ of $\boldsymbol{x}$ is defined by

$$w_S(\boldsymbol{x}) := \sum_{j=1}^{n} s(x_j).$$

For a right linear code C of length n over R, the *minimum socle weight* $d_S(C)$ of C is defined as follows:

$$d_S(C) := \min\{w_S(\boldsymbol{x}) \mid (\boldsymbol{0} \neq) \boldsymbol{x} \in C\}.$$

Lemma 3.15. *Let C be a right linear code of length n and of rank k over R and let A be the $|C| \times n$ array of all codewords in C. Then each column of A corresponds to the following case: the column contains all elements of $R\theta^i$ equally often for some $i \in \{0, 1, \ldots, m-1\}$.*

Proof. Let G be a generator matrix of C. Without loss of generality, we may assume that all elements of the first column of A are in $R\theta^i$. We shall prove the lemma by induction on k.

First we prove the case $k = 1$. We set $G = (x_1, x_2, \ldots, x_n) = (\boldsymbol{x})$. Since R is a chain ring, the ideal generated by $\{x_1, \ldots, x_n\}$ is of the form $R\theta^j$, $j \leq i$. Then C is isomorphic to $R/R\theta^{m-j}$ and each row vector of A is of the form $\boldsymbol{x}(r + R\theta^{m-j})$, $r + R\theta^{m-j} \in R/R\theta^{m-j}$. Now we consider linear maps $\rho_1 : R \to R\,[r \mapsto x_1 r]$ and $\hat{\rho}_1 : R/R\theta^{m-j} \to R[r + R\theta^{m-j} \mapsto x_1 r]$. Since the map $\hat{\rho}_1$ is a homomorphism, the number of times each element $x_i r'$ occurs in the first column of A corresponds to the order $|r + \ker \hat{\rho}_1|$. Therefore each element of $R\theta^i$ occurs in the

$$|\ker \hat{\rho}_1| = |R\theta^{m-i}/R\theta^{m-j}|$$

times equally often in the first column of A.

Next we suppose that the lemma holds in the case $\mathrm{rank}(C) = k - 1$. Assume that G has a form:

$$G = \begin{bmatrix} x_{11} & x_{12} & \cdots & x_{1n} \\ x_{21} & \cdots & \cdots & x_{2n} \\ \vdots & \cdots & \cdots & \vdots \\ x_{k1} & \cdots & \cdots & x_{kn} \end{bmatrix} = \begin{bmatrix} x_{11} & \cdots & x_{1n} \\ & G' & \end{bmatrix} = \begin{bmatrix} \boldsymbol{x}_1 \\ G' \end{bmatrix},$$

where $0 \subseteq x_{11}R \subseteq x_{21}R \subseteq \cdots \subseteq x_{k1}R \subseteq R$. From the assumption, the first column of the $|C'| \times n$ array A' of all codewords of C', where C' is the linear code over R having a generator matrix G', contains all elements of $R\theta^i$ equally often. So the first column of $\boldsymbol{y} + A'$ contains all elements of $R\theta^i$ equally often for all $\boldsymbol{y} \in \boldsymbol{x}_1 R$. Since the array A has the form:

$$\begin{bmatrix} \boldsymbol{y}_1 + A' \\ \boldsymbol{y}_2 + A' \\ \vdots \end{bmatrix},$$

where $\{\boldsymbol{y}_i\}_i = \boldsymbol{x}_1 R$, the first column of A contains all elements of $R\theta^i$ equally often. $\qquad \square$

Then we have the following theorem. A similar result for Lee weights of codes over $\mathbb{Z}_4$ can be found in [12] and corresponds to the special case $R = \mathbb{Z}_4$ in the following result.

Theorem 3.16. *Let C be a right linear code of length n over R. Then we have*

$$|\mathrm{Supp}(C)| = \frac{1}{|C|(q-1)} \sum_{\boldsymbol{x} \in C} w_S(\boldsymbol{x}).$$

Proof. For each $i \in \{0, 1, \ldots, m-1\}$, let n_i be the number of columns of A in which each element of $R\theta^i$ occurs equally often. Then $\sum_{i=0}^{m-1} n_i = |\mathrm{Supp}(C)|$. Therefore we have

$$\sum_{\boldsymbol{x} \in C} w_S(\boldsymbol{x}) = n_0 |C|/|R|\{(q-1) \times (|R| - |\mathrm{Soc}(R)|) + q \times (|\mathrm{Soc}(R)| - 1)\}$$

$$+ n_1 |C|/|R\theta|\{(q-1) \times (|R\theta| - |\mathrm{Soc}(R)|) + q \times (|\mathrm{Soc}(R)| - 1)\} + \cdots$$

$$+ n_{m-1}|C|/|R\theta^{m-1}|\{q \times (|\mathrm{Soc}(R)| - 1)\}$$

$$= |C|\{n_0/q^m((q-1)(q^m - q) + q(q-1)) + n_1/q^{m-1}((q-1)(q^{m-1} - q)$$

$$+ q(q-1)) + \cdots + n_{m-1}/q \times q(q-1)\}$$

$$= |C|(q-1)(n_0 + n_1 + \cdots + n_{m-1})$$

$$= |C|(q-1)|\mathrm{Supp}(C)|. \qquad \square$$

Corollary 3.17. *If C is a right linear code of length n over R, then the rth GHWR of C, $1 \leq r \leq \mathrm{rank}(C)$, satisfies*

$$d_r(C) \geq \left\lceil \frac{(q^r - 1)d_S(C)}{q^r(q-1)} \right\rceil,$$

where $\lceil a \rceil$ denotes the smallest integer greater than or equal to a.

Proof. By Proposition 3.4, we can take an R-submodule D_r of $\mathrm{Soc}(C)$ with $d_r(C) = |\mathrm{Supp}(D_r)|$ and $\mathrm{rank}(D_r) = r$. Since D_r is an $[n, r]$ code over $R/J(R)$, we have $|D_r| = q^r$ From Theorem 3.16, we have

$$|\mathrm{Supp}(D_r)| \geq \frac{|D_r| - 1}{|D_r|(q-1)} d_S(C).$$

$\qquad \square$

References

1. A. Ashikhmin, On generalized Hamming weights for Galois ring linear codes, *Designs, Codes and Cryptography*, **14** (1998) pp. 107–126.
2. S. T. Dougherty and K. Shiromoto, MDR codes over $\mathbb{Z}_k$, *IEEE Trans. Inform. Theory*, **46** (2000) pp. 265–269.
3. A. R. Hammons, P. V. Kumar, A. R. Calderbank, N. J. A. Sloane and P. Solé, The $\mathbb{Z}_4$-linearity of Kerdock, Preparata, Goethals and related codes, *IEEE Trans. Inform. Theory*, **40** (1994) pp. 301–319.
4. T. Helleseth and K. Yang, Further results on generalized Hamming weights for Goethals and Preparata codes over $\mathbb{Z}_4$, *IEEE Trans. Inform. Theory*, **45** (1999) pp. 1255–1258.
5. T. Honold and I. Landjev, Linear codes over finite chain rings, *Electronic Journal of Combinatorics*, **7** (2000), no. 1, Research Paper 11.
6. H. Horimoto and K. Shiromoto, A Singleton bound for linear codes over quasi-Frobenius rings, *Proceedings of the 13th Applied Algebra, Algebraic Algorithms and Error-Correcting Codes* (Hawaii, 1999).
7. H. Horimoto and K. Shiromoto, MDS codes over finite quasi-Frobenius rings, submitted.
8. B. R. McDonald, *Finite rings with identity, Pure and Applied Mathematics*, **28** Marcel Dekker, Inc., New York, 1974.
9. M. A. Tsfasman and S. G. Vladut, Geometric approach to higher weights, *IEEE Trans. Inform. Theory*, **41** (1995) pp. 1564–1588.
10. V. K. Wei, Generalized Hamming weights for linear codes, *IEEE Trans. Inform. Theory*, **37** (1991)pp. 1412–1418.
11. J. A. Wood, Duality for modules over finite rings and applications to coding theory, *American journal of Mathematics*, **121** (1999), 555–575.
12. K. Yang, T. Helleseth, P. V. Kumar and A. G. Shanbhang, On the weights hierarchy of Kardock codes over $\mathbb{Z}_4$. *IEEE Trans. Inform. Theory*, **42** (1996) pp. 1587–1593.
13. K. Yang and T. Helleseth, On the weight hierarchy of Preparata codes over $\mathbb{Z}_4$, *IEEE Trans. Inform. Theory*, **43** (1997) pp. 1832–1842.

Information Rates and Weights of Codes in Structural Matrix Rings

Andrei Kelarev[1] and Olga Sokratova[2]

[1] Faculty of Science and Engineering, University of Tasmania Box 252-37
Hobart, Tasmania 7001, Australia
Andrei.Kelarev@utas.edu.au
http://www.maths.utas.edu.au/People/Kelarev/HomePage.html
[2] Institute of Computer Science, Tartu University J. Liivi 2, 50409 Tartu, Estonia
olga@cs.ut.ee
http://www.cs.ut.ee/people/sokratova/

Abstract. Several efficient error-correcting codes are ideals in certain ring constructions. We consider two-sided ideals in structural matrix rings defined in terms of directed graphs with the set of vertices corresponding to rows and columns, and with edges corresponding to nonzero entries in matrices of the ring. Formulas for Hamming weights of all ideals in structural matrix rings are found and sharp upper bounds for information rates of these ideals are given.

1 Introduction

This paper is a contribution to two directions of research: the investigation of code properties of ideals in ring constructions, and the study of structural matrix rings of graphs.

It is well known that introducing additional algebraic structure results in several advantages for coding applications. For example, linear codes are in general better than arbitrary ones, cyclic codes are better than linear codes, and some most efficient codes have been introduced as ideals of group rings. The additional algebraic structure makes it possible to use a small number of generating elements to store the whole code in computer memory, and to use these generators in faster encoding and decoding algorithms (see [7]). These circumstances have motivated serious attention of several authors to considering ideals of various ring constructions from the point of view of coding applications. We refer to the recent survey [6] and books [5], [8], [9] for earlier results on this topic.

The second direction deals with graphs and their matrix rings. Throughout F is a finite field, the word *graph* means a directed graph without multiple edges but possibly with loops, and $D = (V, E)$ stands for a graph with the set $V = \{1, 2, \ldots, n\}$ of vertices and the set E of edges. Edges of D correspond to the standard elementary matrices of the algebra $M_n(F)$ of all $(n \times n)$-matrices over F. Namely, for $(i, j) \in E \subseteq V \times V$, let $e_{(i,j)} = e_{i,j} = e_{ij}$ be the standard

S. Boztaş and I.E. Shparlinski (Eds.): AAECC-14, LNCS 2227, pp. 151–158, 2001.
© Springer-Verlag Berlin Heidelberg 2001

elementary matrix. Denote by

$$M_D(F) = \bigoplus_{w \in E} F e_w$$

the set of all matrices with nonzero entries corresponding to the edges of the graph D, and zeros in all entries for which there are no edges in D. It is well known and easy to verify that $M_D(F)$ is a subalgebra of $M_n(F)$ if and only if D saitsfies the following property

$$(x, y), (y, z) \in E \Rightarrow (x, z) \in E, \tag{1}$$

for all $x, y, z \in V$. In this case the $M_D(F)$ is called a *structural matrix ring*. Structural matrix rings have been investigated by a number of authors, and many interesting results have been obtained (see, for example, [2], [3], [4], [10], [11], [12], and the monograph [5] for details and references on this direction).

From the point of view of coding theory, the Hamming weights and information rates of ideals are of interest. Indeed, the minimum Hamming weight $w_H(C)$, i.e., the minimum number of nonzero coordinates of elements of the code C in a given basis, is important, because it gives the number of errors a code can detect or correct; and the information rate shows the ratio of the number of message digits, which form the information to be transmitted, to the number of all digits.

Various types of ideals in structural matrix rings have been explored very well in the literature (see, in particular, [10] and [11]). However, properties of ideals essential for coding applications have not been addressed yet. The aim of this note is to find Hamming weights and upper bounds on the information rates of ideals in structural matrix rings. We consider only ideals with Hamming weight greater than one, because codes with weight one cannot detect even a single error.

2   Main Theorems

Our main theorem describes the Hamming weights of all ideals in structural matrix rings (Theorems 1). We also give an exact upper bound on the information rates of ideals with given Hamming weight (Theorem 2).

A few definitions are required for these theorems. The *in-degree* and *out-degree* of a vertex $v \in V$ are defined by

$$\mathrm{indeg}\,(v) = |\{w \in V \mid (w, v) \in E\}|,$$

$$\mathrm{outdeg}\,(v) = |\{w \in V \mid (v, w) \in E\}|.$$

A vertex of D is called a *source* (*sink, isolated vertex*) if $\mathrm{indeg}\,(v) = 0$ and $\mathrm{outdeg}\,(v) > 0$ (respectively, $\mathrm{indeg}\,(v) > 0$, $\mathrm{outdeg}\,(v) = 0$, or $\mathrm{indeg}\,(v) = \mathrm{outdeg}\,(v) = 0$). Denote by $\mathrm{so}(D)$ and $\mathrm{si}(D)$ the sets of all sources and sinks of D, respectively. For each vertex $v \in V$, put

$$\mathrm{so}(v) = \{u \in \mathrm{so}(D) \mid (u, v) \in E\},$$

$$\text{si}(v) = \{u \in \text{si}(D) \mid (v, u) \in E\},$$
$$\overline{V} = V \backslash (\text{so}(D) \cup \text{si}(D)) = \{v \in V \mid \text{indeg}\,(v), \text{outdeg}\,(v) > 0\}.$$

In order to describe all pairs of the information rates and weights of all ideals, it suffices to find maximal information rates of all ideals with each value of the Hamming weight. Denote by $w_{id}(D)$ the maximum number in the set of all minimum Hamming weights of ideals of the ring $M_D(F)$.

Theorem 1. *Let $D = (V, E)$ be a graph defining a structural matrix ring $M_D(F)$. The maximum number in the set of all Hamming weights of ideals of the ring $M_D(F)$ is equal to*

$$w_{id}(D) = \max_{v \in \overline{V}}\{1, |E \cap (\text{so}(D) \times \text{si}(D))|, |\text{si}(v)|, |\text{so}(v)|\}. \qquad (2)$$

For positive integers n, d, denote by $k_d(n)$ the maximum integer k such that there exists a linear (n,k) code with minimum distance d (see [7], [8] or [9]). If there are no codes of this sort, then we put $k_d(n) = 0$.

Theorem 2. *Let $D = (V, E)$ be a graph defining a structural matrix ring $M_D(F)$. For any $1 < d \le w_{id}(D)$, all ideals of the ring $M_D(F)$ with minimum weight d have information rate at most*

$$\frac{1}{|E|} \left[\sum_{v \in \overline{V}} k_d(|\text{si}(v)|) + \sum_{v \in \overline{V}} k_d(|\text{so}(v)|) + k_d(|E \cap (\text{so}(D) \times \text{si}(D))|) \right]. \qquad (3)$$

Note that every structural matrix ring can be regarded as a semigroup ring. Let S be a finite semigroup. Recall that the *semigroup ring* $F[S]$ consists of all sums of the form $\sum_{s \in S} r_s s$, where $r_s \in F$ for all $s \in S$, with addition and multiplication defined by the rules

$$\sum_{s \in S} r_s s + \sum_{s \in S} r'_s s = \sum_{s \in S} (r_s + r'_s)s,$$

$$\left(\sum_{s \in S} r_s s \right) \left(\sum_{t \in S} r'_t t \right) = \sum_{s,t \in S} (r_s r'_t)st.$$

If S is a semigroup with zero θ, then the *contracted semigroup ring* $F_0[S]$ is the quotient ring of $F[S]$ modulo the ideal $F\theta$. Thus $F_0[S]$ consists of all sums of the form $\sum_{\theta \ne s \in S} r_s s$, and all elements of $F\theta$ are identified with zero.

A graph $D = (V, E)$ defines a structural matrix ring if and only if the set

$$S_D = \{\theta\} \cup \{e_{ij} \mid (i, j) \in E\}$$

forms a semigroup, and both of these properties are equivalent to condition (1). Then it is easily seen that the structural matrix ring $M_D(F)$ is isomorphic to the contracted semigroup ring $F_0[S_D]$. Thus our note also continues the investigation of coding properties of ideals in semigroup rings started in [1].

3   Proofs of the Main Theorems

Proof of Theorem 1. First, we show that $M_D(F)$ always has an ideal with Hamming weight given by (2). Consider all possible cases, which may occur in (2).

Case 1: $\max\{1, |E \cap (\text{so}(D) \times \text{si}(D))|, |\text{si}(v)|, |\text{so}(v)| : v \in \overline{V}\} = 1$. In this case the assertion is trivial, since the Hamming weight of the whole ring $M_D(F)$ is equal to 1.

Case 2: $\max\{1, |E \cap (\text{so}(D) \times \text{si}(D))|, |\text{si}(v)|, |\text{so}(v)| : v \in \overline{V}\} = |E \cap (\text{so}(D) \times \text{si}(D))|$. Denote by I the ideal generated in $M_D(F)$ by the element

$$x = \sum_{w \in E \cap (\text{so}(D) \times \text{si}(D))} e_w.$$

It is easily seen that

$$M_D(F) \left(\sum_{w \in E \cap (\text{so}(D) \times \text{si}(D))} F e_w \right) = \left(\sum_{w \in E \cap (\text{so}(D) \times \text{si}(D))} F e_w \right) M_D(F) = 0.$$

Therefore $I = Fx$, and so $w_H(I) = w_H(x) = |E \cap (\text{so}(D) \times \text{si}(D))|$.

Case 3: $\max\{1, |E \cap (\text{so}(D) \times \text{si}(D))|, |\text{si}(v)|, |\text{so}(v)| : v \in \overline{V}\} = |\text{si}(u)|$, for some $u \in V$. Denote by I the ideal generated in $M_D(F)$ by the element

$$y = \sum_{v \in \text{si}(u)} e_{(u,v)}.$$

By the definition of $\text{si}(u)$, we get $y M_D(F) = 0$. Hence

$$I = F \sum_{v \in \text{si}(u)} e_{(u,v)} + \sum_{(u_1,u) \in E} \left(F \sum_{v \in \text{si}(u)} e_{(u_1,v)} \right).$$

Every edge of D occurs at most once in all sums of this expression. Therefore the Hamming weight of I is equal to $|\text{si}(u)|$.

Case 4: $\max\{1, |E \cap (\text{so}(D) \times \text{si}(D))|, |\text{si}(u)|, |\text{so}(u)| : u \in \overline{V}\} = |\text{so}(v)|$, for some $v \in V$. Denote by I the ideal generated in $M_D(F)$ by the element

$$z = \sum_{u \in \text{so}(v)} e_{(u,v)}.$$

The definition of $\text{so}(v)$ yields $M_D(F)z = 0$. Therefore

$$I = F \sum_{u \in \text{so}(v)} e_{(u,v)} + \sum_{(v,v_1) \in E} \left(F \sum_{u \in \text{so}(v)} e_{(u,v_1)} \right).$$

Every edge of D occurs at most once in all sums of this expression. It follows that the Hamming weight of I is equal to $|\text{so}(v)|$.

Thus in all the cases $M_D(F)$ has an ideal with Hamming weight given by (2).

Next, we take any ideal I of $M_D(F)$, and show that it has Hamming weight at most (2). Obviously, we can assume that $I \neq 0$. Choose a nonzero element

$$x = \sum_{(u,v) \in E} x_{(u,v)} e_{(u,v)} \in I,$$

where $x_{(u,v)} \in F$. The following cases are possible:

Case 1: $x_{(u,v)} \neq 0$ for some $u, v \in \overline{V}$. Since $u \in \overline{V}$, there exists u_1 such that $(u_1, u) \in E$. Similarly, $(v, v_1) \in E$ for some $v_1 \in V$. It follows that $1e_{(u_1,v_1)} = e_{(u_1,u)} x e_{(v,v_1)} \in I$. Therefore in this case the Hamming weight of I is equal to 1.

Case 2: $x_{(u,v)} = 0$ for all pairs $u, v \in \overline{V}$, but $x_{(u,v)} \neq 0$ for some $u \in \overline{V}$. Hence $v \in \mathrm{si}(u)$. Moreover, $v' \in \mathrm{si}(u)$ for all $v' \in V$ with $x_{(u,v')} \neq 0$. Since $u \in \overline{V}$, there exists u_1 such that $(u_1, u) \in E$. It follows that

$$e_{(u_1,u)} x \in \sum_{z \in \mathrm{si}(u)} F e_{(u_1,z)}.$$

Therefore the Hamming weight of I is at most $|\mathrm{si}(u)|$.

Case 3: $x_{(u,v)} = 0$ for all pairs $u, v \in \overline{V}$, but $x_{(u,v)} \neq 0$ for some $v \in \overline{V}$. We see that $u \subset \mathrm{so}(v)$. Moreover, $u' \in \mathrm{so}(v)$ for all $u' \in V$ with $x_{(u',v)} \neq 0$. Given that $v \in \overline{V}$, there exists v_1 such that $(v, v_1) \in E$. It follows that

$$x e_{(v,v_1)} \in \sum_{z \in \mathrm{so}(v)} F e_{(z,v_1)}.$$

Therefore the Hamming weight of I is at most $|\mathrm{so}(v)|$.

Case 4: $x_{(u,v)} = 0$ if $u \in \overline{V}$ or $v \in \overline{V}$. Then

$$x \in \sum_{w \in E \cap (\mathrm{so}(D) \times \mathrm{si}(D))} x_w e_w,$$

and so the Hamming weight of x is at most $|E \cap (\mathrm{so}(D) \times \mathrm{si}(D))|$.

Thus in each of these cases the Hamming weight of I does not exceed $w_{id}(D)$ given by (2). This completes the proof. $\qquad \square$

Proof of Theorem 2. Consider any ideal I of the ring $M_D(F)$, which has Hamming weight d, where $1 < d \leq w_{id}(D)$.

Every element $x \in M_D(F)$ has a unique representation in the form $x = \sum_{w \in E} x_w e_w$, where $x_w \in F$. The element x_w is called the *projection* of x on w. Let

$$I_w = \{x_w \mid x \in I\}.$$

For any $S \subseteq E$, denote by I_S the *projection* of I on S, that is the set

$$I_S = \sum_{w \in S} I_w.$$

Let supp(I) be the set of edges w such that $I_w \neq 0$.

If supp(I) $\cap \overline{V} \neq \emptyset$, then I contains an element x such that $x_{(u,v)} \neq 0$ for some $u, v \in \overline{V}$. We can choose u_1, v_1 with $(u_1, u) \in E$, $(v, v_1) \in E$, and get $1e_{(u_1,v_1)} = e_{(u_1,u)}xe_{(v,v_1)} \in I$. This contradicts our choice of d and shows that

$$\mathrm{supp}(I) \subseteq \{(v, \mathrm{si}(v)) \mid v \in \overline{V}\} \cup \{(\mathrm{so}(v), v) \mid v \in \overline{V}\} \cup (E \cap (\mathrm{so}(D) \times \mathrm{si}(D))). \quad (4)$$

Suppose to the contrary that the information rate of I exceeds (3). Then it follows from (4) that one of the following cases occurs.

Case 1: $|\mathrm{supp}(I) \cap \{(u,v) \mid v \in \mathrm{si}(u)\}| \geq k_d(|\mathrm{si}(u)|)$, for some $u \in \overline{V}$. Putting

$$S = \{(u,v) \mid v \in \mathrm{si}(u)\},$$

we get $\dim(I_S) > k_d(|\mathrm{si}(u)|)$. Therefore $w_H(I_S) < d$.

Take any element $z \in I_S$. There exists $x \in I$ such that $z = x_S$. Since $u \in \overline{V}$, there exists $u_0 \in V$ such that $(u_0, u) \in E$. Clearly, the Hamming weight of x_S is equal to the Hamming weight of $e_{(u_0,u)}x_S = e_{(u_0,u)}x \in I$. Therefore $w_H(I) \leq w_H(I_S) < d$. This contradiction shows that the first case is impossible.

Case 2: $|\mathrm{supp}(I) \cap \{(u,v) \mid u \in \mathrm{so}(v)\}| \geq k_d(\mathrm{so}(v))$, for some $v \in \overline{V}$. Putting

$$S = \{(u,v) \mid u \in \mathrm{so}(v)\},$$

we get $\dim(I_S) > k_d(|\mathrm{so}(v)|)$. Hence $w_H(I_S) < d$.

For each $z \in I_S$, there exists $x \in I$ such that $z = x_S$. Since $v \in \overline{V}$, there exists $v_1 \in V$ such that $(v, v_1) \in E$. The Hamming weight of x_S is equal to the Hamming weight of $x_S e_{(v,v_1)} = xe_{(v,v_1)} \in I$. Hence $w_H(I) \leq w_H(I_S) < d$. This contradiction shows that the second case is impossible, either.

Case 3: $|\mathrm{supp}(I) \cap (\mathrm{so}(D) \times \mathrm{si}(D))| \geq k_d((\mathrm{so}(D) \times \mathrm{si}(D)))$. Then it follows that $\dim(I \cap \sum_{v \in \overline{V}} Fe_{(\mathrm{so}(v),v)}) > k_d(|\mathrm{so}(v)|)$; whence $w_H(I) \leq w_H(I \cap \sum_{v \in \overline{V}} Fe_{(\mathrm{so}(v),v)}) < d$. This contradiction completes the proof.   □

4   Special Cases

A *tournament* G is a graph such that, for all distinct $u, v \in G$, either $(u, v) \in E(G)$ or $(v, u) \in E(G)$, but not both. The following proposition shows that for many graphs D the bound (3) is exact.

Proposition 1. *Let $D = (V, E)$ be a graph defining a structural matrix ring $M_D(F)$, and such that every connected component of the subgraph induced in D by the set*

$$\{u \in \overline{V} \mid \mathrm{si}(u) \neq \emptyset\} \quad (5)$$

is a tournament, and every connected component of the subgraph induced in D by the set

$$\{v \in \overline{V} \mid \mathrm{so}(v) \neq \emptyset\} \quad (6)$$

is a tournament. Then, for every $1 < d \leq w_{id}(D)$, the structural matrix ring $M_D(F)$ has an ideal with Hamming weight d and information rate given by (3).

Proof. Let C be a connected component of the subgraph induced in D by the set (5). Every tournament satisfying condition (1) is *acyclic*, i.e., it has no directed cycles. It is well known that every acyclic graph can be topologically ordered. This means that we may reorder the vertices $\{u_1, u_2, \ldots, u_k\}$ of the tournament C so that it has an edge (u_i, u_j) if and only if $i > j$. Put $S_i = \mathrm{si}(u_i)$. It follows from condition (1) that

$$S_1 \subseteq S_2 \subseteq \cdots \subseteq S_k.$$

By induction we can define subspaces L_i of $\sum_{v \in \mathrm{si}(u_i)} F e_{(u_i,v)}$ such that $w(L_i) = d$, $\dim(L_i) = k_d(\sum_{v \in \mathrm{si}(u_i)} F e_{(u_i,v)})$ and $L_1 \subseteq L_2 \subseteq \cdots \subseteq L_k$. Straightforward verification shows that the union of these subspaces is an ideal of $M_D(F)$.

Now, let C be a connected component of the subgraph induced in D by the set (6). Relabel the vertices $\{v_1, v_2, \ldots, v_k\}$ of the tournament C in the opposite direction so that it has an edge (v_i, v_j) if and only if $i < j$. This time we put $S_i = \mathrm{so}(v_i)$. Again it follows that

$$S_1 \subseteq S_2 \subseteq \cdots \subseteq S_k.$$

By induction we can define subspaces L_i of $\sum_{u \in \mathrm{si}(v_i)} F e_{(v_i,u)}$ such that $w(L_i) = d$, $\dim(L_i) = k_d(\sum_{u \in \mathrm{si}(v_i)} F e_{(v_i,u)})$ and $L_1 \subseteq L_2 \subseteq \cdots \subseteq L_k$. Then it is easily seen that the union of these subspaces is an ideal of $M_D(F)$.

Denote by S the sum of these ideals obtained above for all connected components C of the subgraph of D induced by the set (5), together with the sum of ideals given above by all connected components of the subgraph induced in D by the set (6). Then the sum

$$S + \sum_{w \in (E \cap (\mathrm{so}(D) \times \mathrm{si}(D)))} F e_w$$

has the required information rate given in (3). □

The following example shows that the exact values of information rates intricately depend on the structure of the graph D, and for some graphs may be less than the bound (3).

Example 1. The graph $D = (V, E)$ with the set $V = \{1, 2, 3, 4, 5, 6, 7\}$ of vertices and adjacency matrix

$$\begin{bmatrix} 0 & 0 & 0 & 0 & 0 & 0 & 0 \\ 0 & 0 & 0 & 0 & 0 & 0 & 0 \\ 0 & 0 & 0 & 0 & 0 & 0 & 0 \\ 0 & 0 & 0 & 0 & 0 & 0 & 0 \\ 1 & 1 & 1 & 0 & 0 & 0 & 0 \\ 0 & 1 & 1 & 1 & 0 & 0 & 0 \\ 1 & 1 & 1 & 1 & 1 & 1 & 1 \end{bmatrix}$$

satisfies condition (1), and so it defines a structural matrix ring $M_D(F)$. Let $d = 3$, and let $F = GF(2)$. The largest linear subspace with Hamming weight 2 in $F e_{(5,1)} + F e_{(5,2)} + F e_{(5,3)}$ is generated by $e_{(5,1)} + e_{(5,2)} + e_{(5,3)}$, and so it has dimension equal to $k_3(|\operatorname{si}(6)|)$. It contains all vectors of Hamming weight 3. Similarly, the largest linear subspace with Hamming weight 3 in $Fe_{(6,2)} + Fe_{(6,3)} + Fe_{(6,4)}$ is generated by $e_{(6,2)} + e_{(6,3)} + e_{(6,4)}$, and so it has dimension equal to $k_3(|\operatorname{si}(6)|)$, too. However, if we consider the ideal generated by these spaces, it has smaller Hamming weight. Indeed,

$$e_{(7,5)}\big(e_{(5,1)} + e_{(5,2)} + e_{(5,3)}\big) + e_{(7,6)}\big(e_{(6,2)} + e_{(6,3)} + e_{(6,4)}\big) = e_{(7,1)} + e_{(7,4)}.$$

It follows that all ideals of $M_D(F)$ with Hamming weight 3 have information rates strictly less than the upper bound given by (3).

References

1. J. Cazaran, A.V. Kelarev, *Generators and weights of polynomial codes*, Arch. Math. (Basel) 69 (1997), 479–486.
2. S. Dăscălescu, L. van Wyk, *Do isomorphic structural matrix rings have isomorphic graphs?* Proc. Amer. Math. Soc. 124 (1996), no. 5, 1385–1391.
3. S. Dăscălescu, B. Ion, C. Năstăsescu, J. Rios Montes, *Group gradings on full matrix rings*, J. Algebra 220(1999), 709–728.
4. B.W. Green, L. van Wyk, *On the small and essential ideals in certain classes of rings*, J. Austral. Math. Soc. Ser. A 46 (1989), no. 2, 262–271.
5. A.V. Kelarev, "Ring Constructions and Applications", World Scientific, in press.
6. A.V. Kelarev, P. Sole, *Error-correcting codes as ideals in group ring*, Contemporary Mathematics, **273** (2001), 11-18.
7. R. Lidl, G. Pilz, "Applied Abstract Algebra", Springer-Verlag, New York, 1998.
8. V.S. Pless, W.C. Huffman, R.A. Brualdi, "Handbook of Coding Theory", Elsevier, New York, 1998.
9. A. Poli, L. Huguet, "Error-Correcting Codes: Theory and Applications", Prentice-Hall, 1992.
10. L. van Wyk, *Matrix rings satisfying column sum conditions versus structural matrix rings*, Linear Algebra Appl. 249 (1996), 15–28.
11. L. van Wyk, *A link between a natural centralizer and the smallest essential ideal in structural matrix rings*, Comm. Algebra 27 (1999), no. 8, 3675–3683.
12. S. Veldsman, *On the radicals of structural matrix rings*. Monatsh. Math. 122 (1996), no. 3, 227–238.

On Hyperbolic Codes

Olav Geil[1] and Tom Høholdt[2]

[1] Aalborg University, Department of Mathematical Sciences,
Frederik Bajersvej 7G, DK-9220 Aalborg Ø, Denmark
olav@math.auc.dk
[2] Technical University of Denmark, Department of Mathematics,
Bldg 303, DK-2800 Lyngby, Denmark
T.Hoeholdt@mat.dtu.dk

Abstract. We give a new description of the so-called hyperbolic codes from which the minimum distance and the generator matrix are easily determined. We also give a method for the determination of the dimension of the codes and finally some results on the weight hierarchy are presented.

Keywords: Hyperbolic codes, generalized Hamming weights.

1 Introduction

In [9] Saints and Heegard considered a class of codes called hyperbolic cascaded Reed-Solomon codes which can be seen as an improvement of the generalized Reed-Muller codes $\mathrm{RM}_q(r, 2)$. The construction was further generalized by Feng and Rao in [2] to an improvement of the generalized Reed-Muller codes $\mathrm{RM}_q(r, m)$ for arbitrary m. Feng et al. also estimated the minimum distance of the new codes. The codes were further studied in [8] and [5] where the minimum distance was estimated by means of order functions and it was shown using the theory of order domains that the codes are asymptotically bad with respect to the order bound and the codes were renamed hyperbolic codes. In [3] and [4] Feng et al. used the so-called generalized Bezout's theorem to determine the minimum distance and the generalized Hamming weights of several codes and it was realized by Geil and Høholdt in [6] that these results could be obtained by using the so-called footprint from Gröbner basis theory. In this paper we use the footprint to construct a class of codes where the minimum distance is easy to determine, this is done in section 2. In section 3 we then show that these codes are actually the hyperbolic codes, thereby obtaining generator matrices of these, and give a method for the determination of the dimension. It follows that the estimation in [8] of the minimum distance of the hyperbolic codes actually gives the correct minimum distance. Section 4 is devoted to the generalized Hamming weights of the codes and section 5 is the conclusion.

S. Boztaş and I.E. Shparlinski (Eds.): AAECC-14, LNCS 2227, pp. 159–171, 2001.
© Springer-Verlag Berlin Heidelberg 2001

2   A Class of Codes with Known Minimum Distance

In this section we give a new description of a class of codes related to the polynomial ring $\mathbb{F}_q[X_1, \ldots, X_m]$, $m \geq 1$. We determine the minimum distance of the codes. The presentation of the codes relies on the Gröbner basis theoretical concept of a footprint.

Definition 1. *Assume we are given an ideal*

$$I = \langle F_1(X_1, \ldots, X_m), \ldots, F_l(X_1, \ldots, X_m) \rangle \subseteq \mathbb{F}_q[X_1, \ldots, X_m]$$

and a monomial ordering $\prec$ on the set $\mathcal{M}_m$ of monomials in the variables $X_1, \ldots, X_m$. The footprint of I with respect to $\prec$ is given by

$$\Delta_{\prec}(I) := \{M \in \mathcal{M}_m \mid M \text{ is not a leading monomial of any polynomial in } I\} \ .$$

In order to estimate/find the minimum distance of the codes that we are just about to define we will need the following result known as the footprint bound. For a proof of the theorem see [1].

Theorem 1. *Assume we are given an ideal I and a monomial ordering $\prec$ such that $\Delta_{\prec}(I)$ is a finite set. Then the size of $\Delta_{\prec}(I)$ is independent of the actual choice of $\prec$. Let $\overline{\mathbb{F}}_q$ denote the algebraic closure of $\mathbb{F}_q$. The number of common solutions in $(\overline{\mathbb{F}}_q)^m$ of $F_1(X_1, \ldots, X_m), \ldots, F_l(X_1, \ldots, X_m)$ is at most equal to $\#\Delta_{\prec}(I)$. In other words the size of the variety $\mathcal{V}_{\overline{\mathbb{F}}_q}(I)$ satisfies $\#\mathcal{V}_{\overline{\mathbb{F}}_q}(I) \leq \#\Delta_{\prec}(I)$. In particular $\#\mathcal{V}_{\mathbb{F}_q}(I) \leq \#\Delta_{\prec}(I)$ holds.*

Definition 2. *Given a polynomial ring $\mathbb{F}_q[X_1, \ldots, X_m]$ and an indexing $\mathbb{F}_q^m = \{P_1, P_2, \ldots, P_n\}$, where $n = q^m$. Consider the evaluation map*

$$ev : \begin{cases} \mathbb{F}_q[X_1, \ldots, X_m] \to & \mathbb{F}_q^n \\ F & \mapsto (F(P_1), \ldots, F(P_n)) \end{cases} \ .$$

Define the map

$$D : \begin{cases} \mathcal{M}_m \to \mathbb{N}_0 \\ M \mapsto \#\Delta_{\prec}(\langle M, X_1^q, \ldots, X_m^q \rangle) \end{cases}$$

and define the code $E(s) := Span_{\mathbb{F}_q}\{ev(M) \mid M \in \mathcal{M}_m, \ D(M) \leq s\}$.

Note that the value $D(M)$ is easily calculated. It is simply the number of monomials in $\mathcal{M}_m$ that are not divisible by any of the monomials $M, X_1^q, \ldots, X_m^q$.

Remark 1. As $ev(NX_i^q) = ev(NX_i)$ and $D(NX_i^q) \geq D(NX_i)$ for $N \in \mathcal{M}_m$ we need in the definition of $E(s)$ only consider monomials M such that $\deg_{X_i} M < q$, $i = 1, \ldots, m$. It is well-known that the restriction of ev to

$$Span_{\mathbb{F}_q}\{M \in \mathcal{M}_m \mid \deg_{X_i} M < q, \ i = 1, \ldots, m\}$$

is an isomorphism. Hence, for s large enough $E(s) = \mathbb{F}_q^n$.

Definition 3. *We define*

$$\mathcal{M}_m^{(q)}(s) := \{M \in \mathcal{M}_m \mid \deg_{X_i} < q \text{ for } i = 1, \ldots, m, \text{ and } D(M) \le s\} \ .$$

Using this notation we have

$$E(s) = \text{Span}_{\mathbb{F}_q}\{\text{ev}(M) \mid M \in \mathcal{M}_m^{(q)}(s)\} \ .$$

Example 1. In this example we construct the code $E(19)$ related to $\mathbb{F}_3[X_1, X_2, X_3]$. To this end we register in the matrices A_1, A_2 and A_3 below the values

$$\{D(M) \mid M \in \mathcal{M}_3, \deg_{X_i} M < 3, \text{ for } i = 1, 2, 3\} \ .$$

The matrices should be read as follows. The entry in position (i, j) in A_k is $D(X_1^{i-1} X_2^{j-1} X_3^{k-1})$.

$$A_1 = \begin{bmatrix} 0 & 9 & 18 \\ 9 & 15 & 21 \\ 18 & 21 & 24 \end{bmatrix} \qquad A_2 = \begin{bmatrix} 9 & 15 & 21 \\ 15 & 19 & 23 \\ 21 & 23 & 25 \end{bmatrix} \qquad A_3 = \begin{bmatrix} 18 & 21 & 24 \\ 21 & 23 & 25 \\ 24 & 25 & 26 \end{bmatrix}$$

So

$$E(19) = \text{ev}\left(\text{Span}_{\mathbb{F}_3}\{M \mid M \in \mathcal{M}_3^{(3)}(19)\}\right)$$

$$= \text{ev}\left(\text{Span}_{\mathbb{F}_3}\{1, X_1, X_1^2, X_2, X_1 X_2, X_2^2, X_3, X_1 X_3, X_2 X_3, X_1 X_2 X_3, X_3^2\}\right)$$

and $E(19)$ is of dimension $k = 11$. Obviously the code is of length $n = 3^3 = 27$.

A first study of the parameters of the codes reveals the following.

Proposition 1. *The code $E(s)$ is of length $n = q^m$ and minimum distance $d \ge q^m - s$.*

Proof. The first part is obvious. To show the last part we fix an arbitrary monomial ordering $\prec$. A code word in $E(s)$ can be written $\text{ev}(\sum_{i=1}^t \gamma_i M_i)$ where for $i = 1, \ldots, t$, $\gamma_i \in \mathbb{F}_q$, $M_i \in \mathcal{M}_m^{(q)}(s)$ holds. We may without loss of generality assume $M_t \succ M_j$ for $j = 1, \ldots t - 1$ and $\gamma_t \ne 0$. Now

$$\Delta_{\prec}(\langle \sum_{i=1}^t \gamma_i M_i, X_1^q - X_1, \ldots, X_m^q - X_m \rangle) \subseteq \Delta_{\prec}(\langle M_t, X_1^q, \ldots, X_m^q \rangle)$$

and in particular

$$\#\Delta_{\prec}(\langle \sum_{i=1}^t \gamma_i M_i, X_1^q - X_1, \ldots, X_m^q - X_m \rangle) \le s \ .$$

By the footprint bound the maximal number of positions where the codeword is equal to 0 is at most s and the bound $d \ge q^m - s$ follows.

We note that one can also prove Proposition 1 by using the tecniques in [10]. As we shall soon see we can say even more whenever s is chosen properly.

Definition 4. *Define* $S := \{D(M) \mid M \in \mathcal{M}_m, \deg_{X_i} < q,\ i = 1, \ldots, m\}$.

Theorem 2. *For any* $s' \in \mathbb{N}_0$ *there exists a unique* $s \in S$ *such that* $E(s') = E(s)$. *The minimum distance of* $E(s)$ *is given by* $d = q^m - s$.

Proof. The first part follows from the very definition of the codes in combination with Remark 1. To show the last part we need by Proposition 1 only find a code word $\mathrm{ev}(\sum_{i=1}^{t} \gamma_i M_i)$ in $E(s)$ of weight $q^m - s$. That is to say we need only find a nonzero polynomial $\sum_{i=1}^{t} \gamma_i M_i$, ($\gamma_i \in \mathbb{F}_q$, $M_i \in \mathcal{M}_m^{(q)}(s)$ for $i = 1, \ldots, t$) that possesses s different zeros in $\mathbb{F}_q^m$. We first choose any $M \in \mathcal{M}_m^{(q)}(s)$ such that $D(M) = s$. Say $M = X_1^{a_1} \cdots X_m^{a_m}$. Hence

$$s = \#\Delta_{\prec}(\langle M, X_1^q, \ldots, X_m^q \rangle) = q^m - (q - a_1)(q - a_2) \cdots (q - a_m) .$$

Index the elements of $\mathbb{F}_q$ by $\mathbb{F}_q = \{\alpha_1, \ldots, \alpha_q\}$. Consider the polynomial

$$P(X_1, \ldots, X_m) = (X_1 - \alpha_1) \cdots (X_1 - \alpha_{a_1})(X_2 - \alpha_1) \cdots (X_2 - \alpha_{a_2})$$
$$\cdots (X_m - \alpha_1) \cdots (X_m - \alpha_{a_m}) .$$

This polynomial is of the desired form. And the elements in $\mathbb{F}_q^m$ that are not a zero of P are precisely the elements of the form $(\alpha_{i_1}, \alpha_{i_2}, \ldots, \alpha_{i_m})$ with $a_1 < i_1, a_2 < i_2, \ldots, a_m < i_m$. So P has $q^m - (q - a_1)(q - a_2) \cdots (q - a_m) = s$ different zeros and we are through.

Example 2. This is a continuation of Example 1. We have $D(X_1 X_2 X_3) = 19 \in S$. Hence by Theorem 2 the minimum distance of $E(19)$ is given by $d = 3^3 - 19 = 8$.

In the following we will always assume that $s \in S$.

3   Hyperbolic Codes

In [8, p. 922] the so-called hyperbolic codes are considered.

Definition 5. *Let*

$$\mathcal{N}_m^{(q)}(s) := \{X_1^{a_1} \cdots X_m^{a_m} \in \mathcal{M}_m \mid a_i < q, i = 1, \ldots, m,\ \prod_{i=1}^{m} (a_i + 1) < q^m - s\}$$

The hyperbolic codes are now defined as follows.

Definition 6. $Hyp_q(s, m) := \{\underline{c} \in \mathbb{F}_q^n \mid \langle \underline{c}, \mathrm{ev}(M) \rangle = 0 \text{ for all } M \in \mathcal{N}_m^{(q)}(s)\}$.
Here $n = q^m$ *and* $\langle\ ,\ \rangle$ *denotes the standard inner product in* $\mathbb{F}_q^n$.

In [8] the minimum distance of these codes is estimated using the order bound. One gets $d(\text{Hyp}_q(s,m)) \geq q^m - s$. Theorem 2 and the following result proves that this estimate actually is equal to the true minimum distance of the hyperbolic code.

Theorem 3. *Consider* $\mathbb{F}_q[X_1, \ldots, X_m]$ *then* $E(s) = \text{Hyp}_q(s,m)$.

Proof. We first note that if $M \in \mathcal{M}_m^{(q)}(s)$ and $N \in \mathcal{N}_m^{(q)}(s)$ then $\langle \text{ev}(M), \text{ev}(N) \rangle = 0$. To see this let $M = X_1^{a_1} \cdots X_m^{a_m}$ and $N = X_1^{b_1} \cdots X_m^{b_m}$ then

$$\langle \text{ev}(M), \text{ev}(N) \rangle = \sum_{i_1=1}^{q} \alpha_{i_1}^{a_1+b_1} \sum_{i_2=1}^{q} \alpha_{i_2}^{a_2+b_2} \cdots \sum_{i_m=1}^{q} \alpha_{i_m}^{a_m+b_m} \tag{1}$$

where $\mathbb{F}_q = \{\alpha_1, \ldots, \alpha_q\}$. If there exists an i such that $a_i + b_i \not\equiv 0 \mod (q-1)$ then (1) is obviously zero and the same is the case if there exists an i such that $a_i + b_i = 0$, so we only have to consider the case where $a_i + b_i = q - 1$ or $2(q-1)$ for all i. Suppose $a_i + b_i = q - 1$, $i = 1, \ldots, r$ and $a_i + b_i = 2(q-1)$, $i = r+1, \ldots, m$. This of course implies that $a_i = b_i = q - 1$ for $i = r+1, \ldots, m$. Now $q^m - \prod_{i=1}^{m}(q - a_i) \leq s$ so

$$q^m - s \leq \prod_{i=1}^{m}(q - a_i) = \prod_{i=1}^{r}(b_i + 1) = q^{r-m} \prod_{i=1}^{m}(b_i + 1) < q^{r-m}(q^m - s)$$

which is a contradiction, so the case actually never occurs. This proves that $E(s) \subseteq \text{Hyp}_q(s,m)$.

On the other hand we have that

$$\dim(\text{Hyp}_q(s,m))$$

$$= q^m - \#\{(a_1, \ldots, a_m) \mid 0 \leq a_i < q, i = 1, \ldots, m, \; \prod_{i=1}^{m}(a_i + 1) < q^m - s\}$$

$$= \#\{(a_1, \ldots, a_m) \mid 0 < a_i < q, i = 1, \ldots, m, \; \prod_{i=1}^{m}(a_i + 1) \geq q^m - s\}$$

and

$$\dim(E(s))$$

$$= \#\{(a_1, \ldots, a_m) \mid 0 \leq a_i < q, i = 1, \ldots, m, \; D(X_1^{a_1} \cdots X_m^{a_m}) \leq s\}$$

$$= \#\{(a_1, \ldots, a_m) \mid 0 \leq a_i < q, i = 1, \ldots, m, \; \prod_{i=1}^{m}(q - a_i) \geq q^m - s\}$$

where $\dim(C)$ denotes the dimension of the code C. It is clear that the mapping

$$(z_1, \ldots, z_m) \mapsto (q - z_1 - 1, \ldots, q - z_m - 1)$$

is a bijection from

$$\{(a_1, \ldots, a_m) \mid 0 \leq a_i < q, i = 1, \ldots, m, \; \prod_{i=1}^{m}(q - a_i) \geq q^m - s\}$$

to

$$\{(a_1,\ldots,a_m) \mid 0 \le a_i < q, i = 1,\ldots,m, \prod_{i=1}^{m}(a_i+1) \ge q^m - s\}.$$

So the two codes have the same dimension and we have completed the proof.

Remark 2. It follows from Theorem 3 that we now have the generator matrices of the hyperbolic codes.

For $a \in \mathbb{N}$ we define

$$V(m,a) := \#\{(x_1,\ldots,x_m) \mid x_i \in \mathbb{N}, 1 \le x_i \le q, i = 1,\ldots,m, \prod_{i=1}^{m} x_i \le a\}$$

then it follows from above that

$$\dim(\mathrm{Hyp}_q(s,m)) = q^m - V(m, q^m - s - 1).$$

It is not obvious how to get a closed form expression for $V(m,a)$ but since $V(1,a) = \min\{a,q\}$ and $V(m,a) = \sum_{j=1}^{q} V(m-1, \lfloor \frac{a}{j} \rfloor)$ we can easily calculate $V(m,a)$ recursively. One can verify that $V(2,a) = bq + \sum_{j=b+1}^{\min\{a,q\}} \lfloor \frac{a}{j} \rfloor$ where $b := \min\{\lfloor \frac{a}{q} \rfloor, q\}$ and the last sum is zero if $b \ge q$. For $q = 2$ we get $V(m,a) = \sum_{j=0}^{i} \binom{m}{j}$ if $2^i \le a < 2^{i+1}$ in agreement with the fact that the classical binary Reed-Muller codes are hyperbolic codes.

The description in [8] of the hyperbolic codes is based on order domain theory. From the theory in [8] it is clear that the hyperbolic code construction is an improvement of the generalized Reed-Muller code construction in the following sense. For every generalized Reed-Muller code there exists an hyperbolic code of designed minimum distance $d^* = q^m - s$ such that d^* equals the minimum distance of the generalized Reed-Muller code (in this paper we have shown that in general the designed minimum distance d^* equals the true minimum distance of the hyperbolic code). The first code is in some cases of the same dimension as the latter, in other cases of higher dimension. Further there are many more hyperbolic codes related to $\mathbb{F}_q[X_1,\ldots,X_m]$ than there are generalized Reed-Muller codes. We illustrate these observations by an example.

Example 3. We look at the case $q = 64$ and $m = 3$. There are 190 different generalized Reed-Muller codes $\mathrm{RM}_{64}(r,3)$ and 14 224 different hyperbolic codes $\mathrm{Hyp}_{64}(s,3)$. The graphs in Figure 1 are generated by a pointplot routine. Every + corresponds to a generalized Reed-Muller code of the given parameters. The graph marked with a ∘ corresponds to the hyperbolic codes. It appears that given a generalized Reed-Muller code, then in almost all cases there are hyperbolic codes that are of larger minimum distance and are of larger dimension. This was of course one of the reasons for considering the hyperbolic codes in the first place (besides the fact that we get many more codes).

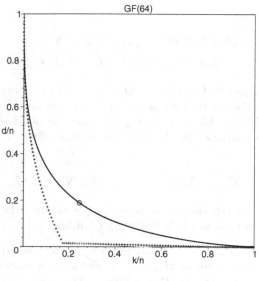

Fig. 1.

It is well-known that generalized Reed-Muller codes are asymptotically bad and it follows from [5, Corollary 2] that the hyperbolic codes are also asymptotically bad since their minimum distance as we have seen equals the order bound.

4   The Generalized Hamming Weights

In this section we are concerned with the generalized Hamming weights of the hyperbolic code. The idea of generalized Hamming weights for a linear code is to generalize the concept of the minimum distance. We have the following definition. Given

$$ U = \{u_1 = (u_{11}, \ldots, u_{1n}), \ldots, u_s = (u_{s1}, \ldots, u_{sn})\} \subseteq \mathbb{F}_q^n $$

define the support of U to be

$$ \mathrm{Supp}(U) := \{i \mid \exists u_t \in U \text{ with } u_{ti} \neq 0\} . $$

Consider a linear code C of dimension k. For $h = 1, \ldots, k$ the hth generalized Hamming weight is defined to be

$$ d_h := \min\{\#\mathrm{Supp}(U) \mid U \text{ is a linear subcode of } C \text{ of dimension } h\} . $$

The set $\{d_1, \ldots, d_k\}$ is called the weight hierarchy for C.

As we will show below the hth generalized Hamming weight of the hyperbolic code $\mathrm{Hyp}_q(m, s)$ is related to the following number.

Definition 7.

$$\eta_h(q, s, m) := \max\{\#\Delta_\prec(\langle M_1, \ldots, M_h, X_1^q, \ldots, X_m^q\rangle)$$
$$| \ M_i \neq M_j \ for \ i \neq j, \ M_i \in \mathcal{M}_m^{(q)}(s) \ for \ i = 1, \ldots, h\} \ .$$

Note that the number $\#\Delta_\prec(\langle M_1, \ldots, M_h, X_1^q, \ldots, X_m^q\rangle)$ is easily calculated. It is simply the number of monomials in $\mathcal{M}_m$ that are not divisible by any of the monomials $M_1, \ldots, M_h, X_1^q, \ldots, X_m^q$. To ease the notation we will in the remaining part of this paper use the following notation. Given $F = \{M_1, \ldots, M_i\}$ then we denote

$$\langle F, X_1^q, \ldots, X_m^q\rangle := \langle M_1, \ldots, M_i, X_1^q, \ldots, X_m^q\rangle \ .$$

To establish the correspondence between $\eta_h(q, s, m)$ and the hth generalized Hamming weight we will need the following definition.

Definition 8. For $M_1, \ldots, M_h \in \mathcal{M}_m$ where $h \geq 2$, let $\gcd(M_1, \ldots, M_h)$ denote the greatest common divisor of $M_1, \ldots, M_h$. For a single element $M_1 \in \mathcal{M}_m$ we write $\gcd(M_1) := M_1$. The set $D = \{M_1, \ldots, M_h\} \subseteq \mathcal{M}_m^{(q)}(s)$ is said to be a dense set related to $Hyp_q(s, m)$ if

$$\{X_1^{b_1} \cdots X_m^{b_m} \in \Delta_\prec(\langle D, X_1^q, \ldots, X_m^q\rangle) \ | \ a_i \leq b_i, i = 1, \ldots, m\} \subseteq \mathcal{M}_m^{(q)}(s) \ ,$$

where $X_1^{a_1} \cdots X_m^{a_m} = \gcd(M_1, \ldots, M_h)$. A set $D = \{M_1, \ldots, M_h\} \subseteq \mathcal{M}_m^{(q)}(s)$ is said to be an optimal set of size h related to $Hyp_q(s, m)$ if $M_i \neq M_j$ for $i \neq j$ and

$$\#\Delta_\prec(\langle D, X_1^q, \ldots, X_m^q\rangle) = \eta_h(q, s, m) \ .$$

Example 4. This is a continuation of Example 1 and 2. The optimal sets of size 2 corresponding to $\mathcal{M}_3^{(3)}(19)$ are $\{X_i X_j, X_1 X_2 X_3\}$, $i \neq j$. Hence $\eta_2(3, 19, 3) = 11$. All these sets are dense as for all choices of $i, j, \ i \neq j$

$$\{X_1^{b_1} X_2^{b_2} X_3^{b_3} \in \Delta_\prec(\langle X_i X_j, X_1 X_2 X_3, X_1^3, X_2^3, X_3^3\rangle) \ | \ 1 \leq b_i, 1 \leq b_j\} = \emptyset \subseteq \mathcal{M}_3^{(3)}(19)$$

holds.

Example 5. Let $M' = X_1^{a_1} \cdots X_m^{a_m} \in \mathcal{M}_m$ be any monomial. Clearly

$$\{X_1^{b_1} \cdots X_m^{b_m} \in \Delta_\prec(\langle M', X_1^q, \ldots, X_m^q\rangle) \ | \ a_i \leq b_i, i = 1, \ldots, m\}$$

is equal to the empty set. Hence, in particular any optimal set of size 1 related to $Hyp_q(s, m)$ is dense.

Having introduced the concept of a dense and optimal set related to $Hyp_q(s, m)$ we are now able to state the following theorem concerning the hth generalized Hamming weight. By Example 5 this theorem is a generalization of Theorem 2.

Theorem 4. *The hth generalized Hamming weight of $Hyp_q(s,m)$ satisfies*

$$d_h \geq q^m - \eta_h(q,s,m) . \tag{2}$$

If a dense optimal set of size h related to $Hyp_q(s,m)$ exists then equality holds in (2).

Proof. To see the first part we argue as in the proof of Proposition 1. Consider an h-dimensional subspace

$$\mathrm{Span}_{\mathbb{F}_q}\{\mathrm{ev}(\sum_{i=1}^{t_1}\gamma_{i1}M_{i1}),\ldots,\mathrm{ev}(\sum_{i=1}^{t_h}\gamma_{ih}M_{ih})\} \subseteq Hyp_q(s,m) .$$

We may without loss of generality assume that all M_{ij} belongs to $\mathcal{M}_m^{(q)}(s)$, that $M_{t_j j} \succ M_{kj}$ and $\gamma_{t_j} \neq 0$ holds for $j = 1, \ldots, h$ and $k = 1, \ldots, t_j - 1$. Further we may assume that $M_{t_1 1}, \ldots, M_{t_h h}$ are different. Now

$$\Delta_\prec((\langle\textstyle\sum_{i=1}^{t_1}\gamma_{i1}M_{i1},\ldots,\sum_{i=1}^{t_h}\gamma_{ih}M_{ih}, X_1^q - X_1,\ldots,X_m^q - X_m\rangle)$$
$$\subseteq \Delta_\prec((\langle M_{t_1},\ldots,M_{t_h}, X_1^q,\ldots,X_m^q\rangle) .$$

Hence, by the footprint bound and the definition of $\eta_h(q,s,m)$ the maximal number of common zeros of

$$\{\sum_{i=1}^{t_1}\gamma_{i1}M_{i1},\ldots,\sum_{i=1}^{t_h}\gamma_{ih}M_{ih}, X_1^q - X_1,\ldots,X_m^q - X_m\}$$

is at most $\eta_h(q,s,m)$. The bound $d_h \geq q^m - \eta_h(q,s,m)$ follows.

To prove the last part of the theorem assume next that we are given a dense optimal set $D_1 = \{M_1,\ldots,M_h\}$ related to $Hyp_q(s,m)$. From above we know

$$d_h \geq q^m - \#\Delta_\prec((\langle D_1, X_1^q,\ldots,X_m^q\rangle) . \tag{3}$$

What remains to be shown is

$$d_h \leq q^m - \#\Delta_\prec((\langle D_1, X_1^q,\ldots,X_m^q\rangle) . \tag{4}$$

Clearly (3) and (4) together proves the theorem. To establish (4) let

$$X_1^{a_1}\cdots X_m^{a_m} = \gcd(M_1,\ldots,M_h)$$

and define

$$D_2 := \{X_1^{b_1}\cdots X_m^{b_m} \in \mathcal{M}_m^{(q)}(s) \mid a_i \leq b_i \text{ for } i = 1,\ldots,m\} .$$

The proof of (4) requires three results. To show the first of these three results first observe that $D_1 \subseteq D_2$. Hence,

$$\Delta_\prec((\langle D_2, X_1^q,\ldots,X_m^q\rangle) \subseteq \Delta_\prec((\langle D_1, X_1^q,\ldots,X_m^q\rangle) .$$

Next observe that by the assumption that D_1 is dense the elements in $\Delta_\prec(\langle D_1, X_1^q, \ldots, X_m^q \rangle)$ that are not in $\Delta_\prec(\langle D_2, X_1^q, \ldots, X_m^q \rangle)$ are all in $\mathcal{M}_m^{(q)}(s)$. And finally observe that by the definition of D_2 there are precisely $\#D_2$ elements in $\mathcal{M}_q^{(m)}(s)$ that are not contained in $\Delta_\prec(\langle D_2, X_1^q, \ldots, X_m^q \rangle)$. Among these the monomials $M_1, \ldots, M_h$. From the above observations we conclude

$$\#\Delta_\prec(\langle D_1, X_1^q, \ldots, X_m^q \rangle) - \#\Delta_\prec(\langle D_2, X_1^q, \ldots, X_m^q \rangle) \leq \#D_2 - h . \tag{5}$$

This is the first result needed to prove (4). Next denote as earlier in this paper $\mathbb{F}_q = \{\alpha_1, \ldots, \alpha_q\}$. Define the map $\mathcal{P} : D_2 \to \mathbb{F}_q[X_1, \ldots, X_m]$ by

$$\mathcal{P}(X_1^{b_1} \cdots X_m^{b_m}) := (X_1 - \alpha_1) \cdots (X_1 - \alpha_{b_1})(X_2 - \alpha_1) \cdots (X_2 - \alpha_{b_2})$$
$$\cdots (X_m - \alpha_1) \cdots (X_m - \alpha_{b_m}) .$$

Now $\mathrm{Span}_{\mathbb{F}_q}\{\mathrm{ev}(\mathcal{P}(M)) \mid M \in D_2\}$ is a $\#D_2$ dimensional subspace of $\mathrm{Hyp}_q(s, m)$. By the very definition of D_2 we have $a_i \leq b_i$ for $i = 1, \ldots, m$. Therefore

$$(X_1 - \alpha_1) \cdots (X_1 - \alpha_{a_1})(X_2 - \alpha_1) \cdots (X_2 - \alpha_{a_2}) \cdots (X_m - \alpha_1) \cdots (X_m - \alpha_{a_m})$$

is a factor of all the polynomials in the image of $\mathcal{P}$. So the size of the support of the $\#D_2$ dimensional subspace $\mathrm{Span}_{\mathbb{F}_q}\{\mathrm{ev}(\mathcal{P}(M)) \mid M \in D_2\}$ is at most $(q - a_1)(q - a_2) \cdots (q - a_m)$. We get

$$\begin{aligned} d_{\#D_2} &\leq (q - a_1)(q - a_2) \cdots (q - a_m) \\ &= q^m - (q^m - (q - a_1)(q - a_2) \cdots (q - a_m)) \\ &= q^m - \#\Delta_\prec(\langle D_2, X_1^q, \ldots, X_m^q \rangle) . \end{aligned} \tag{6}$$

This is the second result needed to prove (4). The third and last result needed to prove (4) and thereby the theorem is the following well-known fact that holds for an arbitrary code C. Namely that

$$d_h \leq d_{h+1} - 1 \tag{7}$$

for any h, $1 \leq h < \dim(C)$. Using (7), (6) and (5) respectively we get

$$\begin{aligned} d_h &\leq d_{\#D_2} - (\#D_2 - h) \\ &\leq q^m - \#\Delta_\prec(\langle D_2, X_1^q, \ldots, X_m^q \rangle) - (\#D_2 - h) \\ &\leq q^m - \#\Delta_\prec(\langle D_1, X_1^q, \ldots, X_m^q \rangle) + (\#D_2 - h) - (\#D_2 - h) \\ &= q^m - \#\Delta_\prec(\langle D_1, X_1^q, \ldots, X_m^q \rangle) \end{aligned}$$

and (4) is established. The proof is completed.

Example 6. This is a continuation of Example 4. By Theorem 4 we have $d_2(E(19)) = 3^3 - \eta_2(3, 19, 3) = 27 - 11 = 16$.

Example 7. Consider the code $\mathrm{Hyp}_5(16, 2)$. The set $D_1 = \{X_2^3, X_1 X_2^2, X_1^2 X_2^2\}$ is a dense and optimal set of size 3. We have $\#\Delta_\prec(\langle D_1, X_1^5, X_2^5 \rangle) = 11$. Hence, by Theorem 4 $d_3 = 5^2 - 11 = 14$. Following the ideas of the proof of Theorem 4 we note that $\gcd(X_2^3, X_1 X_2^2, X_1^2 X_2^2) = X_2^2$ and we therefore consider $D_2 = \{X_2^2, X_2^3, X_1 X_2^2, X_1^2 X_2^2\}$. Now it can be shown that the set D_2 in the proof of Theorem 4 is in general a dense and optimal set related to $\mathrm{Hyp}_q(s, m)$. In particular the set D_2 from this example is a dense and optimal set related to $\mathcal{M}_2^{(5)}(16)$. Now $\#\Delta_\prec(\langle D_2, X_1^5, X_2^5 \rangle) = \#\Delta_\prec(\langle D_1, X_1^5, X_2^5 \rangle) - 1$ is easily seen. We conclude $d_4 = d_3 + 1 = 15$.

We give a special treatment to the second generalized Hamming weight of $\mathrm{Hyp}_q(s, 2)$.

Lemma 1. *Given a hyperbolic code $\mathrm{Hyp}_q(s, 2)$ of dimension at least 2 then there exists an optimal dense set of size 2 related to $\mathcal{M}_2^{(q)}(s)$.*

Proof. Let $O = \{X_1^a X_2^b, X_1^{a+u} X_2^{b-v}\} \subseteq \mathcal{M}_2^{(q)}(s)$ be an optimal set. We will show that either O is dense or another optimal set exists that is dense. Optimality requires that neither $-u, v < 0$ nor $-u, v > 0$ can hold. We therefore without loss of generality assume that $u, v \geq 0$. If $u = 0$ or $v = 0$ then O is dense and we are through. In the remaining part of the proof we therefore assume $u, v \geq 1$. We first assume $u = v = l$. As $X_1^{a+l} X_2^{b-l}$ is contained in $\mathcal{M}_2^{(q)}(s)$ we must have

$$1 \leq l \leq q - 1 - a . \tag{8}$$

Consider

$$\#\Delta_\prec(\langle X_1^a X_2^b, X_1^{a+l} X_2^{b-l}, X_1^q, X_2^q \rangle) = (q^2 - (q - a)(q - b)) - l(q - a - l)$$
$$= (aq + bq - ab) - (q - a)l + l^2 . \tag{9}$$

Now (9) describes a polynomial in l of degree 2 with global minimum in $l = (q - a)/2$. This (possible non integer) value of l is situated precisely in the center of the interval $[1, q - 1 - a]$. Hence, by the assumption that an optimal set $\{X_1^a X_2^b, X_1^{a+l} X_2^{b-l}\}$ exists and by (8) we conclude that $\{X_1^a X_2^b, X_1^{a+1} X_2^{b-1}\}$ is an optimal set. This set clearly is dense. Finally we assume without loss of generality that $0 < v < u$. We have

$$\#\Delta_\prec(\langle X_1^a X_2^b, X_1^{a+u} X_2^{b-v}, X_1^q, X_2^q \rangle)$$
$$= q^2 - (q - (a + u))(q - (b - v)) - u(q - b)$$
$$\leq q^2 - (q - a - u)(q - b + v) - (q - b) - v(q - b) . \tag{10}$$

Consider the dense set

$$O' := \{X_1^{a+u-1} X_2^{b-v}, X_1^{a+u} X_2^{b-v}\} \subseteq \mathcal{M}_2^{(q)}(s) .$$

We have

$$\#\Delta_\prec(\langle X_1^{a+u-1} X_2^{b-v}, X_1^{a+u} X_2^{b-v}, X_1^q, X_2^q \rangle)$$
$$= q^2 - (q - a - u)(q - b + v) - (q - b) - v$$
$$\geq q^2 - (q - a - u)(q - b + v) - (q - b) - v(q - b) \tag{11}$$

where the last inequality follows from the fact $b \leq q - 1$ (this is a necessary condition for $X_1^a X_2^b$ to be contained in $\mathcal{M}_2^{(q)}(s)$). Comparing (10) and (11) we see that if O is optimal then also is O'. The proof is completed.

Proposition 2. *The second generalized Hamming weight of a hyperbolic code* $Hyp_q(s, 2)$ *of dimension at least 2 is given by* $d_2 = q^2 - \eta_2(q, s, 2)$.

Proof. By Theorem 4 and Lemma 1.

Having a dense and optimal set of size h one might hope that it is possible to add a monomial to get a dense and optimal set of size $h + 1$. However, as the following example illustrates this strategy is not in general fruitful.

Example 8. Consider the code $Hyp_7(42, 2)$. Now X_1^6 and X_2^6 are the only monomials in $\mathcal{M}_2^{(7)}(42)$ that are mapped to 42 under the map D. That is $\{X_1^6\}$ and $\{X_2^6\}$ are the only optimal sets of size 1 related to $\mathcal{M}_2^{(7)}(42)$. We want to illustrate that if a set $\{M_1, M_2\}$ is an optimal set of size 2 related to $Hyp_7(42, 2)$ (meaning in particular that $M_1 \neq M_2$) then neither X_1^6 nor X_2^6 can belong to the set. By symmetry it is enough to consider X_1^6. To maximize $\#\Delta_\prec(\langle X_1^6, N, X_1^7, X_2^7 \rangle)$ we must choose $N = X_1^5 X_2^3$. Now $\{X_1^6, X_1^5 X_2^3\}$ is not dense. Using the construction from the proof of Lemma 1 we recognize that $\{X_1^5 X_2^2, X_1^5 X_2^3\}$ is a dense set with

$$\#\Delta_\prec(\langle X_1^5 X_2^2, X_1^5 X_2^3, X_1^7, X_2^7 \rangle) \geq \#\Delta_\prec(\langle X_1^6, X_1^5 X_2^3, X_1^7, X_2^7 \rangle) .$$

By inspection the first number is 39 and the last is 38. By further inspection we find that actually $\{X_1^5 X_2^2, X_1^5 X_2^3\}$ is optimal. We conclude that $d_2(Hyp_7(42, 2)) = 7^2 - 39 = 10$.

Remark 3. In [7] Heijnen and Pellikaan gave a general method for bounding the generalized Hamming weights for a class of duals of evaluation codes and obtained exact values for the generalized Reed-Muller codes. It seems likely that a generalization of their results to a larger class of codes would give our bound (2).

We leave it as an open problem to find the answer to the following question. For which values of q, s, m and h, $1 < h < \dim(Hyp_q(s, m))$ does $Hyp_q(s, m)$ possess a dense and optimal set of size h? We note, that the answer to this question may possible be that all choices of q, s, m and h ensures that a dense and optimal set of size h exists.

5   Conclusion

Using the concept of a footprint from Gröbner basis theory we have given a new description of the hyperbolic codes such that their minimum distance and generator matrix are easily determined. We also presented a method for the determination of the dimension of these codes and gave lower bounds on the generalized Hamming weights with a criterion for having equality in these bounds. Whether it is always possible to meet this criterion is left as an open problem.

References

1. D. Cox, J. Little, and D. O'Shea, *Ideals, Varieties, and Algorithms*, 2nd ed., Springer, Berlin 1997.
2. G.-L. Feng and T.R.N. Rao, "Improved Geometric Goppa Codes, Part I:Basic theory," *IEEE Trans. Inform. Theory*, vol. 41, pp. 1678-1693, Nov. 1995.
3. G.-L. Feng, T. R. N. Rao, G. A. Berg, and J. Zhu, "Generalized Bezout's Theorem and Its Applications in Coding Theory", *IEEE Trans. Inform. Theory*, vol. 43, pp. 1799-1810, Nov. 1997.
4. G.-L. Feng, J. Zhu, X. Shi, and T. R. N. Rao, "The Applications of Generalized Bezout's Theorem to the Codes from the Curves in High Dimensional Spaces", in *Proc. 35th Allerton Conf. Communication, Control and Computing*, pp. 205-214, 1997.
5. O. Geil, "On the Construction of Codes from Order Domains", submitted to *IEEE Trans. Inform. Theory*, June. 2001.
6. O. Geil, and T. Høholdt, "Footprints or Generalized Bezout's Theorem", *IEEE Trans. Inform. Theory*, vol. 46, pp. 635-641, Mar. 2000.
7. P. Heijnen, and R. Pellikaan, "Generalized Hamming weights of q-ary Reed-Muller codes", *IEEE Trans. Inform. Theory*, vol. 44, pp. 181-196, Jan. 1998.
8. T. Høholdt, J. H. van Lint, and R. Pellikaan, "Algebraic Geometry Codes", in *Handbook of Coding Theory*, (V. S. Pless, and W. C. Hufman Eds.), vol 1, pp. 871-961, Elsevier, Amsterdam 1998.
9. K. Saints, and C. Heegard, "On Hyperbolic Cascaded Reed-Solomon codes", *Proc. AAECC-10, Lecture Notes in Comput. Sci.* Vol. 673, pp. 291-303, Springer, Berlin 1993.
10. T. Shibuya and K. Sakaniwa, "A Dual of Well-Behaving Type Designed Minimum Distance," *IEICE Trans. A*, vol. E84-A, no. 2, pp. 647-652, Feb. 2001.

On Fast Interpolation Method
for Guruswami-Sudan List Decoding
of One-Point Algebraic-Geometry Codes

Shojiro Sakata

The University of Electro-Communications,
Department of Information and Communication Engineering,
1-5-1 Chofugaoka Chofu-shi, Tokyo 182-8585, JAPAN
sakata@ice.uec.ac.jp

Abstract. Fast interpolation methods for the original and improved versions of list decoding of one-point algebraic-geometry codes are presented. The methods are based on the Gröbner basis theory and the BMS algorithm for multiple arrays, although their forms are different in the original list decoding algorithm (Sudan algorithm) and the improved list decoding algorithm (Guruswami-Sudan algorithm). The computational complexity is less than that of the conventional Gaussian elimination method.

1 Introduction

List decoding is a kind or rather an extension of bounded-distance decoding. While the conventional bounded-distance decoding aims to find a unique codeword within the error-correction bound, i.e. half the minimum distance, the list decoding attemps to give all the codewords within the Hamming sphere having a given received word as its center and a certain radius greater than half the minimum distance. Sudan [1] invented an algebraic method with polynomial complexity for list decoding of RS codes with coding rate less than $\frac{1}{3}$. Soon later his method was extended to one-point algebraic-geometry (AG) codes by Shokrollahi and Wasserman [2]. Furthermore, Guruswami and Sudan [3] presented an improved version of Sudan algorithm for both RS and one-point AG codes, which is effective even for higher coding rate. We call the original algorithm [1][2] and the improved one [3] Sudan algorithm and Guruswami-Sudan (GS) algorithm, respectively, where GS algorithm is a generalization of Sudan algorithm in the sense that the former is reduced to the latter by taking a special value of its parameter. Either of Sudan and GS algorithms is composed of two main procedures, where the first is to find a kind of interpolation polynomial or function and the second is to factorize the outcome from the first procedure into linear factors over a function field so that one can get all the candidates

---

[1] This work is partly supported by the Science Foundation of the Japanese Educational Ministry under Grant No. 12650368.

S. Boztaş and I.E. Shparlinski (Eds.): AAECC-14, LNCS 2227, pp. 172–182, 2001.

for desired codewords. Both procedures are algebraic and have polynomial complexity, but their original forms have much complexity. Several efficient versions have been investigated, among which are fast interpolation algorithms given by Roth and Ruckenstein [4], Numakami, Fujisawa, and Sakata [5], Olshevsky and Shokrollahi [6], etc., and fast factorization algorithms by Augot and Pecquet [7], Matsumoto [8], Wu and Siegel [9], Gao and Shokrollahi [10], etc. Particularly, a fast interpolation method [5] for GS list decoding of RS codes was given based on a modification of the BMS algorithm [11]. Further, a fast interpolation algorithm [12] for Sudan list decoding of AG codes can be obtained based on the BMS algorithm [13] since the required system of linear equations has a nice structure such as block-Hankel matrix suitable to fast algorithm in these cases. On the other hand, the counterpart for GS list decoding has not been given yet, though some proposal [15] has been offered. Reformulations of GS algorithm for RS codes and AG codes were presented by Nielsen and Høholdt [14] and by Høholdt and Nielsen [15], respectively, based on some results by Feng and Blahut [16].

In this paper we present some new proposals of fast interpolation for GS list decoding of one-point AG codes in the framework of the Høholdt-Nielsen theory [15]. They are inspired by Gröbner basis theory, which has not been referred to in any previous relevant works for list decoding of one-point AG codes, and based on the BMS algorithm [11] for multiple arrays in a fashion different from that used in the fast interpolation of both GS list decoding for RS codes and Sudan list decoding for one-point AG codes.

2 Preliminaries

In this section we present a brief survey of one-point AG codes, particularly Hermitian codes, and GS list decoding algorithm for one-point AG codes according to the formulation by Høholdt and Nielsen [15].

2.1 Hermitian Codes

In this paper we discuss Hermitian codes in most cases, but we can treat any one-point AG codes in a similar fashion. A Hermitian code over a finite field $\mathbf{F}_q$ is defined from a Hermitian curve $\mathcal{X}: x^{q_1+1} + y^{q_1} + y = 0$, where $q = q_1^2$, and we restrict ourselves to consider the case of the field $\mathbf{F}_q$ with characteristic 2, i.e. $q_1 = 2^\sigma$, $q = q_1^2 = 2^{2\sigma}$. The curve is nonsingular and has genus $g = \frac{q_1(q_1-1)}{2}$. For $N = q_1^3$, let $\mathcal{P} = \{P_1, P_2, \cdots, P_N\}$ be all the $\mathbf{F}_q$-rational points of the curve $\mathcal{X}$ except for the infinity point P_∞. For the ideal $I_\mathcal{X} = \langle x^{q_1+1} + y^{q_1} + y \rangle \subset \mathbf{F}_q[x, y]$, we have the coordinate ring $\Gamma(\mathcal{X}) = \mathbf{F}_q[x, y]/I_\mathcal{X}$ and the function field $\mathbf{F}_q(\mathcal{X})$ on the curve $\mathcal{X}$. Then, for a fixed integer m, a Hermitian code is defined as $\mathcal{C} = \{\underline{c} = (c_l) \in \mathbf{F}_q^N | c_l = f(P_l), 1 \leq l \leq N, f \in \mathcal{L}(mP_\infty)\}$ by a linear subspace $\mathcal{L}(mP_\infty)$ of the linear space $\mathcal{L}(\infty P_\infty)$ which is the set of all algebraic functions $f \in \mathbf{F}_q(\mathcal{X})$ having a single pole at the point P_∞. The set $\mathcal{L}(\infty P_\infty)$ can be identified with the ring $R := \Gamma(\mathcal{X})$ in case of the Hermitian curve, and it is spanned by the set of functions $\{x^i y^j | (i,j) \in \Sigma\}$ for a subset $\Sigma := \{(i,j) \in \mathbf{Z}_0^2 | 0 \leq i, 0 \leq j \leq q_1 - 1\}$ of the 2D integral lattice $\mathbf{Z}_0^2$, where we remark that the functions x, y have pole orders $o(x) = -v_{P_\infty}(x) = q_1$, $o(y) = -v_{P_\infty}(y) = q_1 + 1$, respectively. We arrange the basis elements $\phi_1, \phi_2, \cdots$ of $\mathcal{L}(\infty P_\infty)$ in increasing order w.r.t. pole order $o_l = -v_{P_\infty}(\phi_l)$ such that $o_1 = 0 < o_2 < o_3 < \cdots$, where $o_l = l + g - 1$ if $l \geq g$, by Riemann-Roch theorem. This order induces the total order $\leq_T$ over the 2D set Σ as the $(q_1, q_1 + 1)$-weighted order and the multidegree of a function or polynomial $f = \sum_{(i,j) \in \mathrm{Supp}(f)} f_{ij} x^i y^j \in R$ as $\deg(f) := \max_T \mathrm{Supp}(f) \in \Sigma$, which we call the *degree* of f simply. If $n > m \geq 2g - 1$, the code $\mathcal{C}$ has dimension $K := \dim(\mathcal{C}) = m - g + 1$ and minimum distance $d \geq d^* := N - m$, where $\mathcal{L}(m P_\infty) = \langle \phi_1, \phi_2, \cdots, \phi_K \rangle = \langle x^i y^j | (i,j) \in \Sigma, q_1 i + (q_1 + 1)j \leq m \rangle$. For list decoding of Hermitian codes, it is required to find all the codewords $\underline{c} = (c_l) \in \mathcal{C}$ such that $d(\underline{c}, \underline{w}) \leq \tau$ for a given received word $\underline{w} = (w_l) \in \mathbf{F}_q^N$ and an integer τ $(> \lfloor \frac{d^* - 1}{2} \rfloor)$.

2.2 GS List Decoding of Hermitian Codes

For a received word $\underline{w} = (w_l) \in \mathbf{F}_q^N$, we consider the 3D points $\tilde{P}_l = (P_l, w_l) = (\alpha_l, \beta_l, w_l) \in \mathbf{F}_q^3$, $1 \leq l \leq N$, in addition to the 2D points $P_l = (\alpha_l, \beta_l) \in \mathcal{P}$, $1 \leq l \leq N$, on $\mathcal{X}$. Furthermore, we consider the polynomial ring $R[z]$ composed of polynomials with coefficients in R, which is spanned by functions of the form $\Phi_i = \phi_l z^k$, whose order is defined as $\rho(\phi_l z^k) := o_l + km$ [15]. Again we arrange the basis elements Φ_i in increasing order w.r.t. $\rho(\Phi_i)$ such that $\rho(\Phi_0) \leq \rho(\Phi_1) \leq \rho(\Phi_2) \leq \cdots$. Correspondingly, we have the total order $\leq_{\tilde{T}}$ over the 3D set $\tilde{\Sigma} := \Sigma \times \mathbf{Z}_0$ $(\subset \mathbf{Z}_0^3)$ as the $(q_1, q_1 + 1, m)$-weighted order, and the *degree* (i.e. multidegree) of a function or trivariate polynomial $Q(x,y,z) = \sum_{(i,j,k) \in \mathrm{Supp}(Q)} Q_{ijk} x^i y^j z^k$ $(\mathrm{Supp}(Q) \subset \tilde{\Sigma})$ as $\deg(Q(x,y,z)) := \max_{\tilde{T}} \mathrm{Supp}(Q) \in \tilde{\Sigma}$. In $\mathcal{L}(m P_\infty)$, instead of the above-mentioned basis $\{\phi_1, \phi_2, \cdots, \phi_K\}$, we can take another basis called an increasing zero basis $\{\phi_1^{(i)}, \phi_2^{(i)}, \cdots, \phi_K^{(i)}\}$ for each point P_i, $1 \leq i \leq N$, where the functions $\phi_k^{(i)}$ have a zero of increasing multiplicities at the point P_i, i.e. the valuations $v_{P_i}(\phi_k^{(i)})$ (> 0), $1 \leq k \leq K$, are such that $v_{P_i}(\phi_1^{(i)}) < v_{P_i}(\phi_2^{(i)}) < \cdots < v_{P_i}(\phi_K^{(i)})$. (In case of the Hermitian code, we can have $v_{P_i}(\phi_k^{(i)}) = k - 1$, $1 \leq k \leq K'$, for some integer $K' \leq K$, because P_i, $1 \leq i \leq N$, are nonsingular.) Let $\phi_l = \sum_k c_{lk}^{(i)} \phi_k^{(i)}$, $1 \leq l, k \leq K$, where the $K \times K$ matrix $[c_{lk}^{(i)}]$ gives the transformation from the basis $\{\phi_1, \cdots, \phi_K\}$ to the basis $\{\phi_1^{(i)}, \cdots, \phi_K^{(i)}\}$. Then, a function $Q(x,y,z) = Q(P, z) = \sum_{j=0}^J \sum_{k=1}^K Q_{jk} \phi_k z^j$ $(Q_{jk} \in \mathbf{F}_q)$ can be written in the form $\sum_{j=0}^J \sum_{k=1}^K q_{jk}^{(i)} \phi_k^{(i)} z^j$, where $q_{jk}^{(i)} = \sum_{l=1}^K Q_{jl} c_{lk}^{(i)}$. A function $Q(P, z) = \sum_{l=0}^L Q_l \Phi_l$ $(Q_l \in \mathbf{F}_q)$ which is written also as $\sum_{j=0}^J \sum_{k=1}^K q_{jk}^{(i)} \phi_k^{(i)} z^j$ is said to have a zero of multiplicity $\geq s$ at $\tilde{P}_i = (P_i, w_i) = (\alpha_i, \beta_i, w_i)$ if and only if the condition holds

$$q_{jk}^{\prime(i)} := \sum_l q_{lk}^{(i)} \binom{l}{j} w_i^{l-j} = 0, \; j + k \leq s, \tag{1}$$

where we remark that $Q(P, z + w_i) = \sum_j \sum_k q_{jk}^{\prime(i)} \phi_k^{(i)} z^j$. For $L := N\binom{s+1}{2}$, the conditions (1) for all the 3D points $\tilde{P}_i = (P_i, w_i)$, $1 \leq i \leq N$, imply a homogeneous system of L linear equations for the $L + 1$ unknown coefficients Q_l, $0 \leq l \leq L$. Therefore, there exists a function $Q(P, z) \in \langle \Phi_0, \Phi_1, \cdots, \Phi_L \rangle$ having a zero of multiplicity at every point $\tilde{P}_i$, $1 \leq i \leq N$, which is the desired interpolation function in the first stage of GS list decoding of a Hermitian code. In Høholdt-Nielsen's description of GS algorithm based on an analysis of Sudan algorithm by Feng and Blahut [16], given a positive integer s, the integers r_s, t_s are determined as follows:

$$\binom{r_s}{2}m - (r_s - 1)g \leq L < \binom{r_s+1}{2}m - r_s g,$$
$$t_s r_s - g(t_s) \leq L - \left(\binom{r_s}{2}m - (r_s - 1)g\right) < (t_s + 1)r_s - g(t_s + 1), \quad (2)$$

where the function $g(t)$ is defined as $g(t) := t - dim(\mathcal{L}(tP_\infty)) + 1$. Thus we have the following implementation [15] of GS algorithm to find an interpolation function $Q(P, z)$ with $\rho(Q) \leq (r_s - 1)m + t_s$ and finally all the codewords $\underline{c}$ such that $d(\underline{c}, \underline{w}) \leq \tau_s := N - \lfloor \frac{(r_s-1)m+t_s}{s} \rfloor - 1$, where the number of codewords $\underline{c} \in \mathcal{C}$ with $d(\underline{c}, \underline{w}) \leq \tau_s$ is less than r_s.

(i) Construct an increasing zero basis for each l, $1 \leq l \leq N$;
(ii) Find the interpolation function $Q(P, z) \in R[z]$ which has a zero of multiplicity at least s at all points (P_i, w_i), $1 \leq i \leq N$, and whose order is $\rho(Q) \leq (r_s - 1)m + t_s$, i.e. solve the system of linear equations (1);
(iii) Find factors $z - f | Q$ over the function field $\mathbf{F}_q(\mathcal{X})$ or the ring $R[z]$.

Since the complexity $\mathcal{O}(L^3) = \mathcal{O}(N^3 s^6)$ of the Gaussian elimination (ii) is large in the general case ($s \geq 2$), we need a fast interpolation method instead.

3   Fast Interpolation for Sudan and GS List Decoding

Before treating fast interpolation for the more general GS algorithm, we mention that for Sudan algorithm.

3.1   Fast Interpolation for Sudan List Decoding

In case of multiplicity $s = 1$, i.e. for Sudan algorithm, we can have a fast interpolation method by using the BMS algorithm [13], based on the following fact.

Lemma 1 For $\tilde{P}_l = (\alpha_l, \beta_l, w_l)$, $1 \leq l \leq N$, the sets $\tilde{I}^{(l)} := \{F = F(x, y, z) \in R[z] | F(\alpha_l, \beta_l, w_l) = 0\}$, $1 \leq l \leq N$, and $\tilde{I} := \cap_{l=1}^N \tilde{I}^{(l)}$ are ideals in $R[z]$.

A function $F = F(x, y, z) \in R[z]$ can be written in the form $F = \sum_{\tilde{p} \in \text{Supp}(F)} F_{\tilde{p}} X^{\tilde{p}}$, with $\text{Supp}(F) \subset \tilde{\Sigma}$, where $F_{\tilde{p}} \in \mathbf{F}_q$, $X^{\tilde{p}} := x^i y^j z^k$ for $\tilde{p} = (i, j, k) \in \tilde{\Sigma}$. Now, for N 3D arrays $u^{(l)} = (u_{\tilde{p}}^{(l)})$ having the components $u_{\tilde{p}}^{(l)} := \alpha_l^i \beta_l^j w_l^k$, $\tilde{p} = (i, j, k) \in$

$\tilde{\Sigma}$, determined from $\tilde{P}_l = (\alpha_l, \beta_l, w_l)$, $1 \leq l \leq N$, we introduce another 3D array $u = (u_{\tilde{p}})$ over $\mathbf{F}_q$ as their sum, i.e. $u_{\tilde{p}} := \sum_{l=1}^{N} u_{\tilde{p}}^{(l)}$. Then, we have the lemma which implies that we can apply the 3D BMS algorithm [13] to the array u to find a Gröbner basis of the ideal $\tilde{I}$, which contains the desired interpolation polynomial Q as its element.

Lemma 2 $F \in \tilde{I}$ *if and only if the following linear recurrence holds for the given array u:*

$$\sum_{\tilde{q} \in \text{Supp}(F)} F_{\tilde{q}} u_{\tilde{q}+\tilde{p}} = 0, \ ^\forall \tilde{p} \in \tilde{\Sigma}. \tag{3}$$

The BMS algorithm is an iterative algorithm to find a minimal set of polynomials $F = \sum_{\tilde{p} \in \text{Supp}(F)} F_{\tilde{p}} X^{\tilde{p}} \in R[z]$ satisfying the identity (3) up to a certain point $\tilde{p}$, which becomes a Gröbner basis of the ideal $\tilde{I}(u)$ finally after a finite number of iterations of checking components of arrays and updating F at each point $\tilde{p}$ successively (w.r.t. some appropriate total order over $\tilde{\Sigma}$), where the ideal $\tilde{I}(u)$ is defined as the set of all polynomials F satisfying (3) for the array u ($\tilde{I}(u)$ is called the characteristic ideal of u, and on the other hand an array u such that $\tilde{I}(u) = \tilde{I}$ for a given ideal $\tilde{I}$ is called a representative array of $\tilde{I}$ if it exsits [17]).

For the first nonzero pole order $\nu := o_2$, which is equal to q_1 in case of Hermitian codes, the cardinality $\mu := \#F$ of a minimal polynomial set F during the process of the BMS algorithm [13] is equal to $r_1 \nu$, where $\mathcal{O}(r_1) = \mathcal{O}(\sqrt{\frac{N}{K}})$ by (2). Consequently, the computational complexity of the above method is $\mathcal{O}(\mu N^2) = \mathcal{O}(\nu N^{\frac{5}{2}} K^{-\frac{1}{2}})$, which is $\mathcal{O}(N^{\frac{7}{3}})$ in case of Hermitian codes with $\mathcal{O}(K) = \mathcal{O}(N)$, particularly $K \sim cN$ for $c < \frac{1}{3}$. This complexity is the same as that of Olshevsky and Shokrollahi's method [6] and less than $\mathcal{O}(N^3)$ for Gaussian elimination.

The left-hand side of the equation (3) is a component $v_{\tilde{p}}$ of a 3D array $v = (v_{\tilde{p}})$, $\tilde{p} \in \tilde{\Sigma}$, provided that the array $u_{\tilde{p}}$ is defined over the set $2\tilde{\Sigma} := \{\tilde{p} + \tilde{q} | \tilde{p}, \tilde{q} \in \tilde{\Sigma}\}$, where the array v is denoted as $v := F \circ u$. Then, the linear recurrence (3) can be written $F \circ u = 0$, where the right-hand side 0 means the zero array whose components are all zero. For later use, we remember the following definition and the fact from [17]. A set of arrays $u^{(1)}, \cdots, u^{(M)}$ is called a set of representative arrays of an ideal $\tilde{I} \in R[z]$ if and only if $\tilde{I}$ coincides with the characteristic ideal of these arrays defined as $\tilde{I}(u^{(1)}, \cdots, u^{(M)}) := \{F \in R[z] | F \circ u^{(i)} = 0, 1 \leq i \leq M\}$. Given a set of arrays $u^{(1)}, \cdots, u^{(M)}$, by using a modification [11] of the BMS algorithm for multiple arrays, we can find efficiently a minimal set of polynomials F satisfying the identities $F \circ u^{(i)} = 0$ in the form of (3) for the arrays $u^{(i)}$, $1 \leq i \leq M$, which is a Gröbner of the ideal $\tilde{I}(u^{(1)}, \cdots, u^{(M)})$.

3.2   Fast Interpolation for GS List Decoding

Now we extend the valuation $v_{P_l}(f)$, $1 \leq l \leq N$, of $f \in R$ at $P_l \in \mathcal{P}$ to that of any function $F \in R[z]$ at $\tilde{P}_l = (P_l, w_l)$ by $v_{\tilde{P}_l}(f(z - w_l)^k) := v_{P_l}(f) + k$ for $f \in R$.

Thus, for $F = \sum_k f_k(z - w_l)^k$ with $f_k \in R$, we define $v_{\tilde{P}_l}(\sum_k f_k(z - w_l)^k)) := \min_k \{v_{\tilde{P}_l}(f_k(z - w_l)^k)\} = \min_k \{v_{P_l}(f_k) + k\}$. Then, we must find a function $F \in R[z]$ such that $\deg(F)$ is minimum w.r.t. the total order $\leq_{\tilde{T}}$ over $\tilde{\Sigma}$ and $v_{\tilde{P}_l}(F) \geq s_l(= s)$, $1 \leq l \leq N$. Immediately we have the following observations.

(1) For each couple of $\tilde{P}_l$, $1 \leq l \leq N$, and $j \in \mathbf{Z}_0$, the set $\tilde{I}^{(l)}(j) := \{F \in R[z] | v_{\tilde{P}_l}(F) \geq j\}$ is an ideal;

(2) The required interpolation function Q is an element of a Gröbner basis of the ideal $\tilde{I}(s) = \cap_{1 \leq l \leq N} \tilde{I}^{(l)}(s)$.

If we can have some efficient method to obtain a Gröbner basis of an intersection of ideals, we do not need to find all the elements of an increasing zero basis at each point P_l, and neither need to solve a large system of linear equations by Gaussian elimination. The method of obtaining a Gröbner basis of an intersection of two ideals $I_1, I_2 \subset \mathbf{F}[x_1, x_2, \cdots, x_n]$ by the well-known formula $I_1 \cap I_2 = \langle tf_i + (1-t)g_j | f_i \in I_1, g_j \in I_2 \rangle \cap \mathbf{F}[x_1, x_2, \cdots, x_n]$ is not so efficient for our purpose. Before discussing how to find the Gröbner bases of ideals $\tilde{I}^{(l)}(s)$, $1 \leq l \leq N$, and $\tilde{I}(s)$ in $R[z]$, we treat ideals $I \subset R$ in **3.2.1**, and then consider ideals $\tilde{I} \subset R[z]$ in **3.2.2**. Finally, we give a fast algorithm of finding a Gröbner basis of the intersection ideal $\tilde{I}(s)$ in **3.2.3**.

3.2.1 Gröbner Basis of 2D Ideals

First, for $p = (i, j) \in \Sigma$ and P_l, $1 \leq l \leq N$, we define the zero order of p at $P := P_l$, as $o_P(p) := \max\{v_P(f) | f \in R, \deg(f) = p\}$, which is denoted as $o(p)$ simply. Arranging the elements of Σ w.r.t. the total order $<_T$, we have $\Sigma = \{p^{(1)}, p^{(2)}, \cdots\}$, where $p^{(1)} = (0, 0) <_T p^{(2)} <_T \cdots$. Let $f^{(i)} \in R$ be such that $\deg(f^{(i)}) = p^{(i)}$ and $v_P(f^{(i)}) = o(p^{(i)})$, $i = 1, 2, \cdots$. Then, we have a couple of mutually corresponding series $O_P := \{o(p^{(1)}), o(p^{(2)}), \cdots\} \subset \mathbf{Z}_0$ and $F_P := \{f^{(1)}, f^{(2)}, \cdots\} \subset R$. Let $<_P$ be the ordinary partial order over the 2D set $\mathbf{Z}_0^2$. Then, it is easy to see that if $p <_P q$, $o(p) < o(q)$, and that $I_P(j) := \{f \in R | v_P(f) \geq j\}$ is an ideal in R. A subset $\tilde{\Xi} \subset \Sigma$ is said to be *outward* (respectively, *inward*) *stable* (w.r.t. $\leq_P$) if and only if the condition is satisfied: If $p \in \Xi$, $q \in \Sigma$, $p \leq_P$ (respectively, $\geq_P$) q, then $q \in \Xi$. We take the subset $\Sigma_P(j) := \{p \in \Sigma | o(p) \geq j\}$ of Σ, and its complement $\Delta_P(j) := \Sigma \setminus \Sigma_P(j)$. These subsets have the following properties w.r.t. the partial order $\leq_P$ over $\mathbf{Z}_0^2$, i.e. the subset $\Sigma_P(j)$ is outward stable, and $\Delta_P(j)$ is inward stable. This leads us to introduce the subset $S_P(j) := \min_P \Sigma_P(j)$ of $\Sigma_P(j)$, which is the set of all minimal (w.r.t. the partial order $<_P$) points $p \in \Sigma_P(j)$, accompanied with the subset $F_P(j) := \{f \in F_P | \deg(f) \in S_P(j)\}$ of R. Consequently, we have a conclusion about ideals $I_P(j)$ in R.

Theorem 1 $F_P(j)$ *is a Gröbner basis of* $I_P(j)$.

From the above considerations we have the algorithm of finding O_P, F_P iteratively (w.r.t. $\leq_T$) for each $P = (\zeta, \eta)$ and a fixed integer s. Below let $X := x - \zeta$, $Y := y - \eta$.

Algorithm 1 *Step 1:* $j := 1$; $F_P := \{1\}$; $O_P := \{0\}$;
Step 2: $\underline{p} := \underline{p}^{(j+1)} - (1,0)$ *(or* $(0,1)$*)*;
  $f \in \bar{F}_P$ *such that* $\deg(f) = \underline{p}$; $f' := Xf$ *(or* Yf*)*;
  if $v_P(f') \in O_P$ *then*
    by using the procedure mentioned in the proof of Theorem 1 of [15],
    construct $g \in R$ *from* f' *and some* $h \in F_P$ *with* $v_P(h) = v_P(f')$
    such that $v_P(g) \notin O_P$ *and* $v_P(g)$ *is minimum;*
  $F_P := F_P \cup \{g\}$; $O_P := O_P \cup \{v_P(g)\}$;
  else $F_P := F_P \cup \{f'\}$; $O_P := O_P \cup \{v_P(f')\}$;
Step 3: $j := j + 1$; *if* $j \le s$ *then go to Step 2 else stop.*

In the above procedure of [15], the coefficients of the initial terms of Laurent expansions of functions w.r.t. a local parameter t must be calculated. They can be obtained by the following ordinary method in case of Hermitian codes over fields with characteristic 2, i.e. $q_1 = 2^\sigma, q = 2^{2\sigma}$. For $P = (\zeta, \eta)$; $X = x - \zeta, Y = y - \eta$, we can rewrite the curve-defining equation as $Y + Y^{q_1} = aX + bX^{q_1} + X^{q_1+1}$, where $a := \zeta^{q_1}, b := \zeta$, from which it follows that, in the power expansion $Y = \sum_{i \ge 1} c_i X^i$ the coefficients satisfy the equalities: $c_1 = a$, $c_{q_1} + c_1^{q_1} = b$, $c_{q_1+1} = 1$, $c_{q_1 j} + c_j^{q_1} = b\delta_{j1}$. Therefore, $c_j = 0$ except for $c_1 = a$, $c_{q_1^{j-1}(q_1+1)} = 1$, $j \ge 1$, and consequently we have $Y = aX + X^{q_1+1} + X^{q_1(q_1+1)} + X^{q_1^2(q_1+1)} + \cdots$. It is easy to get the expansions of $Y^2, Y^4, \cdots$, and in general those of Y^i, $i \in \mathbf{Z}_0$. We show a simple example of computation by Algorithm 1.

Example 1 Let $q_1 = 2$, $q = 2^2 = 4$. *Then, we have an elliptic curve* $(g = 1)$ *as a special case:* $\mathcal{X} : x^3 + y^2 + y = 0$ *over the field* $\mathbf{F}_4 = \mathbf{F}_2[\alpha]/\langle \alpha^2 + \alpha + 1 \rangle$. *At the infinity point* P_∞, $o(x) = 2$, $o(y) = 3$, *and* $\mathcal{L}(\infty P_\infty) = \langle x^i y^j | 0 \le i, 0 \le j \le 1 \rangle$. *For example, we take* $P = (1, \alpha^2)$, *and let* $X := x - 1$, $Y := y - \alpha^2$. *Then,* $Y + Y^2 = X + X^2 + X^3$, *and thus* $Y = X + X^3 + X^6 + X^{12} + \cdots$. *By applying Algorithm 1 we have the following tables, in which pole orders* o_l, *numbers* l *implying the total order* $\le_T$, *functions* $f^{(l)} \in F_P$, *and valuations* $o(\underline{p}^{(l)}) \in O_P$ *for* $\underline{p}^{(l)} = (i,j)$ *are shown respectively as the* (i,j)-*elements of each (partial) 2D array.*

Pole orders o_l (l)

$i \setminus j$	0	1	2
0	0(1)	3(3)	6(6)
1	2(2)	5(5)	8(8)
2	4(4)	7(7)	
3	6(6)	9(9)	
4	8(8)		

Functions F_P (Valuations O_P)

$i \setminus j$	0	1	2
0	1(0)	$Y + X$(3)	
1	X(1)	$X(Y + X)$(4)	
2	X^2(2)	$X^2(Y + X)$(5)	
3	$X^3 + Y + X$(6)	$(X^3 + 1)Y + X^4 + X^3 + X$(9)	
4	$X^4 + XY + X^2$(7)		

From the table, we have e.g. a Gröbner basis expression of the ideal $I(4) = \langle X^3 + Y + X, XY + X^2, Y^2 + X^2 \rangle$.

The computational complexity of Algorithm 1 is $\mathcal{O}(s^3 N)$. As an alternative, we have another algorithm to find a Gröbner basis of $I_P(j)$ based on the formula

$I_P(j+1) = I_P(j)I_P + I_\mathcal{X}$, which is derived from Proposition 11 mentioned by Matsumoto [8]. Its complexity is $\mathcal{O}(\nu s^2 N)$.

3.2.2 Gröbner Bases of 3D Ideals

Now, as a result of the discussions about ideals I in R in **3.2.1**, we are ready to treat ideals in $R[z]$. Similar to the zero order of a 2D point $\underline{p} = (i,j) \in \Sigma$, we define the zero order of a 3D point $\tilde{p} = (i,j,k) \in \tilde{\Sigma}$ at $\tilde{P} = \tilde{P}_l = (\alpha, \beta, w)$ as $\tilde{o}_{\tilde{P}}(\tilde{p}) := \max\{v_{\tilde{P}}(\tilde{f}) | \deg(\tilde{f}) = \tilde{p}\}$, which is denoted as $\tilde{o}(\tilde{p})$ simply. Arranging the elements of $\tilde{\Sigma}$ w.r.t. the total order $<_{\tilde{T}}$ over $\tilde{\Sigma}$, we have $\tilde{\Sigma} = \{\tilde{p}^{(1)}, \tilde{p}^{(2)}, \cdots\}$, where $\tilde{p}^{(1)} = (0,0,0) <_{\tilde{T}} \tilde{p}^{(2)} <_{\tilde{T}} \cdots$. Let $\tilde{f}^{(i)}$ be such that $\deg(\tilde{f}^{(i)}) = \tilde{p}^{(i)}$ and $v_{\tilde{P}}(\tilde{f}^{(i)}) = \tilde{o}(\tilde{p}^{(i)})$, $i \in \mathbf{Z}_0$. Then, we have a couple of mutually corresponding series $\tilde{O}_{\tilde{P}} := \{\tilde{o}(\tilde{p}^{(1)}), \tilde{o}(\tilde{p}^{(2)}), \cdots\} \subset \mathbf{Z}_0$ and $\tilde{F}_{\tilde{P}} := \{\tilde{f}^{(1)}, \tilde{f}^{(2)}, \cdots\} \subset R[z]$. Again we have a set of concepts for ideals in $R[z]$ similar to those for ideals in R. Thus, we have that if $\tilde{p} <_{\tilde{P}} \tilde{q}$ (w.r.t. the partial order $<_{\tilde{P}}$ over $\tilde{\Sigma}$), $\tilde{o}(\tilde{p}) < \tilde{o}(\tilde{q})$, and that the subset $\tilde{I}_{\tilde{P}}(j) := \{\tilde{f} \in R[z] | v_{\tilde{P}}(\tilde{f}) \geq j\}$ is an ideal. A subset $\tilde{\Xi} \subset \tilde{\Sigma}$ is said to be *outward* (respectively, *inward*) *stable* (w.r.t. $\leq_{\tilde{P}}$) if and only if the condition is satisfied: If $\tilde{p} \in \tilde{\Xi}, \tilde{q} \in \tilde{\Sigma}, \tilde{p} \leq_{\tilde{P}}$ (respectively, $\geq_{\tilde{P}})\tilde{q}$ then $\tilde{q} \in \tilde{\Xi}$. Let $\tilde{\Sigma}_{\tilde{P}}(j) := \{\tilde{p} \in \tilde{\Sigma} | \tilde{o}(\tilde{p}) \geq j\}$, and $\tilde{\Delta}_{\tilde{P}}(j) := \tilde{\Sigma} \setminus \tilde{\Sigma}_{\tilde{P}}(j)$. Then, $\tilde{\Sigma}_{\tilde{P}}(j)$ is outward stable, and $\tilde{\Delta}_{\tilde{P}}(j)$ is inward stable. Taking a subset $\tilde{S}_{\tilde{P}}(j) := \min_{\tilde{P}} \tilde{\Sigma}_{\tilde{P}}(j)$ composed of elements $\underline{p} \in \tilde{\Sigma}_{\tilde{P}}(j)$ which are minimal w.r.t. $\leq_{\tilde{P}}$ and the corresponding subset $\tilde{F}_{\tilde{P}}(j) := \{\tilde{f} \in \tilde{F}_{\tilde{P}} | \deg(\tilde{f}) \in \tilde{S}_{\tilde{P}}(j)\}$, we have a conclusion about ideals $\tilde{I}_{\tilde{P}}(j)$ in $R[z]$.

Theorem 2 $\tilde{F}_{\tilde{P}}(j)$ *is a Gröbner basis of* $\tilde{I}_{\tilde{P}}(j)$, *and* $\tilde{F}_{\tilde{P}}(s) \subseteq \cup_{i=0}^{s}(z-w)^i F_P(s-i)$.

Example 2 *As in the previous example, for* $X = x - \zeta, Y = y - \eta, Z = z - w$, $\tilde{I}_{\tilde{P}}(4) = \langle X^3 + Y + X, XY + X^2, Y^2 + X^2, X^2 Z^2, (Y+X)Z, XZ^3, Z^4 \rangle$ *is a Gröbner basis expression of the ideal* $\tilde{I}_{\tilde{P}}(4)$ *in view of the above-mentioned Gröbner basis of* $I_P(4)$.

3.2.3 Fast Interpolation Method for GS List Decoding

From a Gröbner basis $\tilde{F}^{(l)}(s) := \tilde{F}_{\tilde{P}_l}(s)$ of the ideal $\tilde{I}^{(l)}(s) := \tilde{I}_{\tilde{P}_l}(s)$, we can get a set of νs representative 3D arrays $u^{(l1)}, u^{(l2)}, \cdots, u^{(l\nu)}$ such that $\tilde{I}^{(l)}(s) = \tilde{I}(u^{(l1)}, u^{(l2)}, \cdots, u^{(l\nu)})$, for each l, $1 \leq l \leq N$, where we remember that the definition of representative arrays of an ideal is as written in **3.1**. Since $\tilde{I}(s) = \cap_{l=1}^{N} \tilde{I}^{(l)}(s)$, we can find a Gröbner basis of $\tilde{I}(s)$ by applying the BMS algorithm [11] for multiple arrays iteratively as follows, where we can start the BMS algorithm with the Gröbner basis $\tilde{F}$ of the i-th intersection ideal $\tilde{I} = \cap_{j=1}^{i} \tilde{I}^{(j)}$ obtained in the i-th stage of the following algorithm and update $\tilde{F}$ at each iteration in Step 2 of the $(i+1)$-th stage until all the relevant parts of the arrays have been checked and finally the Gröbner basis of $\tilde{I} \cap \tilde{I}^{(i+1)}$ is obtained.

Algorithm 2   *Step 1:* $\tilde{F} := \tilde{F}^{(1)}$ $(\tilde{I} := \tilde{I}^{(1)})$; $i := 1$;

Step 2: Apply the BMS algorithm to the multiple arrays $u^{(i+1,1)}, u^{(i+1,2)}, \cdots,$
  $u^{(i+1,\nu)}$ *to update* $\tilde{F}$ *so that* $\tilde{F}$ *becomes a Gröbner basis of* $\tilde{I} := \tilde{I} \cap \tilde{I}^{(i+1)}(s)$
  at the final iteration;

Step 3: $i := i + 1$; *if* $i \leq N$ *then go to Step 2 else stop.*

The computational complexity of Algorithm 2 can be estimated roughly as follows. Each ideal $\tilde{I}^{(i)}(s)$ has νs representative arrays, and its Gröbner basis has the delta set $\tilde{\Delta}^{(i)}(s)$ of size $\#\tilde{\Delta}^{(i)}(s) = s^2$, which is equal to the size of each element of $\tilde{F}^{(i)}(s)$. On the other hand, the size of the delta set for a Gröbner basis of the i-th intersection $\cap_{j=1}^{i} \tilde{I}^{(j)}(s)$ can be estimated as is^2, which is not less than the size of each element of $\tilde{F}$, where $\tilde{F}$ is a set of minimal polynomials of a set of partial arrays during the process of Algorithm 2. The number of elements of $\tilde{F}$ is not greater than $r_s\nu$, which is $\mathcal{O}(sN^{\frac{1}{2}}K^{-\frac{1}{2}})$ since $\mathcal{O}(mr_s^2) = \mathcal{O}(Ns^2)$ by (2) and $\mathcal{O}(m) = \mathcal{O}(K)$. Therefore, in Step 2 for each i, we need to update $r_s\nu$ polynomials of size is^2 by s^2 iterations for the νs partial arrays $u^{(i+1,1)}, u^{(i+1,2)}, \cdots, u^{(i+1,\nu)}$, each of size $(i+1)s^2 - is^2 = s^2$. Thus, the total complexity of Algorithm 2 is roughly $\sum_{i=1}^{N} \mathcal{O}(r_s \times is^2 \times s^2 \times \nu s) = \mathcal{O}(\nu^2 s^6 N^{\frac{5}{2}} K^{-\frac{1}{2}})$, which is an overestimation because only a few elements of the current minimal polynomial set $\tilde{F}$ must be checked and updated probably at each of νs iterations of Step 2 for every i. For Hermitian codes the complexity is $\mathcal{O}(s^6 N^{\frac{19}{6}} K^{-\frac{1}{2}})$, which is $\mathcal{O}(s^6 N^{\frac{8}{3}})$ and less than the complexity $\mathcal{O}(s^6 N^3)$ of the Gaussian elimination in case of $K \sim cN$ for $0 < c < 1$. (Remark: Olshevsky and Shokrollahi [6] give neither a method of fast interpolation for GS list decoding of one-point AG codes nor its complexity explicitly.)

4   Concluding Remark

The computational complexity of our fast interpolation method for GS list decoding of one-point AG codes is $\mathcal{O}(\nu^2 s^6 N^{\frac{5}{2}} K^{-\frac{1}{2}})$, which is not reduced to the complexity $\mathcal{O}(\nu N^{\frac{5}{2}} K^{-\frac{1}{2}})$ of the fast interpolation in case of Sudan list decoding ($s = 1$) because the method for GS list decoding has a different form in applying the BMS algorithm for multiple arrays.

References

1. M. Sudan, "Decoding of Reed-Solomon codes beyond the error-correction bound," *J. Complexity*, 13, 180–193, 1997.
2. M.A. Shokrollahi, H. Wasserman, "List decoding of Algebraic-geometric codes," *IEEE Trans. Inform. Theory*, 45, 432–437, 1999.
3. V. Guruswami, M. Sudan, "Improved decoding of Reed-Solomon and algebraic-geometry codes," *IEEE Trans. Inform. Theory*, 45, 1757–1767, 1999.
4. R.M. Roth, G. Ruckenstein, "Efficient decoding of Reed-Solomon codes beyond half the minimum distance," *IEEE Trans. Inform. Theory*, 46, 246–257, 2000.
5. Y. Numakami, M. Fujisawa, S. Sakata, "A fast interpolation method for list decoding of Reed-Solomon codes," *Trans. IEICE*, J83-A, 1309–1317, 2000. (in Japanese)

6. V. Olshevsky, M.A. Shokrollahi, "A displacement approach to efficient decoding of algebraic-geometric codes," Proc. 31st ACM Symp. Theory of Comput., 235–244, 1999.

7. D. Augot, L. Pecquet, "A Hensel lifting to replace factorization in list-decoding of algebraic-geometric and Reed-Solomon codes," *IEEE Trans. Inform. Theory*, 46, 2605–2614, 2000.

8. R. Matsumoto, "On the second step in the Guruswami-Sudan list decoding for AG codes," Tech. Report of IEICE, IT99-75, 65–70, 2000.

9. X.W. Wu, P.H. Siegel, "Efficient list decoding of algebraic-geometric codes beyond the error correction bound," preprint.

10. Sh. Gao, M.A. Shokrollahi, "Computing roots of the polynomials over function fields of curves," in *D. Joyner (Ed.), Coding Theory & Crypt.*, Springer, 214–228, 2000.

11. S. Sakata, "N-dimensional Berlekamp-Massey algorithm for multiple arrays and construction of multivariate polynomials with preassigned zeros," in *T. Mora (Ed.), Appl. Algebra, Algebraic Algorithms & Error-Correcting Codes*, LNCS: 357, Springer, 356–376, 1989.

12. S. Sakata, Y. Numakami, "A fast interpolation method for list decoding of RS and algebraic-geometric codes," presented at the 2000 IEEE Intern. Symp. Inform. Theory, Sorrento, Italy, June 2000.

13. S. Sakata, "Extension of the Berlekamp-Massey algorithm to N dimensions," *Inform. & Comput.*, 84, 207–239, 1990.

14. R. Nielsen, T. Høholdt, "Decoding Reed-Solomon codes beyond half the minimum distance," in *J. Buchmenn, T. Høholdt, T. Stichtenoth, H. Tapia-Recillas (Eds.), Coding Theory, Crypt. & Related Areas*, Springer, 221–236, 2000.

15. T. Høholdt, R. Nielsen, "Decoding Hermitian codes with Sudan's algorithm," preprint.

16. W. Feng, R.E. Blahut, "Some results of the Sudan algorithm" presented at the 1998 IEEE Intern. Symp. Inform. Theory, Cambridge, USA, August 1998.

17. S. Sakata, "On determining the independent point set for doubly periodic arrays and encoding two-dimensional cyclic codes and their duals," *IEEE Trans. Inform. Theory*, 27, 556–565, 1981.

Computing the Genus of a Class of Curves

M.C. Rodríguez-Palánquex[1], L.J. García-Villalba[2], and I. Luengo-Velasco[3]

[1] Sección Departamental de Matemática Aplicada. Escuela Universitaria
de Estadística. UCM. Avda. Puerta de Hierro s/n, 28040 Madrid, Spain
mcrodri@eucmax.sim.ucm.es

[2] Departamento de Sistemas Informáticos y Programación. Facultad de Informática.
UCM. 28040 Madrid, Spain
javiergv@sip.ucm.es

[3] Departamento de Álgebra. Facultad de Ciencias Matemáticas. UCM. 28040
Madrid, Spain
iluengo@eucmos.sim.ucm.es

Abstract. The aim of this paper is to present an exhaustive algebraic
study of a new class of curves, the so-called Quasihermitian curves (that
includes the Hermitian curves), computing its genus. These curves allow to construct good algebraic geometric Goppa codes since they are
absolutely irreducible plane curves with many rational points on $\mathbb{F}_q$.

1   Introduction

Let K be a finite field of characteristic 2, $K = \mathbb{F}_q$ where $q = 2^j$.

For any $a, b \in \mathbb{Z}$, being $a \geq 2$, $b > -a$, $\beta_1, \beta_2 \in K - \{0\}$, we consider the following curves (*Quasihermitian curves*) defined over K given by the homogeneus
equation

$$Y^a Z^b + \beta_1 Y Z^{a+b-1} + \beta_2 X^{a+b} = 0$$

if $b \geq 0$ and if $b < 0$

$$Y^a + \beta_1 Y Z^{a-1} + \beta_2 X^{a+b} Z^{-b} = 0.$$

So, this class of curves is defined by the affine equation

$$y^a + \beta_1 y + \beta_2 x^{a+b} = 0.$$

Quasihermitian curves include some types of curves with many rational points.
If $j = 2j_0$ we have the Hermitian curves [8], which have by equation

$$Y^{2^{j_0}} Z + Y Z^{2^{j_0}} + X^{2^{j_0}+1} = 0$$

and the maximal curves obtained from the affine equation $y^{2^{j_0}} + y = x^m$ where
m is a divisor of $(2^{j_0} + 1)$, see [3].

Curves with many rational points are interesting in Coding theory. In particular, Goppa geometric codes obtained from Hermitian curves have been extensively studied [8] [9]. The special arithmetic properties for the $\mathbb{F}_q$-rational points

S. Boztaş and I.E. Shparlinski (Eds.): AAECC-14, LNCS 2227, pp. 182–191, 2001.
© Springer-Verlag Berlin Heidelberg 2001

of Hermitian curves have allowed to calculate the true minimun distance of such codes.

We begin here a systematic study of properties of Quasihermitian curves and of the Goppa codes obtained from them. In particular, we compute its genus.

We find among these Quasihermitian curves many new maximal curves (see Section 6), i.e. for the non-singular models of these curves the number of $\mathbb{F}_q$-rational points attains the Hasse-Weil upper bound $q+1+2g\sqrt{q}$. So, for example, $Y^2Z^9 + YZ^{10} + X^{11} = 0$ is a maximal curve over $\mathbb{F}_{2^{10}}$ and $Y^4 + YZ^3 + X^3Z = 0$ is a maximal curve over $\mathbb{F}_{2^6}$.

2   Some Previous Definitions and Results

Let C_1 and C_2 be curves (absolutely irreducible projective varieties of dimension 1) defined over K. We shall denote by $g(C_1)$ and $g(C_2)$, the genus of C_1 and C_2, respectively, and by $K(C_1)$ and $K(C_2)$ the respective function fields.

Definition 1. *A non-constant rational map $\phi : C_1 \to C_2$ is said to be purely inseparable, if $K(C_1)$ is a purely inseparable algebraic extension of $\phi^*(K(C_2))$, where ϕ^* is defined by*

$$\phi^* : K(C_2) \to K(C_1)$$
$$f \mapsto f \circ \phi$$

Proposition 1. *(See [5] p.302, and [8] p.127) If $K(C_1)$ is a purely inseparable algebraic extension of $\phi^*(K(C_2))$, then $g(C_1) = g(C_2)$.*

We recall the genus formula of a curve C obtained from the sequence of multiplicities at its infinitely near singular points. The blowing-up at one point process will be also called quadratic transformation (QDT) at such point.

Proposition 2. *(See [1] p.148, [2] p.7, [4] p.124, and [5] p.126) Assume that C is an absolutely irreducible algebraic plane curve of degree h. Let $P_1, P_2, ..., P_N$ be all the singular points of C, $r_{i1}, r_{i2}, ..., r_{in_i}$ $(i = 1, ..., N)$ the multiplicity sequence with respect to blowing-up with center P_i. Then the genus of C is*

$$g(C) = \frac{(h-1)(h-2)}{2} - \sum_{i=1}^{N} \delta(P_i)$$

where, $\forall i = 1, ..., N$, $\delta(P_i) = \sum_{j=1}^{n_i} \frac{r_{ij}(r_{ij}-1)}{2}$.

3   Quasihermitian Curves

Definition 2. *A Quasihermitian curve is the absolutely irreducible projective variety over K given, for any $a, b \in \mathbb{Z}$, $a \geq 2$, $b > -a$, by the homogeneus equation*

$$Y^a Z^b + \beta_1 Y Z^{a+b-1} + \beta_2 X^{a+b} = 0$$

if $b \geq 0$ *or*

$$Y^a + \beta_1 Y Z^{a-1} + \beta_2 X^{a+b} Z^{-b} = 0,$$

if $b < 0$, *where* $\beta_1, \beta_2 \in K - \{0\}$.

Proposition 3. *For any* $a, b \in \mathbb{Z}$, $a \geq 2$ *and* $b > -a$, *the affine curves* C *and* C_1 *with equations, respectively,* $y^a + y + x^{a+b} = 0$ *and* $y_1^a + \beta_1 y_1 + \beta_2 x_1^{a+b} = 0$, *are isomorphic over* $\overline{K}$.

Proof. The rational map $\Phi : \mathbb{A}^2_{\overline{K}} \to \mathbb{A}^2_{\overline{K}}$, $(x, y) \mapsto (\delta x, \epsilon y)$, with δ and $\epsilon \in \overline{K}$ such that $\delta^{a+b} = \frac{\epsilon^a}{\beta_2}$ and $\epsilon^{a-1} = \beta_1$, is an isomorphism. The restricted map verifies $(\Phi|C)(x, y) = (x_1, y_1) \in C_1$ because

$$\epsilon^a y^a + \beta_1 \epsilon y + \beta_2 \delta^{a+b} x^{a+b} = \epsilon^a (y^a + y + x^{a+b}) = 0.$$

Corollary 1. *(See Proposition 1) The curves* C *and* C_1 *have the same genus.*

Proposition 3 and Corollary 1 suggest the following definition.

Definition 3. *For any* $a \geq 2$ *and* $b \geq 0$, *we denote by* $C_{a,a+b}$ *the absolutely irreducible projective curve over* K *given by the homogeneus equation*

$$Y^a Z^b + Y Z^{a+b-1} + X^{a+b} = 0$$

4   Singular Points of $C_{a,a+b}$

If $P = (x, y, z) \in C_{a,a+b}$ is singular, then

$$\begin{cases} y^a z^b + y z^{a+b-1} + x^{a+b} = 0 \\ (a+b) x^{a+b-1} = 0 \\ a y^{a-1} z^b + z^{a+b-1} = 0 \\ b y^a z^{b-1} + (a+b-1) y z^{a+b-2} = 0 \end{cases}$$

We have the following cases, where if a is even, then we write $a = 2^r r_0$ (with r_0 odd), and if a is odd, then we write $a - 1 = 2^s s_0$ (with s_0 odd), see Table 1.

5   Genus

Consider $b \geq 0$, we write $a + b = 2^n b_0$, with b_0 odd, and $\forall \, n' \geq 0$, $m = 2^{n'} b_0$. If $m \geq a$, we are going to consider the curve $C_{a,m}$ (so, the notation $C_{a,m}$ indicates that $m \geq a$), and if $m < a$, the curve $D_{a,m}$ defined by the equation $Y^a + Y Z^{a-1} + X^m Z^{a-m} = 0$. We shall prove that there is a purely inseparable rational map between $C_{a,a+b}$ and $C_{a,m}$, and between $C_{a,a+b}$ and $D_{a,m}$. We begin with the following result.

Table 1. Sing($C_{a,a+b}$):=SINGULAR POINTS of $C_{a,a+b}$

$a = 2$ and $b = 0$	There are none
a is EVEN, $a \neq 2$ and $b = 0$	$(\tau, 1, 0) \ / \ \tau^{r_0} = 1$
a is EVEN and $b = 1$	There are none
a is EVEN and $b \neq 0, 1$	$(0, 1, 0)$
a is ODD and $b = 0, 1$	$(0, \sigma, 1) \ / \ \sigma^{s_0} = 1$
a is ODD and $b \neq 0, 1$	$(0, 1, 0), (0, \sigma, 1) \ / \ \sigma^{s_0} = 1$

Lemma 1. *We consider the affine curve $C_0 : x_0^u + y_0^v + f_0(x_0, y_0) = 0$, where $u, v \in \mathbb{N}$, $u \geq v \geq 2$ and $f_0(x_0, y_0) = \sum_{r,s:us+vr>uv} x_0^r y_0^s$ with $r, s \in \mathbb{N}$ and $\alpha = \gcd(u, v)$ **odd**. Then for the singular point of C_0, $P_0 = (0, 0)$, we have*
$$\delta(P_0) = \frac{(u-1)(v-1)+\alpha-1}{2}$$

Proof. We write $u = n_1 v + r_1$, with $0 \leq r_1 \leq v - 1$ and we distinguish the next cases.

(i) If $r_1 = 0$ then $\alpha = v$ and applying $(n_1 - 1)$ quadratic transformations $(QDTs)$ given by $x_i = x_{i+1}$, $y_i = x_{i+1} y_{i+1}$, we obtain the curve
$$C_{n_1-1} : x_{n_1-1}^v + y_{n_1-1}^v + f_{n_1-1}(x_{n_1-1}, y_{n_1-1}) = 0$$
where $f_{n_1-1}(x_{n_1-1}, y_{n_1-1}) = \sum_{r,s} x_{n_1-1}^{r+(n_1-1)s-(n_1-1)v} y_{n_1-1}^s$.
We note that $\forall i = 1, ..., n_1 - 1$, C_i: $x_i^{u-iv} + y_i^v + f_i(x_i, y_i) = 0$, where $f_i(x_i, y_i) = \sum_{r,s} x_i^{r+is-iv} y_i^s$, with $(r + is - iv)v + (u - iv)s > (u - iv)v$.
Hence we get, from C_{n_1-1}, the nonsingular curve of C_0, $\widetilde{C_0}$. In conclusion we have, by blowing-up, a singularity tree with n_1 points of multiplicity v plus v smooth points on $\widetilde{C_0}$, so $\delta(P_0) = \frac{(v-1)u}{2}$. This result coincides with the Lemma when $v = \alpha$.

(ii) If $r_1 \neq 0$, then $\exists n_2, r_2 \in \mathbb{N}$ such that $v = n_2 r_1 + r_2$. We distinguish again two cases.

(ii.1) If $r_2 = 0$, then $\alpha = r_1$. After $(n_1 - 1)$ $QDTs$ we have the curve
C'_{n_1-1}: $x_{n_1-1}^{v+r_1} + y_{n_1-1}^v + f_{n_1-1}(x_{n_1-1}, y_{n_1-1}) = 0$.
We note that $\forall i = 1, ..., n_1 - 1$, C'_i: $x_i^{(n_1-i)v+r_1} + y_i^v + f_i(x_i, y_i) = 0$, where $f_i(x_i, y_i) = \sum_{r,s} x_i^{r+is-iv} y_i^s$, with $((n_1 - i)v + r_1)s + (r + is - iv)v > ((n_1 - i)v + r_1)v$.
The QDT given by $x_{n_1-1} = x_{n_1}$, $y_{n_1-1} = x_{n_1} y_{n_1}$, transforms the curve C'_{n_1-1} into the curve $C'_{n_1} : x_{n_1}^{r_1} + y_{n_1}^v + f_{n_1}(x_{n_1}, y_{n_1}) = 0$. We are in the case (i), where C'_{n_1} is like C_0 and $v = n_2 r_1$. So we have a singularity tree with n_1 points of multiplicity v, plus n_2 points of multiplicity α, plus α smooth points on $\widetilde{C_0}$. Then
$$\delta(P_0) = \frac{n_1 v(v-1) + n_2 \alpha(\alpha-1)}{2} = \frac{(u-1)(v-1) + \alpha - 1}{2}$$

(ii.2) If $r_2 \neq 0$, then $\exists\, n_3, r_3 \in \mathbb{N} \,/\, r_1 = n_3 r_2 + r_3$. In the case $r_3 = 0$, we have $r_2 = \alpha$, and the sequence of multiplicities is: $\overbrace{v, ..., v}^{n_1}, \overbrace{r_1, ..., r_1}^{n_2}$, $\overbrace{\alpha, ..., \alpha}^{n_3}, \overbrace{1, ..., 1}^{\alpha}$. Then $\delta(P_0) = \frac{n_1 v(v-1) + n_2 r_1 (r_1 - 1) + n_3 \alpha(-1)}{2} = \frac{vu - u - v + \alpha}{2} = \frac{(u-1)(v-1)+\alpha-1}{2}$. So, by applying a finite number of cases, we get $r_i \in \mathbb{N}^*$ such that $r_i = \alpha$ (i.e., $r_{i+1} = 0$). Then $r_{i-1} = n_{i+1} r_i$. Therefore the sequence of multiplicities is

$$\overbrace{v, ..., v}^{n_1}, \overbrace{r_1, ..., r_1}^{n_2}, \overbrace{r_2, ..., r_2}^{n_3}, \overbrace{r_3, ..., r_3}^{n_4}, ..., \overbrace{\alpha, ..., \alpha}^{n_{i+1}}, \overbrace{1, ..., 1}^{\alpha}$$

hence $\delta(P_0) = \frac{vu - u - v + \alpha}{2} = \frac{(u-1)(v-1)+\alpha-1}{2}$.

We write, $\forall a \geq 2,\, b \geq 0,\, a + b = 2^n b_0$ with $n \geq 0,\, b_0 \geq 1$ and b_0 **odd**, and $\forall n' \geq 0,\, m = 2^{n'} b_0$. We consider the following curves

$$-If \quad m \geq a, \quad C_{a,m} : Y^a Z^{m-a} + Y Z^{m-1} + X^m = 0$$

$$-If \quad m < a \quad D_{a,m} : Y^a + Y Z^{a-1} + X^m Z^{a-m} = 0$$

Remark 1. There is a purely inseparable map ϕ between the curves $C_{a,a+b}$ and $C_{a,m}$, and also between $C_{a,a+b}$ and $D_{a,m}$, with degree $2^{n-n'}$ if $n \geq n'$ or $2^{n'-n}$ if $n' > n$.

Remark 2. (See Proposition 1) The curves $C_{a,a+b}$, $C_{a,m}$ and $D_{a,m}$ have the same genus.

We are going to determine of genus of $C_{a,a+b}$ $(b \geq 0)$ and hence, we compute of genus of the curves defined by affine equation

$$y^a + y + x^{a+b} = 0, \forall a \geq 2, \forall b \geq -a.$$

Theorem 1. *For any* $a, b \in \mathbb{N},\ a \geq 2,\ b \geq 0$ *we have*

$$g(C_{a,a+b}) = \frac{s_0(b_0 - 1)}{2} - \alpha_0$$

where $a + b = 2^n b_0$ *with* $n \geq 0$ *and* $b_0 \geq 1$ **odd**; $a - 1 = 2^s s_0$ *with* $s \geq 0$ *and* $s_0 \geq 1$ **odd** *and* $\alpha = \gcd(a, b_0)$, $\alpha = 2\alpha_0 + 1$.

Proof. We consider the seven following cases (see Table 1).

(A) If $a = 2$ and $b = 0$, then $b_0 = 1$, $s_0 = 1$, and $\alpha_0 = 0$. In this case there are no singular points, so $g(C_{2,2}) = 0$, i.e, $g(C_{2,2}) = \frac{s_0(b_0-1)}{2} - \alpha_0 = 0$.

(B) If a is **even**, $a \neq 2$ and $b = 0$ then $a = 2^n b_0$ (i.e., $b_0 = r_0$), $s_0 = a - 1$, $\alpha = \gcd(a, b_0) = b_0$ and $\text{Sing}(C_{a,a}) = \{(\tau, 1, 0)/\tau^{r_0} = 1\}$. We consider the affine curve $1 + z^{a-1} + x^a = 0$. After a translation $x = x_1 + \tau$, $z = z_1$, we may assume that it pass through the origin $(0, 0)$, so

$$1 + z_1^{a-1} + (x_1 + \tau)^a = 0 \Leftrightarrow 1 + z_1^{a-1} + (x_1^{2^n} + \tau^{2^n})^{b_0} = 0 \Leftrightarrow$$

$$\Leftrightarrow z_1^{a-1} + \tau^{2^n(b_0-1)} x_1^{2^n} + p(x_1) = 0, \text{with } \partial p(x_1) \geq 2^n + 1.$$

We have that $\gcd(a - 1, 2^n) = 1$, and by Lemma 1 $\delta((\tau, 1, 0)) = \frac{(a-2)(2^n - 1)}{2}$. Hence $g(C_{a,a}) = \frac{(a-1)(b_0-1)-(b_0-1)}{2} = \frac{s_0(b_0-1)}{2} - \alpha_0$.

(C) If a is **even** and $b = 1$ then $a + 1 = b_0$, $s_0 = a - 1$, and $\alpha_0 = 0$. There are no singular points, and $g(C_{a,a+1}) = \frac{a(a-1)}{2}$, so $g(C_{a,a+1}) = \frac{s_0(b_0-1)}{2} - \alpha_0$.

(D) If a is **even**, $b \neq 0$, $b \neq 1$, then $C_{a,a+b}$ has only one singular point, namely $(0, 1, 0)$. We distinguish two cases: $b_0 > a$ or $b_0 < a$.

(D.1) If $b_0 > a$, we consider the curve C_{a,b_0}, with $b_0 - a \geq 1$.
(D.1.1) If $b_0 - a = 1$, we are in the case **(C)** and then $g(C_{a,a+b}) = g(C_{a,a+1}) = \frac{a(a-1)}{2}$. We have $a + 1 = b_0$, $a - 1 = s_0$ and $\alpha = \gcd(a, b_0) = 1$, then $g(C_{a,a+b}) = \frac{s_0(b_0-1)}{2} - \alpha_0$.
(D.1.2) If $b_0 - a > 1$, then $\text{Sing}(C_{a,a+b}) = \{(0, 1, 0)\}$. For the affine curve $z^{b_0-a} + z^{b_0-1} + x^{b_0} = 0$ we have by Lemma 1

$$\delta((0, 1, 0)) = \frac{(b_0 - a - 1)(b_0 - 1) + \alpha - 1}{2}$$

where $\alpha = \gcd(a, b_0)$. Then

$$g(C_{a,a+b}) = \frac{(b_0-1)(a-1)}{2} - \alpha_0 = \frac{s_0(b_0-1)}{2} - \alpha_0.$$

(D.2) If $b_0 < a$, we consider the curve D_{a,b_0}, $Y^a + Y Z^{a-1} + X^{b_0} Z^{a-b_0} = 0$. Its singular points have to satisfy the relations

$$\begin{cases} X^{b_0-1} Z^{a-b_0} = 0 \\ Z^{a-1} = 0 \\ Y Z^{a-2} + X^{b_0} Z^{a-b_0-1} = 0 \\ Y^a + Y Z^{a-1} + X^{b_0} Z^{a-b_0} = 0 \end{cases}$$

(D.2.1) If $a - b_0 = 1$, then there are no singular points, $a - 1 = b_0 = s_0$, $\alpha = \gcd(a, b_0) = 1$ and

$$g(C_{a,a+b}) = g(D_{a,b_0}) = \frac{(a-1)(a-2)}{2} = \frac{s_0(b_0-1)}{2} - \alpha_0$$

(D.2.2) If $a - b_0 > 1$, the only singular point is $(1, 0, 0)$, and we consider the affine curve: $y^a + yz^{a-1} + z^{a-b_0} = 0$ and $\alpha = \gcd(a, a - b_0)$ is odd, so we have by Lemma 1 $\delta((1, 0, 0)) = \frac{(a-1)(a-b_0-1)+\alpha-1}{2}$. Hence

$$g(C_{a,a+b}) = g(D_{a,b_0}) = \frac{(a-1)(b_0-1)+\alpha-1}{2} = \frac{s_0(b_0-1)}{2} - \alpha_0.$$

(E) If a is **odd**, $b = 0$, then $a = b_0$, $b_0 = 2^s s_0 + 1$, $\alpha = b_0$ and $\mathrm{Sing}(C_{a,a}) = \{(0, \sigma, 1)/\sigma^{s_0} = 1\}$. We consider the affine curve $y^a + y + x^a = 0$, which we do to pass by $(0, 0)$ doing $x = x_1$, $y = y_1 + \sigma$, so

$$(y_1 + \sigma)^a + (y_1 + \sigma) + x_1^a = 0 \Leftrightarrow$$
$$\Leftrightarrow (y_1^{2^s} + \sigma^{2^s})^{s_0}(y_1 + \sigma) + (y_1 + \sigma) + x_1^a = 0 \Leftrightarrow$$
$$\Leftrightarrow x_1^a + \sigma^{2^s(s_0-1)+1} y_1^{2^s} + p(y_1) = 0, \ with \ \partial p(y_1) \geq 2^s + 1.$$

Since $\gcd(a, 2^s) = 1$ and Lemma 1 $\delta((0, \sigma, 1)) = \frac{(a-1)(2^s-1)}{2}$. Hence

$$g(C_{a,a}) = \frac{(a-1)(s_0-1)}{2} = \frac{(s_0-1)(b_0-1)}{2} = \frac{s_0(b_0-1)}{2} - \alpha_0.$$

(F) If a is **odd**, and $b = 1$, then $a + 1 = 2^n b_0$. We distinguish the two possible cases.

(F.1) If $b_0 > a$, we consider the curve C_{a,b_0}, $Y^a Z^{b_0-a} + Y Z^{b_0-1} + X^{b_0} = 0$. As $b_0 - a \geq 2$, then $\mathrm{Sing}(C_{a,b_0}) = \{(0, 1, 0), (0, \sigma, 1)/\sigma^{s_0} = 1\}$. We have the affine curve $z^{b_0-a} + z^{b_0-1} + x^{b_0} = 0$, with $\gcd(b_0 - a, b_0) = \alpha$ which is odd, and then by Lemma 1

$$\delta((0, 1, 0)) = \frac{(b_0 - a - 1)(b_0 - 1) + \alpha - 1}{2}$$

If we now consider the affine curve $y^a + y + x^{b_0} = 0$, and we do $x = x_1$, $y = y_1 + \sigma$, we obtain

$$(y_1 + \sigma)^a + (y_1 + \sigma) + x_1^{b_0} = 0 \Leftrightarrow$$
$$\Leftrightarrow (y_1^{2^s} + \sigma^{2^s})^{s_0}(y_1 + \sigma) + (y_1 + \sigma) + x_1^{b_0} = 0 \Leftrightarrow$$
$$\Leftrightarrow x_1^{b_0} + \sigma^{2^s(s_0-1)+1} y_1^{2^s} + p(y_1) = 0, \ with \ \partial p(y_1) \geq 2^s + 1.$$

As $\gcd(b_0, 2^s) = 1$, by Lemma 1 $\delta((0, \sigma, 1)) = \frac{(b_0-1)(2^s-1)}{2}$. So

$$g(C_{a,a+1}) = \frac{(b_0 - 1)(a - 1 - 2^s s_0 + s_0)}{2} - \alpha_0 = \frac{s_0(b_0 - 1)}{2} - \alpha_0.$$

(F.2) Si $b_0 < a$, we consider the curve D_{a,b_0}, $Y^a + Y Z^{a-1} + X^{b_0} Z^{a-b_0} = 0$. Its singular points have to verify

$$\begin{cases} X^{b_0-1} Z^{a-b_0} = 0 \\ Y^{a-1} + Z^{a-1} = 0 \\ Y^a + Y Z^{a-1} + X^{b_0} Z^{a-b_0} = 0 \end{cases}$$

Then, if $b_0 = 1$, the only singular point is $(1, 0, 0)$; and if $b_0 \neq 1$, the singular points are: $\{(1, 0, 0), (0, \sigma, 1)/\sigma^{s_0} = 1\}$. We distinguish again two cases.

(F.2.1) If $b_0 = 1$, we consider the affine curve $y^a + yz^{a-1} + z^{a-1} = 0$, as $\gcd(a, a-1) = 1$, by Lemma 1 $\delta((1, 0, 0)) = \frac{(a-1)(a-2)}{2}$.

So, $g(C_{a,a+1}) = 0 = \frac{s_0(b_0-1)}{2} - \alpha_0$.

(**F.2.2**) If $b_0 \neq 1$, we consider the afinne curve $y^a + yz^{a-1} + z^{a-b_0} = 0$, and $\gcd(a, a - b_0) = \alpha$, then

$$\delta((1,0,0)) = \frac{(a-1)(a-b_0-1)+\alpha-1}{2}$$

If we consider the affine curve $y^a + y + x^{b_0} = 0$, we have $x = x_1$, $y = y_1 + \sigma$, so we obtain (case (**F.1**)), $x_1^{b_0} + \sigma^{2^s(s_0-1)+1} y_1^{2^s} + p(y_1) = 0$, with $\partial p(y_1) \geq 2^s + 1$ then

$$\delta((0,\sigma,1)) = \frac{(b_0-1)(2^s-1)}{2}$$

Hence

$$g(C_{a,a+1}) = \frac{(a-1)(b_0-1) - s_0(b_0-1)(2^s-1)}{2} - \alpha_0 = \frac{s_0(b_0-1)}{2} - \alpha_0.$$

(**G**) If a is **odd**, $b \neq 0$ and $b \neq 1$. We distinguish the cases: $b_0 = a$, $b_0 > a$ and $b_0 < a$.

(**G.1**) If $b_0 = a$, we consider the curve $C_{a,a}$, which has been already studied in (**E**).

(**G.2**) If $b_0 > a$, we have the curve C_{a,b_0}, $Y^a Z^{b_0-a} + YZ^{b_0-1} + X^{b_0} = 0$, and $\mathrm{Sing}(C_{a,b_0}) = \{(0,1,0),(0,\sigma,1)/\sigma^{s_0} = 1\}$. We consider the affine curve $z^{b_0-a} + z^{b_0-1} + x^{b_0} = 0$, $\gcd(b_0 - a, b_0) = \alpha$ is odd, and by Lemma 1, we have
$\delta((0,1,0)) = \frac{(b_0-a-1)(b_0-1)+\alpha-1}{2}$.
For the affine curve $y^a + y + x^{b_0} = 0$, we do $x = x_1$, $y = y_1 + \sigma$, so we have the affine curve: $(y_1 + \sigma)^a + (y_1 + \sigma) + x_1^{b_0} = 0$, which was studied in (**F.1**), and we know that

$$\delta((0,\sigma,1)) = \frac{(b_0-1)(2^s-1)}{2}$$

Hence $g(C_{a,a+b}) = g(C_{a,b_0}) = \frac{s_0(b_0-1)}{2} - \alpha_0$.

(**G.3**) If $b_0 < a$, we consider the curve D_{a,b_0}, $Y^a + YZ^{a-1} + X^{b_0}Z^{a-b_0} = 0$. Its singular points have to verify

$$\begin{cases} X^{b_0-1}Z^{a-b_0} = 0 \\ Y^{a-1} + Z^{a-1} = 0 \\ YZ^{a-2} + X^{b_0}Z^{a-b_0-1} = 0 \\ Y^a + YZ^{a-1} + X^{b_0}Z^{a-b_0} = 0 \end{cases}$$

(**G.3.1**) If $b_0 = 1$, the only singular point is $(1,0,0)$. We consider the affine curve $y^a + yz^{a-1} + z^{a-1} = 0$. By Lemma 1

$$\delta((1,0,0)) = \frac{(a-1)(a-2)}{2}$$

so, $g(C_{a,a+b}) = g(D_{a,1}) = \frac{(a-1)(a-2)}{2} - \frac{(a-1)(a-2)}{2} = 0$, and since $\alpha = \gcd(a, b_0) = 1$, then $\alpha_0 = 0$, and $g(C_{a,a+b}) = \frac{s_0(b_0-1)}{2} - \alpha_0 = 0$.

(G.3.2) If $b_0 > 1$ then $\text{Sing}(D_{a,b_0}) = \{(1,0,0), (0,\sigma,1)/\sigma^{s_0} = 1\}$. We consider for the point $(1,0,0)$ the affine curve $y^a + yz^{a-1} + z^{a-b_0} = 0$ where $\alpha = \gcd(a, a - b_0)$ is odd. Then, by Lemma 1

$$\delta((1,0,0)) = \frac{(a-1)(a-b_0-1)+\alpha-1}{2}$$

For the points $(0,\sigma,1)$, we consider the affine curve $y^a + y + x^{b_0} = 0$, and we do $x = x_1$ and $y = y_1 + \sigma$. Then, we have the affine curve $(y_1 + \sigma)^a + (y_1 + \sigma) + x_1^{b_0} = 0$. We are in the case **(F.1)** and

$$\delta((0,\sigma,1)) = \frac{(b_0-1)(2^s-1)}{2}$$

Then $g(C_{a,a+b}) = g(D_{a,b_0}) = \frac{s_0(b_0-1)}{2} - \alpha_0$.

Corollary 2. *For any $a, b \in \mathbb{Z}$, $a \geq 2$, $b > -a$, $a+b = 2^n b_0$, if C is the curve the equation $Y^a Z^b + YZ^{a+b-1} + X^{a+b} = 0$, with $b \geq 0$, or $Y^a + YZ^{a-1} + X^{a+b} Z^{-b} = 0$, with $b < 0$, then its genus is*

$$g(C) = \frac{s_0(b_0 - 1)}{2} - \alpha_0$$

where $a + b = 2^n b_0$ with $n \geq 0$ and $b_0 \geq 1$ odd; $a - 1 = 2^s s_0$ with $s \geq 0$ and $s_0 \geq 1$ odd and $\alpha = \gcd(a, b_0)$, $\alpha = 2\alpha_0 + 1$.

6   Examples of Maximal Curves

Let $g(C)$ be the genus of C and let $N(C(\mathbb{F}_q))$ be the number of $\mathbb{F}_q$-rational points of C (i.e., for the non-singular model of C). We present, for example, the following maximal Quasihermitian curves.

$C_{2,5} : Y^2 Z^3 + YZ^4 + X^5 = 0 \qquad g(C_{2,5}) = 2, \quad N(C_{2,5}(\mathbb{F}_{2^4})) = 33$

$C_{2,11} : Y^2 Z^9 + YZ^{10} + X^{11} = 0 \quad g(C_{2,11}) = 5, \quad N(C_{2,11}(\mathbb{F}_{2^{10}})) = 1,345$

$C_{2,13} : Y^2 Z^{11} + YZ^{12} + X^{13} = 0 \quad g(C_{2,13}) = 6, \quad N(C_{2,13}(\mathbb{F}_{2^{12}})) = 4,865$

$C_{3,5} : Y^3 Z^2 + YZ^4 + X^5 = 0 \qquad g(C_{3,5}) = 2, \quad N(C_{3,5}(\mathbb{F}_{2^4})) = 33$

$D_{4,3} : Y^4 + YZ^3 + X^3 Z = 0 \qquad g(D_{4,3}) = 3, \quad N(D_{4,3}(\mathbb{F}_{2^6})) = 113$

and considering that if $a + b = 2^n b_0$, with $b_0 \geq 1$ odd, $n \geq 0$, then there is a purely inseparable rational map ϕ between the curves $C_{a,a+b}$ and $C_{a,2^h b_0}$, with $2^h b_0 \geq a$, and between the curves $C_{a,a+b}$ and $D_{a,2^{h'} b_0}$, with $2^{h'} b_0 < a$ (see Remark 1), we can give other maximal curves from each one of the above curves. For example, with the same genus and the same number of rational points over $\mathbb{F}_{2^4}$ (therefore maximal curves) we have, from $C_{2,5}$, the curves: $C_{2,10}$, $C_{2,20}$ and $C_{2,40}$.

Acknowledgements

This work was supported by DGICYT (*Dirección General de Investigación del Ministerio de Ciencia y Tecnología*) under grants PB97-0284-C02-01, TIC2000-0735 and TIC 2000-0737-C03-02.

During this work J. García-Villalba was with the IBM Research Division K65/C2 at IBM Almaden Research Center, San Jose, California, USA (e-mail: javiervi@almaden.ibm.com). J. García-Villalba would like to express his appreciation to the Programa Complutense del Amo for providing him a grant to stay at IBM Almaden Research Center.

References

1. S. S. Abhyankar. *Algebraic geometry for scientists and engineers*. American Mathematical Society. (1990).
2. D. Le Brigand, J. J. Risler. *Algorithme de Brill-Noether et codage des codes de Goppa*. Publications du laboratoire danalyse numérique. C.N.R.S.
3. A. Cossidente, J. W. P. Hirschfeld, G. Korchmáros, F. Torres. *On plane maximal curves*. Math. AG/9802113. (Feb. 1998).
4. W. Fulton. *Algebraic Curves. An Introduction to Algebraic Geometry*. W.A. Bemjamin, Inc., New York. (1969).
5. R. Hartshorne. *Algebraic geometry*. GTM 76, Springer-Verlag, New York. (1982).
6. M. Namba. *Geometry of projective algebraic curves*. Marcel Dekker, inc. New York. (1984).
7. H. Stichtenoth. *Algebraic Function fields and codes*. Springer-Verlag Berlin Heidelberg. (1993).
8. H. Stichtenoth. *A note on Hermitian codes over $GF(q^2)$*. IEEE Trans. Inf. Theory, vol. 34 (5), pp. 1345-1347. (September 1988).
9. H. J. Tiersma. *Remarks on codes from Hermitian curves*. IEEE Trans. Inform. Theory, vol. IT-33(4). (July, 1987).

Iterations of Multivariate Polynomials
and Discrepancy of Pseudorandom Numbers

Jaime Gutierrez and Domingo Gomez-Perez

Department of Mathematics and Computing,
Faculty of Science, University of Cantabria,
Santander E–39071, Spain
{jaime, domingo}@matesco.unican.es

Abstract. In this paper we present an extension of a result in [2] about a discrepancy bound for sequences of s-tuples of successive nonlinear multiple recursive congruential pseudorandom numbers of higher orders. The key of this note is based on linear properties of the iterations of multivariate polynomials.

1 Introduction

The paper [2] studies the distribution of pseudorandom number generators defined by a recurrence congruence modulo a prime p of the form

$$u_{n+1} \equiv f(u_n, \ldots, u_{n-m+1}) \pmod{p}, \qquad n = m - 1, m, \ldots, \qquad (1)$$

with some *initial values* $u_0, \ldots, u_{m-1}$, where $f(X_1, \ldots, X_m)$ is a polynomial of m variables over the field $\mathbb{F}_p$ of p elements. These nonlinear congruential generators provide a very attractive alternative to linear congruential generators and, especially in the case $m = 1$, have been extensively studied in the literature, see [1] for a survey.

When $m = 1$, for sequences of the largest possible period $t = p$, a number of results about the distribution of the fractions u_n/p in the interval $[0, 1)$ and, more generally, about the distribution of the points

$$\left(\frac{u_n}{p}, \cdots, \frac{u_{n+s-1}}{p} \right) \qquad (2)$$

in the s-dimensional unit cube $[0, 1)^s$ have been obtained, see the recent series of papers [3,5,6,7,8] for more details. In the paper [2], the same method for nonlinear generators of arbitrary order $m > 1$ is presented. In particular, the paper [2] gives a nontrivial upper bound on exponential sums and the discrepancy of corresponding sequences for polynomials of total degree $d > 1$ which have a *dominating term* (see Theorem 1 and Theorem 2 in that paper). As in [2], we say that a polynomial $f(X_1, \ldots, X_m) \in \mathbb{F}_p[X_1, \ldots, X_m]$ has a *dominating term* if it is of the form

$$f(X_1, \ldots, X_m) = a_{d_1 \ldots d_m} X_1^{d_1} \cdots X_m^{d_m} + \sum_{i_1=0}^{d_1-1} \cdots \sum_{i_m=0}^{d_m-1} a_{i_1 \ldots i_m} X_1^{i_1} \cdots X_m^{i_m}$$

S. Boztaş and I.E. Shparlinski (Eds.): AAECC-14, LNCS 2227, pp. 192–199, 2001.

with some integers $d_1 \geq 1, d_2 \geq 0, \ldots, d_m \geq 0$ and coefficients $a_{i_1 \ldots i_m} \in \mathbb{F}_p$ with $a_{d_1 \ldots d_m} \neq 0$. We denote by $\mathcal{DT}$ the class of polynomials having a dominating term.

In this paper we extend Theorem 1 and Theorem 2 of [2] to a very large class of polynomials, including arbitrary polynomials of degree greater than one with respect to the variable X_m, that is, polynomials f with $\deg_{X_m}(f) > 1$. This question appears in [2] as an important open problem. This note is based on properties about composition of multivariate polynomials which could be of independent interest.

The paper is divided into three sections. In Section 2 we study the behaviour of the polynomials under composition. Then Section 3 we extend the result of [2]. Finally, in Section 4 we pose some open problems.

2 Iterations of Multivariate Polynomials

Let $\mathbb{K}$ be an arbitrary field and let f be a polynomial in $\mathbb{K}[X_1, \ldots, X_m]$. As in the paper [2], we consider, for $k = 1, 2, \ldots$, the sequence of polynomials $f_k(X_1, \ldots, X_m) \in \mathbb{K}[X_1, \ldots, X_m]$ by the recurrence relation

$$f_k(X_1, \ldots, X_m) = f(f_{k-1}(X_1, \ldots, X_m), \ldots, f_{k-m}(X_1, \ldots, X_m)),$$

where $f_k(X_1, \ldots, X_m) = X_{1-k}$, for $k = -m + 1, \ldots, 0$.

In this section we will give sufficient conditions for the polynomial f such that the polynomial sequence $f_k, k = -m + 1, \ldots$, is linearly independent. In order to prove this we can suppose, without loss of generality, that $\mathbb{K}$ is an algebraically closed field. A central tool to study this sequence of polynomials is the following ring homomorphism :

$$\phi : \mathbb{K}[X_1, \ldots, X_m] \to \mathbb{K}[X_1, \ldots, X_m]$$

defined as: $\phi(X_1) = f$ and $\phi(X_k) = X_{k-1}$, for $k = 2, \ldots, m$.

Lemma 1. *With the above notations, we have the following:*

- $\phi^j(f_k) = f_{k+j}$, *for $j > 0$ and $k = -m + 1, \ldots, 0, 1, 2, \ldots$.*
- *The polynomial f has degree greater than zero with respect to the variable X_m if and only if ϕ^j is an injective map, for every $j \geq 1$. In particular, the $\{f_r, f_{r+1}, \ldots, f_{r+m-1}\}$ are algebraically independent, for all $r \geq -m + 1$.*

Proof. The proof of the first part it is trivial by the definition of the rinh homomorphism ϕ.

On the other hand, we have that ϕ is injective map if and only if its kernel is trivial, that is, ϕ is injective if and only if

$$\{p \in \mathbb{K}[X_1, \ldots, X_m], \quad \phi(p) = 0\} = \{0\}.$$

If $p \in \mathbb{K}[X_1, \ldots, X_m]$, then $\phi(p) = p(f, X_1, \ldots, X_{m-1})$; so $p = 0$ if and only if $\{X_{m-1}, \ldots, X_1, f\}$ are algebraically independent. If $\deg_{X_m}(f) > 0$ then X_m is algebraically dependent over $\mathbb{K}(f, X_1, \ldots, X_{m-1})$. Consequently $\{X_{m-1}, \ldots, X_1, f\}$ are algebraically independent over $\mathbb{K}$ if and only if we have $\deg_{X_m}(f) > 0$.

Finally, by the first part, we see that $\phi^{r+m}(X_{m-j}) = f_{r+j}$, for $j = 0, \ldots, m-1$. Now, the claim follows by induction on r.   □

We say that a multivariate polynomial $f(X_1, \ldots, X_m) \in \mathbb{K}[X_1, \ldots, X_m]$ is *quasi-linear* in X_m if it is of the form $f = aX_m + g$ where $0 \neq a \in \mathbb{K}$ and $g \in \mathbb{K}[X_1, \ldots, X_{m-1}]$. We denote by $\mathcal{NL}$ the class of *non quasi-linear* in X_m polynomials of degree greater than zero with respect to the variable X_m. So, the class $\mathcal{NL}$ is the set of all polynomials except the polynomials which do not depend on X_m and the *quasi-linear* polynomials.

Lemma 2. *Let f be an element of $\mathcal{NL}$. Then any finite family of the polynomials f_k, $k = -m + 1, \ldots, 0, 1, \ldots$, is linearly independent.*

Proof. We prove it by induction on m. For $m = 1$ it is obvious, because the degree is multiplicative with respect to polynomial composition. Now, we assume that $\deg_{X_m}(f) > 0$ and we suppose that we have a nonzero linear combination:

$$a_r f_r + a_{r+1} f_{r+1} + \cdots + a_{r+s} f_{r+s} = 0, \tag{3}$$

where $a_j \in \mathbb{K}$ and $a_r \neq 0$. We claim that $X_m \in \mathcal{I}$, where $\mathcal{I}$ is the ideal in the polynomial ring $\mathbb{K}[X_1, \ldots, X_m]$, generated by:

$$\mathcal{I} = (X_1, \ldots, X_{m-1}, \bar{f}),$$

with $\bar{f} = f - f(0, \ldots, 0)$.

By Lemma 1, ϕ^{r+m-1} is an injective map and

$$\phi^{r+m-1}(f_{-m+1}) = \phi^{r+m-1}(X_m) = f_r.$$

Applying the inverse of ϕ^{r+m-1} to equation (3), we obtain:

$$a_r X_m + a_{r+1} X_{m-1} + \cdots + a_{r+s} f_{s-m+1} = 0. \tag{4}$$

We show that $\bar{f}_t = f_t - f_t^0 \in \mathcal{I}$, where $f_t^0 = f_t(0, \ldots, 0)$. By the uniqueness of the classical euclidean division

$$f = (X_1 - f_{t-1}^0) g_1 + r_1(X_2, \ldots, X_m)$$

and

$$r_1(X_2, \cdots, X_m) = (X_2 - f_{t-2}^0) g_2 + r_2(X_3, \ldots, X_m).$$

Now, by recurrence, we have:

$$f = (X_1 - f_{t-1}^0) g_1 + \cdots + (X_{m-1} - f_{t-m+1}^0) g_{m-1} + (X_m - f_{t-m}^0) g_m + g_0,$$

where $g_i \in \mathbb{K}[X_i, \ldots, X_m]$, $i = 0, \ldots, m$.

Since, $f_t = f(f_{t-1}, \ldots, f_{t-m})$ we have that $g_0 = f_t^0$. Now, by induction on t, we will show that $\bar{f}_t \in \mathcal{I}$, for $t > 0$. In order to see that, we observe that

$$f_t = f(f_{t-1}, \ldots, f_{t-m})) =$$

$$= \bar{f}_{t-1} g_1(f_{t-1}, \ldots, f_{t-m})) + \cdots + \bar{f}_{t-m} g_m(f_{t-1}, \ldots, f_{t-m})) + g_0.$$

Then, $\bar{f}_t = f_t - g_0 \in \mathcal{I}.$

Using the equation (4), we have:

$$a_r X_m = -a_r^{-1}(a_{r+1} X_{m-1} + \cdots + a_{r+s} f_{s-m+1}).$$

And have just proved that $X_m \in \mathcal{I}$. So, there exist polynomials $w_i \in \mathbb{K}[X_1, \ldots, X_m]$, $i = 1, \ldots, m$, such that

$$X_m = X_1 w_1 + \cdots + X_{m-1} w_{m-1} + \bar{f} w_m,$$

then $X_m = \bar{f}(0, \ldots, 0, X_m) w_m(0, \ldots, 0, X_m)$. As consequence, we can write f as follows:

$$f = X_1 h_1 + \cdots + X_{m-1} h_{m-1} + \alpha X_m + \beta, \tag{5}$$

where $h_i \in \mathbb{K}[X_i, \ldots, X_m]$, $(i = 1, \ldots, m-1)$, $\alpha, \beta \in \mathbb{K}$ and $\alpha \neq 0$. Now, we consider the polynomial

$$H = f(X_1, \ldots, X_{m-1}, Y) - f(X_1, \ldots, X_{m-1}, Z) \in \mathbb{K}[X_1, \ldots, X_{m-1}, Y, Z].$$

We claim there exists a zero $(\alpha_{0,1}, \ldots, \alpha_{0,m-1}, \beta_0, \gamma_0) \in \mathbb{K}^{m+1}$ of the polynomial H, with $\beta_0 \neq \gamma_0$. In order to prove this last claim, we write the polynomial f as univariate polynomial in the variable X_m with coefficients b_j in the polynomial ring $\mathbb{K}[X_1, \ldots, X_{m-1}]$, for $j = 0, \ldots, s$, that is, $f = b_s X_m^s + \cdots + b_1 X_m + b_0$, for $j = 0, \ldots, s$ and $b_s \neq 0$. So,

$$H = b_s(Y^s - Z^s) + \cdots + b_1(Y - Z).$$

If a such zero does not exist, then the zero set of h coincides with the zero set of the polynomial $Y - Z$. Since $Y - Z$ is an irreducible polynomial in $\mathbb{K}[X_1, \ldots, X_{m-1}, Y, Z]$, then by the Nullstellensatz theorem, (see for instance [4]) H is a power of $Y - Z$, i.e., there exists a positive natural number t such that $H = \gamma(Y - Z)^t$, where $0 \neq \gamma \in \mathbb{K}$. We have the following:

$$b_s(Y^s - Z^s) + \cdots + b_1(Y - Z) = \gamma(Y - Z)^t.$$

¿From this polynomial equality, we obtain that $s = t$. Since $\gamma(Y - Z)^s$ is a homogenous polynomial, then $b_s(Y^s - Z^s) = \gamma(Y - Z)^s$. Now, from (5), we get that $s = 1$ and f must be $b_1 X_m + b_0$, that is, f is a quasi-linear polynomial in X_m. By the assumption $f \in \mathcal{NL}$, this is a contradiction.

Finally, we evaluate the left hand of the polynomial equality (4) in the point $P_0 = (\alpha_{0,1}, \ldots, \alpha_{0,m-1}, \beta_0)$, we obtain:

$$a_r \beta_0 + \ldots + a_{r+m-1} \alpha_{0,1} + a_{r+m} f(P_0) + \cdots + a_{r+s} f_{r+s-m}(P_0) = 0. \tag{6}$$

We also evaluate (4) in the point $Q_0 = (\alpha_{0,1}, \ldots, \alpha_{0,m-1}, \gamma_0)$ and we obtain:

$$a_r \gamma_0 + \cdots + a_{r+m-1} \alpha_{0,1} + a_{r+m} f(Q_0) + \cdots + a_{r+s} f_{r+s-m}(Q_0) = 0. \tag{7}$$

We observe that $f_k(P_0) = f_k(Q_0)$ for all $k \geq 0$. Thus, subtracting the equation (7) from the equation (6), we get $a_r(\beta_0 - \gamma_0) = 0$. Again, this is a contradiction and, the result follows.   □

We can also extend the above result to another class of polynomials. We say that a multivariate polynomial $f(X_1, \ldots, X_m) \in \mathbb{K}[X_1, \ldots, X_m]$ of total degree d, has the *dominating variable* X_1 if it is of the form

$$f = a_d X_1^d + a_{d-1} X_1^{d-1} + \cdots + a_0$$

where $d > 0$ and $a_i \in \mathbb{K}[X_2, \ldots, X_m]$, with $a_d \neq 0$. We denote by $\mathcal{DV}$ the class of polynomials having the dominating variable X_1.

Lemma 3. *With the above notations, for polynomial $f \in \mathcal{DV}$ the total degree of the polynomial f_k is d^k, $k = 1, 2, \ldots$. In particular, if $d > 1$, any finite family of the polynomials f_k, $k = -m+1, \ldots, 0, 1, \ldots$, is linearly independent.*

Proof. We prove this statement by induction on k. For $k = 1$ it is obvious.

Now we assume that $k \geq 2$. We have

$$f_k = a_d f_{k-1}^d + a_{d-1}(f_{k-2}, \ldots, f_{k-(m-1)}) f_{k-1}^{d-1} + \cdots + a_0(f_{k-2}, \ldots, f_{k-(m-1)})$$

We remark that for all

$$\deg(a_{d-i}) \leq i, \qquad i = 0, \ldots, d,$$

because $\deg f = d$. Using the induction assumption we obtain

$$\deg(a_{d-i}(f_{k-2}, \ldots, f_{k-(m-1)}) f_{k-1}^{d-i})$$
$$= \deg(a_{d-i}(f_{k-2}, \ldots, f_{k-(m-1)})) + \deg(f_{k-1}^{d-i}) \leq i d^{k-2} + (d-i) d^{k-1},$$

for all $i = 1, \ldots, d$. On the other hand

$$\deg(a_d f_{k-1}^d) \geq \deg(f_{k-1}^d) = d^k.$$

Finally, we observe that $d^k > i d^{k-2} + (d-i) d^{k-1}$ for all $i = 1, \ldots, d$. □

We have the following corollary:

Corollary 1. *If f is a polynomial in $\mathbb{K}[X_1, X_2]$ of total degree greater than one, then any finite family of the polynomials f_k, $k = -m+1, \ldots, 0, 1, \ldots$, is linearly independent.*

Proof. It is an immediate consequence of Lemmas 2 and 3 □

We observe that any polynomial in the class $\mathcal{NL}$ has total degree greater than one. On the other hand, if f is a linear polynomial, the sequence f_k, $k = 1, \ldots$, is obviously linearly dependent.

The following examples illustrate that we have three different classes of multivariate polynomial in m variables. The polynomial $f = X_1^2 + X_2 X_1$ has dominating variable X_1, that is, $f \in \mathcal{DV}$, but it has not a dominating term, $f \notin \mathcal{DT}$. We also have, that f is not a quasi-linear polynomial in X_2. Conversely, $g = X_1 X_2 + 1 \in \mathcal{DT} \bigcap \mathcal{NL}$, but $f \notin \mathcal{DV}$. Finally, $h = X_1^2 + X_2 \in \mathcal{DT} \bigcap \mathcal{DV}$, but $h \notin \mathcal{NL}$.

3   Discrepancy Bound

We denote by $\mathcal{T}$ the union of the three classes $\mathcal{T} = \mathcal{DV} \bigcup \mathcal{DT} \bigcup \mathcal{NL}$.

Following the proof of Theorem 1 in [2], we note that the only condition that they require is the statement of the above results. So, as a consequence of Lemma 2 and 3 and Corollary 1 we have Theorem 1 and Theorem 2 of [2] for polynomials $f(X_1, \ldots, X_m) \in \mathbb{F}_p[X_1, \ldots, X_m]$ with $f \in \mathcal{T}$ if $m > 2$ and for any non-linear polynomial f if $m = 2$.

As in the paper [2], let the sequence (u_n) generated by (1) be purely periodic with an arbitrary period $t \leq p^m$. For an integer vector $\mathbf{a} = (a_0, \ldots, a_{s-1}) \in Z^s$, we introduce the exponential sum

$$S_{\mathbf{a}}(N) = \sum_{n=0}^{N-1} \mathbf{e} \left(\sum_{j=0}^{s-1} a_j u_{n+j} \right),$$

where $\mathbf{e}(z) = \exp(2\pi i z / p)$.

Theorem 1. *Suppose that the sequence (u_n), given by (1) generated by a polynomial $f(X_1, \ldots, X_m) \in \mathbb{F}_p[X_1, \ldots, X_m]$ of the total degree $d \geq 2$ is purely periodic with period t and $t \geq N \geq 1$. If $m = 2$ or $f \in \mathcal{T}$, then the bound*

$$\max_{\gcd(a_0, \ldots, a_{s-1}, p)=1} | S_{\mathbf{a}}(N) | = O\left(N^{1/2} p^{m/2} \log^{-1/2} p \right)$$

holds, where the implied constant depends only on d and s.

As in the paper [2], for a sequence of N points

$$\Gamma = (\gamma_{1,n}, \ldots, \gamma_{s,n})_{n=1}^N$$

of the half-open interval $[0, 1)^s$, denote by Δ_Γ its discrepancy, that is,

$$\Delta_\Gamma = \sup_{B \subseteq [0,1)^s} \left| \frac{T_\Gamma(B)}{N} - | B | \right|,$$

where $T_\Gamma(B)$ is the number of points of the sequence Γ which hit the box

$$B = [\alpha_1, \beta_1) \times \ldots \times [\alpha_s, \beta_s) \subseteq [0, 1)^s$$

and the supremun is taken over all such boxes.

Let $D_s(N)$ denote the discrepancy of the points (2) for $n = 0, \ldots, N - 1$.

Theorem 2. *Suppose that the sequence (u_n), given by (1) generated by a polynomial $f(X_1, \ldots, X_m) \in \mathbb{F}_p[X_1, \ldots, X_m]$ of the total degree $d \geq 2$ is purely periodic with period t and $t \geq N \geq 1$. If $m = 2$ or $f \in \mathcal{T}$, then the bound*

$$D_s(N) = O\left(N^{1/2} p^{m/2} \log^{-1/2} p (\log \log p)^s \right)$$

holds, where the implied constant depends only on d and s.

In particular, Theorems 1 and 2 apply to any *non-linear* with respect to X_1 polynomial. Thus these are direct generalizations of the results of [5].

4   Remarks

We have extended the results of [2] to a very large class of polynomials, including multivariate polynomials f such that $\deg_{X_m}(f) > 1$. The only remain open problem is for a subclass of polynomials of the form $g(X_1, \ldots, X_{m-1}) + aX_m$, where $a \in \mathbb{K}^\times$.

On the other hand, it would be very interesting to extend these results to the case of generators defined by a list of m polynomials of $\mathbb{F}_p[X_1, \ldots, X_m]$:

$$\mathbf{F} = (f_1(X_1, \ldots, X_m), \ldots, f_m((X_1, \ldots, X_m))$$

For each $i = 1, \ldots, m$ we define the sequence of polynomials $f_i^{(k)}(X_1, \ldots, X_m) \in \mathbb{F}_p[X_1, \ldots, X_m]$ by the recurrence relation

$$f_i^{(0)} = f_i, \quad f_i^{(k)}(X_1, \ldots, X_m) = f_i^{(k-1)}(f_1, \ldots, f_m), \qquad k = 0, 1, \ldots.$$

So, for very k, we have the following list of m multivariate polynomials:

$$\mathbf{F^k} = (f_1^k(X_1, \ldots, X_m), \ldots, f_m^k(X_1, \ldots, X_m)).$$

Now, the question is for what general families of polynomials $\mathbf{F}$, for any two numbers r and s with $0 \le r < s$ the polynomials $f_i^r - f_i^s$, $i = 1, \ldots, m$, are linearly independent.

Acknowledgments

This research is partially supported by the National Spanish project PB97-0346.

References

1. J. Eichenauer-Herrmann, E. Herrmann and S. Wegenkittl, *A survey of quadratic and inversive congruential pseudorandom numbers*, Lect. Notes in Statistics, Springer-Verlag, Berlin, **127** (1998), 66–97.
2. F. Griffin, H. Niederreiter and I. Shparlinski, *On the distribution of nonlinear recursive congruential pseudorandom numbers of higher orders,* Proc. the 13th Symp. on Appl. Algebra, Algebraic Algorithms, and Error-Correcting Codes, Hawaii, 1999, Lect. Notes in Comp. Sci., Springer-Verlag, Berlin, 1999, **1719** , 87–93.
3. J. Gutierrez, H. Niederreiter and I. Shparlinski, *On the multidimensional distribution of nonlinear congruential pseudorandom numbers in parts of the period,* Monatsh. Math., **129**, (2000) 31–36.
4. M. Nagata, *Theory of commutative fields,* Translations of Mathematical Monograph, vol. **125**, Amer. Math. Soc., Providence, R.IU., 1993.
5. H. Niederreiter and I. Shparlinski, *On the distribution and lattice structure of nonlinear congruential pseudorandom numbers,* Finite Fields and Their Applications, **5** (1999), 246–253.
6. H. Niederreiter and I. Shparlinski, *On the distribution of inversive congruential pseudorandom numbers modulo a prime power,* Acta Arith., **92**, (2000), 89–98.

7. H. Niederreiter and I. Shparlinski, *On the distribution of pseudorandom numbers and vectors generated by inversive methods*, Appl. Algebra in Engin., Commun. and Computing, **10**, (2000) 189–202.

8. H. Niederreiter and I. E. Shparlinski, 'On the distribution of inversive congruential pseudorandom numbers in parts of the period', *Math. Comp.* (to appear).

Even Length Binary Sequence Families with Low Negaperiodic Autocorrelation

Matthew G. Parker

Code Theory Group, Inst. for Informatikk, HIB,
University of Bergen, Norway
matthew@ii.uib.no
http://www.ii.uib.no/~matthew/MattWeb.html

Abstract. Cyclotomic constructions are given for several infinite families of even length binary sequences which have low negaperiodic autocorrelation. It appears that two of the constructions have asymptotic Merit Factor 6.0 which is very high. Mappings from periodic to negaperiodic autocorrelation are also discussed.

1 Introduction

The Periodic Autocorrelation Function (PACF) of a length N binary sequence, $s(t)$, is,

$$P_s(\omega) = \sum_{t=0}^{N-1} (-1)^{s(t+\omega)-s(t)}, \qquad 0 \le \omega < N \qquad (1)$$

where sequence indices, t, are taken mod N. $s(t)$ has optimal PACF when $|P_s(\omega)| = 1$ if N is odd. For N even, the PACF of $s(t) = 0001$ is $4,0,0,0$, which is perfect as $P_s(\omega) = 0, \forall \omega \ne 0$. But, for N even, $N > 4$, it is conjectured (but not proven) that there is no binary $s(t)$ with perfect PACF. If this conjecture is true then, for N even, $N > 4$, binary $s(t)$ such that $\min_{s(t)}(\max_{1 \le \omega < N} |P_s(\omega)|) = 2$ (4) has best possible PACF, for $4 \nmid N$ ($4|N$), respectively. However, when $s(t)$ is balanced (an equal number of zeros and ones) or almost-balanced ($|\#\text{zeroes} - \#\text{ones}| = 1$) proof of optimality is possible. A recent paper [1] used cyclotomy to construct infinite [1] balanced (almost-balanced) binary sequence families of length $N = 2p$, for certain p prime, with optimal PACF. In this paper we consider the Negaperiodic Autocorrelation Function (NACF) of $s(t)$,

$$Q_s(\omega) = \sum_{t=0}^{N-1} (-1)^{s(t+\omega)-s(t)-\lfloor \frac{t+\omega}{N} \rfloor}, \qquad 0 \le \omega < N \qquad (2)$$

where sequence indices, t, are taken, mod N. For example, the NACF of $s(t) = 110101$ is $Q_s(\omega) = 6, -4, 2, 0, -2, 4$. Binary $s(t)$ has optimal NACF when $|Q_s(\omega)|$

---

[1] 'infinite' means there is no upper limit on N for which the construction is valid.

S. Boztaş and I.E. Shparlinski (Eds.): AAECC-14, LNCS 2227, pp. 200–209, 2001.

$= 1$, $\forall \omega \neq 0$, if N is odd. For even N the NACF of $s(t) = 01$ is $2, 0$ which is perfect as $Q_s(\omega) = 0$, $\forall \omega \neq 0$. But for N even, $N > 2$, we conjecture (but cannot prove) that there is no binary $s(t)$ with perfect NACF. If this conjecture is true then, for N even, $N > 2$, binary $s(t)$ such that $\min_{s(t)}(\max_{1 \leq \omega < N} |Q_s(\omega)|) = 2$, has best possible NACF. We provide constructions for such 'conjectured optimal' sequences, $s(t)$, in Theorems 1 and 2, where $s(t)$ is not necessarily balanced or almost-balanced. [2] We can always define an odd-length binary sequence, $e(t)$, such that $e(t) = s(t) + t \pmod 2$, where $Q_e(\omega) = (-1)^\omega P_s(\omega)$ (Lemma 2), so low odd-length N PACF constructions trivially map to low odd-length N NACF constructions. However most even-length sequences with low NACF cannot be trivially derived from even-length sequences with known PACF, although we do review some useful mappings in Section 5. In this paper cyclotomy is used to construct binary sequence families of even length $N = 2p$ ($N = 4p$) with low NACF for certain p prime. Unlike the sequences of [1], the sequences of this paper are not necessarily balanced or almost-balanced. Sequences with low NACF can be used in spread-spectrum systems in a similar way to sequences with low PACF, and for comparable complexity [9]. The Aperiodic Autocorrelation Function (AACF) of a length N binary sequence, $s(t)$, is,

$$A_s(\omega) = \sum_{t=0}^{N-1} (-1)^{s(t+\omega)-s(t)}, \qquad -N < \omega < N \tag{3}$$

where $s(t) = 0$ for $t < 0$ or $t \geq N$. AACF is the sum and difference of PACF and NACF:

$$
\begin{aligned}
A_s(\omega) &= \tfrac{1}{2}(P_s(\omega) + Q_s(\omega)), & 0 \leq \omega < N \\
A_s(\omega) &= \tfrac{1}{2}(P_s(N-\omega) - Q_s(N-\omega)), & -N \leq \omega < 0
\end{aligned}
\tag{4}
$$

where $|A_s(\omega)| = |A_s(-\omega)|$. It is a well-known open problem to identify lowest possible values of $|A_s(\omega)|$ for a length N sequence, s. 'Golay Merit Factor' (MF) [8] is a common metric used to measure aperiodic optimality of a sequence and is given by,

$$M_s = \frac{N^2}{2 \sum_{\omega=1}^{N-1} |A_s(\omega)|^2} \tag{5}$$

Lower values of $|A_s(\omega)|$ give higher MF. The highest MF for a given length N binary sequence is not known in general. The asymptote, $M_s = 6.0$, $N \to \infty$ is the highest known asymptote for a sequence, s, belonging to an infinite family of binary sequences, where the construction is a cyclic shift (cyclically shifted by approximately $N/4$) of a Legendre or Modified-Jacobi sequence [7,8], although Golay has constructed skewsymmetric binary sequences with MFs generally between 8.00 and 9.00 [3,4,5] up to lengths $N = 100$ or so. The Rudin-Shapiro-based constructions [2,6,11,10], achieve PACF and NACF upper bounds which

---

[2] Computations show that binary $s(t)$ satisfying $\min_{s(t)}(\max_{1 \leq \omega < N} |Q_s(\omega)|) = 2$ exist for all even N up to $N = 38$. This is in contrast to PACF when $4|N$, where computations suggest $\min_{s(t)}(\max_{1 \leq \omega < N} |P_s(\omega)|) = 4$.

appear to be asymptotically of the same order, leading to an asymptotic MF of 3.0.

This paper shows, experimentally, that the constructions of Theorems 1 and 2 also approach $M_s = 6.0$ as $N \to \infty$, and Section 5 argues that this is because these constructions are closely related to Legendre sequences.

2 Construction

Instead of constructing a length N sequence $s(t)$, we construct a length $2N$ sequence $s'(t)$, where $s'(t) = s(t)$, $0 \le t < N$, $s'(t) = s(t) + 1 \pmod 2$, $N \le t < 2N$. The NACF of $s(t)$ and the PACF of $s'(t)$ are related as follows,

$$Q_s(\omega) = \frac{1}{2}P_{s'}(\omega), \qquad 0 \le \omega < N$$

For example, if $s'(t) = 11010111110010100000$ then $s(t) = 1101011111$.
$P_{s'}(\omega) = 20, 0, 4, 0, -4, 0, 4, 0, -4, 0, -20, 0, -4, 0, 4, 0, -4, 0, 4, 0$, so
$Q_s(\omega) = 10, 0, 2, 0, -2, 0, 2, 0, -2, 0$. The constructing method uses cyclotomy, as in [1], to specify a subset C of Z_{2N} to define the characteristic sequence $s'(t)$ of C:

$$s'(t) = \begin{cases} 1, & \text{if } t \in C \\ 0, & \text{otherwise} \end{cases}$$

The PACF is determined by the difference function,

$$d_C(\omega) = |C \cap (C + \omega)|$$

where $C + \omega$ denotes the set $\{c + \omega : c \in C\}$ and '+' denotes addition, mod $2N$. The PACF of $s'(t)$ is then,

$$P_{s'}(\omega) = 2N - 4(|C| - d_C(\omega)) \tag{6}$$

This paper gives constructions for $N = 2p$ and $N = 4p$, p prime. We therefore specify C over Z_{4p} and Z_{8p}. By the Chinese Remainder Theorem (CRT), Z_{rp} is isomorphic to $Z_r \times Z_p$, $\gcd(r, p) = 1$. For $N = rp$, let $C' = \{\{n\} \times C_n \mid C_n \subseteq Z_p^*, 0 \le n < r\}$, $F = \{G \times 0 \mid G \subseteq Z_r\}$, and $C = C' \cup F$. Define $\omega = (\omega_1, \omega_2) \in Z_r \times Z_p$. Then,

$$\begin{aligned} d_C(\omega_1, \omega_2) &= |C \cap (C + (\omega_1, \omega_2))| \\ &= \sum_{k=0}^{r-1} \sum_{n=0}^{r-1} |C_n \cap (C_{k-\omega_1} + \omega_2)| \\ &\quad + |G \cap (G + (w_1, 0))| + \sum_{k=0}^{r-1} |G \cap (k + w_1, C_k + w_2)| \\ &\quad + \sum_{k=0}^{r-1} |(k, C_k) \cap (G + (w_1, w_2))| \end{aligned} \tag{7}$$

From (7) we see that if we know $|C_n \cap (C_m + \omega_2)|$, $\forall n, m, \omega_2 \in Z_p$, and if we can also determine the last three terms involving G, then we can determine $d_C(\omega_1, \omega_2) = d_C(\omega)$, $\forall \omega$, and hence the PACF of $s'(t)$. If we construct C_n from the union of various cyclotomic classes over $GF(p)$, $\forall n$, then $|C_n \cap (C_m + \omega_2)|$ is computable from the cyclotomic numbers over $\mathrm{GF}(p)$. Let D_i be the cyclotomic class of order d, given by,

$$D_i = \{\alpha^i, \alpha^{d+i}, \alpha^{2d+i}, \alpha^{3d+i}, \ldots, \alpha^{p-1-d+i}\}, \qquad 0 \le i < d$$

where α is a primitive generator over $\mathrm{GF}(p)$. Then the cyclotomic number $[i, j]$ of order d over $\mathrm{GF}(p)$ is,

$$[i, j] = |(D_i + 1) \cap D_j| \tag{8}$$

Note that $|C_n \cap (C_m + w_2)| = |w_2^{-1}C_n \cap (w_2^{-1}C_m + 1)|$, $(\bmod\ p)$, for $w_2 \ne 0$. If $C_n = \bigcup_{k \in T_n} D_k$, $T_n \subseteq Z_r$, and $w_2^{-1} \in D_h$, then $w_2^{-1}C_n = \bigcup_{k \in T_n} D_{k+h}$. Therefore,

$$|w_2^{-1}C_n \cap (w_2^{-1}C_m+1)| = |(\bigcup_{k \in T_n} D_{k+h}) \cap (\bigcup_{k \in T_m} D_{k+h}+1)| = \sum_{k \in T_n} \sum_{j \in T_m} [k+h, j+h] \tag{9}$$

i.e. a sum of cyclotomic numbers. We later use cyclotomic numbers to prove the NACF of some of the sequences we construct.

Example 1: $s'(t)$ is described by C comprising F and the C_n which are, in turn, the union of various D_i of order d. Let $2N = rp = 4p$, $d = 2$, and $C_0 = D_0$, $C_1 = D_0$, $C_2 = D_1$, $C_3 = D_1$. Let $G = \{1, 2\}$. Then, for $p = 13$ we can choose $\alpha = 2$ to give $D_0 = \{1, 4, 3, 12, 9, 10\}$ and $D_1 = \{2, 8, 6, 11, 5, 7\}$. Thus, using the CRT, mod 52, we construct the sets, $F = \{13, 26\}$, and,

$$(0, C_0) = \{40\{1, 4, 3, 12, 9, 10\}\} \qquad (1, C_1) = \{13 + 40\{1, 4, 3, 12, 9, 10\}\}$$
$$(2, C_2) = \{26 + 40\{2, 8, 6, 11, 5, 7\}\} \quad (3, C_3) = \{39 + 40\{2, 8, 6, 11, 5, 7\}\}$$

Then $C = (0, C_0) \cup (1, C_1) \cup (2, C_2) \cup (3, C_3) \cup F =$
$\{1, 2, 4, 6, 7, 9, 11, 12, 13, 15, 16, 17, 18, 19, 25, 26, 29, 31, 34, 36, 40, 46, 47, 48, 49, 50\}$.

Therefore, $s'(t) = 0110101101011101111100000110010100101000100000111110$, and

$$P_{s'}(\omega) = 52, 0, 4, 0, -4, 0, 4, 0, -4, 0, 4, 0, -4, 0, 4, 0, -4, 0, 4, 0, -4, 0, 4, 0, -4, 0,$$
$$- 52, 0, -4, 0, 4, 0, -4, 0, 4, 0, -4, 0, 4, 0, -4, 0, 4, 0, -4, 0, 4, 0, -4, 0, 4, 0$$

Finally, the first half of $s'(t)$ is $s(t) = 0110101101011101111000001$, and,

$$Q_s(\omega) = 26, 0, -2, 0, 2, 0, -2, 0, 2, 0, -2, 0, 2, 0, -2, 0, 2, 0, -2, 0, 2, 0, -2, 0, 2, 0$$

Example 1 highlights the following restriction.

Lemma 1. *For $s'(t)$ to satisfy $s'(t + N) = s'(t) + 1$ $(\bmod\ 2)$, $0 \le t < N$, we require that, if $C_n = \bigcup_{i \in T_n} D_i$ $T_n \subseteq Z_d$, then $C_{n+\frac{r}{2}} = \bigcup_{i \notin T_n} D_i$. Moreover, if $j \in G$ $(\notin G)$, then $j + \frac{r}{2}$ $(\bmod\ r) \notin G$, $(\in G)$.*

From Lemma 1 it is sufficient to describe $s(t)$ by defining C_n for $0 \le n < \frac{r}{2}$, and by defining $G' \subset Z_{\frac{r}{2}}$, where $G' = \{g \mid g \in G, g < \frac{r}{2}\}$.

A Compact Description for $s(t)$: $s(t)$ is compactly described by $\mathbf{H} = (G', \{\bigcup_{i \in T_0} D_i\}, \{\bigcup_{i \in T_1} D_i\}, \ldots, \{\bigcup_{i \in T_{\frac{r}{2}-1}} D_i\})$.

So for Example 1 we define $s(t)$ by $\mathbf{H} = (\{1\}, \{D_0\}, \{D_0\})$. Example 1 is taken from Theorem 1 of Section 3 and is a construction for length $N = 2p$ sequences, $s(t)$, with low NACF.

3 Sequences with Low Negaperiodic Autocorrelation

3.1 Symmetries

Two length K sequences, $u(t)$ and $v(t)$ are called 'PACF-equivalent' ('NACF-equivalent') if they have the same distribution of PACF (NACF) magnitudes, and there exist well-defined operations that take $u(t)$ to and from $v(t)$. Such operations are called PACF-equivalent (NACF-equivalent) operations. Before presenting the constructions we first mention some PACF-equivalent operations on $s'(t)$. These translate into NACF-equivalent operations on $s(t)$.

PACF-equivalent operation on $s'(t)$	NACF-equivalent operation on $s(t)$
Cyclic Shift of $s'(t)$	Negacyclic Shift of $s(t)$
Reversal of $s'(t)$	Reversal of $s(t)$
Negation of $s'(t)$	Negation of $s(t)$

The following theorems and conjectures only present constructions for NACF-inequivalent sequences, $s(t)$, and proofs of Theorems 1 and 2 are given at the end of this section.

Theorem 1. *Let $p = 4f + 1$ be prime and $d = 2$. The length $N = 2p$ sequence $s(t)$ has conjectured optimal three-valued out-of-phase negaperiodic autocorrelation, $\{-2, 0, 2\}$, if $\mathbf{H} = (\{1\}, \{D_0\}, \{D_0\})$.*

Theorem 2. *Let $p = 4f + 3$ be prime and $d = 2$. The length $N = 2p$ sequence $s(t)$ has conjectured optimal three-valued out-of-phase negaperiodic autocorrelation, $\{-2, 0, 2\}$, if $\mathbf{H} = (\{0, 1\}, \{D_0\}, \{D_0\})$ or $\mathbf{H} = (\{-\}, \{D_0\}, \{D_0\})$.*

In the following three Conjectures let $\gamma = \{a, b\}\{c, d\}\{e, f\}\{g, h\}$ be short for $\{D_a \cup D_b\}, \{D_c \cup D_d\}, \{D_e \cup D_f\}, \{D_g \cup D_h\}$.

Conjecture 1. Let p be a prime of the form $(n^2 + 1)/2$, $8|(p - 1)$, and $d = 4$. Let $s(t)$ be described by $\mathbf{H} = (G', \gamma)$. Then, for a given γ chosen from Conjecture 1 of Table 1, $\exists \alpha$ and α^{-1} such that the length $N = 4p$ sequence $s(t)$ has near-optimal five-valued out-of-phase negaperiodic autocorrelation $\{-4, -2, 0, 2, 4\}$ or $\{-18, -4, 0, 4, 18\}$, respectively, independent of choice of G'.

Table 1. G' and γ Values for Conjectures 1 and 2

Conjecture 1		Conjecture 2	
G'	γ	G'	γ
$\{2\}$	$\{0,3\}\{1,2\}\{0,1\}\{0,1\}$	$\{0\}$	$\{0,3\}\{1,2\}\{0,1\}\{0,1\}$
$\{0,1,2\}$	$\{1,2\}\{0,3\}\{0,1\}\{0,1\}$	$\{1\}$	$\{1,2\}\{0,3\}\{0,1\}\{0,1\}$
$\{3\}$	$\{2,3\}\{0,1\}\{1,2\}\{1,2\}$	$\{0,2,3\}$	$\{2,3\}\{0,1\}\{1,2\}\{1,2\}$
$\{0,1,3\}$	$\{0,1\}\{2,3\}\{1,2\}\{1,2\}$	$\{1,2,3\}$	$\{0,1\}\{2,3\}\{1,2\}\{1,2\}$

Conjecture 2. Let p be a prime of the form $(n^2 + 1)/2$, $8 \nmid (p - 1)$, and $d = 4$. Let $s(t)$ be described by $\mathbf{H} = (G', \gamma)$. Then, for a given γ chosen from conjecture 2 of Table 1, $\exists \alpha$ and α^{-1} such that the length $N = 4p$ sequence $s(t)$ has near-optimal five-valued out-of-phase negaperiodic autocorrelation $\{-4, -2, 0, 2, 4\}$ or $\{-22, -4, 0, 4, 22\}$, respectively, independent of choice of G'.

Conjecture 3. Let p be a prime of the form $n^2 + 4$, and $d = 4$. Let $s(t)$ be described by $\mathbf{H} = (G', \gamma)$. Then, for a given γ chosen from the left-hand (right-hand) side of Table 2, $\exists \alpha$ and α^{-1} such that the length $N = 4p$ sequence $s(t)$ of H has near-optimal five and seven-valued out-of-phase negaperiodic autocorrelation $\{-4, -2, 0, 2, 4\}$ or $\{-12, -4, -2, 0, 2, 4, 12\}$, respectively, for the single choice of G' from the left-hand (right-hand) side of Table 2.

Table 2. G' and γ Values for Conjecture 3

G'	γ	G'	γ
$\{0\}$	$\{1,3\}\{0,2\}\{0,1\}\{0,1\}$	$\{0,3\}$	$\{0,1\}\{0,2\}\{0,2\}\{0,1\}$
	$\{0,2\}\{1,3\}\{0,1\}\{0,1\}$		$\{0,1\}\{1,3\}\{1,3\}\{0,1\}$
	$\{1,3\}\{0,2\}\{1,2\}\{1,2\}$		$\{1,2\}\{0,2\}\{0,2\}\{1,2\}$
	$\{0,2\}\{1,3\}\{1,2\}\{1,2\}$		$\{1,2\}\{1,3\}\{1,3\}\{1,2\}$

Example 2: A representative sequence of Conjecture 3 is
$H = (\{0,3\}, \{D_0, D_1\}, \{D_0, D_2\}, \{D_0, D_2\}, \{D_0, D_1\})$. Then
$C = \{(0,0) \cup (3,0) \cup (5,0) \cup (6,0) \cup (0, C_0) \cup (1, C_1) \cup (2, C_2) \cup (3, C_3) \cup (4, C_4) \cup (5, C_5) \cup (6, C_6) \cup (7, C_7)\}$, where

$$C_0 = \{D_0 \cup D_1\}, \quad C_1 = \{D_0 \cup D_2\}, \quad C_2 = \{D_0 \cup D_2\}, \quad C_3 = \{D_0 \cup D_1\}$$
$$C_4 = \{D_2 \cup D_3\}, \quad C_5 = \{D_1 \cup D_3\}, \quad C_6 = \{D_1 \cup D_3\}, \quad C_7 = \{D_2 \cup D_3\}$$

Let $p = 29$ and $d = 4$. Using $\alpha = 2$ as a primitive generator, mod 29, $D_0 = \{1, 16, 24, 7, 25, 23, 20\}$, $D_1 = \{2, 3, 19, 14, 21, 17, 11\}$, $D_2 = \{4, 6, 9, 28, 13, 5, 22\}$, $D_3 = \{8, 12, 18, 27, 26, 10, 15\}$. Using the CRT,

$$(0, C_0) = 88\{1, 16, 24, 7, 25, 23, 20, 2, 3, 19, 14, 21, 17, 11\} (\text{mod } 232)$$
$$(1, C_1) = 145 + 88\{1, 16, 24, 7, 25, 23, 20, 4, 6, 9, 28, 13, 5, 22\} (\text{mod } 232)$$
$$\dots \text{etc}$$

Similarly, $F = \{0, 203, 29, 174\}$
Therefore,
$s(t) = 11011000010110111001010011001100111001011011101111100000101$
$\qquad 01010101001111101111000111110010001001000011101001000001000$
    and the NACF of $s(t)$ is,

$116, 2, 0, 2, -4, -2, 0, 2, 4, 2, 0, 2, -4, -2, 0, -2, 4, -2, 0, 2, -4, 2, 0, 2, 4, 2, \dots$ etc

Proof. (of Theorem 1). We wish to compute $d_C(w_1, w_2)$ by evaluating (7) using (8) and (9). For $p = 4f + 1$, $w_2^{-1} \in D_h$ implies $\pm w_2 \in D_{h+1 (\text{mod } 2)}$, and we need this for the last three terms of (7). The cyclotomic numbers of order $d = 2$ for $p = 4f + 1$ are $[0,0] = \frac{p-5}{4}$, $[0,1] = [1,0] = [1,1] = \frac{p-1}{4}$. We have $C_0 = C_1 = D_0$, $C_2 = C_3 = D_1$, $G = \{(1,0),(2,0)\}$. Therefore,

$$
\begin{aligned}
d_C(0,0) &= |C| = 2(p-1) + 2 = 2p \\
d_C(1,0) &= |C_0 \cap C_3| + |C_1 \cap C_0| + |C_2 \cap C_1| + |C_3 \cap C_2| + |G \cap (G + (1,0))| \\
&= |D_0| + |D_1| + 1 = p \\
d_C(2,0) &= 2(|C_0 \cap C_2| + |C_1 \cap C_3|) + |G \cap (G + (2,0))| = 0 + 0 = 0 \\
d_C(3,0) &= d_C(1,0) = p \quad (\text{using } d_C(-w_1, -w_2) = d_C(w_1, w_2)) \\
d_C(0,w_2) &= \sum_{n=0}^{r-1} |C_n \cap (C_n + w_2)| + \sum_{k=0}^{r-1} |G \cap (k, C_k + w_2)| \\
&\quad + \sum_{k=0}^{r-1} |(k, C_k) \cap (G + (0, w_2))| \\
&= [0,0] + [0,0] + [1,1] + [1,1] \\
&\quad + |\{(1,0),(2,0)\} \cap \{(1, C_1 + w_2) \cup (2, C_2 + w_2)\}| \\
&\quad + |\{(1, C_1) \cup (2, C_2)\} \cap \{(1, w_2),(2, w_2)\}| = p - 3 + 1 + 1 = p - 1, \\
&\qquad \text{for } w_2^{-1} \in D_0, \text{ or } w_2^{-1} \in D_1 \\
d_C(1,w_2) &= \sum_{n=0}^{r-1} |C_n \cap (C_{n-1} + w_2)| + \sum_{k=0}^{r-1} |G \cap (k+1, C_k + w_2)| \\
&\quad + \sum_{k=0}^{r-1} |(k, C_k) \cap (G + (1, w_2))| \\
&= [0,1] + [0,0] + [1,0] + [1,1] \\
&\quad + |\{(1,0),(2,0)\} \cap \{(1, C_0 + w_2) \cup (2, C_1 + w_2)\}| \\
&\quad + |\{(2, C_2) \cup (3, C_3)\} \cap \{(2, w_2),(3, w_2)\}| = p - 2 + 2 = p, \\
&\qquad \text{for } w_2^{-1} \in D_0, \text{ or } w_2^{-1} \in D_1
\end{aligned}
$$

similarly $d_C(2, w_2) = p - 1 + 1 + 1 = p + 1$, $d_C(3, w_2) = p - 2 + 2 = p$
for $w_2^{-1} \in D_0$, or $w_2^{-1} \in D_1$

Substituting $d_C(w_1, w_2)$ back into (6) gives the PACF distribution $\{0, 4, -4, N\}$ for $s'(t)$, implying an NACF distribution $\{0, 2, -2\}$ for $s(t)$.   □

Proof. (of Theorem 2) The proof is identical to that of Theorem 1, except that, for $p = 4f + 3$, $w_2^{-1} \in D_h$ implies $w_2 \in D_{h+1(\text{mod } 2)}$, and $-w_2 \in D_h$. Moreover, the cyclotomic numbers of order $d = 2$ for $p = 4f + 3$ are $[0,1] = \frac{p+1}{4}$, $[0,0] = [1,0] = [1,1] = \frac{p-3}{4}$.   □

Conjectures 1 - 3 will hopefully be proved in a similar way to the above, but now cyclotomic numbers of order 4 are required.

4   Asymptotic Merit Factors

By computation, using (5), the constructions of Theorems 1 and 2 give sequences, $s(t)$, with Merit Factor (MF) $M_s \to 6.0$ as $N \to \infty$. Figs 1 and 2 plot MF for increasing prime values, p, for the constructions of Theorems 1 and 2. Very good MFs occur for no negacyclic shift, but Fig 3 presents the best MF over all negacyclic shifts. The highest MF sometimes occurs for a non-zero negacyclic shift. The asymptote of $M_s = 6.0$ is the best known for an infinite construction class of binary sequences [7,8], where cyclically-shifted Legendre and Modified-Jacobi sequences also attain this maximum[3]. Unlike Legendre and Modified-Jacobi sequences, no final shift of the constructed sequences is required to obtain

---

[3] The constructions of [1] appear to asymptote to $M_s = 1.5$ or $M_s = 3.0$

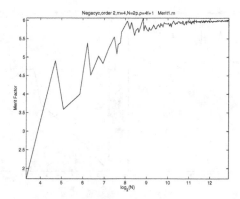

Fig. 1. NegaPeriodic Construction, Theorem 1, $p = 4f + 1$

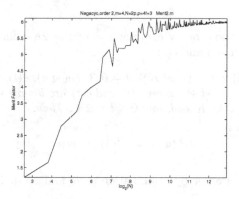

Fig. 2. NegaPeriodic Construction, Theorem 2, $p = 4f + 3$

the asymptote of 6.0. Lemma 3 of the next section shows that the constructions of Theorems 1 and 2 are closely related to Legendre sequences.

5   Mappings Between Periodic and Negaperiodic Autocorrelation

Although the sequence constructions of this paper are new, we also highlight further symmetries that trivially relate PACF and/or NACF coefficient distributions of binary sequences $s(t)$ and $e(t)$, where s and e are not necessarily the same length.

Lemma 2. *Let $e(t) = s(t) + t \pmod 2$, where $s(t)$ and $e(t)$ are binary sequences of length K. Then,*

$$Q_e(\omega) = (-1)^\omega P_s(\omega)$$

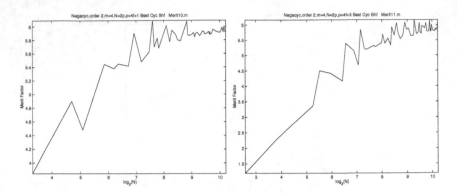

Fig. 3. NegaPeriodic Constructions, $p = 4f + 1$, Theorem 1 (lh), $p = 4f + 3$ Theorem 2 (rh), Best Negacyclic Shift

Proof. Direct inspection, or by examination of the $2K$-point Discrete Fourier Transform (DFT) of $s(t)$ and $e(t)$. □

Lemma 3. *Let* $e(t) = s(t \ (mod \ K))$, $t = 0, 3 \ (mod \ 4)$, $e(t) = s(t \ (mod \ K)) + 1 \ (mod \ 2)$, $t = 1, 2 \ (mod \ 4)$, *where* $s(t)$ *and* $e(t)$ *are binary sequences of length* K *and* $2K$, *respectively,* K *odd, and* $0 \le t < 2K$. *Then,*

$$Q_e(\omega) = 0 \qquad\qquad\qquad \omega \ odd$$
$$Q_e(\omega) = (-1)^{\frac{\omega}{2}} 2P_s(\omega \ (mod \ K)) \ \omega \ even, \qquad 0 \le \omega < 2K$$

Proof. Direct inspection or by examination of K and $2K$-point DFTs of s and e, respectively. □

Example 3: Consider the negated Legendre sequence of length $K = 13$, $s(t) = 1101100001101$. This sequence has PACF
$P_s(\omega) = 13, 1, -3, 1, 1, -3, -3, -3, -3, 1, 1, -3, 1$. $e(t)$ is of length $2K = 26$ and is given by,

$$e(t) = 1101100001101110110000 1101 + 0110011001100110011001 1001 \ (mod \ 2)$$
$$= 1011111000001000101001 0100$$

and $e(t)$ has NACF,

$$Q_e(\omega) = 26, 0, -6, 0, -2, 0, -6, 0, 6, 0, 2, 0, -2, 0, 2, 0, -2, 0, -6, 0, 6, 0, 2, 0, 6, 0$$

$e(t)$ is identical to $s'(t)$ of Example 1 apart from the first bit. In general, an equivalent construction to that of Theorems 1 and 2 for $K = p$ is to make $s(t)$ a negated Legendre sequence, apply Lemma 3, then flip bit 0 or bit K.

Lemma 4. *Let* $e(t) = s(t \ (mod \ K))$, $4 \nmid t$, $e(t) = s(t \ (mod \ K)) + 1 \ (mod \ 2)$, $4|t$, *where* $s(t)$ *and* $e(t)$ *are binary sequences of length* K *and* $4K$, *respectively,* K *odd. Then,*

$$Q_e(\omega) = 0 \qquad\qquad\qquad 4 \nmid \omega$$
$$Q_e(\omega) = 4P_s(\omega \ (mod \ K)) \ 4|\omega, \qquad 0 \le \omega < 4K$$

Proof. Direct inspection or by examination of K and $4K$-point DFTs of s and e, respectively. □

6   Conclusion

This paper has presented new cyclotomic constructions for infinite families of length $N = 2p$ and $N = 4p$ binary sequences with very low negaperiodic autocorrelation. The technique builds length $2N$ sequences with low periodic autocorrelation with the second half the negation of the first half. The desired length N sequence is then simply the first half. Two of the constructions exhibit a Merit Factor approaching 6.0 as N approaches infinity. This is the highest asymptote currently known. A final section highlights further mappings which relate periodic autocorrelation of a binary sequence to the periodic or negaperiodic autocorrelation of another binary sequence.

References

1. Ding, C.,Helleseth, T.,Martinsen, H.M.: New Classes of Binary Sequences with Three-Level Autocorrelation. IEEE Trans. Inform. Theory **47** 1. Jan. (2001) 428–433
2. Golay, M.J.E.: Complementary Series. IRE Trans. Inform. Theory **IT-7** Apr. (1961) 82–87
3. M.J.E.Golay, M.J.E.: Sieves for Low Autocorrelation Binary Sequences. IEEE Trans. Inform. Theory **23** 1. Jan. (1977) 43–51
4. Golay, M.J.E.: The Merit Factor of Long Low Autocorrelation Binary Sequences. IEEE Trans. Inform. Theory **28** 3. May (1982) 543–549
5. Golay, M.J.E.: A New Search for Skewsymmetric Binary Sequences with Optimal Merit Factors. IEEE Trans. Inform. Theory **36** 5. Sept. (1990) 1163–1166
6. Høholdt, T., Jensen, H.E., Justesen, J.: Aperiodic Correlations and the Merit Factor of a Class of Binary Sequences. IEEE Trans. Inform. Theory **31** 4. July (1985) 549–552
7. Jensen, J.M.,Elbrønd Jensen, H.,Høholdt T.: The Merit Factor of Binary Sequences Related to Difference Sets. IEEE Trans. Inform. Theory **37** 3. May (1991) 617–626
8. Høholdt, T.: The Merit Factor of Binary Sequences. Difference Sets, Sequences and their Correlation Properties, A.Pott et a. (eds.), Series C: Mathematical and Physical Sciences, Kluwer Academic Publishers **542** (1999) 227–237
9. Luke, H.D.: Binary Odd-Periodic Complementary Sequences. IEEE Trans. Inform. Theory **43** 1. Jan. (1997) 365–367
10. Parker, M.G., Tellambura, C.: Generalised Rudin-Shapiro Constructions. WCC2001, Workshop on Coding and Cryptography, Paris(France) Jan. 8-12 (2001)
11. Paterson, K.G., Tarokh, V.: On the Existence and Construction of Good Codes with Low Peak-to-Average Power Ratios. IEEE Trans. Inform. Theory **46** 6. Sept. (2000) 1974–1987

On the Non-existence
of (Almost-)Perfect Quaternary Sequences *

Patrice Parraud

Écoles Militaires de St Cyr-Coëtquidan
Centre de Recherche des Écoles de St Cyr-Coëtquidan
F56381 Guer Cedex, France
patrice_parraud@esm-stcyr.terre.defense.gouv.fr

Abstract. We give some new general non-existence results on perfect
and almost-perfect quaternary sequences that are useful for the syn-
chronisation of messages in multiple access communications systems. We
present a conjecture on the non-existence of perfect autocorrelation quaternary sequences.

Keywords: Finite fields – Number theory – Autocorrelation – Binary
and Quaternary Sequences.

1 Introduction

Throughout the paper $s = (s_t)_{t \in \mathbb{N}}$ will denote a complex-valued sequence of
period n, i.e. $s_{t+n} = s_t$ for every t. Associated with it is its generating vector
$(s_0, s_1, \cdots, s_{n-1})$. The periodic autocorrelation of s is classically the coefficients
$\{c_s(u) \mid 0 \leq u \leq n-1\}$ of inner products $c_s(u) = \sum_{t=0}^{n-1} s_t \overline{s_{t+u}}$ where $\bar{s}_t$ denotes
the complex conjugate of s_t and the sum $t+u$ is computed modulo n. The coeffi-
cients $c_s(u)$ with $u \equiv 0 \pmod{n}$ are called "in-phase" coefficients and are equal
to the weight of s, i.e. the number of non-zero elements of s in one period. The
other coefficients $c_s(u)$ for $u \not\equiv 0 \pmod{n}$ are called "out-of-phase" coefficients.
The shift operator σ applied to s is defined by $\sigma(s) = (s_{n-1}, s_0, \cdots, s_{n-2})$, for
all integer $u \pmod{n}$ we have $\sigma^u(s) = (s_{n-u}, \cdots, s_{n-u-1})$.

In this paper, sequences have values in the ring of integers modulo 4 $\mathbb{Z}_4 =
\mathbb{Z}/4\mathbb{Z} = \{0, 1, 2, 3\}$ (called quaternary sequences). These correspond to complex-
valued sequences by the standard isomorphism $x \mapsto i^x$ with $i^2 = -1$. We now
recall some useful basic notations and definitions on the correlation of quaternary
sequences. Let $s = (s_t)$ be a quaternary sequence of period n. Define $m_u =
s - \sigma^u(s)$ for all integer $u \pmod{n}$. The autocorrelation coefficients are equal
for all integer $u \pmod{n}$ to

$$c_s(u) = [\eta_0(m_u) - \eta_2(m_u)] + i[\eta_1(m_u) - \eta_3(m_u)] \qquad (1.1)$$

---

* This work has been done while the author was at Université de Toulon et du Var -
G.R.I.M.(G.E.C.T.) - FRANCE

S. Boztaş and I.E. Shparlinski (Eds.): AAECC-14, LNCS 2227, pp. 210–218, 2001.

where $\eta_l(m_u) = \sharp\{t \in [0, n-1] \mid m_u^t = l\}$ with $l = 0, 1, 2$ or 3. This result is classical and easy to proove using the general definition of the autocorrelation coefficients of a complex-valued sequence.

A periodic quaternary sequence is called *perfect* if all its "out-of-phase" autocorrelation coefficients are equal to zero. Similarly, it is called *almost-perfect* if all its "out-of-phase" autocorrelation coefficients are equal to zero except possibly one.

There are some results regarding the non-existence of periodic quaternary sequences having perfect autocorrelation properties. Chung and Kumar proved in [1] that there exists no perfect autocorrelation quaternary sequences of period 2^k with k greater than 4. Moreover, it is well-known that the existence of a perfect autocorrelation quaternary sequence is equivalent to the existence of a complex circulant Hadamard matrix. In [2], Arasu, Launey and Kumar conjectured that there exists no such matrix of order greater than 16. In this paper, we present new results on the non-existence of perfect autocorrelation quaternary sequences and express as a conjecture that those sequences only exists with period 4, 8 and 16. We establish a similar result concerning the non-existence of almost-perfect autocorrelation quaternary sequences.

2   Perfect Autocorrelation Quaternary Sequences

First, we give a short characterisation of periodic quaternary sequence having perfect autocorrelation properties.

Proposition 2.1. *The quaternary sequence $s = (s_t)$ of period n is perfect if and only if for all integer u (mod n) there exists two integers a and b such that $\eta_0(m_u) = \eta_2(m_u) = a$ and $\eta_1(m_u) = \eta_3(m_u) = b$ with $a + b = n/2$.*

Proof. For all integer u (mod n) we have $c_s(u) = [\eta_0(m_u) - \eta_2(m_u)] + i[\eta_1(m_u) - \eta_3(m_u)] = 0$. But $\eta_0(m_u) + \eta_2(m_u) + \eta_1(m_u) + \eta_3(m_u) = n$. Thus $\eta_0(m_u) = \eta_2(m_u)$ and $\eta_1(m_u) = \eta_3(m_u)$. Finally we get $2\eta_0(m_u) + 2\eta_1(m_u) = n$. $\square$

One of the central theorems on non-existence of perfect quaternary sequences is the following.

Theorem 2.1. *Let $s = (s_t)$ be a quaternary sequence of period n. Define $n_k = \sharp\{t \in [o, n-1] \mid s_t = k\}$ ($k = 0, 1, 2, 3$) as the number of occurencies of k in one generating vector $(s_0, s_1, \cdots, s_{n-1})$ of s. If s is perfect then $n_0(n_0 - 1) + n_1(n_1 - 1) + n_2(n_2 - 1) + n_3(n_3 - 1) - 2n_0n_2 - 2n_1n_3 = 0$.*

Proof. We consider the quaternary $(n-1, n)$-matrix $A = (a_{ut})$ such that the uth row contains the coefficents of $m_u = s - \sigma^u(s)$, that is, $a_{ut} = s_t - s_{t-u}$, where subscripts are computed modulo n. Let N_0, N_1, N_2 and N_3 be the number of 0,1,2 and 3 respectively of this matrix. We calculate N_0, N_1, N_2 and N_3 in two different ways.

$$A = \begin{pmatrix} s_0 - s_{n-1} & s_1 - s_0 & \cdots & s_t - s_{n-t-1} & \cdots & s_{n-1} - s_{n-2} \\ \vdots & \vdots & \vdots & \vdots & \vdots & \vdots \\ s_0 - s_{n-u} & s_1 - s_{n-u+1} & \cdots & s_t - s_{n-t-u} & \cdots & s_{n-1} - s_{n-u-1} \\ \vdots & \vdots & \vdots & \vdots & \vdots & \vdots \\ s_0 - s_1 & s_1 - s_2 & \cdots & s_t - s_{t+1} & \cdots & s_{n-1} - s_0 \end{pmatrix}$$

1) Considering the columns.
The column t of length $n - 1$ is the transpose of the line

$$s_t - s_{n-t-1}, \cdots, s_t - s_{n-t-u}, \cdots, s_t - s_{t+1}$$

where s_t never appears in $\{s_{n-t-1}, \cdots, s_{n-t-u}, \cdots, s_{t+1}\}$.
Let col_0, col_1, col_2 and col_3 be the number of 0,1,2 and 3, respectively, of this column.
If $s_t = 0$ then the column t becomes $-s_{n-t-1} \cdots - s_{n-t-u} \cdots - s_{u+1}$ and then we get $col_0 = n_0 - 1$, $col_1 = n_3$, $col_2 = n_2$ and $col_3 = n_1$.
Similarly we get.
If $s_t = 1$ then $col_0 = n_1 - 1$, $col_1 = n_0$, $col_2 = n_3$ and $col_3 = n_2$.
If $s_t = 2$ then $col_0 = n_2 - 1$, $col_1 = n_1$, $col_2 = n_0$ and $col_3 = n_3$.
If $s_t = 3$ then $col_0 = n_3 - 1$, $col_1 = n_2$, $col_2 = n_1$ and $col_3 = n_0$.
Finally we obtain

$$\begin{cases} N_0 = n_0(n_0 - 1) + n_1(n_1 - 1) + n_2(n_2 - 1) + n_3(n_3 - 1) \\ N_1 = n_0 n_3 + n_1 n_0 + n_2 n_1 + n_3 n_2 \\ N_2 = n_0 n_2 + n_1 n_3 + n_2 n_0 + n_3 n_1 \\ N_3 = n_0 n_1 + n_1 n_2 + n_2 n_3 + n_3 n_0 \end{cases}$$

2) Considering the lines.
The line u of length n is

$$s_0 - s_{n-u}, \cdots, s_t - s_{n-t-u}, \cdots, s_{n-1} - s_{n-u-1}.$$

Let lg_0, lg_1, lg_2 and lg_3 be the number of 0,1,2 and 3 respectively of this line.
Because the quaternary sequence s is perfect and according to proposition 2.1, we have $lg_0 = lg_2 = a$ et $lg_1 = lg_3 = b$ with $a + b = \frac{n}{2}$.
Obviously, it implies that $N_0 = N_2$ and $N_1 = N_3$.
Summing up the equalities in the two cases, we obtain the expected result. $\square$

We now give a result on number theory useful for the next proof.

Theorem 2.2. *Let x be a natural integer and $\prod_p p^{v_p(x)}$ be its prime factor decomposition. A necessary and sufficient condition for x to be the sum of two squares is that for all integer $p \equiv 3 \pmod 4$ the exponent $v_p(x)$ is even.*

Proof. See [5]. $\square$

If we solve the equation $n_0(n_0 - 1) + n_1(n_1 - 1) + n_2(n_2 - 1) + n_3(n_3 - 1) - 2n_0 n_2 - 2n_1 n_3 = 0$ in theorem 21. we find some non-existence conditions on perfect quaternary sequences, summed up in the following theorem.

Theorem 2.3. *Let $s = (s_t)$ be a quaternary sequence of period n and $\prod_p p^{v_p(n)}$ the prime factor decomposition of n. Define $n_k = \sharp\{t \in [0, n-1] \mid s_t = k\}$ ($k = 0, 1, 2, 3$) as the number of occurencies of k in one generating vector $(s_0, s_1, \cdots, s_{n-1})$ of s. If s is perfect then*

1) $n = (n_3 - n_1)^2$ *and* $n \equiv 0 \pmod 4$ *with*

$$n_3 = n_1 + \sqrt{n} \quad n_2 = n_0 = \tfrac{n}{2} - \tfrac{\sqrt{n}}{2} - n_1$$

or

$$n_3 = n_1 - \sqrt{n} \quad n_2 = n_0 = \tfrac{n}{2} + \tfrac{\sqrt{n}}{2} - n_1$$

2) n *is a square and* $n \equiv 0 \pmod{16}$ *with* $n_3 = n_1$ *and*

$$n_2 = \tfrac{n}{2} - n_1 + \tfrac{\sqrt{n}}{4} \quad n_0 = \tfrac{n}{2} - n_1 - \tfrac{\sqrt{n}}{4}$$

or

$$n_2 = \tfrac{n}{2} - n_1 - \tfrac{\sqrt{n}}{4} \quad n_0 = \tfrac{n}{2} - n_1 + \tfrac{\sqrt{n}}{4}$$

3) $n_1 \neq n_3$ *and* $n \neq (n_3 - n_1)^2$ *with for all integer* $p \equiv 3 \pmod 4$ *the exponent* $v_p(n)$ *is even and* $n = a^2 + b^2$ *where a and b are even with*

$$n_3 = n_1 + b \quad n_2 = \tfrac{n}{2} - n_1 - \tfrac{b}{2} + \tfrac{a}{2} \quad n_0 = \tfrac{n}{2} - n_1 - \tfrac{b}{2} - \tfrac{a}{2}$$

or

$$n_3 = n_1 + b \quad n_2 = \tfrac{n}{2} - n_1 - \tfrac{b}{2} - \tfrac{a}{2} \quad n_0 = \tfrac{n}{2} - n_1 - \tfrac{b}{2} + \tfrac{a}{2}$$

and

$$n_3 = n_1 - b \quad n_2 = \tfrac{n}{2} - n_1 + \tfrac{b}{2} + \tfrac{a}{2} \quad n_0 = \tfrac{n}{2} - n_1 + \tfrac{b}{2} - \tfrac{a}{2}$$

or

$$n_3 = n_1 - b \quad n_2 = \tfrac{n}{2} - n_1 + \tfrac{b}{2} - \tfrac{a}{2} \quad n_0 = \tfrac{n}{2} - n_1 + \tfrac{b}{2} + \tfrac{a}{2}$$

Proof. Let $x = n_0$, $y = n_1$, $z = n_2$ and $w = n_3$. Because s is perfect, theorem 21 gives the equation

$$x^2 + y^2 + z^2 + w^2 - x - y - z - w - 2xz - 2yw = 0 \quad (E)$$

with $x + y + z + w = n$ and x, y, z, w integers ranging in $[0, n]$.
We fixe $x = n - y - z - w$ in (E) and obtain

$$4z^2 + 4(y + w - n)z + (2w^2 + 2y^2 - 2nw - 2ny + n^2 - n) = 0$$

which can be viewed as a quadratic polynomial equation with z unknown. Its discriminant is $\Delta = 16(n - (w - y)^2)$. In order to get integer solutions to the equation (E), we need that $\Delta = \delta^2$. Therefore, $z = \tfrac{n}{2} - \tfrac{y}{2} - \tfrac{w}{2} \pm \tfrac{\delta}{8}$ with n, y, w even.

$$\Delta = \delta^2 = 16n - (4(w - y))^2$$

$$\delta^2 + (4(w - y))^2 = 16n$$

Let $X = \delta$ and $Y = 4(w - y)$. We get this new equation

$$X^2 + Y^2 = 16n \quad (E')$$

$\underline{X=0}$: In this case the equation (E') becomes $Y^2 = 16n$ i.e. $(w-y)^2 = n$. If n is not a square then there exists no integer solution to the equation (E') with $X = 0$. If n is a square then we obviously obtain

$$w = y + \sqrt{n} \quad z = x = \frac{n}{2} - \frac{\sqrt{n}}{2} - y \quad or \quad w = y - \sqrt{n} \quad z = x = \frac{n}{2} + \frac{\sqrt{n}}{2} - y$$

$\underline{Y=0}$: In this case the equation (E') becomes $X^2 = 16n$ i.e. $w - y = 0$ with n a square. Therefore, we get $w = y$ and

$$z = \frac{n}{2} - y + \frac{\sqrt{n}}{4} \quad x = \frac{n}{2} - y - \frac{\sqrt{n}}{4} \quad or \quad z = \frac{n}{2} - y - \frac{\sqrt{n}}{4} \quad x = \frac{n}{2} - y + \frac{\sqrt{n}}{4}$$

$\underline{X \neq 0, \quad Y \neq 0}$: Let $16n = 16 \prod_p p^{v_p(n)}$ be the prime factor decomposition of $16n$. Theorem 22 gives a necessary and sufficient condition for $16n$ to be the sum of two squares : for all integer $p \equiv 3 \ (mod \ 4)$ the exponent $v_p(n)$ has to be even. Let $n = a^2 + b^2$ we get $X^2 + Y^2 = 16n = (4a)^2 + (4b)^2$ which yields $X = \delta = \pm 4a$ and $Y = 4(w-y) = \pm 4b$ where a and b play a symetric role. For $w = y + b$, we eventually obtain

$$z = \frac{n}{2} - y - \frac{b}{2} + \frac{a}{2} \quad x = \frac{n}{2} - y - \frac{b}{2} - \frac{a}{2} \quad or \quad z = \frac{n}{2} - y - \frac{b}{2} - \frac{a}{2} \quad x = \frac{n}{2} - y - \frac{b}{2} + \frac{a}{2}$$

and for $w = y - b$

$$z = \frac{n}{2} - y + \frac{b}{2} + \frac{a}{2} \quad x = \frac{n}{2} - y + \frac{b}{2} - \frac{a}{2} \quad or \quad z = \frac{n}{2} - y + \frac{b}{2} - \frac{a}{2} \quad x = \frac{n}{2} - y + \frac{b}{2} + \frac{a}{2}$$

$\square$

It is proved in [1] that there exist no perfect autocorrelation quaternary sequences of period 2^k with k greater than 4. And it is conjectured in [2] that there exists no complex circulant Hadamard matrix of order greater than 16. Our nonexistence theorems on perfect quaternary sequences complete these well-known results. Therefore, it is tempting to express the following conjecture.

> Only perfect periodic quaternary sequences of period 4,8 and 16 exist

To illustrate this conjecture, we sum up in the following table the nonexistence of perfect quaternary sequences of period n with $n \equiv 0 \ mod \ 4$ and $4 \leq n \leq 100$. The notations used are:

† : does not exist by theorem 23.
†k: does not exist by results in [1] with parameter k.
? : non-existence conjectured by results in [2].
⋆ : exists by a exhaustive computer search.

n	4	8	12	16	20	24	28	32	36	40	44	48	52	56	60	64	68
$existence$	⋆	⋆	†	⋆	?	†	†	†4	?	?	†	†	?	†	†	†	†5	?

n	72	76	80	84	88	92	96	100
$existence$	?	†	?	†	†	†	†	?

3   Almost-Perfect Autocorrelation Quaternary Sequences

We recall that an almost-perfect quaternary sequence has all its "out-of-phase" coefficients equal to zero except possibly one. So we begin by a short characterisation of this non-zero autocorrelation coefficient.

Proposition 3.1. *The quaternary sequence* $s = (s_t)$ *of period* n *is almost-perfect if and only if* $n \equiv 0 \pmod 2$ *and all its "out-of-phase" autocorrelation coefficients are equal to zero except one which corresponds to a shift equal to half of period.*

Proof. The proof is obvious for periodical and symetrical reasons of the auto-correlation function. $\square$

Theorem 3.1. *Let* $s = (s_t)$ *be a quaternary sequence of period* n. *Define* $n_k = \sharp\{t \in [o, n-1] \mid s_t = k\}$ $(k = 0, 1, 2, 3)$ *as the number of occurencies of* k *in one generating vector* $(s_0, s_1, \cdots, s_{n-1})$ *of* s. *If* s *is almost-perfect with* $\mid c_s(\frac{n}{2}) \mid = c$ *then*

$$n_0(n_0 - 1) + n_1(n_1 - 1) + n_2(n_2 - 1) + n_3(n_3 - 1) = 2n_0n_2 + 2n_1n_3 \pm c.$$

Proof. We proceed similarly as in theorem 21 with the same notations. We obtain the same result raisoning on columns, i.e.

$$\begin{cases} N_0 = n_0(n_0 - 1) + n_1(n_1 - 1) + n_2(n_2 - 1) + n_3(n_3 - 1) \\ N_1 = N_3 = n_0n_3 + n_1n_0 + n_2n_1 + n_3n_2 \\ N_2 = 2n_0n_2 + 2n_1n_3 \end{cases}$$

The reasoning on lines is a little different because of the line $u = \frac{n}{2}$ which corresponds to the only non-zero autocorrelation coefficient. First, we treat this case separately and then continue with the other lines.

Line $u = \frac{n}{2}$:

Let l_0, l_1, l_2 and l_3 be the number of $0, 1, 2$ and 3, respectively, of this line.

$$\mid c_s(\frac{n}{2}) \mid = c \Leftrightarrow \mid (l_0 - l_2) + i(l_1 - l_3) \mid = c$$

$$\Leftrightarrow (l_0 - l_2)^2 + (l_1 - l_3)^2 = c^2$$

Let $X = l_0 - l_2$, $Y = l_1 - l_3$ et $Z = c$. We get the equation $X^2 + Y^2 = Z^2$. A result on theory numbers (theorem 5.11 in [4]) gives the solutions (called primitive) of this equation.

$X = r^2 - s^2$, $Y = 2rs$, $Z = r^2 + s^2$ with r and s two integers of opposite parity, prime each other and $r > s > 0$.

But, if $n = 2m$ with m odd then $c = n - 2 = 2(m - 1)$ or $c = \frac{n-2}{2} = m - 1$ which are both even.

Moreover, if $n = 4m$ then $c = n - 4 = 4(m - 1)$ or $c = \frac{n-4}{2} = 2(m - 1)$ which are also both even.

Therefore, c is always even and the equation $X^2 + Y^2 = Z^2$ has no solutions. It

implies that either r or s is equal to zero and then $Y = 0$ i.e. $l_1 = l_3$.
The equation is reduced to $X^2 = Z^2$ that is $l_0 - l_2 = \pm c$. First we solve the case $l_0 - l_2 = c$, the other one can be deduced by symmetry.
If $l_0 - l_2 = c$ then the two possible choices are $l_0 = c$ and $l_2 = 0$ or $l_0 = c$ and $l_2 \neq 0$.
If $l_2 = 0$ then $l_1 = l_3 = \frac{n-c}{2}$ else $l_0 = \frac{n+c}{2} - l_1$ and $l_2 = \frac{n-c}{2} - l_1$.
In conclusion for the line $u = \frac{n}{2}$, the four possibilities are the following :

$$l_0 = c \qquad l_2 = 0 \qquad l_1 = l_3 = \frac{n-c}{2}$$
$$l_0 = 0 \qquad l_2 = c \qquad l_1 = l_3 = \frac{n-c}{2}$$
$$l_0 = \frac{n+c}{2} - l_1 \; l_2 = \frac{n-c}{2} - l_1 \; l_1 = l_3$$
$$l_0 = \frac{n-c}{2} - l_1 \; l_2 = \frac{n+c}{2} - l_1 \; l_1 = l_3$$

Line $u \neq \frac{n}{2}$:

We denote by lg_0, lg_1, lg_2 and lg_3 the number of $0, 1, 2$ and 3 respectively of this line.

$$| c_s(i) | = 0 \Leftrightarrow | (lg_0 - lg_2) + i(lg_1 - lg_3) | = 0$$
$$\Leftrightarrow \begin{cases} lg_0 = lg_2 = a \\ lg_1 - lg_3 = b \end{cases} \quad avec \; a + b = \frac{n}{2}$$

Let $\tilde{N}_k$ be the total number of elements equal to k ($0 \leq k \leq 3$) in the matrix without counting the line $u = \frac{n}{2}$.
We have two possibilities for the couples (a, b).
(1) $\forall i, \; i \neq \frac{n}{2} \quad \exists! (a, b) \mid a + b = \frac{n}{2}$ with a and b ranging in $[0 \cdots \frac{n}{2}]$.
In this case, we get $\tilde{N}_0 = \tilde{N}_2 = (n-2)a$ and $\tilde{N}_1 = \tilde{N}_3 = (n-2)b$.
(2) There exist several couples (a_j, b_j) such that $a_j + b_j = \frac{n}{2}$ with an occurency o_j.
In this case we obtain, $\tilde{N}_0 = \tilde{N}_2 = \sum_{j \in J} o_j a_j$ and $\tilde{N}_1 = \tilde{N}_3 = \sum_{j \in J} o_j b_j$ where J counts the number of couples (a_j, b_j) with an occurency equal to o_j.

Finally, we sum up the results for the line $u = \frac{n}{2}$ and the other lines with $N_k = l_k + \tilde{N}_k$ ($0 \leq k \leq 3$).
For the case (1) we have those four possibilities.

$$N_0 = c + (n-2)a \quad N_1 = \frac{n-c}{2} + (n-2)b$$
$$N_2 = 0 + (n-2)a \quad N_3 = \frac{n-c}{2} + (n-2)b$$

$$N_0 = 0 + (n-2)a \quad N_1 = \frac{n-c}{2} + (n-2)b$$
$$N_2 = c + (n-2)a \quad N_3 = \frac{n-c}{2} + (n-2)b$$

$$N_0 = \frac{n+c}{2} - l_1^{n/2} + (n-2)a \quad N_1 = l_1^{n/2} + (n-2)b$$
$$N_2 = \frac{n-c}{2} - l_1^{n/2} + (n-2)a \quad N_3 = l_1^{n/2} + (n-2)b$$

$$N_0 = \frac{n-c}{2} - l_1^{n/2} + (n-2)a \quad N_1 = l_1^{n/2} + (n-2)b$$
$$N_2 = \frac{n+c}{2} - l_1^{n/2} + (n-2)a \quad N_3 = l_1^{n/2} + (n-2)b$$

For the case (2) we have the same equalities replacing $(n-2)a$ and $(n-2)b$ by $\sum_{j \in J} o_j a_j$ and $\sum_{j \in J} o_j b_j$ respectively.

We complete the proof by bringing together the results obtained with the columns and those obtained with the lines.   □

Similarly as in the previous section, if we solve the equation $n_0(n_0 - 1) + n_1(n_1 - 1) + n_2(n_2 - 1) + n_3(n_3 - 1) = 2n_0 n_2 + 2n_1 n_3 \pm c$ stated in theorem 31. we obtain some non-existence conditions on almost-perfect quaternary sequences. The following theorem sums up those results.

Theorem 3.2. *Let $s = (s_t)$ be a quaternary sequence of period n. Define $n_k = \sharp\{t \in [0, n-1] \mid s_t = k\}$ $(k = 0, 1, 2, 3)$ as the number of occurencies of k in one generating vector $(s_0, s_1, \cdots, s_{n-1})$ of s. Let $\mid c_s(\frac{n}{2}) \mid = c$ and $\prod_p p^{v_p(n \pm c)}$ the prime factor decomposition of $n \pm c$. If s is almost-perfect with $\mid c_s(\frac{n}{2}) \mid = c$ then*

1) $n = (n_3 - n_1)^2$ *and* $n \pm c$ *is a square and* $n \pm c \equiv 0 \pmod 4$ *with*

$$n_3 = n_1 + \sqrt{n \pm c} \quad n_2 = n_0 = \frac{n}{2} - \frac{\sqrt{n \pm c}}{2} - n_1$$
$$or$$
$$n_3 = n_1 - \sqrt{n \pm c} \quad n_2 = n_0 = \frac{n}{2} + \frac{\sqrt{n \pm c}}{2} - n_1$$

2) $n_3 = n_1$ *and* $n \pm c$ *is a square and* $n \pm c \equiv 0 \pmod{16}$ *with*

$$n_3 = n_1 \quad n_2 = \frac{n}{2} - n_1 + \frac{\sqrt{n \pm c}}{4} \quad n_0 = \frac{n}{2} - n_1 - \frac{\sqrt{n \pm c}}{4}$$
$$or$$
$$n_3 = n_1 \quad n_2 = \frac{n}{2} - n_1 - \frac{\sqrt{n \pm c}}{4} \quad n_0 = \frac{n}{2} - n_1 + \frac{\sqrt{n \pm c}}{4}$$

3) $n_1 \neq n_3$ *and* $n \neq (n_3 - n_1)^2$ *with for all integers* $p \equiv 3 \pmod 4$ *the exponent* $v_p(n \pm c)$ *is even and* $n \pm c = a^2 + b^2$ *where a and b are even with*

$$n_3 = n_1 + b \quad n_2 = \frac{n}{2} - n_1 - \frac{b}{2} + \frac{a}{2} \quad n_0 = \frac{n}{2} - n_1 - \frac{b}{2} - \frac{a}{2}$$
$$or$$
$$n_3 = n_1 + b \quad n_2 = \frac{n}{2} - n_1 - \frac{b}{2} - \frac{a}{2} \quad n_0 = \frac{n}{2} - n_1 - \frac{b}{2} + \frac{a}{2}$$

and

$$n_3 = n_1 - b \quad n_2 = \frac{n}{2} - n_1 + \frac{b}{2} + \frac{a}{2} \quad n_0 = \frac{n}{2} - n_1 + \frac{b}{2} - \frac{a}{2}$$
$$or$$
$$n_3 = n_1 - b \quad n_2 = \frac{n}{2} - n_1 + \frac{b}{2} - \frac{a}{2} \quad n_0 = \frac{n}{2} - n_1 + \frac{b}{2} + \frac{a}{2}$$

Proof. We proceed similarly as in theorem 23 with the same notations and using theorem 31, however considering $n \pm c$ instead of n in all equations.

□

4   Conclusion

Perfect and almost-perfect periodic quaternary sequences have been the topic of many papers because they are useful in multiple access communications systems. We have presented some new general non-existence theorems about those sequences. In the perfect case, we have been tempted to conjecture that the only such periodic quaternary sequences of period 4, 8 and 16 exist.

References

1. H. CHUNG and P.V. KUMAR - *A new general construction for generalized bent functions* - IEEE Trans. Inform. Theory, **35** : 206–209, 1989.
2. K.T. ARASU and W. LAUNEY and S.L. MA - *On circulant Hadamard matrices* - to appear in Designs, Codes and Cryptography".
3. J. WOLFMANN - *Almost perfect autocorrelation sequences* - IEEE Trans. Inform. Theory, **38** : 1412–1418, 1992.
4. I. NIVEN and H.S. ZUCKERMAN - *The theory of numbers* - John Wiley and Sons, 1980 (4th edition).
5. P. SAMUEL - *Théorie algébrique des nombres* - Hermann, 1971 (2nd edition).

Maximal Periods of $x^2 + c$ in $\mathbb{F}_q$

A. Peinado[1]*, F. Montoya[2], J. Muñoz[2], and A.J. Yuste[3]

[1] Dpto. Ingeniería de Comunicaciones, E.T.S. Ingeniería de Comunicaciones
Universidad de Málaga, Campus de Teatinos - 29071 Málaga, Spain
[2] Dpto. de Tratamiento de la Información y Codificación
Instituto de Física Aplicada (CSIC), C/ Serrano 144, 28006-Madrid, Spain
[3] Dpto. Electrónica, Universidad de Jaen
C/ Alfonso X, 28 - 23700 Linares, Jaen, Spain

Abstract. The orbits produced by the iterations of the mapping $x \mapsto x^2 + c$, defined over $\mathbb{F}_q$, are studied. Several upper bounds for their periods are obtained, depending on the coefficient c and the number of elements q.

Keywords: *Pseudorandom sequence generation, stream ciphers, Pollard generator.*

1 Introduction

Quadratic functions are widely used in Cryptography, defining a great variety of systems [2], [11], [13]. In particular, quadratic functions are used to generate pseudorandom sequences, by means of the iterations of the mapping $x \mapsto x^2$, defined over $\mathbb{Z}_{pq}$, with p, q two distinct odd prime numbers. Many works exist focusing on this topic (see [2], [3], [5], [9]).

However, the mapping $x \mapsto x^2$, is not representative of every quadratic mapping. Hence, none of these results are applicable to the mapping $x \mapsto x^2 + c$, with $c \neq 0$. This mapping is the basis of the Pollard's rho method [12] for integer factorization, which makes use of the orbital structure (tails and cycles) of the functions $f_c \colon \mathbb{Z}_p \to \mathbb{Z}_p$, $f_c(x) = x^2 + c$. In this work, the orbits of the functions $f_c \colon \mathbb{F}_q \to \mathbb{F}_q$, $f_c(x) = x^2 + c$ are analysed and several upper bounds for cycle lengths are obtained, depending on the coefficient c and the number of elements $q = p^n = \#\mathbb{F}_q$, p being an odd prime.

In Section 2, several concepts on functions iteration and quadratic functions are introduced. Section 3 presents the general aspects of the orbital structure of $x \mapsto x^2 + c$, and Section 4 shows the theoretical results about cycle length upper bounds. Finally, Section 5 deals with some experimental results on the cycle length of this function.

---

* This work is supported by CICYT (Spain) under grant TEL98-1020, *Infraestructuras de Seguridad en Internet e Intranets. Aplicación a Redes Públicas y Corporativas.*

S. Boztaş and I.E. Shparlinski (Eds.): AAECC-14, LNCS 2227, pp. 219–228, 2001.

2   Notations and Preliminaries

Let $O(x_0) = \{f^n(x_0) \mid n \in \mathbb{N}\}$ be the f-orbit of an element $x_0 \in \mathbb{F}_q$, with respect to an arbitrary function $f\colon \mathbb{F}_q \to \mathbb{F}_q$, where f^n denotes the n-th iteration of f, or even $O(x_0) = \{x_0, x_1, \ldots, x_{n-1}, x_n, \ldots\}$, $x_n = f(x_{n-1})$. Let $h = h(x)$ be the least positive integer for which an integer k exists such that: i) $0 \leq k < h$, and ii) $x_k = x_h$. The set of elements $x_0, x_1, \ldots, x_{k-1}$ is called the "tail" $T(x_0)$ of the orbit, the set of elements $x_k, x_{k+1}, \ldots, x_{h-1}$ is called the "cycle" $C(x_0)$ of the orbit, and $l(x_0) = l_f(x_0) = h - k$ is the length or period of the cycle (cf. [5], [10, XII]).

Every polynomial $p(X) \in F[X]$ induces, by evaluation, a polynomial function $p\colon F \longrightarrow F$. Proceeding by recurrence on $n \in \mathbb{N}$, we define $p^n(X)$ to be the polynomial obtained by substituting $p(X)$ for X in $p^{n-1}(X)$.

Hence, a quadratic polynomial $p(X) = aX^2 + bX + c \in \mathbb{F}_q[X]$ is linearly conjugated (e.g., see [6]) to a quadratic polynomial $q(X)$ of the form $q(X) = X^2 + k \in \mathbb{F}_q[X]$ (cf. [5]). This relationship allows to simplify the study of quadratic functions, reducing the number of non-vanishing coefficients.

The simplest case of study corresponds to $k = 0$, defining the mapping f as $f\colon \mathbb{Z}_p \to \mathbb{Z}_p$, $f(x) = x^2$. The orbits of this function are completely characterized in [5], where the prime numbers producing cycles of maximal length $(p - 3)/8$, are identified.

We consider now the case $k \neq 0$, defining the mapping f_c as $f_c\colon \mathbb{F}_q \to \mathbb{F}_q$; $f_c(x) = x^2 + c$, $c \in \mathbb{F}_q^*$. The following result shows the impossibility to apply the previous results ($k = 0$) to f_c. Figure 1 illustrates this fact.

Proposition 1. *The function f is not linearly conjugated to f_c, for any $c \in \mathbb{F}_q^*$.*

Proof. The proof is based on the orbital structure of both functions f and f_c. In both cases, there exists a unique element with only one predecessor, *i.e.*, $x = 0$ for the function f, and $x = c$, for the function f_c. Suppose f is conjugated to f_c by means of a permutation polynomial p. Then $p(0) = c$. However, $x = 0$ is an invariant element of f, while c is not in f_c, leading us to a contradiction.

3   Orbits of $x \mapsto x^2 + c$

In this section, general aspects on the orbital structure of f_c are introduced. First, the number of predecessors and succesors of every element $x \in \mathbb{F}_q$ is obtained in the following proposition.

Proposition 2. *Let f_c be the mapping defined by $f_c\colon \mathbb{F}_q \to \mathbb{F}_q$, $f_c(x) = x^2 + c$, with $c \in \mathbb{F}_q^*$. Then, every element $x \in \mathbb{F}_q$ has a unique succesor $f(x)$ and,*

1. *has two predecessors (anti-images) if and only if $x - c$ is a quadratic residue.*
2. *has no predecessor (anti-image) if and only if $x - c$ is not a quadratic residue.*
3. *has one predecessor (anti-image) if and only if $x - c = 0$.*

Proof. It follows directly from the definition of f_c.

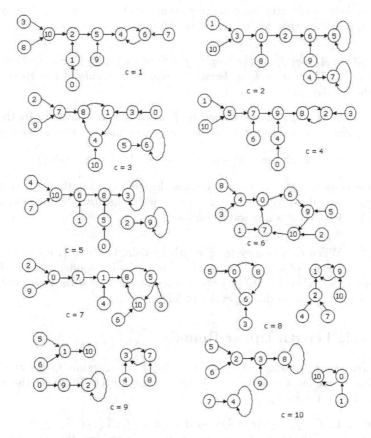

Fig. 1. Orbital spaces of $x \mapsto x^2 + c$, for every $c \in \mathbb{Z}_p^*$, with $p = 11$.

In other words, the element $x = c$ has only one predecessor, but any other element has either two or none predecessors. Unlike the case of function f, the element with only one predecessor ($x = c$) is not invariant ($f(x) = x$). Thus, the maximal cycle length is bounded by $(q+1)/2$, as there exist $(q-1)/2$ quadratic residues. This fact is equivalent to compute the cardinality of $f_c[\mathbb{F}_q]$, which is $\#f_c[\mathbb{F}_q] = (q+1)/2$, since $f_c(x) = f_c(-x)$.

In order to compute the maximal length of the cycles, the following propositions state the conditions for the existence of cycles of length 1 and 2.

Proposition 3. *Let f_c be the mapping defined by $f_c \colon \mathbb{F}_q \to \mathbb{F}_q$, $f_c(x) = x^2 + c$, with $c \in \mathbb{F}_q^*$. Then,*

1. *only one cycle of length 1 exists, if and only if $c = \frac{1}{4}$.*
2. *two cycles of length 1 exist, if and only if $1 - 4c$ is a quadratic residue.*

Proof. It follows directly from the equation $f_c(x) = x$, defining the cycles of length 1, and the expresion of the roots of a quadratic equation.

Proposition 4. *Let f_c be the mapping defined as $f_c\colon \mathbb{F}_q \to \mathbb{F}_q$, $f_c(x) = x^2 + c$ with $c \in \mathbb{F}_q^*$. Then, a cycle of length 2 exists if and only if $1 - 4(c + 1)$ is a quadratic residue.*

Proof. Cycles of length 2 are defined by $f^2(x) = x$, and, therefore, by the equation $x^4 + 2cx^2 - x + c^2 + c = 0$. Since $f(x) - x$ divides $f^2(x) - x$, we have

$$x^4 + 2cx^2 - x + c^2 + c = \left(x^2 - x + c\right)\left(x^2 + x + c + 1\right).$$

Since the equation $x^2 - x + c = 0$ defines the cycles of length 1, the only cycle of length 2 is defined by $x^2 + x + c + 1 = 0$, which has two distinct roots if and only if $1 - 4(c + 1)$ is a quadratic residue.

Remark 1. When only one cycle of length 1 exists, i.e., $c = \frac{1}{4}$, it can be proved that another cycle of length 2 exists if and only if -1 is a quadratic residue. On the other hand, when $1 - 4(c + 1) = 0$, it can also be proved that cycles of length 2 do not exist, thus producing cycles of lentgh 1.

4   Cycle Length Upper Bounds

A theoretical upper bound for cycle lengths of the function f_c is stated in the following theorems. We denote $l_q(c)$ the maximum cycle length of the function f_c, that is, $l_q(c) = \max_{x \in \mathbb{F}_q}(l_{f_c}(x))$.

Theorem 1. *Let f_c be the mapping defined as $f_c\colon \mathbb{F}_q \to \mathbb{F}_q$, $f_c(x) = x^2 + c$, with $c \in \mathbb{F}_q^*$, $p > 9$. Then, if -1 is not a quadratic residue, the cycle length $l_q(c)$ of the function f_c is*

1. $l_q(c) \le (3q + 3)/8$, *if $(q - 3)/4$ is odd.*
2. $l_q(c) \le (3q - 1)/8$, *if $(q - 3)/4$ is even and $-2c$ is a quadratic residue.*
3. $l_q(c) \le (3q + 7)/8$, *if $(q - 3)/4$ is even and $-2c$ is not a quadratic residue.*

Theorem 2. *Let f_c be the mapping defined as $f_c\colon \mathbb{F}_q \to \mathbb{F}_q$, $f_c(x) = x^2 + c$, with $c \in \mathbb{F}_q^*$, $p > 9$. Then, if -1 is a quadratic residue, the cycle length $l_q(c)$ of the function f_c is*

1. $l_q(c) \le (3q + 5)/8$, *if $(q - 1)/4$ is even.*
2. $l_q(c) \le (3q + 1)/8$, *if $(q - 1)/4$ is odd and $-2c$ is a quadratic residue.*
3. $l_q(c) \le (3q + 9)/8$, *if $(q - 1)/4$ is odd and $-2c$ is not a quadratic residue.*

In order to prove these theorems, we need the following previous results.

Lemma 1. *Let $\mathbb{F}_q[i] = \{a + bi \mid a, b \in \mathbb{F}_q\}$, $i^2 = -1$. Consider the norm function $N\colon \mathbb{F}_q[i] \to \mathbb{F}_q$, $N(a + bi) = a^2 + b^2$. Then, we have*

(a) *If -1 is a quadratic residue in $\mathbb{F}_q$, then there are exactly $2q - 1$ elements*
   $a + bi \in \mathbb{F}_q[i]$ satisfying $a^2 + b^2 = 0$ and there are exactly $q - 1$ elements
   $a + bi \in \mathbb{F}_q[i]$ satisfying $a^2 + b^2 = c$ for each $c \in \mathbb{F}_q^*$.
(b) *If -1 is not a quadratic residue in $\mathbb{F}_q$, then there are exactly 1 element*
   $a + bi \in \mathbb{F}_q[i]$ satisfying $a^2 + b^2 = 0$ and there are exactly $q + 1$ elements
   $a + bi \in \mathbb{F}_q[i]$ satisfying $a^2 + b^2 = c$ for each $c \in \mathbb{F}_q^*$

Proof. (a) If -1 is a quadratic residue in $\mathbb{F}_q$, say $-1 = j^2$, then there is a ring isomorphism $\phi\colon \mathbb{F}_q[i] \to \mathbb{F}_q \times \mathbb{F}_q$, given by $\phi(a+bi) = (u, v) = (a+bj, a-bj)$, and the equation $a^2 + b^2 = 0$ is equivalent to the following: $uv = 0$, whose solutions are $(0,0)$; $(u, 0)$, $u \neq 0$; $(0, v)$, $v \neq 0$. Similarly, the equation $a^2 + b^2 = c$, $c \neq 0$, becomes $uv = c$, whose solutions are $(u, c/u)$, with $u \neq 0$, thus proving the first part of the statement.

(b) If -1 is not a quadratic residue in $\mathbb{F}_q$, then $a^2 + b^2 = 0$ with $a \neq 0$, implies $(b/a)^2 = -1$, thus contradicting the asumption. Hence $a = 0$, and similarly $b = 0$. Moreover, in this case, $\mathbb{F}_q[i]$ is a (finite) field. In fact, since $N(xy) = N(x)N(y)$, the relation $xy = 0$ in $\mathbb{F}_q[i]$ implies $N(x)N(y) = 0$, so that $N(x) = 0$ or $N(y) = 0$ and this implies $x = 0$ or $y = 0$, due to the assumption that -1 is not a quadratic residue in $\mathbb{F}_q$. Accordingly, the norm induces a group homomorphism $N\colon \mathbb{F}_q[i]^* = \{a + bi \neq 0\} \to \mathbb{F}_q^*$, which is surjective. In fact, given an element $c \in \mathbb{F}_q$, the number of elements of the mapping $\sigma\colon \mathbb{F}_q \to \mathbb{F}_q$, $\sigma(a) = c - a^2$, is $(q + 1)/2$. As the number of elements of the set $Q = \{b^2 \mid b \in \mathbb{F}_q\}$ is also $(q+1)/2$, we conclude that $(\mathrm{im}\,\sigma) \cap Q$ is not empty. We conclude that every $c \in \mathbb{F}_q^*$ can be written in the form $a^2 + b^2 = c$ and the number of elements $a + bi \in \mathbb{F}_q[i]$ satisfying this condition is exactly equal to $\#\ker N = \#\mathbb{F}_q[i]^*/\#\mathbb{F}_q^* = (q^2 - 1)/(q - 1) = q + 1$.

Proposition 5. *For every $c \in \mathbb{F}_q^*$, $p > 9$, if -1 is not a quadratic residue, we have*

1. *If $(q - 3)/4$ is odd, then there exist $(q + 1)/8$ different pairs (x^2, y^2), such that $c = x^2 + y^2$, with $x^2 \neq y^2$.*
2. *If $(q - 3)/4$ is even and c is a quadratic residue, then there exist $(q + 5)/8$ different pairs (x^2, y^2), such that $c = x^2 + y^2$, with $x^2 \neq y^2$.*
3. *If $(q - 3)/4$ is even and c is not a quadratic residue, then there exist $(q - 3)/8$ different pairs (x^2, y^2), such that $c = x^2 + y^2$, with $x^2 \neq y^2$.*

Proof. From Lemma 1 we have $q+1$ elements $a+bi \in \mathbb{F}_q[i]$ satisfying the equation $a^2 + b^2 = c$, for each $c \in \mathbb{F}_q^*$.

Suppose that $c = b^2$ is a quadratic residue. Hence the pair $(0, b^2)$ is trivially a solution, which can be obtained from four distinct pairs $(0, \pm b)$, $(\pm b, 0)$. Since the elements $a+bi$, $a - bi$, $-a+bi$, and $-a - bi$ produce the same pair (a^2, b^2), there only exist $((q + 1) - 4)/4 = (q - 3)/4$ pairs (a^2, b^2) with $a^2 \neq 0, b^2 \neq 0$.satisfying the equation $a^2 + b^2 = c$.

If $(q - 3)/4$ is even, then the pair (a^2, a^2) is not a solution. In fact, if the pair (a^2, b^2) is a solution, then (b^2, a^2) is also a solution, and the total number of pairs is even. Hence, if we take (a^2, b^2) but not (b^2, a^2), and take the solution $(0, b^2)$, the number of distinct pairs satisfying the equation $a^2 + b^2 = c$ is

$$\frac{(q-3)}{4}\frac{1}{2} + 1 = \frac{q+5}{8}.$$

If $(q-3)/4$ is odd, then it is clear that the pair (a^2, a^2) is a solution. If we apply the restriction imposed in the proposition $(x^2 \neq y^2)$, this solution is not valid. Hence, the number of distinct pairs satisfying the equation $a^2 + b^2 = c$ is

$$\left(\frac{(q-3)}{4} - 1\right)\frac{1}{2} + 1 = \frac{q+1}{8}.$$

Now, we assume that c is not a quadratic residue. Then there are $(q+1)/4$ pairs (a^2, b^2) with $a^2 \neq 0$, $b^2 \neq 0$ satisfying the equation. $a^2 + b^2 = c$. If $(q+1)/4$ is even, the pair (a^2, a^2) is not a solution since $a^2 + a^2 = 2a^2 = c$ leads us to a contradiction. Hence, considering the pair (a^2, b^2) but not (b^2, a^2), the number of distinct pairs satisfying the equation $a^2 + b^2 = c$ is

$$\left(\frac{(q+1)}{4}\right)\frac{1}{2} = \frac{q+1}{8}.$$

If $(q+1)/4$ is odd, then the pair (a^2, a^2) is a solution If we apply the restriction imposed in the proposition $(x^2 \neq y^2)$, this solution is not valid. Hence, considering the pair (a^2, b^2) but not (b^2, a^2), the number of distinct pairs satisfying the equation $a^2 + b^2 = c$ is

$$\left(\frac{(q+1)}{4} - 1\right)\frac{1}{2} = \frac{q-3}{8}.$$

We conclude taking into account that $(q-3)/4 \equiv (q+1)/4 + 1 \pmod 2$.

Proposition 6. *For every $c \in \mathbb{F}_q^*$, $p > 9$, if -1 is a quadratic residue, we have*

1. *If $(q-1)/4$ is even, then $(q-1)/8$ different pairs (x^2, y^2), such that $c = x^2 + y^2$, with $x^2 \neq y^2$.*
2. *If $(q-1)/4$ is odd and c is a quadratic residue, then $(q+3)/8$ different pairs (x^2, y^2), such that $c = x^2 + y^2$, with $x^2 \neq y^2$.*
3. *If $(q-1)/4$ is odd and c is not a quadratic residue, then $(q-5)/8$ different pairs (x^2, y^2), such that $c = x^2 + y^2$, with $x^2 \neq y^2$.*

Proof. From Lemma 1 we have $q-1$ elements $a + bi \in \mathbb{F}_q[i]$ satisfying the equation $a^2 + b^2 = c$, for each $c \in \mathbb{F}_q^*$.

Suppose that $c = b^2$ is a quadratic residue. Hence the pair $(0, b^2)$ is trivially a solution, which can be obtained from four distinct pairs $(0, \pm b)$, $(\pm b, 0)$. Since the elements $a + bi$, $a - bi$, $-a + bi$, and $-a - bi$ produce the same pair (a^2, b^2), there only exist $((q-1) - 4)/4 = (q-5)/4$ pairs (a^2, b^2) with $a^2 \neq 0, b^2 \neq 0$. satisfying the equation $a^2 + b^2 = c$.

If $(q - 5)/4$ is even, then the pair (a^2, a^2) is not a solution. In fact, if the pair (a^2, b^2) is a solution, then (b^2, a^2) is also a solution, and the total number of pairs is even.Hence, if we take (a^2, b^2) but not (b^2, a^2), and take the solution $(0, b^2)$, the number of distinct pairs satisfying the equation $a^2 + b^2 = c$ is

$$\frac{(q - 5)}{4}\frac{1}{2} + 1 = \frac{q + 3}{8}.$$

If $(q - 5)/4$ is odd, then it is clear that the pair (a^2, a^2) is a solution. If we apply the restriction imposed in the proposition $(x^2 \neq y^2)$, this solution is not valid. Hence, the number of distinct pairs satisfying the equation $a^2 + b^2 = c$ is

$$\left(\frac{(q - 5)}{4} - 1\right)\frac{1}{2} + 1 = \frac{q - 1}{8}.$$

Now we assume that c is not a quadratic residue. Then there are $(q - 1)/4$ pairs (a^2, b^2) with $a^2 \neq 0$, $b^2 \neq 0$ satisfying the equation.$a^2 + b^2 = c$. If $(q - 1)/4$ is even then the pair (a^2, a^2) is not a solution since $a^2 + a^2 = 2a^2 = c$ leads us to a contradiction. Hence, considering the pair (a^2, b^2) but not (b^2, a^2), the number of distinct pairs satisfying the equation $a^2 + b^2 = c$ is

$$\left(\frac{(q - 1)}{4}\right)\frac{1}{2} = \frac{q - 1}{8}.$$

If $(q - 1)/4$ is odd, then the pair (a^2, a^2) is a solution. If we apply the restriction imposed in the proposition $(x^2 \neq y^2)$, this solution is not valid. Hence, considering the pair (a^2, b^2) but not (b^2, a^2), the number of distinct pairs satisfying the equation $a^2 + b^2 = c$ is

$$\left(\frac{(q - 1)}{4} - 1\right)\frac{1}{2} = \frac{q - 5}{8}.$$

We conclude taking into account that $(q - 5)/4 \equiv (q - 1)/4 + 1 \pmod 2$.

Proof of theorems 1 and 2. Since $f_c(x) = f_c(-x) = x^2 + c$, the cardinality of $f_c[\mathbb{F}_q]$ is $\#f_c[\mathbb{F}_q] = (q + 1)/2$. On the other hand, the number of elements $y \in f_c[\mathbb{F}_q]$ such that $-y \in f_c[\mathbb{F}_q]$, can be computed solving the following system of equations

$$y = x_1^2 + c$$
$$-y = x_2^2 + c$$

By adding the two equations, we have $x_1^2 + x_2^2 = -2c$, with $x_1^2 \neq x_2^2$. The cardinality of $f_c^2[\mathbb{F}_q]$ can be computed as $\#f_c^2[\mathbb{F}_q] = \#f_c[\mathbb{F}_q] - N_c$, where N_c is the number of pairs (x_1^2, x_2^2), satisfying the equation $x_1^2 + x_2^2 = -2c$. Hence, we can conclude by substituing the values of N_c stated in propositions 5 and 6.

5   Experimental Results

Numerous computations have been carried out to check the validity of the theoretical results, and how far is the previous upper bound to the real maximal cycle length. Most of these results are concerned with $\mathbb{Z}_p$, although similar results are obtained for $\mathbb{F}_q, q = p^n$, with $n > 1$.

As it can be observed in figure 2, the bound in theorem 1 is reached for small prime numbers. However, for $p > 83$ the bound observed in the figure is far from the real value. More precisely, the maximal lengths $l_p(c)$ in the figure are

- $l_p(c) \leq (p-1)/4$, if $p \equiv 1 \pmod 4$.
- $l_p(c) \leq (p-3)/4$, if $p \equiv 3 \pmod 4$.

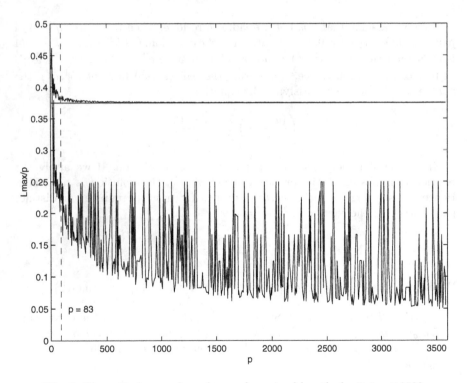

Fig. 2. Theoretical upper bound vs real maximal length, for $7 \leq p \leq 3583$.

Note that this experimental bound implies $\lim_{p\to\infty} l_p(c)/p \leq 1/4 = 0.25$, unlike the theoretical bound which implies $\lim_{p\to\infty} l_p(c)/p \leq 3/8 = 0.375$. Taking into account that the theoretical bound is obtained by computing the cardinality of $f_c^2[\mathbb{Z}_p]$, one could expect an aproximation between theoretical and real bounds by computing the succesives cardinalities of $f_c^i[\mathbb{Z}_p]$, for $i > 2$. However, this approach is not valid in general terms because f_c acts as a permutation over $f_c^2[\mathbb{Z}_p]$ for certain values of p and c, such as, $c = -2$ when 2 is not a quadratic residue in $\mathbb{Z}_p$. In any case, the experimental results obtained computing $\#f_c^4[\mathbb{Z}_p]$ point out that $\#f_c^4[\mathbb{Z}_p]$ is very close to $0.25p$, in many cases.

Considering the prime numbers $p > 83$, the following remarks can be summarized.

Remark 2. As it is known from [5], when p is a 2-safe prime ($p = 2p' + 1$, $p' = 2p'' + 1$, with p, p', p'' distinct odd prime numbers), then the maximal cycle length of the function f is $(p - 3)\, t/2$, where $t = (1/2)^{\text{parity}((p+1)/8)}$. In this case, the maximal cycle length of the function f_c is $(p - 3)/4$ for $c = -2$.

Remark 3. If $p \equiv 1 \,(\mathrm{mod}\,4)$, and f_c has a maximal cycle length of $(p - 1)/4$, then the maximal length of f is always less than $(p - 1)/4$.

Remark 4. If p is a Fermat prime ($p = 2^{2^n} + 1$), then the maximal length of f is 1 [5]. Hence, the maximal length of f_c is greater than that of f.

Remark 5. If p is a Mersenne prime ($p = 2^n - 1$), then the maximal length of f is $n - 1$ [5]. Hence, this length is easily exceeded by the maximal length of f_c, as it can be checked.

Acknowledgements

The authors want to thank to Igor Shparlinski his valuable comments during the redaction process of this paper.

References

1. BACH,E., SHALLIT,J., "Algorithmic Number Theory. Vol I. Efficient algorithms", *The MIT Press*, 1996.
2. BLUM,L., BLUM,M., SHUB,M., "A simple unpredictable pseudorandom number generator", *SIAM Journal on Computing*, **15** (1986), pp. 364-383.
3. GRIFFIN,F., SHPARLINSKI,I., "On the linear complexity profile of the power generator", *IEEE Trans. Inform. Theory*, **46** (2000), pp. 2159-2162.
4. HERNÁNDEZ ENCINAS,L., MONTOYA VITINI,F., MUÑOZ MASQUÉ, J., "Generación de sucesiones pseudoaleatorias mediante funciones cuadráticas en $\mathbb{Z}_{p^n}$, y en su límite proyectivo", Actas de la III Reunión Española de Criptografía, 27-32, (1994).
5. HERNÁNDEZ ENCINAS,L., MONTOYA VITINI,F., MUÑOZ MASQUÉ,J., PEINADO DOMÍNGUEZ,A., "Maximal periods of orbits of the BBS generator", *Proc. 1998 International Conference on Information Security & Cryptology (ICISC'98)*, Seoul, Korea, pp. 71-80, (1998).
6. LANG,S., "Algebra", *Addison-Wesley Publishing Company*, 3rd ed., 1993.
7. LIDL,R., NIEDERREITER,N., "Finite Fields", *Addison-Wesley Publishing Company*, 1983.

8. McELIECE,R., "Finite Fields for computer scientist and engineers". *Kluwer Academic Publishers*, 1987.

9. MONTOYA,F., MUÑOZ,J., PEINADO,A., "Linear complexity of the $x^2(\text{mod}\,p)$ orbits", *Information Processing Letters*, **72** (1999), pp. 3-7.

10. NARKIEWICZ, W., "Polynomial mappings", *Lecture Notes in Math.*, **1600**. Springer, 1995.

11. NYANG,D., SONG,J., "Fast digital signature scheme based on the quadratic residue problem", *Electronics Letters*, **33** (1997), pp. 205-206.

12. POLLARD,J.M., "A Monte Carlo method for factorization", *BIT*, **15** (1975), pp. 331-334.

13. RABIN,M.O., "Digitalized signatures and public key functions as intractable as factorization", *Technical report*, MIT/LCS/TR212, MIT Lab., Comp. Science, Cambridge, Mass, January 1979.

On the Aperiodic Correlation Function of Galois Ring m-Sequences

P. Udaya[1] and S. Boztaş[2]

[1] Dept. of Computer Science and Software Engineering
University of Melbourne, Melbourne, Australia
udaya@cs.mu.oz.au
[2] Dept. of Mathematics, RMIT University,
Melbourne, Australia
serdar@rmit.edu.au

Abstract. We define Gauss-like sums over the Galois Ring $\mathbf{GR}(4,r)$ and bound them using the Cauchy-Schwarz inequality. These sums are then used to obtain an upper bound on the aperiodic correlation function of quadriphase m-sequences constructed from $\mathbf{GR}(4,r)$.
Our first bound δ_1 has a simple derivation and is better than the previous upper bound of Shanbag et. al. for small values of N. We then make use of a result of Shanbag et. al. to improve our bound which gives rise to a bound $\delta_{improved}$ which is better than the bound of Shanbag et. al.
These results can be used as a benchmark while searching for the best phases—termed *auto-optimal* phases—of such quadriphase sequences for use in spread spectrum communication systems. The bounds can also be applied to many other classes of non binary sequences.

Index Terms: Galois Rings, Quadriphase sequences, Aperiodic correlations, Gauss sums over Galois rings

1 Introduction and Motivation

The design of sequences for Code Division Multiple Access (CDMA) communications has been a topic of interest over the last 50 years, starting in the arena of military communications—where the term *spread spectrum* originated since the emphasis then was on spreading the spectrum to 'hide' the transmissions from conventional narrow band receivers or wideband receivers not having access to the correct spreading sequence. Starting in the late 1980s and continuing in the 1990s, the development of mass market public digital cellular radio systems based on CDMA has led to a large increase research activity on all topics related to CDMA. Here we only concentrate on the sequence design aspect and focus on one particular figure of merit for CDMA based systems, the maximum *aperiodic correlation*. Other figures of merit include periodic correlation, partial period correlation, odd correlation and mean-square correlation. For an extensive discussion of these and other issues in CDMA sequence design we refer the reader to the extensive survey article by Helleseth and Kumar in [5]. We are

S. Boztaş and I.E. Shparlinski (Eds.): AAECC-14, LNCS 2227, pp. 229–238, 2001.
© Springer-Verlag Berlin Heidelberg 2001

thus content to provide a brief overview of the reason for being interested in aperiodic correlation of sequences.

For single sequences—as opposed to sequence families—used in pulse compression radar applications, the aperiodic auto-correlation of the sequence, and specifically its maximum off peak magnitude, provides an obvious figure of merit. For CDMA applications, the aperiodic correlation plays a different role of interest. It contributes an additive term to the multiuser interference to which any user is subjected from other users utilizing the same bandwidth. Hence the interest in bounding the aperiodic correlation.

2   Galois Rings and Sequences

In this paper we give an upper bound on the aperiodic correlation function of polyphase sequences constructed from the Galois ring $\mathbf{GR}(p^k, r)$ where p^k is a power of a prime and r is a positive integer. Our result is along the lines of a similar result proved in [6] for binary m-sequences. To obtain this result, we consider Gauss-like sum over $\mathbf{GR}(p^k, r)$—which are essentially Fourier transforms—and bound these sums from above using the Cauchy-Schwartz inequality.

2.1   Definitions

Let $q = p^k$, where p is a prime and $k \geq 2$ is a positive integer. We define $q - ary$ polyphase sequences using a mapping Π from $\mathbf{Z}_q$ to the field of complex numbers which is given by $\Pi : a \mapsto \omega^a$, with $\omega = e^{2\pi\sqrt{-1}/q}$. Clearly Π maps each symbol of $\mathbf{Z}_q$ to a complex q^{th} root of unity.

Hence corresponding to a sequence $M = (m_i : 0 \leq i < N)$ over $\mathbf{Z}_q$, we can define a $q - ary$ polyphase sequence S given by

$$S = (\Pi(m_i) = \omega^{m_i} : 0 \leq i < N). \tag{1}$$

The exponential sum—or the correlation transform—of the sequence S is then defined as

$$\theta(S) = \sum_{i=0}^{N-1} \omega^{m_i}. \tag{2}$$

For the rest of the paper we consider sequences of length N exclusively and omit the N from equations wherever convenient.

The *aperiodic* crosscorrelation function $C_{1,2}(l)$ for two polyphase sequences S^1 and S^2 derived from $\mathbf{Z}_q$ sequences $M^1 = (m_i^1)$ and $M^2 = (m_i^2)$ respectively is defined as

$$C_{1,2}(l) = \begin{cases} \sum_{i=0}^{N-1-l} \omega^{m_i^1 - m_{i+l}^2}, & 0 \leq l \leq N-1 \\ \sum_{i=0}^{N-1+l} \omega^{m_{i-l}^1 - m_i^2}, & 1 - N \leq l < 0 \\ 0, & |l| \geq N. \end{cases} \tag{3}$$

Note that

$$C_{1,1}(0) = C_{2,2}(0) = N \text{ and that } C_{1,2}(l) = C_{1,2}(-l). \tag{4}$$

The *periodic* crosscorrelation function $\theta_{1,2}(.)$ for S^2 and S^2 is defined as

$$\theta_{1,2}(l) = \sum_{i=0}^{N-1} \omega^{m_i^1 - m_{i+l}^2}, \qquad \text{for all } l. \tag{5}$$

Note that $\theta_{1,1}(0) = \theta_{2,2}(0) = N$ and $\theta_{1,2}(l) = \theta_{2,1}^*(-l) = \theta_{1,2}(l+N)$—where z^* denotes the complex conjugate of z—since the period of the sequences is N. Also,

$$\theta_{1,2}(l) = C_{1,2}(l) + C_{1,2}(l-N), \qquad 0 \le l \le N-1.$$

The above equations illustrate the close relationship between the periodic and the aperiodic correlations. For more details on this relationship and its impact on performance, see the above referenced chapter in [5] and the references therein.

2.2   Galois Ring Sequences

Galois rings are the generalizations of Galois fields and have been used widely in the past decade to construct various optimal families of q-ary polyphase sequences [1,9,10,8,3,4,7]. For details on Galois rings we refer the reader to [1,9,10]. Here, we remark that the Galois ring $\mathbf{GR}(p^k, r)$, $r \ge 1$, is a Galois extension of $\mathbf{Z}_{p^k}$, the ring of integers *modulo* p^k, and is isomorphic to the ring $\mathbf{Z}_{p^k}[x]/(f(x))$ where $f(x) \in \mathbf{Z}_{p^k}[x]$ is called a *monic basic irreducible* polynomial of degree r. Let $\mu : \mathbf{Z}_{p^k} \to \mathbf{Z}_p = \mathbf{GF}(p)$ be the (*modulo* p) projection map and extend this map to polynomials over $\mathbf{Z}_{p^k}$ in the natural way. Then, $f(x)$ is a monic basic irreducible polynomial in $\mathbf{Z}_{p^k}[x]$ if $\mu(f(x))$ is a monic irreducible polynomial in $\mathbf{Z}_p[x]$.

Let $\alpha \in \mathbf{GR}(p^k, r)$ be primitive (i.e., an element of multiplicative order $N = p^r - 1$). Without loss of generality α can be taken as one of the roots of $f(x)$.

Then it is natural to define a Galois ring m-sequence M^ν associated with a unit $\nu \in \mathbf{GR}(p^k, r)$ by using the Galois ring trace function $Tr(.)$ defined from $\mathbf{GR}(p^k, r)$ to $\mathbf{Z}_{p^k}$ as

$$M^\nu = (M_i^\nu) = (Tr(\nu\alpha^i) \; : \; 0 \le i < N),$$

where $N = p^r - 1$. The cyclically distinct m-sequences are given by the set

$$\{M^p\} \cup \{M^\nu \; : \; \nu = 1 + \sum_{j=1}^{k-1} p^j \hat{\nu}_j, \; \hat{\nu}_j \in \mathcal{T}\},$$

where $\mathcal{T}$ is the *Teichmuller set* of the Galois ring $\mathbf{GR}(p^k, r)$. $\mathcal{T}$ contains all the powers of the of primitive element α and the zero element, i.e.,

$$\mathcal{T} = \{0, 1, \alpha, \ldots, \alpha^{N-1}\}.$$

This set shares many properties of finite fields. Its nonzero elements are generated by α and it is closed under multiplication. However, *it is not closed under addition.* Note that the sequence M^p is isomorphic to a $\mathbf{GF}(p)$ m-sequence. Hence there are $(p^{kr} - 1)/(p^r - 1)$ cyclically distinct m-sequences over $\mathbf{Z}_p$.

2.3   Gauss Sums of Sequences over $\mathbf{Z}_{p^k}$

We are concerned with deriving upper bounds on the aperiodic correlation of the m-sequences M^ν over $\mathbf{Z}_4$. We follow the method adopted by Sarwate [6] to derive the bound. In the process we require a bound on Gauss like sums or Fourier transform values of these sequences. Let $\Omega_N = \exp(2\pi\sqrt{-1}/N)$ be a complex primitive N^{th} root of unity. Then the Gauss sum or Fourier transform $\hat{S} = (\hat{S}_c)$ of an arbitrary q−ary polyphase sequence $S = (S_i) = \Pi(M) = (w^{m_i})$ is defined as

$$\hat{S}_c = \sum_{i=0}^{N-1} \omega^{m_i}\, \Omega_N^{ic}, \qquad 0 \le c < N.$$

The values $\hat{S}_c$ are referred to as the Fourier transform values and they are related to the polyphase sequence symbols by the inverse Fourier transform

$$S_i = N^{-1} \sum_{c=0}^{N-1} \hat{S}_c\, \Omega_N^{-ic}, \qquad 0 \le i < N.$$

We are now ready to proceed to the proof of the main result of this paper.

3   A Simple Bound on Aperiodic Correlations of Quadriphase $m-$Sequences

In this section we will consider only quadriphase sequences derived from $m-$sequences over $\mathbf{Z}_4$. Let $\alpha \in \mathbf{GR}(4, r)$, r a positive integer, be a primitive element of the multiplicative order $N = 2^r - 1$. The Teichmuller set $\mathcal{T}$ of the Galois ring $\mathbf{GR}(4, r)$ is defined as explained in the previous section. The cyclically distinct quadriphase m-sequences of length N are given by the set

$$\{\Pi(M^\nu)\} = \{(\omega^{Tr(2\alpha^i)})\} \cup \{(\omega^{Tr(\nu\alpha^i)}) \; : \; \nu = 1 + 2\hat{\nu}, \; \hat{\nu} \in \mathcal{T}\}.$$

3.1   Derivation of the Bound

Note that S^2 is a biphase m-sequence and it is well known that $\hat{S}_c^2$ takes the value of $\sqrt{N+1}$ when $c \ne 0$ and takes the value of -1 when $c = 0$ [6,2]. The proof uses the fact that all the phases of binary a m-sequence form an Abelian group under pointwise addition. When $\nu \in \mathcal{T}$, this is not true, and hence we cannot easily bound the transform values. When $c = 0$, $\hat{S}_0^\nu$ is simply the sum of all quadriphase symbols in S^ν. This value has a magnitude $\approx \sqrt{N+1}$ [1,9,10]. To bound the rest of the values we use the Cauchy-Schwartz inequality:

Lemma 1. *If V is a real or complex inner product space, then, for all $x, y \in V$,*

$$| < x, y > | \le ||x|| \, ||y||,$$

where $|| \cdot ||$ denotes the norm of the space which is obtained from the inner product $< \, , \, >$ defined on V via $||x|| = \sqrt{< x, x >}$. Equality holds if and only if one of the vectors x, y is a scalar multiple of the other.

By utilizing the Cauchy-Schwarz inequality we prove the following result.

Theorem 1. *The squared magnitudes of the Fourier coefficients of S^ν satisfy the following inequality*

$$|\hat{S}_k^\nu|^2 \le N(1 + \sqrt{N+1}), \qquad 1 \le k < N.$$

Proof.

$$|\hat{S}_k^\nu|^2 = \sum_{i=0}^{N-1} \omega^{Tr(\nu\alpha^i)} \, \Omega_N^{ik} \sum_{m=0}^{N-1} \omega^{Tr(-\nu\alpha^m)} \, \Omega_N^{-mk}$$

$$= \sum_{i=0}^{N-1} \sum_{m=0}^{N-1} \omega^{Tr(\nu(\alpha^i - \alpha^m))} \, \Omega_N^{k(i-m)}$$

By making the transformation $(i - m) = \tau$,

$$|\hat{S}_k^\nu|^2 = \sum_{\tau=0}^{N-1} \theta(N - \tau) \, \Omega_N^{\tau k},$$

where $\theta(N - \tau)$ is the $(N - \tau)^{th}$ autocorrelation of S^ν.

Then

$$|\hat{S}_k^\nu|^2 = \theta(0) - \theta(1) + \theta(1) + \sum_{\tau=1}^{N-1} \theta(N - \tau) \, \Omega_N^{\tau k}, \qquad (6)$$

It has been shown in [1,9,10], that $|\theta(\tau)|$ is given by

$$|\theta_\nu(l)| = \begin{cases} N, & l \equiv 0 \text{ modulo } N \\ \sqrt{N+1}, & \text{otherwise.} \end{cases} \qquad (7)$$

Also $|\Omega_N^\tau| = 1$, and after using Lemma 1 to bound the last term in (6), we have

$$|\hat{S}_k^\nu|^2 \le N - \sqrt{N+1} + \sqrt{N(N+1)}\sqrt{N}$$

$$\le N(1 + \sqrt{N+1})$$

which proves the theorem.

By using the actual values of $\theta(i)$ we can improve the bound slightly. As in [6], we need some results on the following exponential sums. For any integer c, let $\Gamma(l,c) = \sum_{k=0}^{N-1-l} \Omega_N^{ck}, 0 \le l \le N-1$. Then define $\Gamma_{l,N}$ [6] as

$$\Gamma_{l,N} = N^{-1} \sum_{c=1}^{N-1} |\Gamma(l,c)|.$$

We have the following lemma proved in [6] using a method given by Vinogradov [11].

Lemma 2. *[6, Lemma 1]* $\Gamma_{l,N} < (2/\pi)\ln(4N/\pi)$, *for* $0 \le l \le N-1$

In [6] it is also shown that for $N > 6$, the above bound can be improved to

$$\Gamma_{l,N} < (2/\pi)\ln(4e^{\pi/3}N/3\pi) \tag{8}$$

which reduces the constant in the argument of the logarithm from 1.273.. to 1.209.

Let $\Delta(\nu_1, \nu_2, l, c)$ denote the cross ambiguity function of two m-sequences M^{ν_1} and M^{ν_2}, where

$$\Delta(\nu_1, \nu_2, l, c) = \sum_{i=0}^{N-1} \omega^{M_i^{\nu_1} - M_{i+l}^{\nu_2}} \Omega_N^{ic}, 0 \le c < N. \tag{9}$$

By using Theorem 1 we prove the following lemma.

Lemma 3. *For* $c \ne 0$ *and either* $\nu_1 \ne \nu_2$, *any* l *or* $\nu_1 = \nu_2, l \ne 0$ *modulo* N,

$$|\Delta(\nu_1, \nu_2, l, c)| = \sqrt{N(1 + \sqrt{N+1})}.$$

Proof. We use the fact that m-sequences are closed under pointwise addition or subtraction. Then $\Delta(\nu_1, \nu_2, l, c)$ is equal to the Gauss sum of an appropriate m-sequence. The result then follows from Theorem 1.

We now give the upper bound on the aperiodic crosscorrelation function magnitudes by making use of Lemmas 3 ,2 on the lines of the proof given in [6].

Theorem 2. $|C_{1,2}(l)| < \sqrt{(N+1)} + (2/\pi)\sqrt{N(1 + \sqrt{N+1})}\ \ln(4e^{\pi/3}N/3\pi)$, *for* $l \ne 0$

Proof. In view of (3) and (4) it is sufficient to show the result for $1 \le l \le N-1$. Consider the sum

$$\sum_{c=0}^{N-1} \Delta(\nu_1, \nu_2, l, c)\ \Gamma_{l,c}^*$$

$$= \sum_{c=0}^{N-1} \sum_{i=0}^{N-1} \omega^{M_i^{\nu_1} - M_{i+l}^{\nu_2}}\ \Omega^{ic} \sum_{k=0}^{N-l-1} \Omega^{-kc}$$

$$= \sum_{k=0}^{N-l-1} \sum_{i=0}^{N-1} \omega^{M_i^{\nu_1} - M_{i+l}^{\nu_2}} \sum_{c=0}^{N-1} \Omega^{(i-k)c} \tag{10}$$

$$= N\ C_{1,2}(l),$$

since the inner most sum has value 0 when $i \neq k$ and a value of N when $i = k$. On the other hand the sum can also be written as

$$\sum_{c=0}^{N-1} \Delta(\nu_1, \nu_2, l, c) \, \Gamma_{l,c}^*$$

$$= \Delta(\nu_1, \nu_2, l, 0) \, \Gamma_{l,0}^* + \sum_{c=1}^{N-1} \Delta(\nu_1, \nu_2, l, c) \, \Gamma_{l,c}^* \tag{11}$$

$$= (N - l)\sqrt{(N+1)} + \sum_{c=1}^{N-1} \Delta(\nu_1, \nu_2, l, c) \, \Gamma_{l,c}^*, \quad l \neq 0 (\textit{modulo } N).$$

The function $\Delta(\nu_1, \nu_2, l, 0)$ is the exponential sum of an m-sequence whose value is shown to be equal to $\sqrt{(N+1)}$ [1,9,10]. By combining equations (11), (10) and (8) with Lemma 2, we get the result.

3.2 An Improved Bound

We note that our bound in Theorem 2 can be applied to any set of polyphase sequences provided we have bounds for the Gauss and exponential sums. Here make use of a Gauss sum bound given in [7] for a class of Galois ring sequences. The bound depends on a quantity called the *weighted degree* of the polynomial representing the sequences. Let $f(x)$ ba a polynomial over $\mathbf{GR}(p^k, r)$ with the p-adic expansion

$$f(x) = F_0(x) + pF_1(x) + \cdots + p^{k-1}F_{k-1}(x),$$

where $F_i(x) \in \mathcal{T}[x], 0 \leq i \leq k - 1$ which can be obtained from the p-adic expansion of the coefficients of $f(.)$. Further, we assume that f is nondegenerate, by this we mean that f satisfies the following conditions:

1. $f(0) = 0$,
2. $f \neq 0$ (*modulo p*) and
3. no monomial term in $f(x)$ has degree that is multiple of p.

Let d_j be the degree of $F_j(x), 0 \leq j \leq k - 1$. Then the weighted degree of D of $f(x)$ is defined as

$$D = max\{p^{k-1} \, d_0, p^{k-2} \, d_1, \cdots, d_{k-1}\}$$

By making use of a nondegenerate polynomial f of degree d, many families of optimal polyphase sequences have been defined and studied in [3,7]. Let α be a primitive element of order $N = p^r - 1$ in $\mathbf{GR}(p^k, r)$. Then a sequence associated with a unit $\nu \in \mathbf{GR}(p^k, r)$ and a nondegenerate polynomial f of weighted degree D is given by

$$M^\nu = (M_i^\nu) = (Tr(f(\nu\alpha^i)) \; : \; 0 \leq i < N). \tag{12}$$

Note that when $f(x) = x$, the weighted degree is p^{k-1} and the sequences are m-sequences.

Theorem 3. *[7, Theorem 2] Let $f(x)$ be a nondegenerate polynomial with weighted degree D, Then we have for the Gauss sums of sequences in (12),*

$$|\hat{S}_k^\nu| \le D \sqrt{N+1}, \quad 1 \le k < N.$$

Theorem 4. *[3, Theorem 1] Let $f(x)$ be a nondegenerate polynomial with weighted degree D. Then we have for the exponential sums of sequences in (12),*

$$\theta(S) \le (D-1) \sqrt{N+1}.$$

Note that when $D = 2, p = 2, k = 2$, the above bound is better than the bound in Lemma 3. If we apply the above two bounds to our aperiodic crosscorrelation bound in Theorem 2, we get the following modified bound

Theorem 5. $|C_{1,2}(l)| < (D-1)\sqrt{N+1} + (2/\pi)D\sqrt{N+1} \, \ln(4e^{\pi/3}N/3\pi)$, *for* $l \ne 0$

Proof. The proof runs exactly similar to the proof in Theorem 2 and we use the Gauss and exponential sums in Theorems 3 and 4.

For quadriphase sequences, $D = 2$, and the improved bound then becomes:

$$|C_{1,2}(l)| < \sqrt{N+1} + (4/\pi)\sqrt{N+1} \, \ln(4e^{\pi/3}N/3\pi), \text{ for } l \ne 0. \qquad (13)$$

Clearly the above bound depends on the bound in Theorem 1 and any improvement must come from an improvement to Theorem 1 which is left as an open problem.

4   Conclusions and Comparison of Bounds

In [7], sophisticated techniques are used to obtain an upper bound on the aperiodic correlation function of certain Galois Ring sequences. This bound again depends on the *weighted degree* of the polynomial representing the sequences. Here, we are interested in comparing this bound with our result in Theorems 2 and 5.

The aperiodic crosscorrelation bound in [7] is given by

$$|C_{1,2}(l)| < D\sqrt{N+1}(\ln N + 1), \qquad (14)$$

where D is the weighted degree mentioned above. For m-sequences over $\mathbf{Z}_4$, the weighted D is 2, thus

$$|C_{1,2}(l)| < 2\sqrt{N+1}(\ln N + 1).$$

The Table 1 gives comparison of various bounds for quadriphase m-sequences. Our improved bound in (13) is better than the bound in [7]. Note that from the

Table 1. Comparison of Aperiodic correlation bounds for Quadriphase m-sequences

N	Bound of Theorem2 δ_1	Bound in [7] $\delta_{Shanbhag}$	Improved bound in Theorem5 $\delta_{improved}$	Ratio $\frac{\delta_{improved}}{\delta_{Shanbhag}}$
7	9.87	17.42	10.52	0.604
15	19.98	30.18	18.76	0.622
31	38.80	50.52	31.76	0.629
63	73.69	82.54	52.14	0.632
127	138.05	132.42	83.83	0.633
255	256.23	209.45	132.76	0.634
511	472.17	327.57	207.77	0.634
1023	864.89	507.61	322.12	0.635
2047	1576.01	780.61	495.51	0.635
32767	16641	4126.23	2621.16	0.635
1048575	294323	30439	19346	0.636

Table 1, the improved bound is better than in the bound in [7] by a factor of 0.64. We can write

$$\delta_{improved} < \sqrt{N+1} + (4/\pi)\sqrt{N+1}\ \ln(4e^{\pi/3}N/3\pi)$$
$$\approx \sqrt{N+1} + 1.273\sqrt{N+1}\ (\ln N + 0.19)$$
$$< \sqrt{N+1} + 1.273\sqrt{N+1}\ (\ln N + 1)$$

which is clearly less than $\delta_{Shanbag} = 2\sqrt{N+1}(\ln N + 1)$. In fact using the above upper bound for $\delta_{improved}$ we can write

$$\delta_{Shanbag} - \delta_{improved} > (0.727(\ln N + 1) - 1)\sqrt{N+1}$$

and the right hand side of this expression becomes positive as soon as $N > e$.

However, our simple bound given in Theorem 2 is asymptotically inferior to to the bound in [7].

From Table 1, it is clear that our techniques do not work well for small values of N. In this case it is straightforward to compute $\Gamma_{l,N}$ directly and improve the bounds. We are at present making use of the exact values of $\Gamma_{l,N}$ to find the best phases of these m-sequences with respect aperiodic correlations.

Finally, we remark that, even though we have discussed bounds only for Galois ring sequences, techniques extend easily for other optimal families of polyphase sequences like Kumar-Moreno sequences [12].

References

1. S. Boztaş, R. Hammons, and P. V. Kumar. 4-phase sequences with near-optimum correlation properties. *IEEE, Trans. Inform. Theory*, 38:1101–1113, 1992.

2. S. W. Golomb. *Shift Register Sequences*. Aegean Park Press: California, U.S.A, 1982.
3. P. V. Kumar, T. Helleseth, and A. R. Calderbank. An Upper bound for some Exponential Sums Over Galois Rings and Applications. *IEEE Trans. Inform. Theory*, 41:456–468, 1995.
4. P. V. Kumar, T. Helleseth, A. R. Calderbank, and A. R. Hammons, Jr. Large families of quaternary sequences with low correlation. *IEEE Trans. Inform. Theory*, 42:579–592, 1996.
5. V.S. Pless and W.C. Huffman, editors. *Handbook of Coding Theory, Volume 2*. New York : Elsevier, 1998.
6. D. Sarwate. An Upper Bound on the Aperiodic Autocorrelation Function for a Maximal-Length Sequence. *IEEE Trans. Inform. Theory*, 30:685–687, 1984.
7. A. G. Shanbhag, P. V. Kumar, and T. Helleseth. Upper bound for a hybrid sum over galois rings with application to aperiodic correlation of some q-ary sequences. *IEEE Trans. Inform. Theory*, 42:250–254, 1996.
8. P. Solé. A Quaternary Cyclic Code, and a Familly of Quadriphase Sequences with low Correlation Properties. *Lect. Notes Computer Science*, 388:193–201, 1989.
9. P. Udaya and M. U. Siddiqi. Large Linear Complexity Sequences over $\mathbf{Z}_4$ for Quadriphase Modulated Communication Systems having Good Correlation Properties. *IEEE International Symposium on Information Theory, Budapest, Hungary, June 23-29, 1991.*, 1991.
10. P. Udaya and M. U. Siddiqi. Optimal and Suboptimal Quadriphase Sequences Derived from Maximal Length Sequences over $\mathbf{Z}_4$. *Appl. Algebra Engrg. Comm. Comput.*, 9:161–191, 1998.
11. I.M. Vinogradov. *Elements of Number Theory*. Dover, New York, 1954.
12. P. V. Kumar, and O. Moreno. Prime-phase Sequences with Periodic Correlation Properties Better Than Binary Sequences *IEEE Trans. Inform. Theory*, 37:603–616, 1991.

Euclidean Modules and Multisequence Synthesis

Liping Wang*

State Key Laboratory of Information Security, Graduate School,
University of Science and Technology of China, Beijing 100039, China
A9000@china.com

Abstract. In this paper we extend the concept of Euclidean ring in commutative rings to arbitrary modules and give a special Euclidean $F_q[x]$-module K^n, where F_q is a finite field, n a positive integer and $K = F_q((x^{-1}))$. Thus a generalized Euclidean algorithm in it is deduced by means of $F_q[x]$-lattice basis reduction algorithm. As its direct application, we present a new multisequence synthesis algorithm completely equivalent to Feng-Tzeng' generalized Euclidean synthesis algorithm. In addition it is also equivalent to Mills continued fractions algorithm in the case of the single sequence synthesis.

1 Introduction

Euclidean algorithm and continued fractions technique play important roles in mathematics and other fields. Sequences synthesis problem is also a key problem in coding theory, cryptography and control theory. Many versions of Euclidean algorithm are used to solve such problem [2], [4], [5], [11]. In studying the single sequence synthesis problem, Mills developed a relation between continued fractions algorithm and well-known Berlekamp-Massey algorithm [1], [3], [7], [8]. Since there are so much similarity between Euclidean algorithm and continued fractions, in this paper we extend the concept of Euclidean ring in commutative rings to arbitrary modules. Thus a generalized Euclidean algorithm can be deduced in such modules and so usual Euclidean algorithm and continued fractions algorithm become its special cases. Especially a vector space K^n, where F_q is a finite field, n a positive integer and $K = F_q((x^{-1}))$, is also a Euclidean $F_q[x]$-module. Based on $F_q[x]$-lattice basis reduction algorithm [10], we derive a generalized Euclidean algorithm in it. As its direct application, in Section 3 we present a new multisequence synthesis algorithm. In Section 4 the equivalence with Feng-Tzeng' generalized Euclidean synthesis algorithm [5] is made more explicit. In Section 5 we show that the new synthesis algorithm is also equivalent to the Mills continued fractions algorithm for the single sequence synthesis. Finally, we give our conclusion in Section 6.

* Research supported by NSF under grants No. 19931010 and G 1999035803.

S. Boztaş and I.E. Shparlinski (Eds.): AAECC-14, LNCS 2227, pp. 239–248, 2001.
© Springer-Verlag Berlin Heidelberg 2001

2 Euclidean Modules

In this section we extend the concept of Euclidean ring to arbitrary modules. The chief results is that $F_q[x]$-module K^n is a Euclidean module and so a generalized Euclidean algorithm is deduced in it.

Definition 1. *Let R be a ring with identity and A an R-module. Let A has an equivalence relation denoted by $\sim$. Then A is a Euclidean R-module if there is a function $\phi : A - \{0\} \longrightarrow Z$ such that*
1. if $r \in R$, $\beta \in A$ and $r\beta \neq 0$, then $\phi(\beta) \leq \phi(r\beta)$;
2. if $\alpha, \beta \in A$ and $\beta \neq 0$, then there exist $q \in R$ and $\gamma \in A$ such that

$$\alpha = q\beta + \gamma \tag{1}$$

where $\gamma \not\sim \beta$, or if $\gamma \sim \beta$ then $\phi(\beta) > \phi(\gamma)$.

Example 1. A Euclidean ring R is also a Euclidean R-module under the trivial equivalence relation, i. e. R has two equivalent classes. The set of all nonzero elements is one class and $\{0\}$ is another.

Example 2. The rational number field Q is also a Z-module. The function

$$\phi : \begin{array}{c} Q - \{0\} \\ \alpha = \sum_{i=i_0}^{\infty} a_i \cdot 10^{-i} \end{array} \begin{array}{c} \longrightarrow Z \\ \longmapsto \max\{-i | a_i \neq 0\} \end{array}$$

For $\alpha, \beta \neq 0 \in Q$, then there exist $a \in Z$ and $\gamma \in Q$ such that

$$\alpha = a\beta + \gamma \tag{2}$$

where $\gamma = 0$, or if $\gamma \neq 0$ then $\phi(\beta) > \phi(\gamma)$, and $a = [\frac{\alpha}{\beta}]$, the integer parts of a rational number.

It is easily verified that Q is also a Euclidean Z-module under the trivial equivalence relation. Using the equation (2) repeatedly, we can deduce its Euclidean algorithm in it.

Given $\alpha, \beta \neq 0 \in Q$, we have

$$\alpha = -a_0\beta + \gamma_0 \text{ with } \gamma_0 = 0 \text{ or } \phi(\gamma_0) < \phi(\beta)$$

$$\beta = -a_1\gamma_0 + \gamma_1 \text{ with } \gamma_1 = 0 \text{ or } \phi(\gamma_1) < \phi(\gamma_1)$$

$$\vdots$$

$$\gamma_{k-2} = -a_k\gamma_{k-1} + \gamma_k \text{ with } \gamma_k = 0 \text{ or } \phi(\gamma_k) < \phi(\gamma_{k-1})$$

$$\vdots$$

where $-a_k$ is determined by $\gamma_{k-2}, \gamma_{k-1}$, i. e. $-a_k = [\frac{\gamma_{k-2}}{\gamma_{k-1}}]$.

Given a positive rational number s and let $\alpha = s$ and $\beta = -1$, then module Euclidean algorithm to α and β is continued fractions algorithm to s.

Example 3. Let $K = F_q((x^{-1})) = \{\sum_{i=i_0}^{\infty} a_i x^{-i} | i_0 \in Z, a_i \in F_q\}$. Then K is a Laurent series field. Naturally there exists an action of $F_q[x]$ on K, and so K is also an $F_q[x]$-Module.

Define a map

$$v : \begin{array}{ccc} K - \{0\} & \longrightarrow & Z \\ a(x) = \sum_{i=i_0}^{\infty} a_i x^{-i} & \longmapsto & \min\{i | a_i \neq 0\} \end{array}$$

Analogous to Q, K is a Euclidean $F_q[x]$-module with $\phi = -v$ under the trivial equivalence relation. Thus there is a Euclidean algorithm in it and we will discuss it in Section 5 in detail.

Example 4. In [5] a special Euclidean module was presented. First define an equivalence relation $\sim$ on the ring $F_q[x]$. For a positive integer m and $a(x), b(x) \in F_q[x]$, $a(x) \sim b(x)$ if and only if $\deg(a(x)) \equiv \deg(b(x))$ (mod m). Obviously $\sim$ is a congruence relation on this ring and thus induces a partition into $m + 1$ congruence classes. It is easily verified that $F_q[x]$ is a Euclidean $F_q[x^m]$-module under the equivalence relation $\sim$ and $\phi(a(x)) = \deg(a(x))$. Hence there are corresponding Euclidean algorithm in it, see [5].

Next we give an important example.

Let $K = F_q((x^{-1}))$ and n a positive integer, then K^n is a vector space with rank n. Naturally there exists an action of $F_q[x]$ on K^n, and so K^n is an $F_q[x]$-Module. In addition we frequently use three important functions. First v is extended to a function on K^n, written as V.

$$V : K^n - \{0\} \longrightarrow Z$$
$$\beta = (b_i(x))_{0 \leq i \leq n-1} \longmapsto \min\{v(b_i(x)) | 0 \leq i \leq n-1\}$$

Define a projection.

$$\theta_k : \begin{array}{ccc} K^n & \longrightarrow & F_q^n \\ \beta = (b_i(x))_{0 \leq i \leq n-1} & \longmapsto & (b_{i,k})_{0 \leq i \leq n-1} \end{array}$$

where $b_i(x) = \sum_{j=j_0}^{\infty} b_{i,j} x^{-j}$, $0 \leq i \leq n - 1$, for $k \in Z$. For we often use $\theta_{V(\beta)}(\beta)$, so it is simply denoted $\theta(\beta)$.

Besides define $\pi_i : F_q^n \longrightarrow F_q$ by $(a_0, \cdots, a_{n-1}) \mapsto a_i$ for $0 \leq i \leq n - 1$.

Define $n + 1$ classes on K^n. The first class is denoted by $[1, *, \cdots, *]^{(0)} = \{\beta \in K^n | \pi_0(\theta(\beta)) \neq 0\}$. The second class is denoted by $[0, 1, *, \cdots, *]^{(1)} = \{\beta \in K^n | \pi_0(\theta(\beta)) = 0, \pi_1(\theta(\beta)) \neq 0\}$, $\cdots$, the n-th class is denoted by $[0, \cdots, 0, 1]^{(n-1)} = \{\beta \in K^n | \pi_j(\theta(\beta)) = 0$, for all $j, 0 \leq j \leq n - 2, \pi_{n-1}(\theta(\beta)) \neq 0\}$. The last class has only one element 0. For any vector $\beta \in K^n$, it belongs to one and only one class. We use $\beta \sim \gamma$ if β and γ are in same class. First we have

Lemma 1. *Let $\beta^{(0)}, \cdots, \beta^{(n-1)}$ be n nonzero vectors belonging to distinct classes in K^n. Then they are $F_q[x]$-linearly independent.*

Analogous to the division algorithm in rings, consider the case in K^n. Given two nonzero vectors $\alpha, \beta \in K^n$, $\alpha \sim \beta \in [0, \cdots, 0, 1, *, \cdots, *]^{(u)}$ with $0 \leq u \leq n - 1$ and $V(\alpha) \leq V(\beta)$, then we have

$$\alpha = \frac{\pi_u(\theta(\alpha))}{\pi_u(\theta(\beta))} x^{V(\beta)-V(\alpha)} \beta + \delta_1 \tag{3}$$

with $V(\delta_1) \geq V(\alpha)$.

If $\delta_1 \sim \beta$ and $V(\delta_1) > V(\beta)$, or $\delta_1 \not\sim \beta$, then the process terminates. Otherwise, then $V(\delta_1) > V(\alpha)$ and continue the above process.

$$\delta_1 = \frac{\pi_u(\theta(\delta_1))}{\pi_u(\theta(\beta))} x^{V(\beta)-V(\delta_1)} \beta + \delta_2 \tag{4}$$

If $\delta_2 \not\sim \beta$, or $\delta_2 \sim \beta$ and $V(\delta_2) > V(\beta)$, then terminates. Otherwise repeat the above process till we get

$$\delta_{k-1} = \frac{\pi_u(\theta(\delta_{k-1}))}{\pi_u(\theta(\beta))} x^{V(\beta)-V(\delta_{k-1})} \beta + \delta_k \tag{5}$$

where $\delta_k \not\sim \beta$, or $\delta_k \sim \beta$ and $V(\delta_k) > V(\beta)$.

Since the value of $V(\delta_i)$ strictly increases for $1 \leq i \leq k$, then such process can be finished in finite steps. Thus we have

$$\alpha = \beta\left(\frac{\pi_u(\theta(\alpha))}{\pi_u(\theta(\beta))} x^{V(\beta)-V(\alpha)} + \cdots + \frac{\pi_u(\theta(\delta_{k-1}))}{\pi_u(\theta(\beta))} x^{V(\beta)-V(\delta_{k-1})}\right) + \gamma \tag{6}$$

where $\gamma = \delta_k$ and γ satisfies the above conditions. Therefore we have

Theorem 1. *Let two vectors α and $\beta \neq 0$ in K^n, $\alpha \sim \beta$ and $V(\alpha) \leq V(\beta)$, then there uniquely exist $q(x) \neq 0 \in F_q[x]$ and $\gamma \in K^n$ such that*

$$\alpha = q(x)\beta + \gamma \tag{7}$$

where $V(\gamma) > V(\beta)$ if $\beta \sim \gamma$.

Since $V(\beta) - V(\alpha) > V(\beta) - V(\delta_1) > \cdots > V(\beta) - V(\delta_k)$, we also have

Corollary 1. *With same notation as the above theorem. Then*

$$\deg(q(x)) = V(\beta) - V(\alpha) \ . \tag{8}$$

It is easily verified that K^n is a Euclidean $F_q[x]$-module under the above equivalence relation and $\phi(\beta) = -V(\beta)$ for $\beta \in K^n$.

By repeated use of Theorem 1, we can get a multidivisor form as follows.

Theorem 2. *Let $\beta^{(0)}, \cdots, \beta^{(t-1)}$, $1 \leq t \leq n$, be t nonzero vectors which belong to distinct classes in K^n. Given a vector $\alpha \in K^n$, $\alpha \sim \beta^{(u)}$, for some u, $0 \leq u \leq t-1$, and $V(\alpha) \leq V(\beta^{(u)})$. Then there exist unique elements $q^{(0)}(x), \cdots, q^{(t-1)}(x) \in F_q[x]$, $\gamma \in K^n$ such that*

$$\alpha = \sum_{h=0}^{t-1} q^{(h)}(x)\beta^{(h)} + \gamma \tag{9}$$

where $\gamma \not\sim \beta^{(h)}$ for all h, $0 \leq h \leq t-1$, or $V(\gamma) > V(\beta^{(j)})$ if $\gamma \sim \beta^{(j)}$ with some j, $0 \leq j \leq t-1$.

By repeated application of Theorem 2, we introduce a generalized Euclidean algorithm in this module.

Theorem 3. *Given t nonzero vectors $\beta_1^{(0)}, \cdots, \beta_1^{(t-1)} \in K^n$ which belong to t distinct equivalent classes, $1 \leq t \leq n$, and $\alpha_1 \sim \beta_1^{(u_0)}$, $V(\alpha_1) \leq V(\beta_1^{(u_0)})$, we obtain the following series of equations*

$$\alpha_j = \sum_{h=0}^{t-1} q_j^{(h)}(x)\beta_j^{(h)} + \gamma_j \tag{10}$$

where $V(\gamma_j) > V(\beta_j^{(u_j)})$ if $\gamma_j \sim \beta_j^{(u_j)}$.

$$\alpha_{j+1} = \beta_j^{(u_j)}, \beta_{j+1}^{(u_j)} = \gamma_j, \beta_{j+1}^{(h)} = \beta_j^{(h)} \text{ for } h \neq u_j \tag{11}$$

for $j = 1, 2, \cdots$, until some $j = k$ such that $\gamma_k \nsim \beta_k^{(h)}$ for all $0 \leq h \leq t - 1$.

By Euclidean algorithm we can get the greatest common divisor in rings, by the above theorem we have

Corollary 2. *The submodule generated by $\beta_1^{(0)}, \cdots, \beta_1^{(t-1)}, \alpha$ is the submodule generated by $\beta_k^{(0)}, \cdots, \beta_k^{(t-1)}, \gamma_k$.*

Remark 1. If $t = n - 1$ and $\beta_1^{(0)}, \cdots, \beta_1^{(t-1)}, \alpha$ are $F_q[x]$-linearly independent, by Lemma 1 and Corollary 2 $\beta_k^{(0)}, \cdots, \beta_k^{(t-1)}, \gamma_k$ are $F_q[x]$-linearly independent and they are a reduced basis of the $F_q[x]$-lattice $\Lambda(\beta_1^{(0)}, \cdots, \beta_1^{(n-2)}, \alpha)$. In [10] given a basis of a lattice, there are finite steps to obtain its reduced basis. Our generalized Euclidean algorithm can be consider as its special form and so we can obtain the required result in finite steps.

For convenience, we rewrite the above equtions and define

$$\beta_0^{(u_0)} = \alpha_1, \beta_0^{(h)} = \beta_1^{(h)}, \text{ for } h \neq u_0 \text{ and } \gamma_0 = \beta_1^{(u_0)}.$$

For $j > 1$, we have

$$\alpha_j = \beta_{j-1}^{(u_{j-1})}, \beta_{j-1}^{(u_{j-1})} = \gamma_{j-1}, \beta_j^{(h)} = \beta_{j-1}^{(h)} \text{ for } h \neq u_{j-1}.$$

Then (10) can be rewritten as

$$\gamma_j = (-q_j^{(u_{j-1})}(x))\gamma_{j-1} + \beta_{j-1}^{(u_{j-1})} + \sum_{h=0,h\neq u_{j-1}}^{t-1} (-q_j^{(h)}(x))\beta_{j-1}^{(h)} . \tag{12}$$

3   Multisequence Synthesis Algorithm

In this section we apply module Euclidean algorithm to the multisequence synthesis problem, which is to find a shortest linear recurrence satisfied by given multiple sequences. First we formulate the problem.

Let $a^{(h)} = (a_0^{(h)}, \cdots, a_{N-1}^{(h)})$, $0 \leq h \leq m - 1$, be m sequences, each of length N, over a finite field F_q. A nonzero polynomial $q(x) = \sum_{j=0}^{d} c_j x^j$ is called an annihilating polynomial of $a^{(0)}, \cdots, a^{(m-1)}$ if

$$c_d a_k^{(h)} + c_{d-1} a_{k-1}^{(h)} + \cdots + c_0 a_{k-d}^{(h)} = 0 \tag{13}$$

for all k and h, $d \le k \le N - 1, 0 \le h \le m - 1$. When $c_d = 1$, it is called a characteristic polynomial. A minimal polynomial is defined by a characteristic polynomial with minimum degree. The multisequences synthesis problem of $a^{(0)}, \cdots, a^{(m-1)}$ is to find one of their minimal polynomials.

In [12] a new multisequence synthesis algorithm (LBRMS) was presented by means of $F_q[x]$-lattice basis reduction algorithm. We review it simply. Let $a^{(h)}(x) = \sum_{j=0}^{N-1} a_j^{(h)} x^{-j-1}$ be the formal negative-power series of $a^{(h)}$, $0 \le h \le m - 1$, and $q(x)$ a polynomial over F_q of degree d. By Lemma 1 [12], then $q(x)$ is an annihilating polynomial of $a^{(0)}, \cdots, a^{(m-1)}$ if and only if for each h, $0 \le h \le m - 1$, there exists a unique polynomial $p^{(h)}(x) \in F_q[x]$ such that

$$v(q(x) \cdot a^{(h)}(x) - p^{(h)}(x)) > N - d .\tag{14}$$

Therefore the multisequence synthesis problem of $a^{(0)}, \cdots, a^{(m-1)}$ is reduced to finding a monic polynomial of least degree satisfying (14).

Set $\alpha = (a^{(0)}(x), \cdots, a^{(m-1)}(x), x^{-N-1})_{m+1}$, $e_0 = (1, 0, \cdots, 0)_{m+1}$, $\cdots$, and $e_{m-1} = (0, \cdots, 0, 1, 0)_{m+1}$. Obviously $e_0, \cdots, e_{m-1}, \alpha$ are $F_q[x]$-linearly independent. Therefore they span a free $F_q[x]$-submodule of K^{m+1}, i. e. an $F_q[x]$-lattice $\Lambda(e_0, \cdots, e_{m-1}, \alpha)$ with rank $m + 1$. In detail,

$$\Lambda(e_0, \cdots, e_{m-1}, \alpha) = \{q(x) \cdot \alpha + \sum_{i=0}^{m-1} p_i(x) e_i | \ p_i(x), q(x) \in F_q[x], 0 \le i \le m - 1\}.$$

And a function $\eta : \Lambda(e_0, \cdots, e_{m-1}, \alpha) \longrightarrow F_q[x]$ is defined by $\beta \mapsto b_m(x) x^{N+1}$ with $\beta = (b_0(x), \cdots, b_m(x))$. If $\omega_0, \cdots, \omega_m$ is a reduced basis with $V(\omega_0) \ge \cdots \ge V(\omega_m)$, setting

$$S = \{\beta \in \Lambda(\omega_0, \cdots, \omega_m) | \pi_m(\theta(\beta)) \ne 0\}\tag{15}$$

then $c\eta(\omega_s)$, where c is a constant such that $c\eta(\omega_s)$ is monic and s satisfies $0 \le s \le m - 1$, $\omega_s \in S$ and $\omega_j \notin S$ for $j < s$, is a minimal polynomial.

Apply the module Euclidean algorithm to $\alpha, e_1, \cdots, e_m$, and we can present a new multisequence synthesis algorithm as follows.

Algorithm 1:
Input: m sequences $a^{(0)}, \cdots, a^{(m-1)}$, each of length N, over a finite field F_q.
Output: a minimal polynomial $m(x)$.

1. Set $\gamma_0 = \alpha, \beta_0^{(0)} = e_{m-1}, \beta_0^{(1)} = e_{m-2}, \cdots, \beta_1^{(m-1)} = e_0, j = 0$.
2. $j \leftarrow j + 1$.
3. Apply the generalized Euclidean algorithm for $\gamma_0, \beta_0^{(0)}, \cdots, \beta_0^{(m-1)}$, i. e.

$$\gamma_j = (-q_j^{(u_{j-1})}(x))\gamma_{j-1} + \beta_{j-1}^{(u_{j-1})} + \sum_{h=0, h\ne u_{j-1}}^{m-1} (-q_j^{(h)}(x))\beta_{j-1}^{(h)}.\tag{16}$$

Set $\gamma_{j-1} \in [0, \cdots, 0, 1, *, \cdots, *]^{(u_{j-1})}$, $\beta_j^{(u_{j-1})} = \gamma_{j-1}$, and $\beta_j^{(h)} = \beta_{j-1}^{(h)}$ for all $h \ne u_{j-1}$.

4. until the process terminates, i. e. $j = k$ such that $\gamma_k \not\sim \beta_k^{(h)}$, for all h, $h = 0, 1, \cdots, m - 1$.

5. $c\eta(\gamma_k)$ is a minimal polynomial of $a^{(0)}, \cdots, a^{(m-1)}$, where c is a constant such that $c\eta(\gamma_k)$ is monic.

By Corollary 2 we know

$$\Lambda(e_0, \cdots, e_{m-1}, \alpha) = \Lambda(\beta_k^{(0)}, \cdots, \beta_k^{(m-1)}, \gamma_k) . \tag{17}$$

Furthermore, we have

Theorem 4. $\gamma_k \in [0, \cdots, 0, 1]^{(m)}$.

According to Lemma 1 and Theorem 4, we know $\beta_k^{(0)}, \cdots, \beta_k^{(m-1)}, \gamma_k$ is a reduced basis for the lattice $\Lambda(\alpha, e_0, \cdots, e_{m-1})$ and γ_k is the unique element in $\beta_k^{(0)}, \cdots, \beta_k^{(m-1)}, \gamma_k$ such that the m-th component is not zero. Thus $c\eta(\gamma_k)$ is a minimal polynomial.

Theorem 5. *With same notation as the above. Then for all j, $0 \le j \le k - 1$,*

$$\deg(\eta(\gamma_j)) < \deg(\eta(\gamma_{j+1})) \tag{18}$$

$$\deg(\eta(\gamma_j)) + V(\gamma_j) < \deg(\eta(\gamma_{j+1})) + V(\gamma_{j+1}) \tag{19}$$

The proof is omitted because of the space limit.

Remark 2. The module Euclidean algorithm also can be thought as a slight modifications about $F_q[x]$-lattice basis reduction algorithm, i. e. one step in Euclidean algorithm is several steps in LBRMS such that the degree of the polynomial in every step strictly increases.

4   Comparison with Feng-Tzeng' Algorithm

In [5] Feng-Tzeng presented a generalized Euclidean algorithm and applied it to solving multisequence synthesis problem, also see Appendix. In this section we demonstrate the equivalence between the two Euclidean modules and so the equivalence between the two synthesis algorithms is made explicit.

The polynomial ring $F_q[x]$ is also a Euclidean $F_q[x^m]$-module under the special equivalence relation in Example 4 [5]. Define a function

$$\sigma : F_q[x] \longrightarrow K^m$$
$$f(x) = \sum_{j=0}^d c_j x^j \mapsto (a_0(x), \cdots, a_{m-1}(x))$$

where for all i, $0 \le i \le m - 1$,

$$a_i(x) = \sum_{\substack{j = 0 \\ j = mq_j + r_j, 0 \le r_j \le m - 1, i = m - 1 - r_j}}^{d} c_j x^{-(N-1-q_j)-1}$$

We easily get

Theorem 6. *With the same notation as the above. Then*
(1) σ *is a group monomorphism.*
(2) *for arbitrary* $q(x), f(x) \in F_q[x]$, $\sigma(q(x^m)f(x)) = q(x)\sigma(f(x))$.

The initial conditions in Feng-Tzeng' algorithm are give, also see Appendix. From Theorem 6 we have $\sigma(r_0(x)) = (a^{(0)}(x), \cdots, a^{(m-1)}(x))$, $\sigma(b_0^{(m-1)}(x)) = (1, 0, \cdots, 0)_m$, $\cdots$, $\sigma(b_0^{(0)}(x)) = (0, \cdots, 0, 1)_m$. Furthermore we have

Theorem 7. *For* $0 \leq j \leq k$, *we have*
(1) $\gamma_j = (\sigma(r_j(x)), U_j(x)x^{-N-1})$.
(2) $\beta_j^{(h)} = (\sigma(b_j^{(h)}(x)), V_j^{(h)}(x)x^{-N-1})$ *for all* h, $0 \leq h \leq m - 1$.

In addition, we consider the terminating condition. In [5] the algorithm terminates when we reach the step k such that

$$\deg(r_{j-1}(x)) \geq \deg(U_{j-1}(x)) \text{ for } 1 \leq j \leq k \tag{20}$$

$$\deg(r_k(x)) < \deg(U_k(x^m)) \tag{21}$$

At this step because of $N + 1 - \lceil \frac{\deg(r_k(x))}{m} \rceil > N + 1 - \deg(U_k(x))$, we have $\gamma_k \in [0, \cdots, 0, 1]^{(m)}$ and $\gamma_j \notin [0, \cdots, 0, 1]^{(m)}$ for all j, $0 \leq j \leq k - 1$.

Therefore Algorithm 1 is completely equivalent to Feng-Tzeng' generalized Euclidean algorithm and our representation method is more convenient.

5 Comparison with Mills Algorithm

In this section we show that Algorithm 1 is equivalent to Mills continued fractions algorithm when applied to solving the single sequence synthesis problem.

Given a sequence $a = (a_0, \cdots, a_{N-1})$, then $a(x) = \sum_{i=0}^{N-1} a_i x^{-i-1} \in K$ is the formal power-negative series of a. Setting $\gamma_0 = \alpha_0 = (a(x), x^{-N-1})$ and $\gamma_{-1} = \beta_0^{(1)} = (1, 0)$, we begin to execute the generalized Euclidean algorithm.

$$\gamma_{-1} = q_1(x)\gamma_0 + \gamma_1, \text{ where } V(\gamma_1) > V(\gamma_0) \text{ and } \gamma_1 \sim \gamma_0$$

$$\gamma_0 = q_2(x)\gamma_1 + \gamma_2$$

$$\vdots$$

$$\gamma_{k-2} = q_k(x)\gamma_{k-1} + \gamma_k$$

untill $\gamma_k \nsim \gamma_{k-1}$ the algorithm terminates.

For each i, $-1 \leq i \leq k$, set $Q_i(x) = \eta(\gamma_i) \in F_q[x]$ and $R_i \in K$ denotes the first component of γ_i. Since there exists the unique polynomial $P_i(x)$ such that $v(Q_i(x)a(x) - P_i(x)) > 0$, we have $R_i = Q_i(x)a(x) - P_i(x)$. Thus the initial conditions are:

$$Q_0 = 1, Q_{-1} = 0, P_0 = 0, P_{-1} = 1, R_0 = a(x), R_{-1} = -1.$$

Hence the sequences $\{P_i(x)\}$, $\{Q_i(x)\}$,$\{R_i\}$ satisfy the same recursion.

$$P_{i-2}(x) = q_i(x)P_{i-1}(x) + P_i(x) \tag{22}$$

$$Q_{i-2}(x) = q_i(x)Q_{i-1}(x) + Q_i(x) \tag{23}$$

$$R_{i-2} = q_i(x)R_{i-1} + R_i \tag{24}$$

If $0 \leq i \leq k - 1$, then $\gamma_i \in [1, *]$ and so $V(\gamma_i) = v(R_i)$. Thus we rewrite the equation (24), then

$$\frac{R_{i-2}}{R_{i-1}} = q_i(x) + \frac{R_i}{R_{i-1}} . \tag{25}$$

Since $v(R_i) > v(R_{i-1})$, we have $q_i(x) = [\frac{R_{i-2}}{R_{i-1}}]$, where $[\ ]$ denotes the integer part of the power-negative series, i. e. if $f(x) = \sum_{j=0, c_0 \neq 0}^{\infty} c_i x^{d-j}$, then

$$[f(x)] = \begin{cases} \sum_{j=0}^{d} c_j x^{d-j} & \text{if } d \geq 0 \\ 0 & \text{if } d = 0 \end{cases}$$

The above process is actually the continued fractions technique and so Algorithm 1 is equivalent to Mills continued fractions algorithm for the single sequence synthesis.

6   Conclusion

In [12] we presented a multisequence synthesis algorithm (LBRMS) by means of $F_q[x]$-lattice basis reduction algorithm. Since the generalized Euclidean algorithm in this paper is only its slight modifications, Algorithm 1, Feng-Tzeng' generalized Euclidean algorithm and Mills algorithm can be derived from LBRMS. In addition, the new concept of Euclidean module makes the continued fractions technique become a special Euclidean algorithm. Therefore our module Euclidean algorithm can also be referred as the generalization of the continued fractions algorithm.

Appendix: Feng-Tzeng' Generalized Euclidean Synthesis Algorithm

Input: $a^{(0)}, \cdots, a^{(m-1)}$, length of N, over a finite field F_q.
Output: a minimal polynomial of $a^{(0)}, \cdots, a^{(m-1)}$.

1. Set $r_0(x) = \sum_{h=0}^{m-1} x^{m-1-h} \sum_{i=0}^{N-1} a_i^{(h)} x^{m(N-1-i)}$, $U_0(x) = 1$.
   $b_0^{(h)}(x) = x^{Nm+h}, V_0^{(h)}(x) = 0$ for all h, $0 \leq h \leq m - 1$, and $j = 0$.
2. $j \leftarrow j + 1$.

3. Calculate $r_j(x)$ by generalized Euclidean algorithm, i. e.

$$r_j(x) = (-Q_j^{(u_{j-1})}(x^m))r_{j-1}(x) + b_{j-1}^{(u_{j-1})}(x) + \sum_{h=0, h \neq u_{j-1}}^{m-1} (-Q_j^{(h)}(x^m))b_{j-1}^{(h)}(x)$$

Let $u_j = \deg(r_{j-1}(x)) \bmod m$,
$b_j^{(u_{j-1})}(x) = r_{j-1}(x)$, and $b_j^{(h)}(x) = b_{j-1}^{(h)}(x)$ for all $h \neq u_{j-1}$.

4. Find $U_j(x)$ from $U_{j-1}(x)$ and $V_{j-1}^{(h)}(x)$ so that

$$U_j(x) = (-Q_j^{(u_{j-1})}(x))U_{j-1}(x) + V_{j-1}^{(u_{j-1})}(x) + \sum_{h=0, h \neq u_{j-1}}^{m-1} (-Q_j^{(h)}(x))V_{j-1}^{(h)}(x)$$

Let $V_j^{(u_{j-1})}(x) = U_{j-1}(x)$, and $V_j^{(h)}(x) = V_{j-1}^{(h)}(x)$ for all $h \neq u_{j-1}$.

5. If $\deg(r_j(x)) \geq \deg(U_j(x^m))$, go back to 2. Otherwise, go to 6.
6. Let $k = j$. Then $cU_k(x)$ is a shortest length LFRS, where c is a constant such that $cU_k(x)$ is monic.

References

1. Berlekamp, E. R.: Algebraic Coding Theory. New York: McGrawHill (1968)
2. Camion, P. : An iterative Euclidean algorithm. LNCS **365** (1987) 88–128.
3. Cheng, U. J.: On the continued fraction and Berlekamp's algorithm. IEEE Trans. Inform. Theory, vol. **IT-30** (1984) 541–544
4. Dornstetter, J. L.: On the equivalence between Berlekamp's algorithm and Euclid's algorithm. IEEE Trans. Inform. Theory, vol. **IT-33** (1987) 428–431
5. Feng, G. L., Tzeng, K. K.: A generalized Euclidean algorithm for multisequence shift-register synthesis. IEEE Trans. Inform. Theory, vol. **IT-35** (1989) 584–594
6. Hungerford, T. W.: Algebra, Springer-Verlag (1974)
7. Massey, J. L.: Shift-register synthesis and BCH decoding. IEEE Trans. Inform. Theory, vol. **IT-15** (1969) 122–127
8. Mills, W. H.: Continued fractions and linear recurrence. Math. Comp. vol. **29** (1975) 173–180
9. Norton, G. H.: On the minimal realizations of a finite sequence. J. Symb. Comput. **20** (1995) 93–115
10. Schmidt, W. M.: Construction and estimation of bases in function fields. J. Number Theory **39** (1991) 181–224
11. Sugiyama, Y., Kasahara, M., Hirasawa, S., Namekawa, T.: A method for studying the key equation for decoding Goppa codes. Inform. Contr. vol. **27** (1975) 87–99
12. Wang, L. P., Zhu, Y. F.: F[x]-lattice basis reduction algorithm and multisequence synthesis. Science in China, in press
13. Wang, L. P., Zhu, Y. F.: Multisequence synthesis over an integral ring. Cryptography and Lattices (2001) in press

On Homogeneous Bent Functions

Chris Charnes[1,2], Martin Rötteler[2], and Thomas Beth[2]

[1] Department of Computer Science and Software Engineering,
University of Melbourne,
Parkville, Vic, 3052, Australia.
charnes@cs.mu.oz.au

[2] Institut für Algorithmen und Kognitive Systeme
Universität Karlsruhe
Am Fasanengarten 5
D-76128 Karlsruhe, Germany
{roettele,eiss_office}@ira.uka.de

Dedicated to Hanfried Lenz on the occasion of his 85th birthday

Abstract. A new surprising connection between invariant theory and the theory of bent functions is established. This enables us to construct Boolean function having a prescribed symmetry given by a group action. Besides the quadratic bent functions the only other known homogeneous bent functions are the six variable degree three functions constructed in [14]. We show that these bent functions arise as invariants under an action of the symmetric group on four letters. Extending to more variables we apply the machinery of invariant theory to construct previously unknown homogeneous bent functions of degree three in 8 and 10 variables. This approach gives a great computational advantage over the unstructured search problem. We finally consider the question of linear equivalence of the constructed bent functions.

Keywords: Bent functions, invariant theory, design theory.

1 Introduction

Recently an interesting class of six variable bent functions which are invariant under S_4, the symmetric group on four letters, was found in a computer enumeration by Qu, Seberry, and Pieprzyk [14]. The algebraic normal form of these functions is homogeneous of degree three. The search for homogeneous bent functions that have some degree of symmetry was motivated by cryptographic applications. Loosely speaking, the symmetry property ensures that in repeated evaluations of the functions (such as in cryptographic algorithms) partial evaluations can be reused. As a consequence cryptographic algorithms which are designed using such symmetric Boolean functions have a fast implementation. The application of these ideas to the design of *hashing functions* are discussed in [14].

The relation between the bentness of Boolean functions and their algebraic normal forms seems not to be completely understood at present. For instance,

S. Boztaş and I.E. Shparlinski (Eds.): AAECC-14, LNCS 2227, pp. 249–259, 2001.
© Springer-Verlag Berlin Heidelberg 2001

Carlet [3] characterized the algebraic normal forms of the six variable bent functions. But this characterization does not provide a method of obtaining symmetric bent functions.

The fact that the sought after bent functions have homogeneous normal form and have some symmetry suggest that they could be studied in the context of *invariant theory*. This possibility was already observed in [4] where the connection between the six variable bent functions and the maximal *cliques* of a certain graph [13] was established. This correspondence was used to describe the action of S_4 on this class of functions.

In this paper we demonstrate that it is indeed possible to use invariant theory to construct homogeneous bent functions. In fact it is possible to specify the symmetry group and then to search for those Boolean functions which possess this symmetry. In many cases (see Sections 5.1 and 5.2) this leads to a considerable reduction of the size of the search space. By this method we found previously unknown eight variables homogeneous bent functions of degree 3. Moreover, these functions have a concise description in terms of certain *designs* and graphs. We expect that this connection will play a role in elucidating the structure of the algebraic form of these functions.

2   Background

In this section we briefly recall the basic definitions and properties of Boolean functions. Our main object of interest is the ring

$$\mathcal{P}_n := \mathrm{GF}(2)[x_1, \ldots, x_n]/(x_1^2 - x_1, \ldots x_n^2 - x_n)$$

of Boolean polynomials in n variables over the finite field $\mathrm{GF}(2)$. The ring $\mathcal{P}_n$ is graded, i. e., it has a direct sum decomposition which is induced by the degrees of the Boolean polynomials.

To a vector $v \in \mathrm{GF}(2)^n$ we associate the monomial $x^v := x_1^{v_1} \cdots x_n^{v_n} \in \mathcal{P}_n$. The one's complement of v is denoted by $\bar{v}$. Now let $f \in \mathcal{P}_n$ be written in algebraic normal form (see [12]): $f = \sum_{v \in \mathrm{GF}(2)^n} \alpha_v x^v$. The degree of f is defined by $\deg(f) := \max\{\mathrm{wgt}(v) : \alpha_v \neq 0\}$, where $\mathrm{wgt}(v)$ denotes the Hamming weight of v. Then $\mathcal{P}_n = \bigoplus_{i=0}^n V_i$ where V_i is the the vector space of all *homogeneous* polynomials with respect to this degree function. Clearly $\dim(V_i) = \binom{n}{i}$, corresponding to the fact that there are $2^{\binom{n}{0}} \cdot 2^{\binom{n}{1}} \cdots 2^{\binom{n}{n}} = 2^{2^n}$ Boolean functions in n variables.

Definition 1. *The Fourier transform of a Boolean function $f \in \mathcal{P}_n$ is the function $F : \mathrm{GF}(2)^n \to \mathbb{C}$ defined by*

$$F(s) := \frac{1}{\sqrt{2^n}} \sum_{v \in \mathrm{GF}(2)^n} (-1)^{v \cdot s + f(v)},$$

for each $s \in \mathrm{GF}(2)^n$. Here $x \cdot y := \sum_{i=1}^n x_i y_i$ is the inner product on $\mathrm{GF}(2)^n$.

Boolean functions whose Fourier spectrum has constant absolute values are particularly interesting. This is possible only when the number n of variables is even.

Definition 2. *A Boolean function f is called bent if $|F(s)| = 2^{n/2}$ holds for all $s \in \mathrm{GF}(2)^n$.*

Bent functions have been studied extensively for the last 30 years and there is a large literature which deals with these functions. Amongst which we mention the works of Dillon [6], Rothaus [16], Carlet [3], and Dobbertin [8].

3   Invariant Theory

In this section we briefly recall some basic definitions and results of polynomial invariant theory. Let $\mathrm{GL}(n, K)$ denote the group of invertible $n \times n$ matrices with entries in the field K (in case of a finite field $\mathrm{GF}(q)$ of q elements we write $\mathrm{GL}(n, q)$ instead). For each group $G \leq \mathrm{GL}(n, K)$ we define an operation on the polynomial ring $R_n := K[x_1, \ldots, x_n]$ in the following way: for $g \in G$ and $f \in R_n$ we set

$$f^g := f((x_1, \ldots, x_n) \cdot g),$$

i. e., the elements of G operate via K-linear coordinate changes. We are interested in the fixed points under this operations, i. e., the polynomials which satisfy $f^g = f$ for all $g \in G$. These polynomials are called invariant polynomials or simply invariants. The set of all invariants forms a ring which is usually denoted by $K[x_1, \ldots, x_n]^G$. In the complex case $K = \mathbb{C}$ invariant theory has had important applications in the theory of error-correcting codes; cf. Gleason's theorem [12,15].

Example 1 (Symmetric polynomials). The symmetric group $\mathbf{S}_n$ acts naturally on n variable polynomials by permuting their variables. It is well-known (see, e. g., [11], p. 13) that the *elementary symmetric polynomials*

$$
\begin{aligned}
\sigma_{1,n} &:= & x_1 + x_2 + \ldots + x_n \\
\sigma_{2,n} &:= & x_1 x_2 + x_1 x_3 + \ldots + x_{n-1} x_n \\
&\vdots & \vdots \\
\sigma_{n,n} &:= & x_1 x_2 \cdots x_n
\end{aligned}
$$

generate the ring $K[x_1, \ldots, x_n]^{\mathbf{S}_n}$ for all fields K. Invariant theory and bent functions already meet at this point, since it is known [12] that the quadratic elementary symmetric polynomials $\sigma_{2,n}$ are bent functions if interpreted as elements of the ring $\mathcal{P}_n$. Savický has been shown in [17] that $\sigma_{2,n}$ is essentially[1] the only class of bent functions invariant under the full symmetric group.

---

[1] the linear and constant terms can be chosen so that there are four bent functions of degree 2 in n variables which are invariant under the full symmetric group.

The invariants of degree i form a K-vector space of dimension d_i. In order to gain some information about the number of invariants of a fixed degree we use the generating function of the (graded) ring R_n^G and is defined in terms of the d_i by

$$P_G(z) := \sum_{i \geq 0} d_i z^i$$

in the ring of formal power series.

In the context of invariant theory this generating function $P_G(z)$ is called the *Molien series* (see [19]). Since for a finite group G the invariant ring is finitely generated, $P_G(z)$ is in fact a rational function. In the non-modular case $P_G(z)$ can be computed quite elegantly as the following theorem shows.

Theorem 1 (T. Molien, 1897). *Let G be a finite subgroup of $\mathrm{GL}(n, K)$ and suppose that* $\mathrm{char}(K)$ *does not divide* $|G|$. *Then the following identity holds:*

$$P_G(z) = \frac{1}{|G|} \sum_{g \in G} \frac{1}{\det(1 - zg)}.$$

In case of a permutation group action we use the following additional result (cf. [18, Proposition 4.2.4]) which is the basis our computational approach.

Theorem 2. *Let G be a finite group, X be a finite G-set, and K be a field. Then the Molien series of $K[X]^G$ is given by $P_G(z) = \sum_{i \geq 0} d(i, X, G) z^i$, with certain coefficients $d(i, X, G)$ which depend on i and the action of G on X but not on K.*

For a more complete account of invariant theory see for example [18] or [19]. In what follows we will apply the theory of invariants to the setting where $K = \mathrm{GF}(2)$, and G is a subgroup of S_n which acts naturally by permuting the variables.

4   Homogeneous Bent Functions in Six Variables

A as preliminary step we give an alternative characterization of the results obtained in [14] using invariant theory. To describe the Boolean functions in question we require the following graph which was used in [4].

Definition 3 (Nagy [13]). *Let $\Gamma_{(n,k)}$ be the graph whose vertices correspond to the $\binom{n}{k}$ unordered subsets of size k of a set $\{1, \ldots, n\}$. Two vertices of $\Gamma_{(n,k)}$ are joined by an edge whenever the corresponding k-sets intersect in a subset of size one.*

The bent functions $\{f_i(x_1, \ldots, x_6)\}$ considered in [14] are parametrized by the (maximal) cliques of size four of $\Gamma_{(6,3)}$.

Theorem 3 ([4]). *The thirty homogeneous bent functions in six variables listed in [14] are in one to one correspondence with the complements of the cliques $\{C_i\}$ of $\Gamma_{(6,3)}$.*

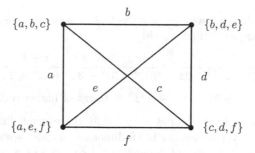

Fig. 1. A labeled clique $\mathcal{C}$ in $\Gamma_{(6,3)}$

There is a canonical labeling of the edges of the graph in Figure 1 which is induced by the vertex labels (3-sets) of the clique.

Theorem 4 ([4]). *The automorphism group $S_4 = \langle\alpha,\beta,\gamma\rangle$ of a clique $\mathcal{C}$ is generated by the involutions:* $\alpha = (a,d)(c,e)$, $\beta = (a,b)(d,f)$, *and* $\gamma = (a,c)(d,e)$.

In Theorem 4 generator α corresponds to a labeling of the edges induced in Figure 1 by the subgraph $| \ | \times$. Similarly, β corresponds to the subgraph $\lceil\_\rfloor$ and γ to the subgraph $\diagdown\diagup$.

This theorem leads us to an important observation which yields the connection to invariants: the automorphism group $S_4 = \langle\alpha,\beta,\gamma\rangle$ also is the stabilizer (taken in S_6) of the polynomial $f \in R_n$ corresponding to the clique $\mathcal{C}$, i.e., $\mathrm{Stab}_{S_6}(f) = \{g \in S_6 : f^g = f\} = \langle\alpha,\beta,\gamma\rangle \cong S_4$.

Hence it is natural to look for polynomials which are invariant under this group action in order to find bent functions. Using this idea we now show that there are homogeneous invariants in $\mathrm{GF}(2)[x_1,\ldots,x_6]$ of degree 3 which are bent functions.

We first define a permutation group on the letters $\{1,\ldots,6\}$ by specializing $a := 3$, $b := 1$, $c := 6$, $d := 4$, $e := 5$, $f := 2$. This yields the group

$$S_4 := \langle(3,4)(5,6),(1,3)(2,4),(3,6)(4,5)\rangle \leq S_6,$$

i.e., the symmetric group of order 4, in the permutation representation obtained as the automorphisms of the clique (this S_4 is the symmetry group of the bent function given in [14, eq. (11)]).

We consider the invariant ring $\mathrm{GF}(2)[x_1,\ldots,x_6]^{S_4}$ of this group action over the field $\mathrm{GF}(2)$. Since the characteristic of the prime field divides the order of the group, the investigation of this ring falls into the domain of modular invariant theory [18]. However, in our case the situation is considerably simplified since we are dealing with a permutation representation and hence Theorem 2 applies. For the investigation of the general case of a symmetry group $G \leq \mathrm{GL}(n,2)$ with an associated invariant ring $\mathrm{GF}(2)[x_1,\ldots,x_n]^G$ the many facets of modular invariant theory have to be taken into account.

In the six variable case we compute the Molien series of G using the computer algebra system **Magma** [2] and obtain

$$P_G(z) = \frac{z^8 - z^7 + z^6 + z^4 + z^2 - z + 1}{z^{14} - 2z^{13} + 2z^{10} + 2z^9 - 3z^8 - 3z^6 + 2z^5 + 2z^4 - 2z + 1}$$

$$= 1 + z + 3z^2 + 6z^3 + 11z^4 + \text{ terms of higher order.}$$

The coefficient of z^3 in this series shows that there are 6 linear independent invariants of degree 3. Intersecting the 6-dimensional space spanned by these invariants with the vector space of dimension $\binom{6}{3}$ spanned by all squarefree monomials of degree 3 gives the vector space spanned by the polynomials

$$
\begin{aligned}
a_1 := {}& x_1x_2x_3 + x_1x_2x_4 + x_1x_2x_5 + x_1x_2x_6 + x_1x_3x_4 + x_1x_5x_6 + \\
& x_2x_3x_4 + x_2x_5x_6 + x_3x_4x_5 + x_3x_4x_6 + x_3x_5x_6 + x_4x_5x_6, \\
a_2 := {}& x_1x_3x_5 + x_1x_4x_6 + x_2x_3x_6 + x_2x_4x_5, \\
a_3 := {}& x_1x_3x_6 + x_1x_4x_5 + x_2x_3x_5 + x_2x_4x_6.
\end{aligned}
$$

Enumerating all elements in the vector space spanned by a_1, a_2 and a_3 we find two homogeneous bent functions

$$
\begin{aligned}
b_1 := {}& x_1x_2x_3 + x_1x_2x_4 + x_1x_2x_5 + x_1x_2x_6 + x_1x_3x_4 + x_1x_3x_6 + \\
& x_1x_4x_5 + x_1x_5x_6 + x_2x_3x_4 + x_2x_3x_5 + x_2x_4x_6 + x_2x_5x_6 + \\
& x_3x_4x_5 + x_3x_4x_6 + x_3x_5x_6 + x_4x_5x_6, \\
b_2 := {}& x_1x_2x_3 + x_1x_2x_4 + x_1x_2x_5 + x_1x_2x_6 + x_1x_3x_4 + x_1x_3x_5 + \\
& x_1x_4x_6 + x_1x_5x_6 + x_2x_3x_4 + x_2x_3x_6 + x_2x_4x_5 + x_2x_5x_6 + \\
& x_3x_4x_5 + x_3x_4x_6 + x_3x_5x_6 + x_4x_5x_6.
\end{aligned}
$$

Remark 1.   – The search space has now been considerably reduced by using the symmetry group S_4: instead of searching 2^{20} Boolean polynomials we need only consider 2^3 polynomials.

– To a bent function f in $n = 2k$ variables there corresponds a bent function $\tilde{f}$ (called the *Fourier transform* of f, cf. [3]) defined by

$$F(s) = 2^k(-1)^{\tilde{f}(s)}$$

for each $s \in \mathrm{GF}(2)^n$. We observe that the bent functions b_1 and b_2 are Fourier transforms of each other. The fact that b_1 and b_2 are related to each other by a Fourier transform follows from [6, Remark 4.8]: a homogeneous bent function f of degree k contains the monomial x^v iff $\tilde{f}$ contains the monomial $x^{\overline{v}}$ where $\overline{v}$ is the one's complement of v.

– The pairing of bent functions induced by the Fourier transform is the same as that induced by the mapping which takes a 3-set $\{a, b, c\}$ to its complement in $\{1 \ldots, 6\}$. This is an automorphism of order two of the graph $\Gamma(6,3)$, whose full automorphism group is $S_6 \times C_2$; see [4].

– We remark that b_2 is the bent function given in [14, eq. (11)]. The stabilizer of b_1 respectively b_2 in S_6 is precisely the S_4 we have started from. Therefore, all bent functions can be obtained either from b_1 or from b_2 by application of a transversal T of $S_4 \setminus S_8$. So we reproduced the result observed in [14] that there are $720/24 = 30$ bent functions.

5   Homogeneous Bent Functions in Eight Variables

We now consider the eight variable homogeneous bent functions. The question of the existence of such functions was raised in [14].

Using the computer algebra system Magma [2] it is possible to generate a system of representatives for the conjugacy classes of subgroups of the symmetric group S_8. For each representative subgroup G of a conjugacy class we computed the vector space of invariants of degree 3.

We found that two groups of orders six and seven, produced homogeneous bent functions of degree 3 which we describe in the following two subsections. The search was performed over the whole subgroup lattice of S_8 with exception of some groups of small order. We had to restrict the size of the search space of Boolean functions to be less than or equal to 2^{14} which rules out a number of subgroups, the largest of which was of order 24. In the extreme case if G was chosen to be identity group this would lead to the space of *all* 2^{56} homogeneous polynomials of degree 3.

5.1   Sixfold Symmetry, Designs and Graphs

The cyclic group $C_6 = \langle (1, 2)(3, 4, 5, 6, 7, 8) \rangle \leq S_8$ gives the invariant bent function

$$
\begin{aligned}
f_{32} = {} & x_1x_3x_6 + x_2x_4x_7 + x_1x_5x_8 + x_2x_3x_6 + x_1x_4x_7 + x_2x_5x_8 + \\
& x_1x_3x_8 + x_2x_3x_4 + x_1x_4x_5 + x_2x_5x_6 + x_1x_6x_7 + x_2x_7x_8 + \\
& x_1x_4x_6 + x_2x_5x_7 + x_1x_6x_8 + x_2x_3x_7 + x_1x_4x_8 + x_2x_3x_5 + \\
& x_3x_4x_6 + x_4x_5x_7 + x_5x_6x_8 + x_3x_6x_7 + x_4x_7x_8 + x_3x_5x_8 + \\
& x_3x_4x_5 + x_4x_5x_6 + x_5x_6x_7 + x_6x_7x_8 + x_3x_7x_8 + x_3x_4x_8 + \\
& x_3x_5x_7 + x_4x_6x_8.
\end{aligned}
$$

We have listed the 32 terms of f_{32} according to the orbits of C_6 acting on the monomials (representing each orbit by one row). In the remaining parts of this paper we shall adopt this convention of representing bent functions.

With respect to the action of C_6 we found eight invariant bent functions. The number of terms in these functions varies: four functions have 32 terms, and four functions have 24 terms. The invariant bent function

$$
\begin{aligned}
f_{24} = {} & x_1x_3x_6 + x_2x_4x_7 + x_1x_5x_8 + x_2x_3x_6 + x_1x_4x_7 + x_2x_5x_8 + \\
& x_1x_3x_8 + x_2x_3x_4 + x_1x_4x_5 + x_2x_5x_6 + x_1x_6x_7 + x_2x_7x_8 + \\
& x_1x_4x_6 + x_2x_5x_7 + x_1x_6x_8 + x_2x_3x_7 + x_1x_4x_8 + x_2x_3x_5 + \\
& x_3x_4x_6 + x_4x_5x_7 + x_5x_6x_8 + x_3x_6x_7 + x_4x_7x_8 + x_3x_5x_8
\end{aligned}
$$

is rather special in that the monomials comprising it correspond to the blocks of a $t - (v, k, \lambda)$ design (cf. [1]), where $t = 1$, $v = 8$, $k = 3$, and $\lambda = 9$.

It also has a description in terms of the Nagy graph $\Gamma(8, 3)$ (cf. Section 4). The four orbits of C_6 on g (displayed above), can be represented as the following subgraphs of $\Gamma(8, 3)$: each orbit of six monomials is a disjoint union of two triangles (or 3-cliques) in the complement of $\Gamma(8, 3)$. Another way of seeing this is in terms of the Johnson graph $J(3, 2)$ [9].

Moreover, the property $f_{24} \leq f_{32}$ holds, i.e., the monomials of f_{24} are contained in the monomials of f_{32}. The difference $f_{32} - f_{24}$ is the polynomial

$$x_3x_4x_5 + x_4x_5x_6 + x_5x_6x_7 + x_6x_7x_8 + x_3x_7x_8 + x_3x_4x_8 + x_3x_5x_7 + x_4x_6x_8$$

which factorizes as $(x_3 + x_6)(x_4 + x_7)(x_5 + x_8)$. Nevertheless, f_{24} and f_{32} are linearly equivalent, cf. Section 6.

5.2 Sevenfold Symmetry

The cyclic group $C_7 = \langle (1,2,3,4,5,6,7) \rangle \leq S_8$ gives the invariant bent function

$$
\begin{aligned}
f_{35} = \; & x_1x_2x_3 + x_2x_3x_4 + x_3x_4x_5 + x_4x_5x_6 + x_5x_6x_7 + x_1x_6x_7 + x_1x_2x_7 + \\
& x_1x_2x_4 + x_2x_3x_5 + x_3x_4x_6 + x_4x_5x_7 + x_1x_5x_6 + x_2x_6x_7 + x_1x_3x_7 + \\
& x_1x_2x_5 + x_2x_3x_6 + x_3x_4x_7 + x_1x_4x_5 + x_2x_5x_6 + x_3x_6x_7 + x_1x_4x_7 + \\
& x_1x_3x_5 + x_2x_4x_6 + x_3x_5x_7 + x_1x_4x_6 + x_2x_5x_7 + x_1x_3x_6 + x_2x_4x_7 + \\
& x_1x_4x_8 + x_2x_5x_8 + x_3x_6x_8 + x_4x_7x_8 + x_1x_5x_8 + x_2x_6x_8 + x_3x_7x_8.
\end{aligned}
$$

Overall there are twelve bent functions which are invariant under C_7. Again, the bent function

$$
\begin{aligned}
f_{28} = \; & x_1x_2x_3 + x_2x_3x_4 + x_3x_4x_5 + x_4x_5x_6 + x_5x_6x_7 + x_1x_6x_7 + x_1x_2x_7 + \\
& x_1x_2x_4 + x_2x_3x_5 + x_3x_4x_6 + x_4x_5x_7 + x_1x_5x_6 + x_2x_6x_7 + x_1x_3x_7 + \\
& x_1x_2x_5 + x_2x_3x_6 + x_3x_4x_7 + x_1x_4x_5 + x_2x_5x_6 + x_3x_6x_7 + x_1x_4x_7 + \\
& x_1x_4x_8 + x_2x_5x_8 + x_3x_6x_8 + x_4x_7x_8 + x_1x_5x_8 + x_2x_6x_8 + x_3x_7x_8
\end{aligned}
$$

has the property $f_{28} \leq f_{35}$. The 28 terms of f_{28} are characteristically split into two subsets of size 7 and 21 respectively according to whether the monomials contain the variable x_8 or not. We observe that the monomials in the set of size 21 correspond to a $1 - (7,3,9)$ design (the monomials comprising the set of size 7 can also be considered as a $1 - (7,2,2)$ design). While the monomials of the difference

$$f_{35} - f_{28} = x_1x_3x_5 + x_2x_4x_6 + x_3x_5x_7 + x_1x_4x_6 + x_2x_5x_7 + x_1x_3x_6 + x_2x_4x_7$$

correspond to a $1 - (7,3,3)$ design.

The bent function f_{35} can also be described as follows. There are five orbits of C_7 acting on the monomials of f_{35}. The graph (cf. Definition 3) corresponding to each orbit is a 7-cycle subgraph of either: the Johnson graph $J(8,3)$, the Nagy graph $\Gamma(8,3)$, or the complement of the Nagy graph.

6 Linear Equivalence

In this section we consider the problem of sorting into equivalence classes the previously found bent functions. Recall that two functions $f, g \in \mathcal{P}_n$ are called linearly equivalent if there is an *affine* transformation between f and g. This means that there is a matrix $A \in \mathrm{GL}(n,2)$ and a vector $v \in \mathrm{GF}(2)^n$ such that $f^{(A,v)} := f(A \cdot (x_1, \ldots, x_n)^t + v) = g$ (cf. [12]). In what follows we make some remarks about the equivalence problem.

- One of the 2-Sylow subgroups of GL(8, 2) of order 2^{28} is the group consisting of all upper triangular matrices. There is a transformation A which belongs to this 2-Sylow subgroup and acts on the variables as

$$x_1 \mapsto x_1 + \sum_{i=3}^{8} x_i, \quad x_2 \mapsto x_2 + \sum_{i=3}^{8} x_i, \quad x_k \mapsto x_k \quad \text{for } k = 3 \ldots 8. \quad (1)$$

  This transformation establishes that $f_{24}^A = f_{32}$.
- As stated in Section 5 the bent functions invariant under the groups C_6 and C_7 occur in pairs (f, g) where $g \leq f$. With respect to this group action the pairing is preserved and hence the pairs form a single orbit of length six. Combining this with the transformation (1) we see that all homogeneous bent functions in $\mathcal{P}_8$ that have symmetry group C_6 form a single orbit under the action of GL(8, 2).
- Following Hou [10] we compute the rank $r_3(f)$ of a homogeneous Boolean function f of degree 3. We associate to the given homogeneous polynomial

$$f = \sum_{\substack{v \in \mathrm{GF}(2)^n: \\ \mathrm{wgt}(v)=3}} \alpha_v x^v$$

a binary $n \times \binom{n}{2}$ matrix. In [10] cosets of the second order Reed-Muller code are considered, but in our case f has already the required form so we can compute the rank of this matrix directly. Hou has shown [10] that $r_3(f)$ is preserved under the action of the affine group. A computation establishes that $r_3(f_{32}) = 6$ and $r_3(f_{35}) = 7$, thus showing that f_{32} and f_{35} are not linearly equivalent.

7   Conclusions and Outlook

Bent functions are of fundamental importance in cryptographic applications. We have produced for the first time homogeneous bent functions of degree three in 8 and 10 variables. We have demonstrated a connection between these functions and 1-designs as well as certain graphs.

We have demonstrated that the language of invariant theory is a natural setting for the construction of homogeneous bent functions which have symmetry. This setting has proven to be of great computational advantage in the search for such bent functions.

Some open problems naturally suggest themselves from our investigation.

- So far we have restricted ourselves to subgroups of S_n permuting the n variables. It seems possible that other subgroups of GL(n, 2) not consisting entirely of permutations could feature in the search for bent functions.
- Another open question is how the bent functions found fit into the known classes of bent functions [3,8].
- We would hope that the observed connections between designs and homogeneous bent functions will provide further insight to the problem of constructing families of bent functions.

258   C. Charnes, M. Rötteler, and T. Beth

Acknowledgements

We would like to thank the following people who provided us with helpful discussions and remarks: Claude Carlet, Markus Grassl, Jozef Pieprzyk, Larry Smith, Andrew Thomason, Vladimir Tonchev.

References

1. Th. Beth, D. Jungnickel, and H. Lenz. *Design Theory*, volume I. Cambridge University Press, 2nd edition, 1999.
2. W. Bosma, J.J. Cannon, and C. Playoust. The Magma algebra system I: The user language. *J. Symb. Comp.*, 24:235–266, 1997.
3. C. Carlet. Two new classes of bent functions. In *Advances in Cryptology-Eurocrypt'93*, volume 765 of *Lecture Notes in Computer Science*, pages 77–101. Springer, 1993.
4. C. Charnes. A class of symmetric bent functions. (submitted), 2000.
5. J. H. Conway, R. T. Curtis, S. P. Norton, and R. A. Wilson. *Atlas of Finite Groups*. Clarendon Press, Oxford, 1985.
6. J. Dillon. A survey of bent functions. *NSA Technical Journal*, Special Issue:191–215, 1972.
7. J. Dixon and B. Mortimer. *Permutation Groups*, volume 163 of *Graduate Texts in Mathematics*. Springer, 1996.
8. H. Dobbertin. Construction of bent functions and balanced Boolean functions with high nonlinearity. In B. Preneel, editor, *Fast Software Encryption*, volume 1008 of *Lecture Notes in Computer Science*, pages 61–74, 1995.
9. C. Godsil. *Algebraic Combinatorics*. Chapman and Hall, New York, London, 1993.
10. X. Hou. Cubic bent functions. *Discrete Mathematics*, 189:149–161, 1998.
11. I. G. MacDonald. *Symmetric Functions and Hall Polynomials*. Oxford University Press, 1979.
12. F. J. MacWilliams and N. J. A. Sloane. *The Theory of Error-Correcting Codes*. North-Holland, Amsterdam, 1977.
13. Z. Nagy. A certain constructive estimate of the Ramsey number (in Hungarian). *Mat. Lapok*, 1972.
14. C. Qu, J. Seberry, and J. Pieprzyk. On the symmetric property of homogeneous Boolean functions. In J. et al. Pieprzyk, editor, *Proceedings of the Australian Conference on Information Security and Privacy (ACISP)*, volume 1587 of *Lecture Notes in Computer Science*, pages 26–35. Springer, 1999.
15. E. M. Rains and N. J. A. Sloane. *Handbook of Coding Theory*, volume I, chapter Self-dual codes, pages 177–294. Elsevier, 1998.
16. O. S. Rothaus. On "bent" functions. *Journal of Combinatorial Theory, Series A*, 20:300–305, 1976.
17. P. Savický. On the bent Boolean functions that are symmetric. *Europ. J. Combinatorics*, 15:407–410, 1994.
18. L. Smith. *Polynomial Invariants of Finite Groups*. A. K. Peters, 1995.
19. B. Sturmfels. *Algorithms in Invariant Theory*. Springer, 1993.

A   Homogeneous Bent Functions in Ten Variables

The cyclic group $C_9 = \langle (1,2,3,4,5,6,7,8,9) \rangle \leq S_{10}$ gives the following invariant bent function (we have abbreviated i for the variable x_i and replaced 10 by 0).

$$123 + 234 + 345 + 456 + 567 + 678 + 789 + 189 + 129 +$$
$$125 + 236 + 347 + 458 + 569 + 167 + 278 + 389 + 149 +$$
$$126 + 237 + 348 + 459 + 156 + 267 + 378 + 489 + 159 +$$
$$140 + 250 + 360 + 470 + 580 + 690 + 170 + 280 + 390 +$$
$$147 + 258 + 369$$

Like the functions displayed in Section 5 this function has a characteristic decomposition into sets of sizes 9 and 63 respectively according to whether the monomials contain the variable x_{10} or not. The set of size 63 forms a $1 - (9,3,21)$ design.

For more information about homogeneous bent functions in ten variables the reader is referred to the web page

http://avalon.ira.uka.de/home/roetteler/bent.html

In Table 1 we have displayed the maximal subgroups of the symmetric group S_{10}, see [5] and [7] regarding the maximal subgroups of S_n and A_n. For each of the groups given in Table 1 we performed a search over the lattice of subgroups in order to find homogeneous bent functions of degree 3. This search was not exhaustive in that some groups of large index in the respective maximal subgroup could not be searched. The maximal order of a subgroup which could not be searched is given in the fourth column.

The full symmetric group S_{10} and the alternating group A_{10} do not produce invariant bent functions of degree 3. We do not have to take into account the subgroups of the alternating group since we are searching over the whole subgroup lattices of the other maximal subgroups of S_{10}.

Table 1. Summary of the search over the subgroups of S_{10}

Group name	Size	No. bents	Max. order
$S_1 \times S_9$	362880	72	288
$S_2 \times S_8$	80640	0	288
$S_3 \times S_7$	30240	0	288
$S_4 \times S_6$	17280	32	288
$S_2 \wr S_5$	3840	32	96
$S_5 \wr S_2$	28800	32	288
$P\Gamma L(2,9)$	1440	0	16

Partially Identifying Codes
for Copyright Protection

Sylvia Encheva[1] and Gérard Cohen[2]

[1] HSH, Bjørnsonsg. 45, 5528 Haugesund, Norway,
sbe@hsh.no
[2] ENST, 46 rue Barrault, 75634 Paris Cedex 13, France,
cohen@enst.fr

Abstract. If Γ is a q-ary code of length n and $a, b, ..., e$ are codewords, then c is called a descendant of $a, b, ..., e$ if $c_i \in \{a_i, b_i, ..., e_i\}$ for $i = 1, ..., n$. Codes Γ with the property that coalitions of a limited size have restrictions on their descendants are studied. Namely, we consider codes with the following partial identification property, referred to later as 2- secure frameproof codes (2-SFPC): any two non intersecting coalitions of size at most 2 have no common descendant.

Index Terms: Secure frameproof codes, copyright protection, designs.

1 Introduction

The ability to resolve ownership disputes and copyright infringement is difficult in the worldwide digital age. There is an increasing need to develop techniques that protect the owner of digital data. Digital watermarking is a technique used to embed a known piece of digital data within another one. The embedded piece acts as a fingerprint for the owner, allowing the protection of copyright, authentication of the data, and tracing of illegal copies.

A publisher embeds a unique fingerprint pattern into each distributed copy of a document, keeping a database of sold copies and their corresponding fingerprints. If, later on, an illegally distributed copy is discovered, he may trace that copy back to the offending user by comparing its fingerprint to the database. We consider the attack which results when two users collude and compare their independently marked copies. Then they can detect and locate the differences, and combine their copies into a new one whose fingerprint differs from all the users'.

Codes were introduced in [2] (see also [11]) as a method of "digital fingerprinting" which prevents a coalition of a given size from forging a copy with no member of the coalition being caught, or from framing an innocent user.

The outline of the paper is as follows. Definitions and basic results are presented in Section 2. In Section 3, we first prove that Sylvester type matrices are 2-secure frameproof codes. Although, they can accommodate a limited number of users, their length 2^i make them suitable for practical applications and concatenation.

S. Boztaş and I.E. Shparlinski (Eds.): AAECC-14, LNCS 2227, pp. 260–267, 2001.
© Springer-Verlag Berlin Heidelberg 2001

Section 4 is devoted to q-ary 2-secure frameproof codes. In particular we show that an equidistant code with $2d > n$ is a 2-SFPC, thus weakening the sufficient condition $4d > 3n$ in [3] and [12]. In the last section, we show how to combine two secure frameproof codes to build larger ones.

2   Definitions and Basic Results

We use the notation of [11] for fingerprinting issues and of [6] for codes and Hadamard matrices. We identify a vector with its *support*, set of its non-zero positions. For any positive real number x we shall denote by $\lfloor x \rfloor$ its integer part and by $\lceil x \rceil$ the smallest integer at least equal to x.

A set $\Gamma \subseteq GF(q)^n$ is called an (n, M, d)-*code* if $|\Gamma| = M$ and the minimum Hamming distance between two of its elements (codewords) is d.

Suppose $\mathcal{C} \subseteq \Gamma$. For any position i define the *projection*

$$P_i(\mathcal{C}) = \bigcup_{a \in \mathcal{C}} a_i.$$

Define the *feasible set* of $\mathcal{C}$ by:

$$F(\mathcal{C}) = \{x \in GF(q)^n : \forall i, x_i \in P_i(\mathcal{C})\}.$$

The feasible set $F(\mathcal{C})$ represents the set of all possible n-tuples (descendants) that could be produced by the coalition $\mathcal{C}$ by comparing the codewords they jointly hold. Observe that $\mathcal{C} \subseteq F(\mathcal{C})$ for all $\mathcal{C}$, and $F(\mathcal{C}) = \mathcal{C}$ if $|\mathcal{C}| = 1$.

If two non-intersecting coalitions can produce the same descendant, it will be impossible to trace with certainty even one guilty user. This motivates the following definition from [11].

Definition 1. *An (n, M)- code Γ is called a s-secure frameproof code (s-SFPC for short) if, for every couple of coalitions $\mathcal{C}, \mathcal{C}' \subseteq \Gamma$ such that $|\mathcal{C}| \leq s, |\mathcal{C}'| \leq s$ and $\mathcal{C} \cap \mathcal{C}' = \emptyset$, we have $F(\mathcal{C}) \cap F(\mathcal{C}') = \emptyset$.*

The previous property can be rephrased as follows when $q = 2$:

For any ordered $2s$-tuple of codewords written as columns, there is a coordinate where the $2s$-tuple $(1..10..0)$ of weight s or its complement occurs. For $s = 2$ this turns out to have been studied in another context under the name of "separation" (see, e.g., [4], [5], [10]).

In this paper, we consider the case $s = 2$.

3   Binary 2-Secure Frameproof Codes

Theorem 1. *[1] If H_n is an n times n Hadamard matrix with $n > 1$, then n is even and for any two distinct rows of H there are precisely $\frac{n}{2}$ columns in which the entries in the two rows agree. Further, if $n > 2$ then n is divisible by 4, and for any three distinct rows of H there are precisely $\frac{n}{4}$ columns in which the entries in all three rows agree.*

Let H_n be a Sylvester type matrix of order $n = 2^i$ [6]. If $+1$'s are replaced by 1's and -1's by 0's (note that this is not the classical way, but insures that the all 0 vector is a codeword), H_n is changed into the binary Hadamard matrix A_n [6], which is an $(n = 2^i, M = 2^i, d = 2^{i-1})$ code.

Theorem 2. *The matrix A_n is an $(n = 2^i, M = 2^i)$ 2-SFPC.*

Proof. Set $n = 4m$ with $m = 2^{i-2}$. By [6] Ch. 2, since any two rows in A_{4m} are orthogonal, they agree in $2m$ places and differ in $2m$ places. Any such two rows, different from the all 0 vector, contain m columns of the four possible types $(11)^T, (00)^T, (01)^T, (10)^T$.

We distinguish between two cases, according to whether or not one of the four codewords is the first row of A_{4m} (the all 0 vector).

Case 1. Let $c_1, c_2, c_3, c_4 \in A_{4m}$ be all different from the first row in A_{4m}. Then A_{4m} will not be 2-SFPC if the following occurs:

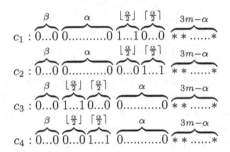

where the number of columns $(00)^T$ in any two rows of A_{4m} is $\alpha + \beta = m$, a '*' being indifferently a 0 or 1. If α is an odd number, then $[c_2, c_4]$ have less than $\alpha + \beta - 1 = m - 1$ columns $(00)^T$, a contradiction. Thus α is even, say $\alpha = 2\alpha'$. Then A_{4m} will fail to be 2-SFPC if the following occurs:

$$
\begin{array}{l}
c_1 : \overbrace{0...0}^{\beta}\,\overbrace{0..........0}^{2\alpha'}\,\overbrace{1...1}^{\alpha'}\overbrace{0...0}^{\alpha'}\,\overbrace{0........0}^{2m-\beta-3\alpha'}\,\overbrace{1.........1}^{2m-\beta-3\alpha'}\,\overbrace{1.........1}^{2m-\beta-3\alpha'}\,\overbrace{1.........1}^{2m-\beta-3\alpha'} \\
c_2 : \overbrace{0...0}^{\beta}\,\overbrace{0..........0}^{2\alpha'}\,\overbrace{0...0}^{\alpha'}\overbrace{1...1}^{\alpha'}\,\overbrace{1.........1}^{2m-\beta-3\alpha'}\,\overbrace{0........0}^{2m-\beta-3\alpha'}\,\overbrace{1.........1}^{2m-\beta-3\alpha'}\,\overbrace{1.........1}^{2m-\beta-3\alpha'} \\
c_3 : \overbrace{0...0}^{\beta}\,\overbrace{1...1}^{\alpha'}\overbrace{0...0}^{\alpha'}\,\overbrace{0..........0}^{2\alpha'}\,\overbrace{1.........1}^{2m-\beta-3\alpha'}\,\overbrace{1.........1}^{2m-\beta-3\alpha'}\,\overbrace{0........0}^{2m-\beta-3\alpha'}\,\overbrace{1.........1}^{2m-\beta-3\alpha'} \\
c_4 : \overbrace{0...0}^{\beta}\,\overbrace{0...0}^{\alpha'}\overbrace{1...1}^{\alpha'}\,\overbrace{0..........0}^{2\alpha'}\,\overbrace{1.........1}^{2m-\beta-3\alpha'}\,\overbrace{1.........1}^{2m-\beta-3\alpha'}\,\overbrace{1.........1}^{2m-\beta-3\alpha'}\,\overbrace{0........0}^{2m-\beta-3\alpha'}
\end{array}
$$

Columns in $[c_1, c_2, c_3, c_4]$ are written this way up to position $\beta + 4\alpha'$ since any two rows in A_{4m} have columns $(00)^T$ in m positions. The number of $(01)^T$ and $(10)^T$ columns are equal, otherwise two of the rows c_1, c_2, c_3, c_4 would have more than m columns $(00)^T$. Therefore any two of the rows c_1, c_2, c_3, c_4 must contain m columns $(11)^T$ from position $\beta + 4\alpha' + 1$ on, since any two of them do not contain columns $(11)^T$ on the first $\beta + 4\alpha'$ positions. We now upperbound by $4m$ the number of columns containing at least one 0, getting $\beta + 4\alpha' + 4(2m - \beta - 3\alpha') \leq 4m$, or $3\beta + 8\alpha' \geq 4m$. On the other hand, for any two rows, the number of $(00)^T$ columns is $m = \beta + 2\alpha'$. This leads to $2\alpha' = m, \beta = 0$, which is impossible since the first column of A_{4m} contains only 0's.

Case 2 now. Suppose that c_1 is the first row in A_{4m}, i.e. $wt(c_1) = 0$. Then A_{4m} will not be 2-SFPC if the following occurs.

$$
\begin{array}{c}
\overset{m}{\overbrace{}}\ \overset{m}{\overbrace{}}\ \overset{m}{\overbrace{}}\ \overset{m}{\overbrace{}} \\
c_1 : 0...0\ 0...0\ 0...0\ 0...0 \\
\overset{m}{\overbrace{}}\ \overset{m}{\overbrace{}}\ \overset{m}{\overbrace{}}\ \overset{m}{\overbrace{}} \\
c_2 : 0...0\ 0...0\ 1...1\ 1...1 \\
\overset{m}{\overbrace{}}\ \overset{m}{\overbrace{}}\ \overset{m}{\overbrace{}}\ \overset{m}{\overbrace{}} \\
c_3 : 0...0\ 1...1\ 0...0\ 1...1 \\
\overset{m}{\overbrace{}}\ \overset{m}{\overbrace{}}\ \overset{m}{\overbrace{}}\ \overset{m}{\overbrace{}} \\
c_4 : 01..1\ 0...0\ 10..0\ 1...1
\end{array}
$$

The row c_4 should have a 0 in the first position since all rows of A_{4m} do. The support of c_4 in positions $\{m+1, m+2, ..., 2m\}$ has size 0 by the assumption that A_{4m} is not 2-SFPC. In this case $[c_2, c_4]$ will have $m + 1$ columns $(00)^T$, again reaching a contradiction. Therefore A_{4m} is a 2-SFPC.   □

4   On Equidistant 2-Secure Frameproof Codes

Positions where two codewords coincide are denoted $\overbrace{*...*}$, and positions where they have different coordinates are denoted $\overbrace{\frown\frown}$. We now improve on the sufficient condition $(d/n) > 1 - (1/s^2)$ from [3] and [12] for a code to be a s-SFPC, in the special case of equidistant codes and $s = 2$.

Proposition 1. *Let C be an equidistant q-ary code with $dist(c_i, c_j) = d$ and length n. If $2d > n$, then C is 2-SFPC.*

Proof. Suppose C is not 2-SFPC. Then w.l.o.g. we may assume that there are four codewords $c_1, c_2, c_3, c_4 \in C$ such that c_1, c_2 and c_3, c_4 can produce a common descendant. Let

c_1, c_2 coincide on positions $1, 2, ..., x + \alpha + y = n - d$,
c_1, c_3 coincide on positions $1, 2, ..., x + \alpha$; $x + \alpha + y + 1, ..., x + \alpha + 2y$,
c_2, c_3 coincide on positions $1, 2, ..., x + \alpha$; $x + \alpha + 2y + 1, ..., x + \alpha + 3y$,
c_1, c_4 coincide on positions $x+1, ..., x+\alpha+y+m_1$; $x+\alpha+3y+1, ..., 2x+\alpha+3y-m_1$,
c_2, c_4 coincide on positions $x + 1, ..., x + \alpha + y$; $x + \alpha + 2y + 1, ..., x + \alpha + y + m_2$;
$2x + 3y + \alpha - m_1 + 1, ..., 3x + \alpha + 3y - m_1 - m_2 = n$.

Thus the 4 times n array, obtained from c_1, c_2, c_3, c_4, may be described as follows

$$
\begin{array}{l}
c_1 : \overbrace{*...*}^{x} \overbrace{*...*}^{\alpha} \overbrace{*...*}^{y} \overbrace{*.*}^{m_1} \overbrace{*.*}^{y-m_1} \overbrace{....}^{m_2} \overbrace{....}^{y-m_2} \overbrace{*...*}^{x-m_1} \overbrace{....}^{x-m_2} \\
c_2 : \overbrace{*...*}^{x} \overbrace{*...*}^{\alpha} \overbrace{*...*}^{y} \overbrace{....}^{m_1} \overbrace{....}^{y-m_1} \overbrace{*.*}^{m_2} \overbrace{*.*}^{y-m_2} \overbrace{....}^{x-m_1} \overbrace{*...*}^{x-m_2} \\
c_3 : \overbrace{*...*}^{x} \overbrace{*...*}^{\alpha} \overbrace{....}^{y} \overbrace{*.*}^{m_1} \overbrace{*.*}^{y-m_1} \overbrace{*.*}^{m_2} \overbrace{*.*}^{y-m_2} \overbrace{....}^{x-m_1} \overbrace{....}^{x-m_2} \\
c_4 : \overbrace{....}^{x} \overbrace{*...*}^{\alpha} \overbrace{*...*}^{y} \overbrace{*.*}^{m_1} \overbrace{....}^{y-m_1} \overbrace{*.*}^{m_2} \overbrace{....}^{y-m_2} \overbrace{*...*}^{x-m_1} \overbrace{*...*}^{x-m_2} .
\end{array}
$$

Combining $\alpha + m_1 + m_2 = n - d$, ($c_3, c_4$ coincide on $n - d$ positions) and $\alpha + x + y = n - d$, (c_1, c_2 coincide on $n - d$ positions) we obtain $m_1 + m_2 = x + y$. The last equality together with

$$m_1 \le x, m_1 \le y, m_2 \le x, m_2 \le y$$

leads to $x = y = m_1 = m_2$. Looking at the positions where c_1, c_3 coincide we get

$$x + y + \alpha = 2y + \alpha = n - d.$$

Replacing $2y$ by d gives $\alpha + 2d = n$. This is possible only for $\alpha = 0, 2d = n$. By assumption $2d > n$, therefore C is 2-SFPC.   □

Note that the bound $2d > n$ cannot be improved in general: there exist equidistant codes with $2d = n$ which are not 2-SFPC (take for example the 4 vectors of weight 1, length 4 and at distance 2 apart).

Let's now look at how designs can provide such equidistant 2-SFPC.
A parallel class in a $2 - (v, k, \lambda)$ design with $v \equiv 0 \pmod{k}$ is a set of v/k pairwise disjoint blocks. A 2-design with parameters v, k, λ, r, b is resolvable if the block set can be partitioned into r disjoint parallel classes. Any such partition is called a resolution. A resolvable design is affine resolvable (or affine) if any two blocks that are not in the same parallel class meet in a constant number (say μ) points.
An equidistant $(n, M = qt, d)_q$ code is optimal if

$$d = \frac{nt(q-1)}{qt - 1}.$$

Proposition 2. *[9] An optimal equidistant q-ary code with parameters*

$$n = \frac{q^2\mu - 1}{q - 1}, M = q^2\mu, d = q\mu$$

exists if and only if there exists an affine design with parameters

$$v = q^2\mu, k = q\mu, \lambda = \frac{q\mu - 1}{q - 1}, r = \frac{q^2\mu - 1}{q - 1}, b = \frac{q^3\mu - q}{q - 1}.$$

Proposition 3. *If A is an incidence matrix of an affine $2 - (v, k, \lambda)$ design, corresponding to a ternary optimal equidistant code, then A is a 2-SFPC.*

Proof. Let A be as described in the proposition and let C be a corresponding ternary optimal equidistant code. The existence of C follows by Proposition 2. By Proposition 1, C is 2-SFPC. Thus for any $c_1, c_2, c_3, c_4 \in C$ there is at least one coordinate j such that $\{c_1^j, c_2^j\} \neq \{c_3^j, c_4^j\}$. In the case $q = 3$ it implies that at least one the two equalities $c_1^j = c_2^j$ or $c_3^j = c_4^j$ occur.
A may be obtained from C by replacing any symbol i in C by the i-th row of I_3. Thus we are sure that for any four rows in A there is a coordinate where one of the columns $(1100)^T$ or $(0011)^T$ occurs, i.e. A is a 2-SFPC. $\qquad\square$

Example 1. The unique 2-(9,3,1) design [9] is an equidistant code with $2d = n$ and it is a 2-SFPC. The corresponding ternary code is 2-SFPC with parameters $(4, 9, 3)$.

5   Combining Codes

New secure frameproof codes may be obtained via separating hash families (SHF), which are a more general object introduced in [11] (allowing for different size coalitions w_1 and w_2 in Definition 1).

Definition 2. *Let n', m', w_1, w_2 be positive integers such that $n' \geq m'$. An $(n', m', \{w_1, w_2\})$-separating hash family is a set of functions $\mathcal{F}$, such that $|Y| = n', |X| = m', f : Y \to X$ for each $f \in \mathcal{F}$, and for any $C_1, C_2 \subseteq \{1, 2, ..., n\}$ such that $|C_1| = w_1, |C_2| = w_2$ and $C_1 \cap C_2 = \emptyset$, there exists at least one $f \in \mathcal{F}$ such that*

$$\{f(y) : y \in C_1\} \cap \{f(y) : y \in C_2\} = \emptyset.$$

The notation $SHF(n; n', m', \{w_1, w_2\})$ will be used to denote an $(n', m', \{w_1, w_2\})$-separating hash family with $|\mathcal{F}| = n$.

Furthermore, in [12], a general concatenation construction of SHF is presented.

Theorem 3. *Suppose there exist $SHF(n'; M, n_0, \{w_1, w_2\})$ and $SHF(n''; n_0, m', \{w_1, w_2\})$. Then there exists an $SHF(n'n''; M, m', \{w_1, w_2\})$.*

The next results provide a way to combine secure frameproof codes, like those obtained in the previous sections, into new SFPC. These constructions give codes with reasonably good number of codewords in some special cases.

Theorem 4. *[7] Let $M = M_1.M_2$, with $M_2 \geq M_1$ and M_2 is not divisible either by 2 or by 3. Suppose $C_1(n_1, M_1)$ and $C_2(n_2, M_2)$ are binary 2-SFPC. Then there is a 2-SFPC (n, M) code with $n = n_1 + 4n_2$.*

Let $\mathbf{S}$ be the rightward cyclic shift operator on n-tuples and $\mathbf{v}$ be a binary p-tuple. In the v-representation of $GF(p)$, the element i is represented by the p-tuple $\mathbf{S}^i(\mathbf{v})$, the ith rightward cyclic shift of $\mathbf{v}$ [8].

Lemma 1. *[8] If p is a Mersenne prime and v is a binary m-sequence of length p, or more generally if $(p-1)/2$ is odd and v is a Legendre sequence, then the v-representation of $GF(p)$ yields an equidistant $(p, p, (p+1)/2)$ binary code.*

Example 2. Suppose we take $C_1 = C_2$ where $M_1 = 11, n_1 = 11$, then $M = 121, n = 55$. Applying Theorem 4, with $C_1(11, 11)$ and $C_2(55, 121)$, leads to a 2-SFPC with $M = 1331, n = 231$.

Take now $M_1 = M_2 = 13, n_1 = n_2 = 13$, then $M = 169, n = 65$. Applying Theorem 4 a second time with $M_1 = 13, n_1 = 13$ and $M_2 = 169, n_1 = 65$ leads to a 2-SFPC with $M = 1859, n = 273$.

Binary 2-SFPC may also be obtained from p-ary 2-SFPC exploiting ideas from [8], whose notation we follow. When the positive integers m and n are relatively prime, the Chinese remainder theorem (CRT) specifies a one-to-one correspondence between m times n arrays $A = \{a(i,j)\}$ with entries from an arbitrary alphabet and $m.n$-tuples $\mathbf{b} = [b_0, ..., b_{mn-1}]$ over the same alphabet, where

$$b_i = a(i \bmod m, i \bmod n),$$

"$i \bmod m$" denoting the remainder when i is divided by m. The next result is the main theorem in [8]. We omit its proof.

Theorem 5. *Let p be a prime and let V be a p-ary (n, M, d) equidistant code such that $\gcd(n, p) = 1$. Let v be a binary p-tuple with Hamming weight $wt(v)$ where $0 < wt(v) < p$. Let each codeword $\mathbf{c} = [c_0, c_1, ..., c_{n-1}]$ in V determine a p times n array A such that the ith column of*
A is the transpose of the p-tuple that is the v-representation of the ith component of $\mathbf{c}$, and let $\mathbf{b}$ be the binary N-tuple (where $N = np$) that corresponds to the array A by the CRT correspondence. Then the set of N-tuples $\mathbf{b}$ corresponding to the codewords $\mathbf{c}$ of V form a binary $(np, M, d(p+1)/2)$ equidistant code.

Acknowledgments

The authors wish to thank a referee for his useful comments.

References

1. Th. Beth, D. Jungnickel and H. Lenz, *Design Theory*, Wissenschaftsverlag, Berlin, (1985).
2. D. Boneh and J. Shaw, "Collusion-secure fingerprinting for digital data", *Springer-Verlag LNCS* 963 (1995), pp. 452-465 (Advances in Cryptology-Crypto '95).

3. B. Chor, A. Fiat and M. Naor, "Tracing traitors", *Springer-Verlag LNCS* 839 (1994), pp. 257-270 (Advances in Cryptology-Crypto '94).

4. A.D. Friedman, R.L. Graham and J.D. Ullman, "Universal single transition time asynchronous state assignments", *IEEE Trans. Comput.* vol.18 (1969) pp. 541-547.

5. J. Körner and G. Simonyi, "Separating partition systems and locally different sequences", *SIAM J. Discrete Math.*, vol. 1, pp. 355-359, 1988.

6. F. J. MacWilliams and N. J. A. Sloane, *The Theory of Error-Correcting Codes*, North-Holland, Amsterdam, (1977).

7. T. Nanya and Y. Tohma, "On Universal Single Transition Time Asynchronous State Assignments", *IEEE Trans. Comput.* vol. 27 (1978) pp. 781-782.

8. Q.A. Nguyen, L. Györfi and J. Massey, "Constructions of Binary Constant-Weight Cyclic Codes and Cyclically Permutable Codes", *IEEE Trans. on Inf. Theory*, vol. 38, no. 3, (1992), pp. 940-949.

9. V.S. Pless, W.C. Huffman - Editors, *Handbook of Coding Theory*, Elsevier, Amsterdam (1998).

10. Yu.L. Sagalovitch, "Separating systems", *Problems of Information Transmission*, Vol. 30 (2) (1994), pp. 105-123.

11. D.R. Stinson, Tran Van Trung and R. Wei, "Secure Frameproof Codes, Key Distribution Patterns, Group Testing Algorithms and Related Structures", *J. Stat. Planning and Inference*, vol. 86 (2)(2000), pp. 595-617.

12. D.R. Stinson, R. Wei and L. Zhu, "New Constructions for Perfect Hash Families and Related Structures Using Combinatorial Designs and Codes", *J. of Combinatorial Designs*, vol. 8 (2000), pp. 189-200.

On the Generalised Hidden Number Problem
and Bit Security of XTR

Igor E. Shparlinski

Department of Computing, Macquarie University
Sydney, NSW 2109, Australia
igor@comp.mq.edu.au   http://www.comp.mq.edu.au/~igor/

Abstract. We consider a certain generalisation of the hidden number
problem which has recently been introduced by Boneh and Venkatesan.
We apply our results to study the bit security of the XTR cryptosystem
and obtain some analogues of the results which have been known for the
bit security of the Diffie-Hellman scheme.

1 Introduction

Let p be a prime. We denote by $\mathbb{F} = \mathbb{F}_p$ and $\mathbb{K} = \mathbb{F}_q$ the fields of p and $q = p^m$ elements, respectively, where $m \geq 1$ is integer.

As usual we assume that $\mathbb{F}$ is represented by the elements $\{0, \ldots, p-1\}$.

For integers s and $r \geq 1$ we denote by $\lfloor s \rfloor_r$ the remainder of s on division by r. We also use $\log z$ to denote the binary logarithm of $z > 0$.

Here we study a variant of the *hidden number problem* introduced in 1996 by Boneh and Venkatesan [1,2]. This problem can be stated as follows: recover an unknown element $\alpha \in \mathbb{F}$ such that for polynomially many known random $t \in \mathbb{F}$ approximations to the values of $\lfloor \alpha t \rfloor_p$ are known.

It has turned out that for many applications the condition that t is selected uniformly at random from $\mathbb{F}$ is too restrictive. Examples of such applications include the bit security results for the Diffie-Hellman, Shamir and several other cryptosystems [7,8] and rigorous results on attacks (following the heuristic arguments of [9,18]) on the DSA and DSA-like signature schemes [5,19,20].

It has been systematically exploited in the aforementioned papers [5,7,8,19,20] that the method of [1] can be adjusted to the case when t is selected from a sequence which has some uniformity of distribution property. Thus, these papers have employed bounds of various exponential sums which are natural tools to establish the corresponding uniformity of distribution property.

In particular, the case when t is selected from a small subgroup of $\mathbb{F}^*$ has been studied in [7] and used to generalise (and correct) some results of [1] about the bit security of the Diffie-Hellman key. The results of [7] are based on bounds of exponential sums with elements of subgroups of $\mathbb{F}^*$, namely on Theorem 3.4 and Theorem 5.5 of [12].

Here, motivated by applications to the recently introduced XTR cryptosystem, we consider a similar problem for subgroups in the *extension field* $\mathbb{K}$. Let

$$\mathrm{Tr}(z) = z + z^p + \ldots + z^{p^{m-1}}$$

S. Boztaş and I.E. Shparlinski (Eds.): AAECC-14, LNCS 2227, pp. 268–277, 2001.

be the trace of $z \in \mathbb{K}$ in $\mathbb{F}$, see [16] for this and other basic notions of the theory of finite fields. Also, let $\mathcal{G} \subseteq \mathbb{K}^*$ be a subgroup of the multiplicative group $\mathbb{K}^*$. We consider the following question: recover a number $\alpha \in \mathbb{K}$ such that for polynomially many known random $t \in \mathcal{G}$, approximations to the values of $\lfloor \mathrm{Tr}(\alpha t) \rfloor_p$ are known. Then we apply our results to obtaining a statement about the bit security of XTR [14,15], see also [3,24,26]. Unfortunately analogues of the bounds of exponential sums of Theorem 3.4 and Theorem 5.5 of [12] are not known for non-prime fields. Thus for subgroups of $\mathbb{K}^*$ we use a different method, which is based on bounds of [4,6] for the number of solutions of certain equations in finite fields. Unfortunately it produces much weaker results.

For a prime p and $k \geq 0$ we denote by $\mathrm{MSB}_{k,p}(x)$ any integer u such that

$$\left| \lfloor x \rfloor_p - u \right| \leq p/2^{k+1}. \tag{1}$$

Roughly speaking $\mathrm{MSB}_{k,p}(x)$ gives k most significant bits of x, however this definition is more flexible and suits better to our purposes. In particular we remark that k in the inequality (1) need not be integer. We remark that here the notion of most significant bits is tailored to modular residues and does not match the usual definition for integers.

Throughout the paper the implied constants in symbols 'O' depend on m and occasionally, where obvious, may depend on the small positive parameter ε; they all are effective and can be explicitly evaluated.

2   Distribution of Trace

The following bound on the number of zeros of sparse polynomials is a version of a similar result from [4,6]. We present it in the form given in [24].

Lemma 1. *Let $g \in \mathbb{K}^*$ be of multiplicative order T and let $s \geq 2$ be an integer. For elements $a_1, \ldots, a_s \in \mathbb{K}^*$ and s integers $e_1, \ldots, e_s$ we denote by W the number of solutions of the equation*

$$\sum_{i=1}^{s} a_i g^{e_i u} = 0, \qquad u \in [0, T-1].$$

Then $W \leq 3T^{1-1/(s-1)} E^{1/(s-1)}$, where $E = \min_{1 \leq i \leq s} \max_{j \neq i} \gcd(e_j - e_i, T)$.

For $\gamma \in \mathbb{K}$ and integers r and h we denote by $N_\gamma(\mathcal{G}, r, h)$ the number of solutions of the congruence

$$\mathrm{Tr}(\gamma t) \equiv r + y \pmod{p}, \qquad t \in \mathcal{G}, \; y = 0, \ldots, h-1.$$

From Lemma 1 one immediately derives an upper bound on $N_\gamma(\mathcal{G}, r, h)$.

Lemma 2. *For any $\gamma \in \mathbb{K}^*$ and any subgroup $\mathcal{G} \subseteq \mathbb{K}^*$,*

$$N_\gamma(\mathcal{G}, r, h) \leq 3h|\mathcal{G}|^{1-1/m} D^{1/m},$$

where $D = \max_{1 \leq \nu < m} \gcd(p^\nu - 1, |\mathcal{G}|)$.

Proof. Let g be a generator of $\mathcal{G}$, thus g is of order $T = |\mathcal{G}|$. Then for every $a \in \mathbb{F}$ the congruence

$$\mathrm{Tr}(\gamma t) \equiv a \pmod{p}, \qquad t \in \mathcal{G},$$

is equivalent to the congruence

$$\sum_{i=1}^{s} a_i g^{e_i u} = 0, \qquad u \in [0, t-1],$$

with $s = m+1$, $e_i = p^{i-1}$, $a_i = \gamma^{e_i}$ for $i = 1, \ldots, s-1$ and $a_s = -a$, $e_s = 0$. Because T is a divisor of $q-1$ we see that $\gcd(p, T) = 1$. Thus by Lemma 1 the number of solutions of the above congruence does not exceed $3T^{1-1/m} D^{1/m}$ for each $a \in \mathbb{F}$. Considering $a \in [r, r+h-1]$ we derive the result. $\square$

3   Lattices

As in [1,2], our results rely on rounding techniques in lattices. We therefore review a few related results and definitions.

Let $\{\mathbf{b}_1, \ldots, \mathbf{b}_s\}$ be a set of linearly independent vectors in $\mathbb{R}^s$. The set

$$L = \{\mathbf{z} \ : \ \mathbf{z} = c_1 \mathbf{b}_1 + \ldots + c_s \mathbf{b}_s, \quad c_1, \ldots, c_s \in \mathbb{Z}\}$$

is called an *s-dimensional full rank lattice* with *basis* $\{\mathbf{b}_1, \ldots, \mathbf{b}_s\}$.

It has been remarked in Section 2.1 of [17], then in Section 2.4 of [21] and then in Section 2.4 of [22] that the following statement holds, which is somewhat stronger than that usually used in the literature. It follows from the modification of [23] of the lattice basis reduction algorithm of [13] and some result of [10].

For a vector $\mathbf{u}$, let $\|\mathbf{u}\|$ denote its *Euclidean norm*.

Lemma 3. *There exists a deterministic polynomial time algorithm which, for a given s-dimensional full rank lattice L and a vector $\mathbf{r} \in \mathbb{R}^s$, finds a vector $\mathbf{v} \in L$ with*

$$\|\mathbf{v} - \mathbf{r}\| \leq \exp\left(O\left(\frac{s \log^2 \log s}{\log s}\right)\right) \min\{\|\mathbf{z} - \mathbf{r}\|, \quad \mathbf{z} \in L\}.$$

Let $\omega_1, \ldots, \omega_m$ be a fixed basis of $\mathbb{K}$ over $\mathbb{F}$.

For an integer $k \geq 0$ and $d \geq 1$ elements $t_1, \ldots, t_d \in \mathbb{K}$, we denote by $\mathcal{L}_k(t_1, \ldots, t_d)$ the $d+m$-dimensional lattice generated by the rows of the following $(d+m) \times (d+m)$-matrix

$$\begin{pmatrix} p & \ldots 0 & 0 & \ldots 0 \\ 0 & \ldots 0 & 0 & \ldots 0 \\ \vdots & \ddots & \vdots & \vdots \\ 0 & \ldots p & 0 & \ldots 0 \\ \mathrm{Tr}(\omega_1 t_1) & \ldots \mathrm{Tr}(\omega_1 t_d) & 1/2^{k+1} & \ldots 0 \\ \vdots & \vdots & \vdots & \ddots \vdots \\ \mathrm{Tr}(\omega_m t_1) & \ldots \mathrm{Tr}(\omega_m t_d) \ 0 & \ldots 1/2^{k+1} \end{pmatrix}. \qquad (2)$$

Lemma 4. *Let p be a sufficiently large n-bit prime number and let $\mathcal{G}$ be a subgroup of $\mathbb{K}^*$ of cardinality T with $T \geq q^\rho$ and $\max_{1 \leq \nu < m} \gcd(p^\nu - 1, |\mathcal{G}|) \leq T^{1-\tau}$ for some constants $\rho, \tau > 0$. Then for any $\varepsilon > 0$, $\eta = (1 - \rho\tau + \varepsilon)n + 6$, and $d = \lceil 2m/\varepsilon \rceil$ the following statement holds. Let*

$$\alpha = \sum_{j=1}^{m} a_j \omega_j \in \mathbb{K}, \qquad a_1, \ldots, a_m \in \mathbb{F},$$

be a fixed element of $\mathbb{K}$. Assume that $t_1, \ldots, t_d \in \mathcal{G}$ are chosen uniformly and independently at random. Then with probability $P \geq 1 - q^{-1}$ for any vector $\mathbf{s} = (s_1, \ldots, s_d, 0, \ldots, 0)$ with

$$\left(\sum_{i=1}^{d} \left(\mathrm{Tr}(\alpha t_i) - s_i \right)^2 \right)^{1/2} \leq 2^{-\eta} p,$$

all vectors

$$\mathbf{v} = (v_1, \ldots, v_d, v_{d+1}, \ldots, v_{d+m}) \in \mathcal{L}_k (t_1, \ldots, t_d)$$

satisfying

$$\left(\sum_{i=1}^{d} (v_i - s_i)^2 \right)^{1/2} \leq 2^{-\eta} p,$$

are of the form

$$\mathbf{v} = \left(\left\lfloor \left| \sum_{j=1}^{m} b_j \mathrm{Tr}\,(\omega_j t_1) \right| \right\rfloor_p, \ldots, \left\lfloor \left| \sum_{j=1}^{m} b_j \mathrm{Tr}\,(\omega_j t_d) \right| \right\rfloor_p, b_1/2^{k+1}, \ldots, b_m/2^{k+1} \right)$$

with some integers $b_j \equiv a_j \pmod{p}$, $j = 1, \ldots, m$.

Proof. As in [1] we define the modular distance between two integers r and l as

$$\mathrm{dist}_p(r, s) = \min_{b \in \mathbb{Z}} |r - l - bp| = \min \left\{ \lfloor r - l \rfloor_p, p - \lfloor r - l \rfloor_p \right\}.$$

We see from Lemma 2 that for any $\beta \in \mathbb{K}$ with $\beta \neq \alpha$ the probability $P(\beta)$ that

$$\mathrm{dist}_p (\mathrm{Tr}(\alpha t), \mathrm{Tr}(\beta t)) \leq 2^{-\eta+1} p$$

for $t \in \mathbb{K}$ selected uniformly at random is

$$P(\beta) \leq 3 \left(2^{-\eta+2} p + 1 \right) T^{-\tau/m} \leq 2^{-\eta+5} p T^{-\tau/m} \leq 2^{-\eta+5} p^{1-\rho\tau}$$
$$\leq 2^{-\eta-(1-\rho\tau)n+5} = 2^{-\varepsilon n - 1} \leq p^{-\varepsilon} = q^{-\varepsilon/m},$$

provided that p is large enough.

Therefore, choosing $d = \lceil 2m/\varepsilon \rceil$, for any $\beta \in \mathbb{K}$ we obtain

$$\Pr \left[\forall i \in [1, d] \mid \mathrm{dist}_p (\mathrm{Tr}(\alpha t_i), \mathrm{Tr}(\beta t_i)) \leq 2^{-\eta+1} p \right] = P(\beta)^d \leq q^{-\varepsilon d/m} \leq q^{-2},$$

where the probability is taken over $t_1, \ldots, t_d \in \mathcal{G}$ chosen uniformly and independently at random.

From here, we derive

$$\Pr\left[\forall \beta \in \mathbb{K}\backslash\{\alpha\}, \ \forall i \in [1,d] \ | \ \mathrm{dist}_p\left(\mathrm{Tr}(\alpha t_i), \mathrm{Tr}(\beta t_i)\right) \leq 2^{-\eta+1}p\right] \leq q^{-1}.$$

The rest of the proof is identical to the proof of Theorem 5 of [1]. Indeed, we fix some $t_1, \ldots, t_d \in \mathcal{G}$ with

$$\min_{\beta \in \mathbb{K}\backslash\{\alpha\}} \ \max_{i \in [1,d]} \ \mathrm{dist}_p\left(\mathrm{Tr}(\alpha t_i), \mathrm{Tr}(\beta t_i)\right) > 2^{-\eta+1}p. \tag{3}$$

Let $\mathbf{v} \in \mathcal{L}_k(t_1, \ldots, t_d)$ be a lattice point satisfying

$$\left(\sum_{i=1}^{d}(v_i - s_i)^2\right)^{1/2} \leq 2^{-\eta}p.$$

Since $\mathbf{v} \in \mathcal{L}_k(t_1, \ldots, t_d)$, there are integers $b_1, \ldots, b_m, z_1, \ldots, z_d$ such that

$$\mathbf{v} = \left(\sum_{j=1}^{m} b_j \mathrm{Tr}(\omega_j t_1) - z_1 p, \ldots, \sum_{j=1}^{m} b_j \mathrm{Tr}(\omega_j t_d) - z_d p, \frac{b_1}{2^{k+1}}, \ldots, \frac{b_m}{2^{k+1}}\right).$$

If $b_j \equiv a_j \pmod{p}$, $j = 1, \ldots, m$, then for all $i = 1, \ldots, d$ we have

$$\sum_{j=1}^{m} b_j \mathrm{Tr}(\omega_j t_i) - z_i p = \left\lfloor \left|\sum_{j=1}^{m} b_j \mathrm{Tr}(\omega_j t_i)\right| \right\rfloor_p = \mathrm{Tr}(\alpha t_i),$$

since otherwise there would be $i \in \{1, \ldots, d\}$ such that $|v_i - s_i| > 2^{-\eta}p$.

Now suppose that $b_j \not\equiv a_j \pmod{p}$ for some $j = 1, \ldots, m$. Put

$$\beta = \sum_{j=1}^{m} b_j \omega_j.$$

In this case we have

$$\left(\sum_{i=1}^{d}(v_i - s_i)^2\right)^{1/2}$$

$$\geq \max_{i \in [1,d]} \ \mathrm{dist}_p\left(\sum_{j=1}^{m} b_j \mathrm{Tr}(\omega_j t_i), s_i\right)$$

$$\geq \max_{i \in [1,d]} \left(\mathrm{dist}_p\left(\mathrm{Tr}(\alpha t_i), \sum_{j=1}^{m} b_j \mathrm{Tr}(\omega_j t_i)\right) - \mathrm{dist}_p(s_i, \mathrm{Tr}(\alpha t_i))\right)$$

$$\geq \max_{i \in [1,d]} \left(\mathrm{dist}_p\left(\mathrm{Tr}(\alpha t_i), \mathrm{Tr}(\beta t_i)\right) - \mathrm{dist}_p(s_i, \mathrm{Tr}(\alpha t_i))\right)$$

$$> 2^{-\eta+1}p - 2^{-\eta}p = 2^{-\eta}p$$

that contradicts our assumption. As we have seen, the condition (3) holds with probability exceeding $1 - q^{-1}$ and the result follows. $\qquad\square$

4   Trace Approximation Problem

Using Lemma 4 in the same way as Theorem 5 of [1] is used in the proof of Theorem 1 of that paper, we obtain

Theorem 1. *Let p be a sufficiently large n-bit prime number and let $\mathcal{G}$ be a subgroup of $\mathbb{K}^*$ with $|\mathcal{G}| \geq q^\rho$ and $\max_{1 \leq \nu < m} \gcd(p^\nu - 1, |\mathcal{G}|) \leq |\mathcal{G}|^{1-\tau}$ for some positive constants $\rho, \tau > 0$. Then for any $\varepsilon > 0$, $k = \lceil(1 - \rho\tau + \varepsilon)n\rceil$ and $d = \lceil 4m/\varepsilon \rceil$ the following statement holds. There exists a deterministic polynomial time algorithm $\mathcal{A}$ such that for any $\alpha \in \mathbb{K}$ given $2d$ values $t_i \in \mathcal{G}$ and $s_i = \mathrm{MSB}_{k,p}(\mathrm{Tr}(\alpha t_i))$, $i = 1, \ldots, d$, its output satisfies*

$$\Pr_{t_1, \ldots, t_d \in \mathcal{G}} [\mathcal{A}(t_1, \ldots, t_d; s_1, \ldots, s_d) = \alpha] \geq 1 - q^{-1}$$

if $t_1, \ldots, t_d$ are chosen uniformly and independently at random from $\mathcal{G}$.

Proof. We follow the same arguments as in the proof Theorem 1 of [1] which we briefly outline here for the sake of completeness. We refer to the first d vectors in the matrix (2) as p-vectors and we refer to the other m vectors as trace-vectors. Let

$$\alpha = \sum_{j=1}^{m} a_j \omega_j \in \mathbb{K}, \qquad a_1, \ldots, a_m \in \mathbb{F}.$$

We consider the vector $\mathbf{s} = (s_1, \ldots, s_d, s_{d+1}, \ldots, s_{d+m})$ where

$$s_{d+j} = 0, \qquad j = 1, \ldots, m.$$

Multiplying the jth trace-vector of the matrix (2) by a_j and subtracting a certain multiple of the jth p-vector, $j = 1, \ldots, m$, we obtain a lattice point

$$\mathbf{u}_\alpha = (u_1, \ldots, u_d, a_1/2^{k+1}, \ldots, a_m/2^{k+1}) \in \mathcal{L}_k(t_1, \ldots, t_d)$$

such that

$$|u_i - s_i| \leq p2^{-k-1}, \qquad i = 1, \ldots, d+m,$$

where $u_{d+j} = a_j/2^{k+1}$, $j = 1, \ldots, m$. Therefore,

$$\|\mathbf{u}_\alpha - \mathbf{s}\| \leq (d+m)^{1/2} 2^{-k-1} p.$$

Let

$$\eta = (1 - \rho\tau + \varepsilon/2)n + 6.$$

By Lemma 3 (used with a slightly rougher constant $2^{(d+m)}$) in polynomial time we can find $\mathbf{v} = (v_1, \ldots, v_d, v_{d+1}, \ldots, v_{d+m}) \in \mathcal{L}_k(t_1, \ldots, t_d)$ such that

$$\|\mathbf{v} - \mathbf{s}\| \leq 2^{d+m} \min\{\|\mathbf{z} - \mathbf{s}\|, \quad \mathbf{z} \in \mathcal{L}_k(t_1, \ldots, t_d)\}$$
$$\leq 2^{d+m-k-1}(d+m)^{1/2}p \leq 2^{-k+O(1)}p \leq 2^{-\eta-1}p,$$

provided that p is large enough. We also have

$$\left(\sum_{i=1}^{d} (u_i - s_i)^2\right)^{1/2} \le d^{1/2} 2^{-k-1} p \le 2^{-\eta-1} p.$$

Therefore,

$$\left(\sum_{i=1}^{d} (u_i - v_i)^2\right)^{1/2} \le 2^{-\eta} p.$$

Applying Lemma 4 (with $\varepsilon/2$ instead of ε), we see that $\mathbf{v} = \mathbf{u}_\alpha$ with probability at least $1 - q^{-1}$, and therefore the components $a_1, \ldots, a_m$ of α can be recovered from the last m components of $\mathbf{v} = \mathbf{u}_\alpha$. $\square$

5   Applications to XTR

We start with a brief outline of the XTR settings, concentrating only on the details which are relevant to this work.

Let $m = 6$, thus $\mathbb{K} = \mathbb{F}_{p^6}$. We also consider another field $\mathbb{L} = \mathbb{F}_{p^2}$, thus we have a tower of extensions $\mathbb{F} \subseteq \mathbb{L} \subseteq \mathbb{K}$. Accordingly, we denote by $\mathrm{Tr}_{\mathbb{K}/\mathbb{L}}(u)$ and $\mathrm{Tr}_{\mathbb{L}/\mathbb{F}}(v)$ the trace of $u \in \mathbb{K}$ in $\mathbb{L}$ and the trace of $v \in \mathbb{L}$ in $\mathbb{F}$. In particular, $\mathrm{Tr}_{\mathbb{L}/\mathbb{F}}\left(\mathrm{Tr}_{\mathbb{K}/\mathbb{L}}(u)\right) = \mathrm{Tr}(u)$ for $u \in \mathbb{K}$.

The idea of XTR is based on the observation that for some specially selected element $g \in \mathbb{K}^*$, which we call *the XTR generator* of prime multiplicative order $l > 3$ such that

$$l | p^2 - p + 1, \tag{4}$$

one can efficiently compute $\mathrm{Tr}_{\mathbb{K}/\mathbb{L}}(g^{xy})$ from the values of x and $\mathrm{Tr}_{\mathbb{K}/\mathbb{L}}(g^y)$ or, alternatively from the values of y and $\mathrm{Tr}_{\mathbb{K}/\mathbb{L}}(g^x)$. This allows us to reduce the size of messages to exchange (namely, just $\mathrm{Tr}_{\mathbb{K}/\mathbb{L}}(g^x)$ and $\mathrm{Tr}_{\mathbb{K}/\mathbb{L}}(g^y)$ rather than g^x and g^y) in order to create a common XTR key $\mathrm{Tr}_{\mathbb{K}/\mathbb{L}}(g^{xy})$.

As it follows from Theorem 24 of [26] (see also [3,14]) any polynomial time algorithm to compute $\mathrm{Tr}_{\mathbb{K}/\mathbb{L}}(g^{xy})$ from g^x and g^y can be used to construct a polynomial time algorithm to compute g^{xy} from the same information. In [24] the same result has been obtained with an algorithm which compute $\mathrm{Tr}_{\mathbb{K}/\mathbb{L}}(g^{xy})$ only for a positive proportion of pairs g^x and g^y. Furthermore, the same results hold even for algorithms which compute only $\mathrm{Tr}(g^{xy})$. We recall that any element $v \in \mathbb{L}$ can be represented by a pair $\left(\mathrm{Tr}_{\mathbb{L}/\mathbb{F}}(v), \mathrm{Tr}_{\mathbb{L}/\mathbb{F}}(\vartheta v)\right)$ where ϑ is a root of an irreducible quadratic polynomial over $\mathbb{F}$. Thus $\mathrm{Tr}(g^{xy})$ is a part of the representation of $\mathrm{Tr}_{\mathbb{K}/\mathbb{L}}(g^{xy})$. In fact the same result holds for $\mathrm{Tr}(\omega g^{xy})$ with any fixed $\omega \in \mathbb{K}^*$.

Thus the above results suggest that breaking XTR is not easier than breaking the classical Diffie–Hellman scheme.

Here we obtain one more result of this kind and show that even computing a certain positive proportion of bits of $\mathrm{Tr}(g^{xy})$ from $\mathrm{Tr}_{\mathbb{K}/\mathbb{L}}(g^x)$ and $\mathrm{Tr}_{\mathbb{K}/\mathbb{L}}(g^y)$ is as hard as breaking the classical Diffie–Hellman scheme. In fact we prove a stronger statement that computing a certain positive proportion of bits of $\mathrm{Tr}_{\mathbb{F}_{p^6}/\mathbb{F}_p}(g^{xy})$ from the values of g^x and g^y is as hard as computing g^{xy} from these values.

We remark that although this result is analogous to those known for the bit security of the Diffie–Hellman scheme [1,7] it is much weaker due to lack of non-trivial estimates of the appropriate exponential sums.

For a positive integer k we denote by $\mathcal{O}_k$ the oracle such that for any given values of g^x and g^y, it outputs $\mathrm{MSB}_{k,p}(\mathrm{Tr}(g^{xy}))$.

Theorem 2. *Let p be a sufficiently large n-bit prime number and let the order of the XTR generator satisfy the inequality $l \geq p^\lambda$. Then for any $\varepsilon > 0$ and $k = \lceil (1 - \lambda/6 + \varepsilon)n \rceil$, there exists a polynomial time algorithm which, given the values of $U = g^u$ and $V = g^v$, where $u, v \in [0, \ldots, l-1]$, makes $\lceil 24/\varepsilon \rceil$ calls of the oracle $\mathcal{O}_k$ and computes g^{uv} correctly with probability at least $1 - p^{-6}$.*

Proof. The case $u = 0$ is trivial. Now assume that $1 \leq u \leq l - 1$. Then $g_u = g^u$ is an element of multiplicative order l (because l is prime).

One easily verifies that (4) implies $\gcd(p^\nu - 1, l) = 1$, $\nu = 1, \ldots, 5$.

Select a random element $r \in [0, l - 1]$. Applying the oracle $\mathcal{O}_k$ to U and $V_r = g^{v+r} = Vg^r$ we obtain

$$\mathrm{MSB}_{k,p}\left(\mathrm{Tr}\left(g^{u(v+r)}\right)\right) = \mathrm{MSB}_{k,p}\left(\mathrm{Tr}(g^{vu}t)\right)$$

where $t = g_u^r$.

Selecting $d = \lceil 24/\varepsilon \rceil$ such elements $r_1, \ldots, r_d \in [0, l - 1]$ uniformly and independently at random we can now apply Theorem 1 with $\alpha = g^{uv}$, $m = 6$, $\rho = \lambda/6$, $\tau = 1$ and the group $\mathcal{G}$ generated by g_u (which coincides with the group generated by g). $\square$

In particular, if l is of order p^2, then we see that about 84% of the bits of $\mathrm{Tr}(g^{xy})$ (or about 42% of the bits which are needed to encode the private key $\mathrm{Tr}_{\mathbb{F}_{p^6}/\mathbb{F}_{p^2}}(g^{xy})$) are as hard as g^{xy}.

6   Remarks

We remark that because we assume m to be fixed the lattices which arise in our setting are of fixed dimension. Therefore one can use even exact algorithms to find the closest vector [11] instead of approximate lattice basis reduction based algorithms as we have done in Lemma 3. On the other hand, using Lemma 3 (or other similar statements) has an additional advantage that they allow us to study this problem when m grows (slowly) together with p because one can take ε as a slowly decreasing to zero function of p.

The method of this work is similar to that of [25], where a variant of Lemma 1 has been used for the problem of finding an m-sparse polynomial

$$f(X) = \alpha_1 X^{e_1} + \ldots + \alpha_m X^{e_m} \in \mathbb{F}[X]$$

from approximate values of $\lfloor f(t) \rfloor_p$ at polynomially many points $t \in \mathbb{F}$ selected uniformly at random.

Finally, probably the most challenging problem is to obtain a similar result for smaller values of k, say for $k = O(\log^{1/2} p)$ as it is known for the Diffie–Hellman scheme [1,7]. In fact for subgroups $\mathcal{G} \subseteq \mathbb{K}^*$ of order $T \geq q^{1/2+\varepsilon}$ this can easily be done using the bound

$$\max_{\gamma \in \mathbb{K}^*} \left| \sum_{t \in \mathcal{G}} \exp\left(2\pi i \mathrm{Tr}\left(\gamma t\right)/p\right) \right| \leq q^{1/2}$$

of exponential sums which is nontrivial only for such "large" subgroups, for example, see Theorem 8.78 in [16] (combined with Theorem 8.24 of the same work) or the bound (3.15) in [12]. However these subgroups are too large to be useful for applications to XTR which has been our principal motivation. On the other hand, Theorem 3.4 and Theorem 5.5 of [12] are nontrivial for much smaller subgroups but apply only to prime fields. To be more precise, Theorem 5.5 of [12] can be extended to composite fields but it does not appear to be enough for our applications. It seems only to imply that for infinitely many (rather than for all or almost all) pairs of primes (p, l) satisfying the above XTR constraints, about $\log^{1/2} p$ bits of $\mathrm{Tr}\left(g^{xy}\right)$ are as hard as g^{xy}.

Acknowledgement

The author thanks Frances Griffin for careful reading the manuscript and helpful remarks.

References

1. D. Boneh and R. Venkatesan, *Hardness of computing the most significant bits of secret keys in Diffie–Hellman and related schemes*, Lect. Notes in Comp. Sci., Springer-Verlag, Berlin, **1109** (1996), 129–142.
2. D. Boneh and R. Venkatesan, *Rounding in lattices and its cryptographic applications*, Proc. 8th Annual ACM-SIAM Symp. on Discr. Algorithms, ACM, NY, 1997, 675–681.
3. A. E. Brouwer, R. Pellikaan and E. R. Verheul, *Doing more with fewer bits*, Lect. Notes in Comp. Sci., Springer-Verlag, Berlin, **1716** (1999), 321–332.
4. R. Canetti, J. B. Friedlander, S. Konyagin, M. Larsen, D. Lieman and I. E. Shparlinski, *On the statistical properties of Diffie–Hellman distributions*, Israel J. Math., **120** (2000), 23–46.
5. E. El Mahassni, P. Q. Nguyen and I. E. Shparlinski, *The insecurity of Nyberg–Rueppel and other DSA-like signature schemes with partially known nonces*, Proc. Workshop on Lattices and Cryptography, Boston, MA, 2001 (to appear).
6. J. B. Friedlander, M. Larsen, D. Lieman and I. E. Shparlinski, *On correlation of binary M-sequences*, Designs, Codes and Cryptography, **16** (1999), 249–256.
7. M. I. González Vasco and I. E. Shparlinski, *On the security of Diffie-Hellman bits*, Proc. Workshop on Cryptography and Computational Number Theory, Singapore 1999, Birkhäuser, 2001, 257–268.

8. M. I. González Vasco and I. E. Shparlinski, *Security of the most significant bits of the Shamir message passing scheme,* Math. Comp. (to appear).

9. N. A. Howgrave-Graham and N. P. Smart, *Lattice attacks on digital signature schemes,* Designs, Codes and Cryptography, (to appear).

10. R. Kannan, *Algorithmic geometry of numbers,* Annual Review of Comp. Sci., **2** (1987), 231–267.

11. R. Kannan, *Minkowski's convex body theorem and integer programming,* Math. of Oper. Research, **12** (1987), 231–267.

12. S. V. Konyagin and I. Shparlinski, *Character sums with exponential functions and their applications,* Cambridge Univ. Press, Cambridge, 1999.

13. A. K. Lenstra, H. W. Lenstra and L. Lovász, *Factoring polynomials with rational coefficients,* Mathematische Annalen, **261** (1982), 515–534.

14. A. K. Lenstra and E. R. Verheul, *The XTR public key system,* Lect. Notes in Comp. Sci., Springer-Verlag, Berlin, **1880** (2000), 1–19.

15. A. K. Lenstra and E. R. Verheul, *Key improvements to XTR,* Lect. Notes in Comp. Sci., Springer-Verlag, Berlin, **1976** (2000), 220–233.

16. R. Lidl and H. Niederreiter, *Finite fields,* Cambridge University Press, Cambridge, 1997.

17. D. Micciancio, *On the hardness of the shortest vector problem,* PhD Thesis, MIT, 1998.

18. P. Q. Nguyen, *The dark side of the Hidden Number Problem: Lattice attacks on DSA,* Proc. Workshop on Cryptography and Computational Number Theory, Singapore 1999, Birkhäuser, 2001, 321–330.

19. P. Q. Nguyen and I. E. Shparlinski, *The insecurity of the Digital Signature Algorithm with partially known nonces,* Preprint, 2000, 1–26.

20. P. Q. Nguyen and I. E. Shparlinski, *The insecurity of the elliptic curve Digital Signature Algorithm with partially known nonces,* Preprint, 2001, 1–16.

21. P. Q. Nguyen and J. Stern, *Lattice reduction in cryptology: An update,* Lect. Notes in Comp. Sci., Springer-Verlag, Berlin, **1838** (2000), 85–112.

22. P. Q. Nguyen and J. Stern, 'The two faces of lattices in cryptology', *Proc. Workshop on Lattices and Cryptography, Boston, MA, 2001*, Springer-Verlag, Berlin, (to appear).

23. C. P. Schnorr, *A hierarchy of polynomial time basis reduction algorithms,* Theor. Comp. Sci., **53** (1987), 201–224.

24. I. E. Shparlinski, *Security of polynomial transformations of the Diffie–Hellman key,* Preprint, 2000, 1–8.

25. I. E. Shparlinski, *Sparse polynomial approximation in finite fields,* Proc. 33rd ACM Symp. on Theory of Comput., Crete, Greece, July 6-8, 2001, 209–215.

26. E. R. Verheul, *Certificates of recoverability with scalable recovery agent security,* Lect. Notes in Comp. Sci., Springer-Verlag, Berlin, **1751** (2000), 258–275.

CRYPTIM: Graphs as Tools
for Symmetric Encryption

Vasyl Ustimenko

Dept. of Mathematics and Computing Science
University of the South Pacific
ustimenk@manu.usp.ac.fj

Abstract. A combinatorial method of encryption is presented. The general idea is to treat vertices of a graph as messages and walks of a certain length as encryption tools. We study the quality of such an encryption in case of graphs of high girth by comparing the probability to guess the message (vertex) at random with the probability to break the key, i.e. to guess the encoding walk. In fact the quality is good for graphs which are close to the Erdös bound, defined by the Even Cycle Theorem. We construct special linguistic graphs of affine type whose vertices (messages) and walks (encoding tools) could be both naturally identified with vectors over $GF(q)$, and neighbors of the vertex defined by a system of linear equations. For them the computation of walks has a strong similarity with the classical scheme of linear coding. The algorithm has been implemented and tested.

Keywords: cryptography, constructive combinatorics, data communication, networks, security, privacy, e-commerce, virtual campus.

1   Introduction

The current work is motivated by security concerns on transmitting data across a University of the South Pacific (USP) intranet, called USPNet, (http://www.usp.ac.fj) designed and implemented to cater for a distance mode tele-education and associated administration - as an e-commerce application. USPNet is expected to facilitate teaching by making education accessible to and helping remove the 'tyranny of distance' between the twelve geographically remotely distributed member countries of the USP.

We have developed a prototype for CRYPTIM, a system to encrypt text and image data for transmission over the USPNet. The prototype is being used for investigation, evaluation and demonstration of the potential of a new encryption scheme. It has been implemented as a software package and trialled with the USPNet intranet.

2   Cryptosystem Requirements

Assume that an unencrypted message, *plaintext*, which can be text or image data, is a string of bits. It is to be transformed into an encrypted string or *ci-*

S. Boztaş and I.E. Shparlinski (Eds.): AAECC-14, LNCS 2227, pp. 278–286, 2001.
© Springer-Verlag Berlin Heidelberg 2001

phertext, by means of a cryptographic algorithm and a *key*: So that the recipient can read the message, encryption must be *invertible*.

Conventional wisdom holds that in order to defy easy decryption, a cryptographic algorithm should produce seeming chaos: that is, ciphertext should look and test random. In theory an eavesdropper should not be able to determine any significant information from an intercepted ciphertext. Broadly speaking, attacks to a cryptosystem fall into 2 categories: *passive attacks*, in which adversary monitors the communication channel and *active attacks*, in which the adversary may transmit messages to obtain information (e.g. ciphertext of chosen plaintext).

Passive attacks are easier to mount, but yields less. Attackers hope to determine the plaintext from the ciphertext they capture; an even more successful attacks will determine the key and thus comprise the whole set of messages.

An assumption first codified by Kerckhoffs in the nineteen century is that the algorithm is known and the security of algorithm rests entirely on the security of the key.

Cryptographers have been improving their algorithms to resist the following list of increasingly aggressive attacks:

i). *ciphertext only* – the adversary has access to the encrypted communications;

ii). *known plaintext* – the adversary has some plaintext and its corresponding ciphertext;

iii). *chosen text* – the adversary chooses the plaintext to be encrypted or the adversary picks the ciphertext to be decrypted (chosen ciphertext) or adversary chooses the plaintext to be encrypted depending on ciphertext received from previous requests (adaptive chosen plaintext).

Chosen-text attacks are largely used to simplify analysis of a cryptosystem.

In our system, we have considered a symmetric approach based on [8], [9] and [11]. It gives the method to develop a family of optimal algorithms, which are able to work efficiently with long keys and long messages. They have theoretically, a universal flexibility in the sense of sizes of the text and the key. Some of them are variations of *one time pad algorithms*, but others are "multi time pad" algorithms and can be even resistant to attacks of type (ii). Within the framework of the project "CRYPTIM" (abbreviation for enCRYPtion of Text and Image data), and supported by the USP Research Committee, algorithms have been implemented for usage in USPNet. Through this work, we are testing the resistance of the "multi time pad" algorithm to attacks of type (iii) and present the first results from such tests.

3   Walks on Graphs of Large Girth as Encoding Tools

One of the classical models of the procedure for encoding data is to present the information to be sent as a variety of n-tuples over the finite Galois field $GF(q)$. We have to "encode" our message x by taking an affine transformation $y = Ax + b$, where A is a certain matrix and b is another n-tuple.

Our proposal is based on the combinatorial method of construction of linear and nonlinear codes, which has a certain similarity with the classical scheme above.

Let Γ be a k-regular graph and $V(\Gamma)$ is the set of its vertices. Let us refer to the sequence $\rho = (v_1, v_2, \cdots, v_n)$, where $v_i \in V(\Gamma)$, $v_i \neq v_{i+2}$, $i = 1, \cdots, v_{n-2}$, and $v_i \Gamma v_{i+1}$, $i = 1, \cdots, n-1$, and $v_\rho = v_n$ as *encoding sequence* and *encoded vertex* of $v = v_1$. Clearly for $u = v_\rho$ there is sequence μ of length s such that $u_\mu = v$. We refer to μ as decoding sequence for v_ρ and write $\mu = \rho^{-1}$.

In the case of vertex transitive graphs set of all *encoding sequences* of certain length starting from the chosen vertex v_0 may be considered as the set of possible keys. To apply the key μ from this set to the vertex v means taking the last vertex of walk μ^g where g is the graph automorphism moving v_0 to v. In case of *parallelotopic graphs* defined below there exists a combinatorial way of description keys in a uniform way, which does not depend on starting vertex (or message).

The girth $g = g(\Gamma)$ of a graph Γ is the length of the shortest cycle in the graph.

If the length of the encoding sequence ρ of the k-regular graph Γ of girth $g = g(\Gamma)$ is less then g, then $v_\rho \neq v$ for any vertex v.

If one knows the length $t \leq g/2$ of the decoding sequence the probability of generating the correct message applying the encoding sequence at random is $1/(k(k-1)^{t-1})$. In this case the algorithm is $k(k-1)^t$ secure. We will use the term *graph encryption scheme* for the pair (Γ, t). It is reasonable to consider the following class of parallelotopic graphs.

Let Γ be a bipartite graph with partition sets P_i, $i = 1, 2$ (inputs and outputs) . Let M be a disjoint union of finite sets M_1 and M_2. We say that Γ is a bipartite parallelotopic graph over (M_1, M_2) if there exists a function $\pi : V(\Gamma) \to M$ such that if $p \in P_i$, then $\pi(p) \in M_i$ and for every pair (p, j), $p \in P_i$, $j \in M_i$, there is a unique neighbour u with given $\pi(u) = j$.

It is clear that the bipartite parallelotopic graph Γ is a $(|M_1|, |M_2|)$ - biregular graph.

So parallelotopic graph is just bipartite graph with special colourings for inputs outputs into $|M_1|$ and M_2 colours, respectively, such that for each vertex there exists a unique neighbour of the any given colour.

We refer also to the function π in the definition of bipartite parallelotopic graph also as a labelling. We will often omit the term "bipartite", because all our graphs are bipartite. In case of encryption scheme of bipartite graph we will use one of the partition set (inputs) as the textspace.

Linguistic graphs:

Let M^t be the Cartesian product of t copies of the set M. We say that the graph Γ is a *linguistic graph* over the set M with parameters m, k, r, s if

Γ is a bipartite parallelotopic graph over (V_1, V_2), $M_1 = M^r$, $M_2 = M^s$ with the set of points $I = M^m$ (inputs) and set of lines $O = M^k$ (outputs). (i.e. M^m and M^k are the partition sets of Γ). It is clear that $m + r = k + s$.

We use the term linguistic coding scheme for a pair (Γ, n), where Γ is linguistic graph and $n < g$ is the length of encoding sequences.

We choose a bipartite graph in the definition above because regular trees are infinite bipartite graphs and many biregular finite graphs of high girth can be obtained as their quotients (*homomorphic images*).

Using linguistic graphs our messages and coding tools are words over the *alphabet M* and we can use the usual matching between real information and vertices of our graph. In case of $M = GF(q)$ the similarity with the linear coding is stronger, because of our messages and keys are tuples over the $GF(q)$.

4   Absolutely Optimal Schemes from Graphs of Large Girth

One time pads. whose keys and strings of random bits at least as long as a message itself, achieve the seeming impossibility: an eavesdropper is not able to determine any significant information from an intersected ciphertext. The simplest classical example: if p_i is the i-th bit of the plaintext, k_i is the i-th bit of the key, and c_i is the first bit of the ciphertext, then $c_i = p_i + k_i$, where $+$ is exclusive or, often written XOR, and is simply addition modulo 2. One time pads must be used exactly once: if a key is ever reused, the system becomes highly vulnerable.

It is clear that encryption scheme above is irresistible to attacks of type (ii).

Families of one time pads can be constructed for the case, when the key space and the message space have the same magnitude. For theoretical studies of cryptographic properties of graph Γ we will always look at encryption scheme (Γ, t) , where $t = [g/2]$ and g is the girth of Γ.

Let Γ_i be an *absolutely optimal* family of graphs, i.e. family of graphs such that the ratio $p_{\text{key}}(i)/p_{\text{mes}}(i)$ of probabilities $p_{\text{key}}(i)$ and $p_{\text{mes}}(i)$ to guess the encoding sequence and to guess the message in the scheme (Γ_i, t_i), respectively, goes to 1 when i is growing.

The constructions of *absolutely optimal* families of schemes of high girth of increasing degree are connected with studies of some well-known problems in Extremal Graph Theory (see [2]). Let $ex(v, n)$ be, as usual, the greatest number of edges (size) in a graph on v vertices, which contains no cycles C_3, C_4, ..., C_n.

From Erdös' Even Cycle Theorem and its modifications [2] it follows that

$$ex(v, 2k) \le Cv^{1+1/k} \qquad\qquad 1$$

where C is a positive constant.

It is easy to see that the magnitude of the extremal family of regular graphs of given girth and of unbounded degree have to be on the Erdös upper bound (2.1). This bound is known to be sharp precisely when $k = 2, 3$, and 5. Thus the problem of constructing absolutely optimal families of high girth is a difficult one. It has been shown in [10] that the incidence graphs of simple groups of Lie type of rank 2 can be used as absolutely optimal encryption schemes with certain resistance to attacks of kind (i), examples of families of absolutely optimal coding schemes of parallelotopic graphs of girth 6, 8, 12 have been considered. Let us look at some of them.

Example 1

Let $P = \{(x_1, x_2, x_3, x_4, x_5) | x_i \in GF(q)\}$, $L = \{[y_1, y_2, y_3, y_4, y_5] | y_i \in GF(q)\}$. Let us define a bipartite graph I as: $(a, b, c, d, e) I[x, y, z, u, v]$ if and only if

$$y - b = xa$$
$$z - 2c = -2xb$$
$$u - 3d = -3xc$$
$$2v - 3e = 3zb - 3yc - ua$$

Input (a, b, c, d, e) and output $[x, y, z, u, v]$ are connected by edge in graph I iff the conditions above hold.

From the equations above, it follows that $\pi : \pi((x_1, x_2, x_3, x_4, x_5)) = x_1$ and $\pi([y_1, y_2, y_3, y_4, y_5]) = y_1$ is a labelling for the parallelotopic graph I.

It can be shown that for $\mathrm{char} GF(q) > 3$ the girth of this graph is at least 12. Directly from the equations above we can get that I defines the linguistic coding scheme with parameters $(1, 1, 5, 5)$ of affine type over $GF(q)$. It is clear that in case of encoding tuples of length 5 we get $p_{\text{key}} = 1/q(q-1)^4$, $p_{\text{mes}} = 1/q^5$ and $I = I_5(q)$ is an absolutely optimal family of linguistic graphs.

Example 2

Let $GF(q^2)$ be the quadratic extension of $GF(q)$ and $x \to x^q$ be the Frobenius automorphism of $GF(q^2)$. Let $P = \{(x_1, x_2, x_3) | x_1 \in GF(q), x_2 \in GF(q^2), x_3 \in GF(q)\}$, $L = \{[y_1, y_2, y_3] | y_1 \in GF(q^2), y_2 \in GF(q^2), y_3 \in GF(q)\}$. Let us define the bipartite graph $I = I_3(q)$ as: $(a, b, c) I[x, y, z]$ if and only if

$$y - b = xa$$
$$z - c = ay + ay^q.$$

It is clear that rules $\pi((x_1, x_2, x_3)) = x_1$ and $\pi([y_1, y_2, y_3]) = y_1$ define the parallelotopic scheme of affine type over the $GF(q)$ (but not over the $GF(q^2)$). Its parameters are $(1, 2, 4, 5)$. It can be shown that the girth of $I = I_3(q)$ is at least 8. It is easy to check that $I_3(q)$ is a family of linguistic absolutely optimal graphs.

Example 1 gives us families of graphs with sizes on the Erdös bound, and Example 2 gives examples of graphs with the sizes on the similar bound for biregular graphs of given degree. Both examples above are special induced subgraphs in the incidence graph of geometries finite simple groups of Lie type of rank 2, which are also form a families of absolutely optimal graphs. Such incidence geometries are not even a parallelotopic graphs, but there is an effective way to compute walks, based on the possibility of embedding of geometry into related Lie algebra [10].

For known absolutely optimal schemes of high girth with the resistance to attacks of type (ii) girth is ≤ 16. The problem of breaking the key is equivalent to solution of the system of nonlinear equations of degree $d(g)$ depending on the girth g. Absolutely optimal schemes of this kind can be used as blocks of larger coding schemes.

5   Optimal Schemes of Unbounded Girth

It is known that one time pads are impractical because in real life we need to deal with large amounts of information. A reasonable strategy is to consider the weaker requirement then equality of dimensions d_{key} of key space and dimension d_{pt} of plain text space. Let us consider the family of graphs Γ_i of increasing girth g_i such that for corresponding coding scheme $(\Gamma_i, t = [g_i/2])$ $\lim (p(i)_{\text{key}})^c / p(i)_{\text{mes}} = 1$, $i \to \infty$ where c is the constant which does not depend on i.

In this situation we say that the schemes of Γ_i form an *optimal family* of schemes. It is easy to check that in case of the optimal family of schemes corresponding to graphs of degree l_i and unbounded girth g_i we have

$$g_i \geq \gamma \log_{l_i - 1}(v_i) \qquad\qquad 2$$

The last formula means that Γ_i, $i = 1, \ldots$ form an infinite family of graphs of large girth in the sense of N. Biggs [1].

A few examples of such families are known (see [1] and [6]).

We have $\gamma \leq 2$, because of (1), but no family has been found for which $\gamma = 2$. Bigger γs correspond to more secure coding schemes. A. Lubotzky (see [7]) conjectured that $\gamma \leq 4/3$.

6   Folders

In practice for the encryption of large data by graph schemes (Γ, t) we need to lift the requirement that sizes of key space and text space are "close", t can be much smaller than half of girth Γ.

For the purpose of convenient encoding by graphs of "potentially infinite" text over a finite alphabet (like the External alphabet of a Turing machine), we need an infinite family of parallelotopic graphs of increasing girth, with a hereditary property: we can add a new part of text, and encode the entire text in a larger graph in such a way that the encoding of the initial part will be the same. This leads to the idea of a *folder* of parallelotopic graphs.

A surjective homomorphism $\eta : \Gamma_1 \to \Gamma_2$ of bipartite parallelotopic graphs Γ_i. $i = 1, 2$ with labelings π_1 and π_2, respectively, such that $\pi_2(\eta(v)) = \pi_1(v)$ is referred to as *parallelotopic morphism* of graphs.

A folder F is a family Γ_j, $j = 1, 2, \ldots$ of graphs and homomorphisms $t_{i,j}$ satisfying the following properties.

(P_1) The Γ_i are parallelotopic (or bipartite parallelotopic) graphs over a finite set M with local labellings denoted by π.

(P_2) For any pair i, j of positive integers, $i > j$, there is a parallelotopic morphism $t_{i,j}$ from Γ_i to Γ_j.

(P_3) $t_{i,j} \circ t_{j,k} = t_{i,k}$ for $i > j > k$ (commutative properties)

Let us assume the existence of the projective limit Γ of Γ_i. We refer to Γ as the cover of folder Γ_i.

If Γ is a forest we refer to the folder as a *free parallelotopic* folder. It is clear that in this case the Γ_i, $i = 1, \ldots$ form an infinite family of graphs of unbounded girth. There is a canonical parallelotopic morphism $t_i : \Gamma \to \Gamma_i$. If T is a connected component of the forest Γ then $t_i(T)$ is a connected component of $t_i(\Gamma)$ and family $t_i(T)$ is a free folder with the cover T.

Remark.

Let Γ_i be a free folder over the $GF(q)$, where the cover Γ is a q-regular tree. We could construct the "Theory of Γ_i-codes" in which the distance in the graph Γ_i would play the role of a Hamming metric in the classical case of linear codes. Of course, the Hamming metric is distance-transitive, i.e., for each k the automorphism group acts transitively on pairs of vectors at a distance k. The distance in the graph Γ_i may not be distance transitive, but we have an "asymptotical" distance transitivity, because of the distance transitivity of the tree Γ and the fact that $\lim(\Gamma_i) = \Gamma$.

The following statements justify the definition of folders.

Theorem 1 (see [9]) There exists a free folder of k-regular parallelotopic graphs for any $k > 2$.

Theorem 2 (see [9]) There exists a free folder of q-regular linguistic graphs of affine type over $GF(q)$ for any prime power $q > 2$.

In fact, Theorem 1 follows from Theorem 2 because we may always consider k -regular induced parallelotopic subgraphs defined by $\pi(v) \in K$, $|K| = k < q$ of the linguistic graph over $GF(q)$.

Explicit constructions of folders satisfying requirements of Theorem 2 will be presented in the next section.

7   Concluding Remarks

Explicit constructions of an optimal folders of linguistic graphs over the $M = GF(q)$ with good complexity of computation of walks had been considered in [9].

We are exploring one of them, which is the free folder of q-regular linguistic graphs $L_n(q)$ such that input $(x_1, x_2, \ldots, x_n) = (x)$ and output $[y_1, y_2, \ldots, y_n] = [y]$ are neighbors if $x_i - y_i = x_{k(i)}y_{s(i)}$ for $2 \leq i \leq n$, where $k(i) < i$, $s(i) \leq i$ and n can be any number, $\pi(x) = x_1$, $\pi([y] = y_1$.

Extremal properties of this family the reader can find in [6].

In fact the parallelotopic morphism of $L_n(q)$ onto $L_m(q)$, $n > m$ induced by canonical projecture of n-dimensional vector space onto m-dimensional. Each graph $L_n(q)$ is similar to the graph from Example 1 above.

Of course we are not computing the adjacency matrix, but have two affine operators $N(\alpha, (x))$ and $N(\alpha, [y])$ - compute the neighbour of (x) and $[y]$ with the first component α.

Let us compare our encryption with the following popular scheme of linear encryption:

We treat our message as a polynomial $f(x)$ over $GF(q)$ (our tuple is an array of coefficients of $f(x)$). The linear coding procedure is just a multiplication of our

$f(x)$ of degree $n-1$ by a polynomial $g(x)$, $\deg(g(x)) = t$, $t > 0$. Thus, y is just an array of coefficients of the polynomial $F(x) = f(x)g(x)$, $m = \deg F(x) = n+t-1$.

It is clear that this symmetric encoding is irresistible to attacks of type (ii) and sizes of plaintext and ciphertext are different. Counting of operation in case of equal dimensions of the plaintext and the ciphertext for the classical scheme as above and our scheme corresponding to $L_n(q)$ where q is a prime shows that our encryption is faster.

We are exploring a straightforward approach to look at what kind of finite automaton (roughly graph) we need for encryption.

The development of prototype CRYPTIM allows us to test the resistance of the algorithm above to attacks of different time. Our initial results from such tests show that the results are encouraging ([4]). Let us consider, for example, case of $p = 127$ (size of ASCII alphabet minus "delete" character). Let $t(k, l)$ be time (in seconds) we need to encrypt (or decrypt because of symmetry) file, size of which is k kilobites with password of length l (key space roughly 2^{7l})) by a Pentium II. Then some values of $t(k, l)$ can be presented by the following matrix

k\ l	9	13	17	21	15
1	1	1	1	2	2
2	2	3	3	4	4
3	4	6	8	9	11
4	16	16	23	30	33
5	22	27	35	44	52
6	38	54	64	88	105

The proposed algorithm is robust and compares well with the performance of some existing algorithms, at least in case of $L_n(q)$ over the prime q.

Acknowledgements

The work on CRYPTIM is supported by a research grant from the University of the South Pacific. Useful discussions were held with colleagues at the 1998 Joint Algebra Colloquium of Imperial College, Queen Mary and King's College (London) and Brian Hartley Seminar at (joint math. seminar of Manchester University and UMIST). Insightful comments were also received from participants at the 2000 Conference on Algebraic Combinatorics (Santa Cruz, California). Software developers of Synopsis Computer Corporation and Quickturn Company (California).

References

1. N.L. Biggs, *Graphs with large girth*, Ars Combinatoria, 25C (1988), 73–80.
2. B. Bollobás, *Extremal Graph Theory*, Academic Press.
3. Don Coppersmith, *The Data Encryption Standard (DES) and its strength against attacks, IBM J. Res Dev.*, 38 (1994), 243-250.

4. S. Fonua, V. Ustimenko and D. Sharma, *A System for Secure data Transmission within Corporate Networks: A Case Study of USPNet* MSc Thesis - in progress, Department of Mathematics and Computing, 2000.

5. S. Landau. *Standing the Test of Time:The Data Encryption Standard,* Notices of the AMS, March 2000, pp 341-349.

6. F. Lazebnik, V. A. Ustimenko and A. J. Woldar, *A New Series of Dense Graphs of High Girth,* Bull (New Series) of AMS, v.32, N1, (1995), 73-79.

7. A. Lubotzky, *Discrete Groups, Expanding graphs and Invariant Measures,* Progr. in Math., 125, Birkhoiser, 1994.

8. V. A. Ustimenko, *Random Walks on special graphs and Cryptography,* AMS Meeting, Louisville, March , 1998.

9. V. A. Ustimenko, *Coordinatization of regular tree and its quotients,* In the volume "Voronoi's Impact in Modern Science": (Proceedings of Memorial Voronoi Conference, Kiev, 1998), Kiev, IM AN Ukraine, July, 1998, pp. 125 - 152.

10. V. A. Ustimenko, *On the Varieties of Parabolic Subgroups, their Generalizations and Combinatorial Applications,* Acta Applicandae Mathematicae 52 (1998): pp. 223–238.

11. V. Ustimenko and D. Sharma, *Special Graphs in Cryptography* in Proceedings of 2000 International Workshop on Practice and Theory in Public Key Cryptography (PKC 2000), Melbourne, December 1999.

12. V. Ustimenko and D. Sharma, *CRYPTIM: a system to encrypt text and image data in E-commerce for the South Pacific,* Proceedings of International ICSC Symposium on Multi Agents and Mobile Agents in Virtual Organizations and E-Commerce (MAMA 2000). December 11 -13, Wollongong, Australia.

An Algorithm for Computing Cocyclic Matrices Developed over Some Semidirect Products

V. Álvarez, J.A. Armario, M.D. Frau, and P. Real*

Dpto. Matemática Aplicada I, Universidad de Sevilla, Avda. Reina Mercedes s/n
41012 Sevilla, Spain,
{valvarez,armario,mdfrau,real}@us.es

Abstract. An algorithm for calculating a set of generators of representative 2-cocycles on semidirect product of finite abelian groups is constructed, in light of the theory over cocyclic matrices developed by Horadam and de Launey in [7,8]. The method involves some homological perturbation techniques [3,1], in the homological correspondent to the work which Grabmeier and Lambe described in [12] from the viewpoint of cohomology. Examples of explicit computations over all dihedral groups D_{4t} are given, with aid of MATHEMATICA.

1 Introduction

Let G be a group, U a trivial G-module. Functions $\psi\colon G \times G \to U$ which satisfy $\psi(a,b)\psi(ab,c) = \psi(b,c)\psi(a,bc)$, $a,b,c \in G$ are called 2-cocycles [19]. A cocycle is a coboundary $\delta\alpha$ if it is derived from a set mapping $\alpha\colon G \to U$ having $\alpha(1) = 1$ by $\delta\alpha(a,b) = \alpha(a)^{-1}\alpha(b)^{-1}\alpha(ab)$. For each G and U, the set of cocycles forms an abelian group $Z^2(G,U)$ under pointwise multiplication, and the coboundaries form a subgroup $B^2(G,U)$. Two cocycles ψ and ψ' are cohomologous if there exists a coboundary $\delta\alpha$ such that $\psi' = \psi \cdot \delta\alpha$. Cohomology is an equivalence relation and the cohomology class of ψ is denoted $[\psi]$. It follows that the quotient group $Z^2(G,U)/B^2(G,U)$ consisting of the cohomology classes, forms an abelian group $H^2(G,U)$, which is known as the second cohomology group of G with coefficients in U. For each $n \geq 0$ one may define the cocycle analogous in dimension n (n-cocycle). In spite of the important role played by cocycles in Algebraic Topology, Representation Theory and Quantum Systems, the problem of explicitly determining a full representative set of n-cocycles for given G and U does not appear to have been traditionally studied by cohomologists, at least, till the last decade.

A 2-cocycle ψ is naturally displayed as a cocyclic matrix (associated to ψ, developed over G); that is, a $|G| \times |G|$ square matrix whose rows and columns are indexed by the elements of G (under some fixed ordering) and whose entry

---

* All authors are partially supported by the PAICYT research project FQM–296 from Junta de Andalucía and the DGESIC research project PB98–1621–C02–02 from Education and Science Ministry (Spain).

S. Boztaş and I.E. Shparlinski (Eds.): AAECC-14, LNCS 2227, pp. 287–296, 2001.
© Springer-Verlag Berlin Heidelberg 2001

in position (g, h) is $\psi(g, h)$. This notion was fruitfully used by Horadam and de Launey [6,7,15] proving some interesting connections between combinatorial design theory and 2-cocycles, as well as connections between coding theory and 2-cocycles. It is also apparent that cocyclic matrices, associated with cocycles with coefficients in $\mathbb{K}_2 = \{-1, 1\}$, account for large classes of so-called Hadamard matrices [8], and may consequently provide an uniform approach to the famous Hadamard conjecture.

These facts have yield that over the past decade considerable effort has been devoted to computations of cocycles and cocyclic matrices. Using classical methods involving the Universal Coefficient Theorem, Schur multipliers, inflation and transgression, two algorithms for finding 2-cocycles representing 2-dimensional cohomology classes can be worked out. The first one [7,8] applies to an abelian group G and the second [10] over groups G for which the word problem is solvable.

Horadam and de Launey's method is based on an explicit version of the well-known Universal Coefficient Theorem, which provides a decomposition of the second cohomology group into the direct sum of two summands,

$$H^2(G, U) \cong Ext(G/[G, G], U) \oplus Hom(H_2(G), U).$$

These connections make possible the translation of cocyclic development onto a (co)homological framework.

This link becomes stronger noting the "Bar construction" [19] related to G. It is a DG-module, which consists of the $\mathbb{Z}$–modules

$$M_0(G) = \mathbb{Z}, \quad M_m(G) = < [g_1, \ldots, g_m] : g_i \in G, 1 \leq i \leq m >,$$

and differential ∂,

$$\partial_1([g_1]) = 0, \quad \partial_{m+1}([g_1, \ldots, g_{m+1}]) = (-1)^{m+1}([g_1, \ldots g_m]) +$$

$$+ ([g_2, \ldots, g_{m+1}]) + \sum_{i=1}^{m} (-1)^i ([g_1, \ldots, g_i g_{i+1}, \ldots, g_{m+1}]).$$

The quotient $Ker(\partial_m)/Im(\partial_{m+1})$ is known to be the mth integral homology group of G, $H_m(G)$. Let $R_2(G)$ denote the quotient $M_2(G)/Im(\partial_3) \supseteq H_2(G)$.

Taking into account what ∂_3 means, it is readily checked that the map

$$\phi : Z^2(G, \mathbb{K}_2) \rightarrow Hom(R_2(G), \mathbb{K}_2)$$
$$h \mapsto \phi(h)$$

such that

$$\phi(h) \left(\sum_{(a,b) \in G \times G} \lambda_{(a,b)}(a, b) + Im(\partial_2) \right) = \sum_{(a,b) \in G \times G} \lambda_{(a,b)} h[a, b]$$

defines an isomorphism between the set of 2-cocycles and $Hom(R_2(G), \mathbb{K}_2)$ [7].

The problem of computing a set of generators for 2-cocycles hence translates to the problem of determining a set of *coboundary*, *symmetric* and *commutator* generators, such that

$$Z^2(G, \mathbb{K}_2) \cong B^2(G, \mathbb{K}_2) \oplus Ext_{\mathbb{Z}}(G/[G, G], \mathbb{K}_2) \oplus Hom(H_2(G), \mathbb{K}_2).$$

A minimal set for symmetric generators may be calculated from a primary invariant decomposition of $G/[G, G] \cong H_1(G)$, as a Kronecker product of back negacyclic matrices [7]. A minimal set for coboundary generators is derived from the multiplication table of G by means of linear algebra manipulations. But it is far from clear how to get a minimal set for commutator generators, in general. One should try to compute the second homology group of G by means of $(M_2, \partial_2, \partial_3)$. Indeed, ∂_2 is not needed for finite groups G, since $H_2(G)$ is a direct sum of finite cyclic groups as it is the case. This procedure is not suitable in practice, since matrices involved are large in most cases.

On the other hand, Flannery calculates these summands as the images of certain embeddings which are complementary, called inflation and transgression. Calculation of representative 2-cocycles associated to $Ext(G/[G, G], U)$ (inflation) is canonical. However, calculation of a complement of the image by the embeddings of inflation in $H^2(G, U)$ as the image of transgression is not canonical, anyway. As a matter of fact, it depends on the choice of a Schur complement. This is a potential source of difficulties in computation of representative 2-cocycles associated with elements of $Hom(H_2(G), U)$. This method has already been implemented in [11], using the symbolic computational system MAGMA.

Using a far different approach, Grabmeier and Lambe present in [12] alternate methods for calculating representative 2-cocycles for all finite p–groups from the point of view of Homological Perturbation Theory [13,14,20]. The computer algebra system AXIOM has been used in order to make calculations in practice.

Here we present a method for explicitly determining a full set of representative 2-cocycles for the elements of the second cohomology group $H^2(G, \mathbb{Z})$ where G is $\mathbb{Z}_r \times_\chi \mathbb{Z}_s$. All general statements given in this paper are applicable to any semidirect product of finite abelian groups, but for simplicity in the exposition, for this class, only the case $\mathbb{Z}_r \times_\chi \mathbb{Z}_s$ will be presented.

Our method could be seen as a mixture of both the algorithms given by Flannery in [10] and Grabmeier–Lambe in [12]. Indeed, we compute representative 2-cocycles proceeding from $Hom(H_2(\mathbb{Z}_r \times_\chi \mathbb{Z}_s), \mathbb{K}_2)$. This alternate method is based on some Homological Perturbation techniques developed in the work of authors [3,1] on the determination of "homological models"(those differential graded modules hG with $H_n(G) = H_n(hG)$, see [4] for instance), for semidirect products of finite abelian groups with group action. The algorithm is straightforward enough to be programmed in any computer algebra system, as we have done in MATHEMATICA[2].

The main steps are to define functions and $F : M_2(\mathbb{Z}_r \times_\chi \mathbb{Z}_s) \to V_2$ and $d_i : V_i \to V_{i-1}$, where V_i are certain "perturbed" simple algebras. These will be defined in such a way that for any representative 2-cycle z in the quotient $\ker d_2/\mathrm{Im}\, d_3$, the elevation of z through F will define a representative 2-cocycle. This is the homology analogous to the work of Grabmeier–Lambe in [12].

It should be noted that explicit formulae for representative 2-cocycles of $H^2(\mathbb{Z}_r \times_\chi \mathbb{Z}_s, \mathbb{Z}_2)$ are given in [21]. The approach explained in this paper covers the more general case of any semidirect product of finite abelian groups.

Similar algorithms may be considered to reach many other settings, progressing from any finite group with known homological model.

2 The Algorithm

Let $\mathbb{Z}_r \times_\chi \mathbb{Z}_s$ be a semidirect product, χ a group action such that

$$(a_1, b_1) \cdot (a_2, b_2) = (a_1 + \chi(b_1, a_2), b_1 + b_2), \quad a_1, a_2 \in \mathbb{Z}_r,\ b_1, b_2 \in \mathbb{Z}_s.$$

Let consider the following auxiliary sets

$$V_2 = \mathbb{Z}[x^2, xy, y^2], \quad V_3 = \mathbb{Z}[x^3, x^2y, xy^2, y^3],$$

$$B_2 = \{[n, m] \otimes [\,] : 1 \le n, m < r\} \cup \{[n] \otimes [m] : 1 \le n < r, 1 \le m < s\} \cup$$
$$\cup \{[\,] \otimes [n, m] : 1 \le n, m < s\},$$

$$B_3 = \{[n, m, k] \otimes [\,] : 1 \le n, m, k < r\} \cup \{[n, m] \otimes [k] : 1 \le n, m < r, 1 \le k < s\} \cup$$
$$\cup \{[n] \otimes [m, k] : 1 \le n < r, 1 \le m, k < s\} \cup \{[\,] \otimes [n, m, k] : 1 \le n, m, k < s\}.$$

We will define $\mathbb{Z}$–linear functions $g_3 : V_3 \to B_3$, $f_i : B_i \to V_i$ for $i = 2, 3$, $\phi_2 : B_2 \to B_3$, $\rho_3 : B_3 \to B_2$, $d_3 : V_3 \to V_2$ and $f_\infty : B_2 \to V_2$. Let

$$g_3(x^3) = ([1, 1, 1] + \cdots + [1, r-1, 1]) \otimes [\,],$$
$$g_3(x^2y) = ([1, 1] + \cdots + [1, r-1]) \otimes [1],$$
$$g_3(xy^2) = [1] \otimes ([1, 1] + \cdots + [1, s-1]),$$
$$g_3(y^3) = [\,] \otimes ([1, 1, 1] + \cdots + [1, r-1, 1])$$

$$f_2([n, m] \otimes [\,]) = x^2, \text{ if } n + m \ge r,$$
$$f_2([n] \otimes [m]) = (nm)\, xy,$$
$$f_2([\,] \otimes [n, m]) = y^2, \text{ if } n + m \ge s,$$

$$f_3([n, m, k] \otimes [\,]) = k\, x^3, \text{ if } n + m \ge r,$$
$$f_3([n, m] \otimes [k]) = k\, x^2y, \text{ if } n + m \ge r,$$
$$f_3([n] \otimes [m, k]) = n\, xy^2, \text{ if } m + k \ge s,$$
$$f_3([\,] \otimes [n, m, k]) = k\, y^3, \text{ if } n + m \ge s,$$

$$\phi_2([n, m] \otimes [\,]) = -([1, 1, m] + \cdots [1, n-1, m]) \otimes [\,]$$
$$\phi_2([n] \otimes [m]) = -([1, 1] + \cdots [1, n-1]) \otimes [m] + [n] \otimes ([1, 1] + \cdots [1, m-1]),$$
$$\phi_2([\,] \otimes [n, m]) = -[\,] \otimes ([1, 1, m] + \cdots [1, n-1, m])$$

$$\rho_3([n, m] \otimes [k]) = [\chi(k, n), \chi(k, m)] \otimes [\,] - [n, m] \otimes [\,],$$
$$\rho_3([n] \otimes [m, k]) = [n] \otimes [k] - [\chi(m, n)] \otimes [k],$$

$$D_3(x^2y) = r\, xy,$$
$$D_3(xy^2) = -s\, xy.$$

These morphisms are understood to be zero otherwise. Let define

$$d_3 = D_3 + f_2\rho_3 g_3 - f_2\rho_3\phi_2\rho_3 g_3 + f_2(\rho_3\phi_2)^2\rho_3 g_3 - \cdots,$$

and $f_\infty : B_2 \to V_2$,

$$f_\infty = f_2 - f_2\rho_3\phi_2 + f_2(\rho_3\phi_2)^2 - \cdots$$

Geometric series of these types converge to define a map, as it is proved in the more general setting of generalized semidirect products of finite abelian groups in [1]. The fact is that ρ_* decreases the dimension on the second component, and ϕ_* either increments the dimension only on the first component or decreases the value of the element in the second component. Hence the composition $\phi_{i-1}\rho_i$ becomes nilpotent.

Notice that the sets B_i defined above consist of the products

$$B_i = \bigoplus_{0 \le j \le i} (M_j(\mathbb{Z}_r) \otimes M_{i-j}(\mathbb{Z}_s)).$$

There is a connecting map $F_2 : M_2(\mathbb{Z}_r \times_\chi \mathbb{Z}_s) \to B_2$, so that

$$F_2[(a_1, b_1), (a_2, b_2)] = [\,]\otimes[b_2, b_1] + 2[\chi(b_2, a_2)]\otimes[b_1] + 2[\chi(b_2, a_2), \chi(b_2 b_1, a_1)]\otimes[\,] -$$

$$- [\chi(b_2 b_1 b_2, a_2), \chi(b_2 b_1 b_2 b_1, a_1)] \otimes [\,] - [\chi(b_2 b_2, a_2)] \otimes [b_1].$$

Theorem 1. *Assume the notation above.*

1. $H_2(\mathbb{Z}_r \times_\chi \mathbb{Z}_s) = H_2(V_2)$, *which is computed from* d_3.
2. *The map* $F = f_\infty \circ F_2 : M_2(\mathbb{Z}_r \times_\chi \mathbb{Z}_s) \to V_2$ *induces an isomorphism in homology, such that for any* $z \in H_2(V_2)$ *the elevation of* z *through* F *defines a cocyclic matrix over* $\mathbb{Z}_r \times_\chi \mathbb{Z}_s$.

In [1] the authors find a homological model for semidirect products of finite abelian groups. In particular, attending to the groups $\mathbb{Z}_r \times_\chi \mathbb{Z}_s$, it is proved that $H_2(M_2(\mathbb{Z}_r \times_\chi \mathbb{Z}_s) = H_2(V_2)$. Moreover F is shown to induce an isomorphism in homology.

Nevertheless the formula for F is not explicitly given there, since it is complicated to give an explicit formula for f_∞ in the general case of semidirect products of groups.

It is a remarkable fact that for every finite group G, $H_2(G)$ is a finite abelian group [5]. This way, it is only needed d_3 in order to compute $H_2(V_2)$ by means of Veblen's algorithm [22].

This process consists in calculating the integer Smith normal form D of the matrix M representing d_3 with regards to basis $\mathcal{B} = \{x^3, x^2y, xy, y^3\}$ and $\mathcal{B}' = \{x^2, xy, y^2\}$.

Let $\mathcal{U} = \{u_1, u_2, u_3, u_4\}$ and $\mathcal{V} = \{v_1, v_2, v_3\}$ define these change basis, such that $D_{\mathcal{U},\mathcal{V}} = PM_{\mathcal{B},\mathcal{B}'}Q$, for appropriated change basis matrices P and Q.

Now we explain what we mean with "elevate z through F".

We want to determine all cocyclic matrices over $\mathbb{Z}_r \times_\chi \mathbb{Z}_s$. That is, all representative 2-cocycles of $\mathbb{Z}_r \times_\chi \mathbb{Z}_s$. Thus it suffices to calculate which x in $M_2(\mathbb{Z}_r \times_\chi \mathbb{Z}_s)$ are shown to give non trivial homological information in $H_2(V_2)$.

For each generator z in $H_2(V_2)$, the *elevation of z through F* relates to the set of elements in $M_2(\mathbb{Z}_r \times_\chi \mathbb{Z}_s)$ which projects onto z with 2-homological information. This can be achieved in two single elevations: one from V_2 to B_2, the other from B_2 to $M_2(\mathbb{Z}_r \times_\chi \mathbb{Z}_s)$.

¿From the theorem above, an algorithm for calculating representative 2-cocycles may be derived in a straightforward manner.

Notice that map F should be called the *universal 2–cochain*, following Grabmeier–Lambe's notation in [12].

Algorithm 1 INPUT DATA: *a semidirect product $\mathbb{Z}_r \times_\chi \mathbb{Z}_s$.*

Step 1. *Compute $d_3 : V_3 \to V_2$, the differential of the homological model of $\mathbb{Z}_r \times_\chi \mathbb{Z}_s$ in dimension 3.*

Step 2. *Compute $H_2(\mathbb{Z}_r \times_\chi \mathbb{Z}_s)$ and representative cycles from d_3.*

Step 3. *Elevate the representative cycles from $H_2(\mathbb{Z}_r \times_\chi \mathbb{Z}_s)$ to $M_2(\mathbb{Z}_r \times_\chi \mathbb{Z}_s)$ via F.*

OUTPUT DATA: *Set of commutator generators for a basis of cocyclic matrices over $\mathbb{Z}_r \times \mathbb{Z}_s$.*

It should be taken into account that **Step 2** often requires to compute the Smith normal form of the matrix corresponding to d_3, which is always of size 4×3, independently of indexes r and s of the factors. This is the fundamental improvement in the calculus of the commutator generators, since the size of matrices which arises from the complex (M_*, ∂_*) depends on the order of the group (the matrix corresponding to operator ∂_3 is of size $(rs)^3 \times (rs)^2$ for the semidirect product $\mathbb{Z}_r \times_\chi \mathbb{Z}_s$).

It may be possible to extend the Theorem 1 and its associated algorithm to other certain families of groups, with homological models already known, such as central extensions [18], finitely generated torsion free nilpotent groups [16], metacyclic groups [17] and many others. It is only needed to find explicit formulae for the analogous of maps F_2 and F.

3 An Example: Dihedral Groups $D_{2t \cdot 2}$

In this section we apply Algorithm 1 in the particular case of dihedral groups. A MATHEMATICA program is used, which authors provide in [2].

It should be noted that dihedral groups $D_{t \cdot 2}$ for odd values of t do not provide 2-homological information, since $H_2(D_{t \cdot 2})$ is known to be zero in this case.

Let $D_{2t \cdot 2} = \{(0,0), (1,0), \ldots, (2t-1,0), (1,1), \ldots, (2t-1,1)\}$,

$$\chi(0,n) = n, \qquad \chi(1,n) = 2t - n, \qquad \forall n \in \mathbb{Z}_{2t}.$$

An explicit formula for F can be worked out for these groups, so that if we define $\lambda : \mathbb{Z}_k \times \mathbb{Z}_k \to \mathbb{Z}_2$, $k \geq 2$, as $\lambda[x, y] = 1$ if $x + y \geq k$ and 0 otherwise, it is readily checked that

$$F[(a_1, b_1), (a_2, b_2)] = b_1 b_2 y^2 + 2b_1 \chi(b_2, a_2) xy + 2b_1(\chi(b_2, a_2) - 1)x^2 +$$

$$+ 2\lambda[\chi(b_2, a_2), \chi(b_2 b_1, a_1)]x^2 - \lambda[\chi(b_1, a_2), a_1]x^2 - a_2 b_1 xy - b_1(a_2 - 1)x^2.$$

Let consider the cases $t = 1$, $D_{2 \cdot 2} = \{(0, 0), (1, 0), (0, 1), (1, 1)\}$,
$t = 2$, $D_{4 \cdot 2} = \{(0, 0), (1, 0), (2, 0), (3, 0), (0, 1), (1, 1), (2, 1), (3, 1)\}$,
and $t = 6$, $D_{12 \cdot 2} = \{(0, 0), (1, 0), \ldots, (11, 0), (0, 1), (1, 1), \ldots, (11, 1)\}$.

Step 1. Compute d_3.

$d(V_3)$	$t = 1$	$t = 2$	$t = 6$
x^3	0	0	0
$x^2 y$	$2xy$	$2x^2 + 4xy$	$10x^2 + 12xy$
xy^2	$-2xy$	$-2x^2 - 4xy$	$-10x^2 - 12xy$
y^3	0	0	0

Step 2. Compute $H_2(D_{2t \cdot 2})$ and representative cycles from d_3.
In order to compute $H_2(D_{2t \cdot 2})$ in the cases $t = 1, 2, 6$, it is useful to calculate the Smith normal form $D_t = P_t M_t Q_t$ of the matrix M_t associated to d_3, with basis change matrices P_t and Q_t, respectively. In these cases,

	$t = 1$	$t = 2$	$t = 6$
D_t	$\begin{pmatrix} 2 & 0 & 0 \\ 0 & 0 & 0 \\ 0 & 0 & 0 \\ 0 & 0 & 0 \end{pmatrix}$	$\begin{pmatrix} 2 & 0 & 0 \\ 0 & 0 & 0 \\ 0 & 0 & 0 \\ 0 & 0 & 0 \end{pmatrix}$	$\begin{pmatrix} 2 & 0 & 0 \\ 0 & 0 & 0 \\ 0 & 0 & 0 \\ 0 & 0 & 0 \end{pmatrix}$
Q_t	$\begin{pmatrix} 0 & 1 & 0 \\ 1 & 0 & 0 \\ 0 & 0 & 1 \end{pmatrix}$	$\begin{pmatrix} 1 & -2 & 0 \\ 0 & 1 & 0 \\ 0 & 0 & 1 \end{pmatrix}$	$\begin{pmatrix} -1 & -6 & 0 \\ 1 & 5 & 0 \\ 0 & 0 & 1 \end{pmatrix}$

Hence, $H_2(D_{2t \cdot 2}) = \mathbb{Z}_2$ for $t = 1, 2, 6$ and the representative cycle is the first element in the new basis $\mathcal{U}$ of $\mathbb{Z}[V_2]$.
In order to translate to the basis $\mathcal{B}$ of $\mathbb{Z}[V_2]$ the homological information which $H_2(D_{2t \cdot 2})$ provides, it suffices to select the odd entries of each of the columns of Q_t corresponding to each representative cycle in the basis $\mathcal{U}$ (that is, to select which elements of $\mathbb{Z}[V_2]$ with regards to basis $\mathcal{B}$ have an odd entry in the position corresponding to a representative cycle with coordinates in basis $\mathcal{U}$). The homological information is concentrated in elements with coordinates $(-, n, -)_{\mathcal{B}}$ for odd values of n in the case $t = 1$, in elements $(n, -, -)_{\mathcal{B}}$ for odd values of n in the case $t = 2$, and in elements $(n, m, -)_{\mathcal{B}}$ for n, m of distinct parity in the case $t = 6$.

Step 3. Elevate the representative cycles from $H_2(D_{2t\cdot 2})$ to $M_2(D_{2t\cdot 2})$ via F. It suffices to detect which elements of $M_2(D_{2t\cdot 2})$ are carried out via F to elements $(-, n, -)_B$ for odd n ($t = 1$), $(n, -, -)_B$ for odd n ($t = 2$), and $(n, m, -)_B$ for n, m of distinct parity ($t = 6$). These elements indicate the positions in the $|D_{2t\cdot 2}| \times |D_{2t\cdot 2}|$ commutator cocyclic generator matrix which are not trivial.

In the case $t = 1$, we obtain the following elements:

$$[(0, 1), (1, 0)], [(1, 1), (1, 0)], [(0, 1), (1, 1)], [(1, 1), (1, 1)].$$

For $t = 2$,

$$[(1, 0), (3, 0)], [(1, 0), (3, 1)], [(2, 0), (2, 0)], [(2, 0), (3, 0)], [(2, 0), (2, 1)],$$
$$[(2, 0), (3, 1)], [(3, 0), (1, 0)], [(3, 0), (2, 0)], [(3, 0), (3, 0)], [(3, 0), (0, 1)],$$
$$[(3, 0), (2, 1)], [(3, 0), (3, 1)], [(0, 1), (2, 0)], [(0, 1), (2, 1)], [(1, 1), (1, 0)],$$
$$[(1, 1), (2, 0)], [(1, 1), (1, 1)], [(1, 1), (3, 1)], [(2, 1), (1, 0)], [(2, 1), (1, 1)],$$
$$[(3, 1), (1, 0)], [(3, 1), (3, 0)], [(3, 1), (1, 1)], [(3, 1), (3, 1)].$$

In the case $t = 6$, the elements which are carried out via F to elements $(n, m, -)_B$ for n, m of distinct parity are those $[(a_1, b_1), (a_2, b_2)]$ such that

$$\begin{cases} b_1 = 0,\ a_1 + a_2 > 11; \\ \text{or} \\ b_1 = 1,\ a_1 < a_2. \end{cases}$$

Output data: set of commutator generators for a basis of cocyclic matrices over $D_{2t\cdot 2}$, $t = 1, 2, 6$. Assuming $K^2 = 1$, we obtain

$t = 1$	$t = 2$	$t = 6$
$\begin{pmatrix} A_1 & A_1 \\ B_1 & B_1 \end{pmatrix}$	$\begin{pmatrix} A_2 & A_2 \\ B_2 & B_2 \end{pmatrix}$	$\begin{pmatrix} A_6 & A_6 \\ B_6 & B_6 \end{pmatrix}$

where

$$A_1 = \begin{pmatrix} 1 & 1 \\ 1 & 1 \end{pmatrix}, \quad B_1 = \begin{pmatrix} 1 & K \\ 1 & K \end{pmatrix}, \quad A_2 = \begin{pmatrix} 1 & 1 & 1 & 1 \\ 1 & 1 & 1 & K \\ 1 & 1 & K & K \\ 1 & K & K & K \end{pmatrix}, \quad B_2 = \begin{pmatrix} 1 & 1 & K & 1 \\ 1 & K & K & 1 \\ 1 & K & 1 & 1 \\ 1 & K & 1 & K \end{pmatrix},$$

$$A_6 = \begin{pmatrix} 1 & 1 & \cdots & 1 & 1 \\ 1 & 1 & \cdots & 1 & K \\ \vdots & \vdots & & \vdots & \vdots \\ 1 & 1 & \cdots & K & K \\ 1 & K & \cdots & K & K \end{pmatrix}, \quad B_6 = \begin{pmatrix} 1 & K & \cdots & K & K \\ 1 & 1 & \cdots & K & K \\ \vdots & \vdots & & \vdots & \vdots \\ 1 & 1 & \cdots & 1 & K \\ 1 & 1 & \cdots & 1 & 1 \end{pmatrix}.$$

Note that A_6 is usually called *back negacyclic*.

In general, it may be proved that for $t > 2$ the computation of $H_2(D_{2t\cdot 2})$ reduces to the matrices

$$D_t = \begin{pmatrix} 2 & 0 & 0 \\ 0 & 0 & 0 \\ 0 & 0 & 0 \\ 0 & 0 & 0 \end{pmatrix} \quad \text{and} \quad Q_t = \begin{pmatrix} -1 & -t & 0 \\ 1 & t-1 & 0 \\ 0 & 0 & 1 \end{pmatrix}$$

so that $H_2(D_{2t \cdot 2}) = \mathbb{Z}_2$ and the homological information is concentrated in elements with coordinates $(n, m, -)_B$ for n, m of distinct parity.

Hence, the set of commutator generators for a basis of cocyclic matrices over $D_{2t \cdot 2}$ reduces to $\left\{ \begin{pmatrix} A_t & A_t \\ B_t & B_t \end{pmatrix} \right\}$, where A_t is the correspondant back negacyclic matrix and B_t consists in the matrix whose rows are the ones of A_t displayed in reverse order.

It should be noted that the cocyclic matrices over dihedral groups have already been found from Flannery's techniques in [10].

Remark 1. The case $t = 2$ is also studied in [7], where the commutator generator is said to be

$$
\begin{pmatrix}
1 & 1 & 1 & 1 & 1 & 1 & 1 & 1 \\
1 & 1 & 1 & B & B & 1 & 1 & 1 \\
1 & 1 & B & B & B & 1 & 1 & B \\
1 & B & B & B & B & 1 & B & B \\
1 & 1 & 1 & 1 & 1 & 1 & 1 & 1 \\
1 & B & B & B & B & 1 & B & B \\
1 & 1 & B & B & B & 1 & 1 & B \\
1 & 1 & 1 & B & B & 1 & 1 & 1
\end{pmatrix}
$$

with $B^2 = 1$.

Both matrices differ in the (Hadamard) product of a coboundary generator C and a symmetric generator S, which are

$$
C = \begin{pmatrix}
1 & 1 & 1 & 1 & 1 & 1 & 1 & 1 \\
1 & A & 1 & A & A & A & 1 & 1 \\
1 & 1 & 1 & 1 & A & 1 & A & 1 \\
1 & A & 1 & A & A & 1 & 1 & A \\
1 & 1 & A & 1 & 1 & 1 & A & 1 \\
1 & A & 1 & 1 & A & 1 & 1 & 1 \\
1 & A & A & A & A & A & 1 & A \\
1 & 1 & 1 & A & A & 1 & 1 & 1
\end{pmatrix}, \quad
S = \begin{pmatrix}
1 & 1 & 1 & 1 & 1 & 1 & 1 & 1 \\
1 & D & 1 & D & 1 & D & 1 & D \\
1 & 1 & 1 & 1 & 1 & 1 & 1 & 1 \\
1 & D & 1 & D & 1 & D & 1 & D \\
1 & 1 & 1 & 1 & 1 & 1 & 1 & 1 \\
1 & D & 1 & D & 1 & D & 1 & D \\
1 & 1 & 1 & 1 & 1 & 1 & 1 & 1 \\
1 & D & 1 & D & 1 & D & 1 & D
\end{pmatrix},
$$

with $A = D = -1$.

The matrix C arises from any of the set map $\alpha_k : D_{4 \cdot 2} \to \mathbb{K}_2$, $k \in \{1, -1\}$,

$$\alpha(0, 0) = 1, \quad \alpha(1, 0) = -1, \quad \alpha(2, 0) = -1, \quad \alpha(3, 0) = 1,$$

$$\alpha(0, 1) = k, \quad \alpha(1, 1) = k, \quad \alpha(2, 1) = k, \quad \alpha(3, 1) = -k.$$

References

1. V. Álvarez, J.A. Armario, M.D. Frau and P. Real. Homology of semidirect products of finite abelian groups with group action. *Preprint* of Dpto. Matemática Aplicada I, University of Seville (Spain, 2001).

2. V. Álvarez, J.A. Armario, M.D. Frau and P. Real. Homology of semidirect products of groups: algorithms and applications. *Preprint* of Dpto. Matemática Aplicada I, University of Seville (Spain, 2001).

3. V. Álvarez, J.A. Armario, P. Real. On the homology of semi-direct products of groups. Colloquium on Topology, Gyula (Hungary, 1998).

4. J.A. Armario, P. Real and B. Silva. On p-minimal homological models of twisted tensor products of elementary complexes localized over a prime. *Contemporary Mathematics*, **227**, 303–314, (1999).

5. K.S. Brown. Cohomology of groups. *Graduate Texts in Math.*, **87**, Springer-Verlag, New York (1982).

6. W. de Launey and K.J. Horadam. A weak difference set construction for higher-dimensional designs. *Designs, Codes and Cryptography*, **3**, 75–87, (1993).

7. W. de Launey and K.J. Horadam. Cocyclic development of designs, *J. Algebraic Combin.*, **2** (3), 267–290, 1993. Erratum: *J. Algebraic Combin.*, (1), pp. 129, 1994.

8. W. de Launey and K.J. Horadam. Generation of cocyclic Hadamard matrices. *Computational algebra and number theory* (Sydney, 1992), volume **325** of *Math. Appl.*, 279–290. Kluwer Acad. Publ., Dordrecht, (1995).

9. D.L. Flannery. Transgression and the calculation of cocyclic matrices. *Australas. J. Combin.*, **11**, 67–78, (1995).

10. D.L. Flannery. Calculation of cocyclic matrices. *J. of Pure and Applied Algebra*, **112**, 181–190, (1996).

11. D.L. Flannery and E.A. O'Brien. Computing 2-cocycles for central extensions and relative difference sets. *Comm. Algebra*, **28(4)**, 1939–1955, (2000).

12. J. Grabmeier, L.A. Lambe. Computing Resolutions Over Finite p-Groups. *Proceedings ALCOMA'99*. Eds. A. Betten, A. Kohnert, R. Lave, A. Wassermann. *Springer Lecture Notes in Computational Science and Engineering*, Springer-Verlag, Heidelberg, (2000).

13. V.K.A.M. Gugenheim and L.A. Lambe. Perturbation theory in Differential Homological Algebra, I. *Illinois J. Math.*, **33**, 556–582, (1989).

14. V.K.A.M. Gugenheim, L.A. Lambe and J.D. Stasheff. Perturbation theory in Differential Homological Algebra II, *Illinois J. Math.*, **35** (3), 357–373, (1991).

15. K.J. Horadam and A.A.I. Perera. Codes from cocycles. *Lecture Notes in Computer Science*, volume **1255**, 151–163, Springer–Verlag, Berlin–Heidelberg–New York, (1997).

16. J. Huebschmann. Cohomology of nilpotent groups of class 2. *J. Alg.*, **126**, 400–450, (1989).

17. J. Huebschmann. Cohomology of metacyclic groups. *Transactions of the American Mathematical Society*, **328** (1), 1–72, (1991).

18. L.A. Lambe and J.D. Stasheff. Applications of perturbation theory to iterated fibrations. *Manuscripta Math.*, **58**, 367–376, (1987).

19. S. Mac Lane. Homology. *Classics in Mathematics* Springer-Verlag, Berlin, (1995). Reprint of the 1975 edition.

20. P. Real. Homological Perturbation Theory and Associativity. *Homology, Homotopy and Applications*, **2**, 51–88, (2000).

21. K. Tahara. On the Second Cohomology Groups of Semidirect Products. *Math Z.*, **129**, 365–379, (1972).

22. O. Veblen. Analisis situs. *A.M.S. Publications*, **5**, (1931).

Algorithms for Large Integer Matrix Problems

Mark Giesbrecht[1], Michael Jacobson, Jr.[2], and Arne Storjohann[1]

[1] Ontario Research Centre for Computer Algebra, University of Western Ontario,
London, ON, N6A 5B7, Canada
{mwg,storjoha}@scl.csd.uwo.ca
[2] Dept. of Computer Science, University of Manitoba,
Winnipeg, MB, R3T 2N2, Canada
jacobs@cs.umanitoba.ca

Abstract. New algorithms are described and analysed for solving various problems associated with a large integer matrix: computing the Hermite form, computing a kernel basis, and solving a system of linear diophantine equations. The algorithms are space-efficient and for certain types of input matrices — for example, those arising during the computation of class groups and regulators — are faster than previous methods. Experiments with a prototype implementation support the running time analyses.

1 Introduction

Let $A \in \mathbb{Z}^{n \times (n+k)}$ with full row-rank be given. The lattice $\mathcal{L}(A)$ is the set of all $\mathbb{Z}$-linear combinations of columns of A. This paper describes new algorithms for solving the following problems involving $\mathcal{L}(A)$: computing the Hermite basis, computing a kernel basis, and given an integer vector b, computing a diophantine solution x (if one exists) to the linear system $Ax = b$.

By Hermite basis of A we mean the unique lower-triangular matrix $H \in \mathbb{Z}^{n \times n}$ such that $\mathcal{L}(H) = \mathcal{L}(A)$ and each off-diagonal entry is nonnegative and strictly smaller than the positive diagonal entry in the same row. A kernel for A is an $N \in \mathbb{Z}^{(n+k) \times k}$ such that $\mathcal{L}(N) = \{v \in \mathbb{Z}^{n+k} \mid Av = 0\}$. The problem of computing H and N often occurs as a subproblem of a larger number-theoretic computation, and the input matrices arising in these applications often have some special properties. The algorithms we give here are designed to be especially efficient for an input matrix $A \in \mathbb{Z}^{n \times (n+k)}$ which satisfies the following properties:

- A is sparse. More precisely, let μ be the number of nonzero entries in A. Then $\mu = O(n^{1+\epsilon})$ for some $0 \le \epsilon < 1$.
- The dimension k of the kernel is small compared with n.
- Let l be the smallest index such that the principal $(n-l) \times (n-l)$ submatrix of the Hermite basis H of $\mathcal{L}(A)$ is the identity. Then l is small compared with n.

Sparse input-matrices which satisfy these conditions on k and l are typical in computations for computing class groups and regulators of quadratic fields

S. Boztaş and I.E. Shparlinski (Eds.): AAECC-14, LNCS 2227, pp. 297–307, 2001.
© Springer-Verlag Berlin Heidelberg 2001

using the algorithm described in [4,7]. The diagonal elements of the Smith form of the matrix yield the elementary divisors of the class group (i.e., they give the class group as a product of cyclic groups), and the kernel (in the case of real quadratic fields) is used to compute the regulator. In practice, the number of diagonal elements of the Hermite basis which are not one is rarely larger than the rank of the class group. Since class groups are often cyclic or very close to being cyclic (as predicted by the Cohen-Lenstra heuristics [1]), l is small as well. Thus, the algorithms described in this paper are especially effective for these types of input.

Many algorithms have been proposed for computing the Hermite basis; for a survey we refer to [12]. The algorithm proposed in [12] — which is deterministic and computes a unimodular transformation-matrix, but does not exploit the sparsity of A or the fact that l may be small — requires about $O(n^4 (\log \|A\|)^2)$ bit operations where $\|A\| = \max_{ij} |A_{ij}|$. Moreover, that algorithm requires intermediate storage for about $O(n^3 (\log \|A\|))$ bits. The algorithm we propose computes H in an expected number of about $O(\mu n^2 (\log \|A\|) + n^3 (\log \|A\|)^2 (l^2 + k \log \|A\|))$ bit operations. When A is sparse and k and l are small compared to n we essentially obtain an algorithm which requires about $O(n^3 (\log \|A\|)^2)$ bit operations. Moreover, the algorithm requires intermediate space for only about $O(n^2 \log \|A\|)$ bits, for both sparse and dense input matrices. However, in practice, when A is sparse the storage requirements are reduced by a factor of two.

Table 1. Running times: A constant size entries and $k < n$.

Section	Word operations	Type
§3 Permutation conditioning	$O(n^3)$	LV
§4 Leading minor computation	$O(\mu n^2 (\log n))$	LV
§5 Lattice conditioning	$O(kn^3 (\log n)^3)$	DET
§6 Kernel basis computation	$O(k^2 n^3 (\log n)^2)$	DET
§7 Hermite basis computation	$O(kn^3 (\log n)^3 + l^2 n^3 (\log n)^2)$	DET
§8 System solving	$O(n^3 (\log n)^2)$	DET

For the analyses of our algorithms we assume we are working on a binary computer which has words of length ω, and if we are working with an input matrix $A \in \mathbb{Z}^{n+(n+k)}$, that ω satisfies

$$\omega > \max \left(6 + \log \log \left((\sqrt{n} \|A\|)^n \right), 1 + \log(2(n^2 + n)) \right). \tag{1}$$

Primes in the range $2^{\omega - 1}$ and 2^ω are called *wordsize* primes. We assume that a wordsize prime can be chosen uniformly and randomly at unit cost. Complexity results will be given in terms of word operations. For a more thorough discussion of this model see the text [13].

The computation is divided into a number of phases. The first three phases (described in Sections 3, 4 and 5) can be viewed as precomputation. Once these are complete, computing a kernel and Hermite basis, as well as solving diophantine systems involving A, can be accomplished deterministically in the running times indicated in Table 1.

The first phase – permutation conditioning – is to find a wordsize prime p for which A has full row-rank modulo p and permute the columns via a permutation matrix P such that the principal $n \times n$ submatrix B_1 has generic rank-profile: $B = AP = [B_1|B_2]$. The inverse modulo p of B_1 is also computed during this phase.

The second phase – leading minor computation – is to compute the determinant d of B_1. This is the only phase where we exploit the possible sparseness of A to get a better asymptotic running-time bound. In practice, we use Wiedemann's algorithm modulo a collection of distinct primes; this is easy to parallelize.

The third phase – lattice conditioning – is to compute a $Q \in \mathbb{Z}^{k \times n}$ which is used to compress the information from the columns of B_2 with B_1 to obtain a single $n \times n$ matrix $B_1 + B_2 Q$ from which the Hermite basis of B can be recovered.

2   Preliminaries

We recall the notion of a recursive and iterated inverse. Let R be a commutative ring with identity.

Recursive Inverse

Suppose that $A \in \mathsf{R}^{n \times n}$ enjoys the special property that each principal minor is invertible over R. The recursive inverse is a data structure that requires space for only n^2 ring elements but gives us the inverse of all principal minors of A. By "gives us" the inverse we mean that we can compute a given inverse×vector or vector×inverse product in quadratic time — just as if we had the inverse explicitly.

For $i = 1, \ldots, n$ let A_i denote the principal $i \times i$ submatrix of A. Let d_i be the i-th diagonal entry of A. For $i = 2, \ldots, n$ let $u_i \in \mathsf{R}^{1 \times (i-1)}$ and $v_i \in \mathsf{R}^{(i-1) \times 1}$ be the submatrices of A comprised of the first $i - 1$ entries in row i and column i, respectively. In other words, for $i > 1$ we have

$$A_i = \left[\begin{array}{c|c} A_{i-1} & v_i \\ \hline u_i & d_i \end{array}\right].$$

The *recursive inverse* of A is the expansion

$$A^{-1} = V_n D_n U_n \cdots V_2 D_2 U_2 V_1 D_1 U_0 D_0, \tag{2}$$

where V_i, D_i and U_i are $n \times n$ matrix defined as follows. For $i = 1, 2, \ldots, n$ let $B_i = \mathrm{diag}(A_i^{-1}, I_{n-i}) \in \mathsf{R}^{n \times n}$. Then

$$B_1 = \left[\begin{array}{c|c} d_1^{-1} & \\ \hline & I_{n-1} \end{array}\right],$$

and for $i > 1$ we have

$$
B_i =
\overset{V_i}{\left[\begin{array}{c|c|c}
I_{i-1} & -B_{i-1}v_i & \\
\hline
 & 1 & \\
\hline
 & & I_{n-i}
\end{array}\right]}
\overset{D_i}{\left[\begin{array}{c|c|c}
I_{i-1} & & \\
\hline
 & (d_i - v_i u_i)^{-1} & \\
\hline
 & & I_{n-i}
\end{array}\right]}
\overset{U_i}{\left[\begin{array}{c|c|c}
I_{i-1} & & \\
\hline
 & -u_i & 1 \\
\hline
 & & I_{n-r}
\end{array}\right]}
B_{i-1}.
$$

The expression (2) for A^{-1} as the product of structured matrices has some practical advantages in addition to giving us the inverse of all principal submatrices. Suppose that A is sparse, with $O(n^{1+\epsilon})$ entries for some $0 \le \epsilon < 1$. Then the V_i will also be sparse and $A^{-1}v$ or $v^T A^{-1}$ for a given $v \in \mathsf{R}^{n \times 1}$ can be computed in $n^2/2 + O(n^{1+\epsilon})$ ring operations.

Iterated Inverse

Now, let $U \in \mathsf{R}^{n \times k}$ and $V \in \mathsf{R}^{k \times n}$ be given in addition to A. Suppose the perturbed matrix $A + UV$ is invertible. The iterated inverse is a data structure that gives us $(A + UV)^{-1}$ but requires only $O(n^2 k)$ ring operations to compute if we already have the inverse of A.

For $i = 0, 1, 2, \ldots, k$ let U_i and V_i be the submatrices of U and V comprised of the principal i columns and rows, respectively. Let u_i and v_i be the i-th column and row of U and V, respectively. Note that $u_i v_i$ is an $n \times n$ matrix over R while while $v_i u_i$ is a 1×1 matrix over R. For $i = 0, 1, \ldots, n$ suppose that $(A + U_i V_i)$ is invertible, and let $B_i = (A + U_i V_i)^{-1}$. Then $B_0 = A^{-1}$ and for $i > 0$ we have

$$
B_i = (I + \bar{u}v_i)B_{i-1} \quad \text{where} \quad \bar{u}_i = -1/(1 + v_i B_{i-1} u_i)B_{i-1}u_i \in \mathsf{R}^{n \times 1}.
$$

The vector $\bar{u}_i$ can be computed using B_{i-1} in $O(n^2 + ni)$ ring operations. Thus, if we start with B_0, we can compute the *iterated inverse* expansion

$$
(A + UV)^{-1} = (I + \bar{u}_k v_k) \cdots (I + \bar{u}_2 v_2)(I + \bar{u}_1 v_1)A^{-1}
$$

in $O(n^2 k + nk^2)$ ring operations. Using the iterated inverse, we can compute $(A + UV)^{-1}u$ or $u^T (A + UV)^{-1}$ for a given $u \in \mathsf{R}^{n \times 1}$ using $O(n^2 + nk)$ ring operations. Note that for our applications k is typically much smaller than n.

3 Permutation Conditioning

Let $A \in \mathbb{Z}^{n \times (n+k)}$ be given. Choose random wordsize primes in succession until a prime p is found for which A has full rank modulo p. The rank check is performed using gaussian elimination. The lower bound (1) on ω (the word length on the computer) ensures such a prime will be found in an expected constant number of iterations. Once a good prime is found, we can also compute a $(n + k) \times (n + k)$ permutation matrix P such that each principal submatrix of AP is nonsingular modulo p. Let $B = AP$. Let C be the modulo p recursive inverse of the principal $n \times n$ submatrix of B. We call the tuple (B, P, C, p) a *permutation conditioning* of A. Producing a permutation conditioning requires an expected number of $O(n^3 + n^2(\log \|A\|))$ word operations.

4   Computation of Leading Minor

Let (B, P, C, p) be a permutation conditioning of $A \in \mathbb{Z}^{n+(n+k)}$. Let B_1 be the principal $n \times n$ submatrix of B. Let μ be a bound on the number of nonzero entries in B_1 and let $d = \det B_1$. For a wordsize prime p, the image $d \bmod p$ can be computed in an expected number of $O(\mu(n + (\log \|A\|)))$ word operations using the method of Wiedemann [14]. Hadamard's bound gives $|d| \leq (\sqrt{n}\|A\|)^n$, so if we have images for at least $\lceil n(\log_2 \sqrt{n}\|A\|)/(\omega - 1)\rceil + 1 = O(n(\log n + \log \|A\|))$ distinct primes we can compute d using Chinese remaindering. We obtain the following.

Proposition 1. *The principal $n \times n$ minor of B can be computed using an expected number of $O(\mu n^2 (\log n + \log \|A\|) + \mu n (\log \|A\|)^2)$ word operations.*

Now assume we have computed $d = \det B_1$. Let $v \in \mathbb{Z}^{n \times 1}$ be the n-th column of I_n. Then the last entry of $B_1^{-1} dv$ will be the determinant of the principal $(n-1) \times (n-1)$ submatrix of B_1. The vector $B_1^{-1} dv$ is computed in $O(n^3 (\log n + \log \|A\|)^2)$ word operations using p-adic lifting as described in [2]. Because we have the recursive inverse of B_1, we get the following:

Proposition 2. *Let a permutation conditioning (B, P, C, p) together with the principal $t \times t$ minor of B be given, $t > 1$. Then the determinant of the principal $(t - 1) \times (t - 1)$ minor of B can be computed in $O(n^3 (\log n + \log \|A\|)^2)$ word operations.*

5   Lattice Conditioning

Let a permutation conditioning (B, P, C, p) of $A \in \mathbb{Z}^{n \times (n+k)}$ be given. Write $B = [\,B_1 | B_2\,]$ where B_1 is $n \times n$. Assume $d = \det B_1$ is also given. Recall that $\det \mathcal{L}(B)$ is the product of diagonal entries in the Hermite basis of B.

Definition 1. *A lattice conditioning of B is a tuple (Q, W, c) such that:*

- $Q \in \mathbb{Z}^{k \times n}$,
- $\gcd(c, pd^2) = \det \mathcal{L}(B)$ *where* $c = \det(B_1 + B_2 Q)$,
- W *is the modulo p iterated inverse of* $B_1 + B_2 Q$.

The purpose of a lattice conditioning is to compress the information from the extra columns B_2 into the principal n columns. Note that

$$[\,B_1 | B_2\,] \left[\begin{array}{c|c} I_n & \\ \hline Q & I_k \end{array}\right] = [\,B_1 + B_2 Q | B_2\,]$$

where the transforming matrix is unimodular. The condition $\gcd(c, pd^2) = \det \mathcal{L}(B)$ on c means that we can neglect the columns B_2 when computing the Hermite basis of B. Note that the condition $\gcd(c, d^2) = \det \mathcal{L}(B)$ would also suffice, but using the modulus pd^2 ensures that $B_1 + B_2 Q$ is nonsingular modulo p.

We have the following result, which follows from the theory of modulo d computation of the Hermite form described in [3], see also [12, Proposition 5.14]. Let (Q, W, c) be a lattice conditioning of B. Then

Lemma 1. $\mathcal{L}\left(\left[\,B_1 + B_2 Q | d^2 I\,\right]\right) = \mathcal{L}(B)$.

The algorithm to compute a lattice conditioning is easiest to describe recursively. Let $\bar{B}$ and $\bar{B}_2$ be the matrices B and B_2, respectively, but with the last column removed. Assume we have recursively computed a lattice conditioning $(\bar{Q}, \bar{W}, \bar{c})$ for $\bar{B}$. Let u be the last column of B. We need to compute a $v \in \mathbb{Z}^{1 \times n}$ such that $\gcd(c, pd^2)$ is minimized, where $c = \det(B_1 + \bar{B}_2 \bar{Q} + uv)$. Using the iterated inverse $\bar{W}$, compute $\bar{u} = (B_1 + \bar{B}_1 \bar{Q})^{-1} \bar{c} u$ using linear p-adic lifting. This costs $O(n^3 (\log n + \log \|A\|)^2)$ word operations. It is easy to derive from elementary linear algebra that $c = \bar{c} + v\bar{u}$. We arrive at the problem of computing v such that

$$\gcd(\bar{c} + v_1 \bar{u}_1 + v_2 \bar{u}_2 + \cdots v_n \bar{u}_n, d^2) = \gcd(\bar{c}, \bar{u}_1, \bar{u}_2, \ldots, \bar{u}_n, d^2). \qquad (3)$$

This problem, the "modulo N extended gcd problem" with $N = pd^2$, is studied in [11]. From [6] we know that there exists a v with entries bounded in magnitude by $O((\log d)^2)$. We may assume (by induction) the same bound for entries in $\bar{Q}$. Then $\|B_1 + \bar{B}_2 \bar{Q}\| = O(n(\log d)^2 (\log \|A\|))$ and Hadamard's bound gives that $\max(d, \bar{c}, \|\bar{u}\|) = O(n(\log n + \log \|A\|))$.

Lemma 2. *A solution $v \in \mathbb{Z}^{1 \times n}$ to the modulo pd^2 extended gcd problem (3) which satisfies $\|v\| = O((\log d)^2)$ can be computed in $O(n^2 (\log n + \log \|A\|)^2 + n^3 (\log n + \log \|A\|)^3)$ word operations.*

We obtain the following result.

Proposition 3. *Let a permutation conditioning (B, P, C, p) for $A \in \mathbb{Z}^{n+(n+k)}$ together with the principal $n \times n$ minor d of B be given. Suppose that $k < n$. Then a lattice conditioning (Q, W, c) for (B, P, C, p) which satisfies $\|Q\| = O((\log d)^2)$ can be computed in $O(kn^3 (\log n + \log \|A\|)^3)$ word operations.*

In practice, the code fragment below will compute a suitable $v \in \mathbb{Z}^{n \times 1}$ and c quickly. Correctness is easy to verify.

```
c ← c̄;    g ← gcd(c, pd²);
for i from 1 to n do
    v[i] ← 0; g ← gcd(g, ū[i]);
    while gcd(c, pd²) ≠ g do c ← c + ū[i];    v[i] ← v[i] + 1
```

6   Kernel Basis Computation

Let a permutation conditioning (B, P, C, p) of $A \in \mathbb{Z}^{n+(n+k)}$ be given. Write $B = \left[\,B_1 | B_2\,\right]$ where $B_1 \in \mathbb{Z}^{n \times n}$. Assume $d = \det B_1$ is also given. We want to compute a basis of the kernel of A, i.e., an $N \in \mathbb{Z}^{(n+k) \times k}$ such that $\mathcal{L}(N) = \{v \in \mathbb{Z}^{n+k} \mid Bv = 0\}$. Noting that $AN = 0$ if and only if $BP^{-1}N = 0$ shows it will be sufficient to compute a kernel basis of B.

The construction given in the next fact is classical. The bound is also easy to derive. See for example [12].

Fact 1. *Let* $X = B_1^{\text{adj}} B_2$ *and let* H *be the trailing* $k \times k$ *submatrix of the Hermite basis of* $\begin{bmatrix} B_1 | B_2 \\ \hline I \end{bmatrix}$. *Then a kernel basis for* B *is given by* $N = \begin{bmatrix} -XH(1/d) \\ \hline H \end{bmatrix}$. *Moreover,* $||N|| \le (\sqrt{n}||A||)^n$.

A happy feature of the basis given by Fact 1 is that it is canonical; it is the only basis which has trailing $k \times k$ submatrix in Hermite form. Suppose we had some other kernel basis $\bar{N}$ for B. Then we could construct H by transforming the trailing $k \times k$ block of $\bar{N}$ to Hermite form. We will use this observation in our construction of N. Recover X by solving the matrix system $B_1 X = dB_2$ using linear p-adic lifting. Let $M = \begin{bmatrix} -X \\ \hline dI \end{bmatrix} \in \mathbb{Z}^{(n+k) \times k}$. Then $BM = 0$. The following observation is well known.

Fact 2. *Let* $M \in \mathbb{Z}^{(n+k) \times k}$ *have rank* k *and satisfy* $BM = 0$. *If* $G \in \mathbb{Z}^{k \times k}$ *is such that* $\mathcal{L}(G^T) = \mathcal{L}(N^T)$ *then* MG^{-1} *is a basis for the kernel for* B.

Compute the Hermite basis G^T of M^T. Then MG^{-1} is a basis for the kernel of B. In particular dG^{-1} is integral and has each diagonal entry a divisor of d. Recover H by computing the Hermite form of of dG^{-1}. Recovering G and H is accomplished using the modulo d algorithm as described in [3] or [5]. The cost is $O(nk^2)$ operations with integers bounded in length by $\log|d| = O(n(\log n + \log||A||))$ bits, or $O(n^3k^2(\log n + \log||A||)^2)$ word operations. This also bounds the cost of constructing X and post-multiplying X by $H(1/d)$.

Proposition 4. *Let a permutation conditioning* (B, P, C, p) *for* $A \in \mathbb{Z}^{n+(n+k)}$ *together with the principal* $n \times n$ *minor of* B *be given. Then a kernel basis for* A *can be computed in* $O(k^2 n^3(\log n + \log||A||)^2)$ *word operations.*

7   Hermite Basis Computation

Recall that l is the minimal index such that the principal $(n - l) \times (n - l)$ submatrix of the Hermite basis of A is the identity. Our result is:

Proposition 5. *Let a permutation conditioning* (B, P, C, p) *for* $A \in \mathbb{Z}^{n+(n+k)}$ *together with the principal* $n \times n$ *minor* d *of* B *be given. Suppose* $k < n$. *Then the Hermite basis of* A *can be computed in in* $O(kn^3(\log n + \log||A||)^3 + l^2 n^3(\log n + \log||A||)^2)$ *word operations.*

Proof. (Sketch) Let $\bar{B}$ be the first l rows of $B_1 + QB_2$. Write $\bar{B}$ as $[\bar{B}_1 | \bar{B}_2]$ where $\bar{B}_1$ is $(n - l) \times (n - l)$. Find $\bar{d} = \det \bar{B}_1$ using $l - 1$ applications of Proposition 2. Let $\bar{C}$ be the recursive inverse of $\bar{B}_1$. (Note that we get $\bar{C}$ for free from C.) Compute a lattice conditioning $(\bar{Q}, \bar{W}, \bar{c})$ for $(\bar{B}, I_{n+k}, \bar{C}, p)$. Then $\gcd(\bar{c}, p\bar{d}^2) = 1$. Furthermore:

$$\begin{matrix} B \\ \begin{bmatrix} \bar{B}_1 | \bar{B}_2 \\ \hline * \ | \ * \end{bmatrix} \end{matrix} \begin{bmatrix} I_{n-l} & \\ \hline Q & I_{k+l} \end{bmatrix} \begin{bmatrix} (\bar{B}_1 + \bar{B}_2\bar{Q})^{-1}\bar{c} & \\ \hline & I_{k+l} \end{bmatrix} = \begin{bmatrix} \bar{c}I_{n-l} | \bar{B}_2 \\ \hline * \ | \ * \end{bmatrix}.$$

where the transformed matrix on the right can be computed in $O(kn^3(\log n + \log\|A\|)^2)$ word operations using p-adic lifting. By an extension of Lemma 1, the Hermite basis of this matrix augmented with d^2I will be the Hermite basis of B. The basis is computed using $O(nl^2)$ operations with integers bounded in length by $\log|d| = O(n(\log n + \log\|A\|))$ bits.

8   System Solving

Our result is:

Proposition 6. *Let the following (associated to an $A \in \mathbb{Z}^{n+(n+k)}$) be given:*

- *a permutation conditioning (B, P, C, p),*
- *the principal $n \times n$ minor d of B, and*
- *a lattice conditioning (Q, W, c) for (B, P, C, p) which satisfies $\|Q\| = O((\log d)^2)$.*

Then given a column vector $b \in \mathbb{Z}^{n+k}$, a minimal denominator solution to the system $Ax = b$ can be computed in $O(n^3(\log n + \log\|A\|)^2)$ word operations.

Proof. The technique is essentially that used in [9]; we only give the construction here. Write B as $B = \begin{bmatrix} B_1 | B_2 \end{bmatrix}$ where B_1 is $n \times n$. Compute $v = B_1^{-1}db$ and $w = (B_1 + B_2Q)^{-1}cb$. Find $s, t \in \mathbb{Z}$ such that $sd + tc = \gcd(d, c)$. Then

$$x = sP\begin{bmatrix} I_n \\ \hline \end{bmatrix} v + tP\begin{bmatrix} I_n \\ \hline Q \end{bmatrix} w$$

is a solution to $Ax = b$ with minimal denominator.

Note that there exists a diophantine solution to the system if and only if the minimal denominator is one.

9   Massaging and Machine Word Lifting

The algorithms in previous sections make heavy use of p-adic lifting to solve linear systems. For efficiency, we would like to always choose p to be a power of two. That is, $p = 2^\omega$ where ω is the length of a word on the particular architecture we are using, for example $\omega = 32$, 64, 128. Then the lion's share of computation will involve machine arithmetic.

Unfortunately, the input matrix A may not have full rank modulo two, causing the permutation conditioning described in Section 2 to fail. In this section we show how to transform A to a "massaged" matrix B of the same dimension as A but such that all leading minors of B are nonsingular modulo two. The massaged B can then be used as input in lieu of A.

The construction described here is in the same spirit as the Smith form algorithm for integer matrices proposed by [8] and analogous to the massaging process used to solve a linear polynomial system described in [10].

Definition 2. *A* massaging *of A is tuple* (B, P, G, C) *such that:*

- $G \in \mathbb{Z}^{n \times n}$ *is in Hermite form with each diagonal entry a power of two,*
- $G^{-1}A$ *is an integer matrix of full rank modulo two,*
- $(B, P, C, 2^\omega)$ *is a permutation conditioning of A.*

Now we describe an algorithm to compute a massaging. Let $\bar{A}$ be the submatrix of A comprised of the first $n-1$ rows. Recursively compute a massaging $(\bar{P}, \bar{G}, \bar{B}, \bar{C})$ for $\bar{A}$. Write $\bar{B} = [\bar{B}_1 | \bar{B}_2]$ where $\bar{B}_1$ has dimension $(n-1) \times (n-1)$. Let $b = [b_1 | b_2]$ be the last row of A where b_1 has dimension $n-1$. Consider the over-determined linear system $x[\bar{B}_1 | \bar{B}_2] = [b_1 | b_2]$. This system is necessarily inconsistent since we assumed that A has full row rank. But for maximal t, we want to compute an $x \in \{0, 1, \ldots, 2^t - 1\}^{n-1}$ such that $xB_1 \equiv b_1 \bmod 2^t$, $xB_2 \equiv b_2 \bmod 2^{t-1}$ and $xB_2 \not\equiv b_2 \bmod 2^t$. At the same time find an elementary permutation matrix E such that the first component of $(b_2 - xB_2)E$ is not divisible by 2^t. The computation of x and E is accomplished using linear p-adic lifting with $p = 2^\omega$; for a description of this see [2] or [9]. Set

$$G = \left[\begin{array}{c|c} \bar{G} & \\ \hline x & 2^t \end{array}\right], \quad P = \bar{P}\left[\begin{array}{c|c} I_{n-1} & \\ \hline & E \end{array}\right], \quad B = \left[\begin{array}{c|c} I_{n-1} & \\ \hline & 1/2^t \end{array}\right]\left[\begin{array}{c|c} I_{n-1} & \\ \hline -x & 1 \end{array}\right]\left[\begin{array}{c} B \\ \hline b \end{array}\right].$$

Update the recursive inverse to produce C as described in Section 2.

We now estimate the complexity of computing a massaging. By Hadamard's bound, $\log_2 \det G \le n(\log_2 \sqrt{n} + \log_2 \|A\|)$ which gives the worst-case bound

$$\lceil n + n \log_2(\sqrt{n}\|A\|)/\omega\rceil = O(n(\log n + \log \|A\|))$$

on the number of lifting steps. This a worst-case factor of only $O(\log n + \log \|A\|)$ more lifting steps than required to compute only a permutation conditioning.

The only quibble with massaging is that entries in B might be larger than entries in A. Recall that the parameter l is used to denote the smallest index such that the Hermite basis of A has principal $(n-l) \times (n-l)$ submatrix the identity. Then entries in the first $n-l$ rows of B are bounded by $\|A\|$. The bound

$$\|G^{-1}\| \le (l+1)^{(l+1)/2} \tag{4}$$

is easy to derive. It follows that $\|G^{-1}B\| \le n(l+1)^{(l+1)/2}\|A\|$. We remark that the bound (4) is pessimistic but difficult to improve substantially in the worst case. It is an unfortunate byproduct of the fact that the ring $\mathbb{Z}$ is archimedian. In practice, $\|G^{-1}\|$ is much smaller.

10   Implementation and Execution

All the algorithms described in the previous sections have been implemented in C using the GNU MP large integer package. While the implementation is still experimental, preliminary results are very encouraging for computing the determinant, kernel and Hermite form of matrices with the small k and l.

We have employed this code on matrices generated during the computation of class groups and regulators of quadratic fields using the algorithm described in [7]. This algorithm uses the index-calculus approach and is based largely on the self-initializing quadratic-sieve integer-factorization algorithm. As in the factoring algorithm, the matrices generated are very sparse, with on the order of only 0.5% of entries nonzero.

The kernel of the matrix is required to compute the regulator of a real quadratic field. In practice, only a few vectors in the kernel are sufficient for this purpose, so the dimension of the kernel is small. As noted earlier, the expected number of diagonal elements of the Hermite basis which are not 1 is also small. The algorithms described in this paper are especially effective for this type of input.

Timings

The following table summarizes some of the execution timings on input as described above. Times are in hours and minutes.

Input				Timings HH:MM				
n	$n+k$	l	$\% \mu$	Massaging	Det	Cond	Kernel	Hermite
6000	6178	1	.373	00:14	05:50	02:40	–	00:03
6000	6220	1	.460	00:17	06:33	03:10	–	00:03
5000	5183	0	.542	00:09	07:55	00:02	02:50	–
6000	6181	0	.473	00:15	27:15	00:04	05:07	–
8600	8908	0	.308	00:38	20:30	00:14	19:15	–
10500	10780	0	.208	01:09	68:06	00:15	36:40	–

All computations were performed on 866Mhz Pentium III processors with 256Mb of RAM. Machine word lifting was used. The times for the determinant computation represent total work done; each determinant was computed in parallel on a cluster of ten such machines.

The first two rows in the table correspond to input matrices from the computation of the class groups of two imaginary quadratic orders. In this case, there is no regulator and hence the kernel does not have to be computed. The remaining examples all arise from real quadratic fields. The Hermite basis was trivial for all theses examples, a fact which was immediately detected once the lattice determinant had been computed. The second example and the last example correspond to quadratic orders with 90 and 101 decimal-digit discriminants, respectively. These are the largest discriminants for which the class group and regulator have been computed to date.

For comparison, previous methods described in [7], and run on a 550Mhz Pentium, required 5.2 days to compute the determinant and Hermite form of the 6000 × 6220 matrix. The 6000 × 6181 matrix required 12.8 days of computing time to find the determinant, Hermite form and kernel on the same machine. Computation of the Hermite form of the 10500 × 10780 matrix required 12.1 days. In this latter case, the computation of the kernel was not possible without the new methods described in this paper.

References

1. H. Cohen and H. Lenstra, Jr. Heuristics on class groups of number fields. In *Number Theory, Lecture notes in Math.*, volume 1068, pages 33–62. Springer-Verlag, New York, 1983.

2. J. D. Dixon. Exact solution of linear equations using p-adic expansions. *Numer. Math.*, 40:137–141, 1982.

3. P. D. Domich, R. Kannan, and L. E. Trotter, Jr. Hermite normal form computation using modulo determinant arithmetic. *Mathematics of Operations Research*, 12(1):50–59, 1987.

4. J. L. Hafner and K. S. McCurley. A rigorous subexponential algorithm for computation of class groups. *J. Amer. Math. Soc.*, 2:837–850, 1989.

5. C. S. Iliopoulos. Worst-case complexity bounds on algorithms for computing the canonical structure of finite abelian groups and the Hermite and Smith normal forms of an integer matrix. *SIAM Journal of Computing*, 18(4):658–669, 1989.

6. H. Iwaniec. On the problem of Jacobsthal. *Demonstratio Mathematica*, 11(1):225–231, 1978.

7. M. J. Jacobson, Jr. *Subexponential Class Group Computation in Quadratic Orders*. PhD thesis, Technischen Universität Darmstadt, 1999.

8. F. Lübeck. On the computation of elementary divisors of integer matrices. *Journal of Symbolic Computation*, 2001. To appear.

9. T. Mulders and A. Storjohann. Diophantine linear system solving. In S. Dooley, editor, *Proc. Int'l. Symp. on Symbolic and Algebraic Computation: ISSAC '99*, pages 281–288. ACM Press, 1999.

10. T. Mulders and A. Storjohann. Rational solutions of singular linear systems. In C. Traverso, editor, *Proc. Int'l. Symp. on Symbolic and Algebraic Computation: ISSAC '00*, pages 242–249. ACM Press, 2000.

11. A. Storjohann. A solution to the extended gcd problem with applications. In W. W. Küchlin, editor, *Proc. Int'l. Symp. on Symbolic and Algebraic Computation: ISSAC '97*, pages 109–116. ACM Press, 1997.

12. A. Storjohann. *Algorithms for Matrix Canonical Forms*. PhD thesis, ETH – Swiss Federal Institute of Technology, 2000.

13. J. von zur Gathen and J. Gerhard. *Modern Computer Algebra*. Cambridge University Press, 1999.

14. D. Wiedemann. Solving sparse linear equations over finite fields. *IEEE Trans. Inf. Theory*, IT-32:54–62, 1986.

On the Identification of Vertices and Edges Using Cycles

Iiro Honkala[1] *, Mark G. Karpovsky[2], and Simon Litsyn[3]

[1] Department of Mathematics, University of Turku 20014 Turku, Finland,
honkala@utu.fi
[2] College of Engineering, Boston University, Boston, MA 02215, USA
markkar@bu.edu
[3] Department of Electrical Engineering–Systems,
Tel-Aviv University, Ramat-Aviv 69978, Israel
litsyn@eng.tau.ac.il

Abstract. The subgraphs C_1, C_2, ..., C_k of a graph G are said to identify the vertices (resp. the edges) of G if the sets $\{j : v \in C_j\}$ (resp. $\{j : e \in C_j\}$) are nonempty for all the vertices v (edges e) and no two are the same. We consider the problem of minimizing k when the subgraphs C_i are required to be cycles or closed walks. The motivation comes from maintaining multiprocessor systems, and we study the cases when G is the binary hypercube, or the two-dimensional p-ary space with respect to the Lee metric.

Keywords: Identification, cycle, binary hypercube, Hamming distance, Lee metric, graph, multiprocessor system.

1 Introduction

Assume that G is a finite, undirected graph, and that each vertex (node) contains a processor and each edge represents a connection (dedicated communication link) between two processors. We wish to maintain the system, and consider the case in which at most one of the processors (or alternatively, at most one of the connecting wires between the processors) is not working. We can send test messages and route them through this network in any way we choose. What is the smallest number of messages we have to send if based on which messages safely come back (i.e., the idea is that the messages are routed to eventually reach the starting point) we can tell which vertex (resp. edge) is broken (if any)? See [4], [25], [24].

A sequence $v_0 e_1 v_1 e_2 \ldots e_n v_n$ of vertices v_i in G and edges $e_i = (v_{i-1}, v_i)$ in G is called a *walk*. If $v_0 = v_n$, it is a *closed walk*. We would like to find a collection of closed walks $C_1, C_2, \ldots, C_k$ that together contain all the vertices (resp. edges) and moreover, the sets $\{j : v \in C_j\}$ (resp. $\{j : e \in C_j\}$) are all different. We denote the minimum cardinality k by $V^*(G)$ (resp. $E^*(G)$). Since a walk may

---

* Research supported by the Academy of Finland under grant 44002.

S. Boztaş and I.E. Shparlinski (Eds.): AAECC-14, LNCS 2227, pp. 308–314, 2001.

contain the same vertex and the same edge more than once, the minimum k remains the same even if we only require that the C_i's are connected.

For technical reasons ([4], [25]), we would like our closed walks to be *cycles*, i.e., closed walks $v_0 e_1 v_1 \ldots v_n$, $n \geq 3$, where $v_i \neq v_j$ whenever $i \neq j$, except that $v_0 = v_n$. We denote the minimum cardinality k in this second variant by $V(G)$ (resp. $E(G)$).

In Section 2 we assume that G is the binary hypercube $\mathbb{F}_2^n$, where $\mathbb{F}_2 = \{0, 1\}$. Its vertices are all the binary words in $\mathbb{F}_2^n$, and edge set consists of all pairs of vertices connecting two binary words that are Hamming distance one apart. We denote by $d(\mathbf{x}, \mathbf{y})$ the Hamming distance between the vectors $\mathbf{x}, \mathbf{y} \in \mathbb{F}_2^n$ and $w(\mathbf{x})$ the number of ones in $\mathbf{x}$. In Section 3 we consider the p-ary space $\mathbb{Z}_p^2$ with respect to the Lee metric.

The results of this paper are from [11], where more detailed proofs can be found. Various other identification problems have been considered, e.g., in [4], [16], [17], [18], [24], [25] and in [2], [3], [5], [6],[7], [8], [9], [10], [12], [13], [14], [19], [20], [21], [22].

2   Binary Hypercubes

In this section we assume that G is the binary hypercube $\mathbb{F}_2^n$ and denote $V(G)$ and $E^*(G)$ by $V(n)$ and $E^*(n)$.

For arbitrary sets, we have the following trivial identification theorem:

Theorem 1. *A collection A_1, A_2, $\ldots$, A_k of subsets of an s-element set S is called identifying, if for all $x \in S$ the sets $\{i : x \in A_i\}$ are nonempty and different. Given s, the smallest identifying collection of subsets consists of $\lceil \log_2(s+1) \rceil$ subsets.* □

Of course, both the vertex and edge identification problems are special cases of this problem, and for the binary hypercube with $k = 2^n$ vertices we get the lower bound

$$V(n) \geq \lceil \log_2(2^n + 1) \rceil = n + 1.$$

Theorem 2. $V(n) = n + 1$ *for all $n \geq 2$.*

Proof. We construct $n + 1$ cycles all starting from the all-zero vector $\mathbf{0}$. Let C_0 be any cycle starting from $\mathbf{0}$ which visits the all-one vector $\mathbf{1}$. Given i, $1 \leq i \leq n$, let C_i be a cycle which visits exactly once all the points whose i-th coordinate equals 0: this is simply an $(n-1)$-dimensional Gray code (see, e.g., [23, p. 155]).

These $n + 1$ cycles together have the required property: a point x lies in C_i if and only if $x_i = 0$. The cycle C_0 guarantees that also the all-one vector lies in at least one cycle. □

Consider now the edge identification problem using closed walks. The number of edges is clearly $n2^{n-1}$, and by Theorem,

$$E^*(n) \geq \lceil \log_2(n2^{n-1} + 1) \rceil = n + \lfloor \log_2 n \rfloor.$$

There is a construction [11] which uses $n + \lfloor \log_2 n \rfloor + 2$ closed walks, and we therefore get the following theorem.

Theorem 3. $n + \lfloor \log_2 n \rfloor \leq E^*(n) \leq n + \lfloor \log_2 n \rfloor + 2.$

3 The Case $G = \mathbb{Z}_p^2$ with Respect to the Lee Metric

In this section G is the set $\mathbb{Z}_p^n$ with respect to the Lee metric, i.e., two vertices $(x_1, x_2, \ldots, x_n)$ and $(y_1, y_2, \ldots, y_n)$ are adjacent if and only if $x_j - y_j = \pm 1$ for a unique index j and $x_i = y_i$ for all $i \neq j$.

We denote $V(G)$, $V^*(G)$ and $E^*(G)$ by $V(p, n)$, $V^*(p, n)$ and $E^*(p, n)$. The simple proof of the following theorem can be found in [11].

Theorem 4. $V^*(p, 1) = E^*(p, 1) = \lceil p/2 \rceil$ for all $p \geq 5$.

Consider the case $n = 2$. The graph can be drawn as a $p \times p$ grid (cf. Figure 1). For each i, denote by $(1, i)$, $(2, i)$, $\ldots$, (p, i) the vertices on the i-th horizontal row from the bottom. We operate on the coordinates modulo p. Each vertex (i, j) is adjacent to the four vertices $(i - 1, j)$, $(i + 1, j)$, $(i, j - 1)$ and $(i, j + 1)$. For instance, $(1, p)$ and $(1, 1)$ are adjacent.

Theorem 5. $\lceil 2 \log_2 p \rceil \leq V(p, 2) \leq 2 \lceil \log_2(p + 1) \rceil + 1$ for all $p \geq 4$.

Proof. Let D be the cycle $(1, p)$, $(2, p)$, $(2, p - 1)$, $\ldots$, moving alternately one step to the right and one step down until it circles back to $(1, p)$; and let E (resp. F) be the cycle obtained from D by shifting it down by one step (resp. two steps); cf. Figure 1 where $p = 12$.

Take $k = \lceil \log_2 p \rceil$, and let **A** be the $k \times p$ matrix whose i-th column is the binary representation of i. For $p = 12$, we have

$$\mathbf{A} = \begin{pmatrix} 0\,0\,0\,0\,0\,0\,0\,0\,1\,1\,1\,1 \\ 0\,0\,0\,0\,1\,1\,1\,1\,0\,0\,0\,0 \\ 0\,0\,1\,1\,0\,0\,1\,1\,0\,0\,1\,1 \\ 0\,1\,0\,1\,0\,1\,0\,1\,0\,1\,0\,1 \end{pmatrix}.$$

From each row A_i of **A** we form a cycle B_i as follows (cf. Figure 2):

The cycle starts from the vertex $(1, 1) \in D$, and moves to $(1, p)$ and $(2, p)$.

Assume that we currently lie in $(j, p - j + 2) \in D$.

– If the j-th bit of A_i is 0, then we take one step down and one step to the right to $(j + 1, p - j + 1) \in D$.

– If the j-th bit of A_i is 1, then we move up along the j-th column and circle round to the point $(j, p - j + 1)$ and take one step to the right to $(j + 1, p - j + 1) \in D$.

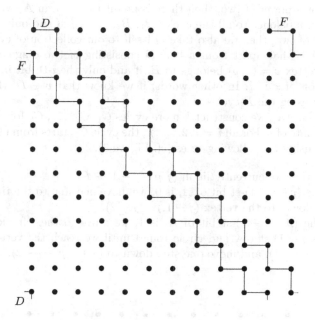

Fig. 1. The cycles D and F.

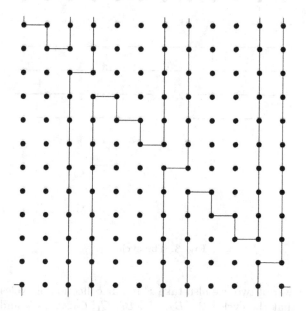

Fig. 2. The cycle B_3.

If p is not a power of two, then there is no all 1 column in $\mathbf{A}$, which implies that a vertex v belongs to all the cycles B_1, B_2, ..., B_k if and only if $v \in D$. If p is a power of two, then we also take D itself to our collection of cycles.

The idea is that apart from a belt D consisting of two diagonals, we can say that a vertex $v = (x, y)$ belongs to B_i if and only the i-th bit in the binary representation of x is 1. In other words, if we know that $v \notin D$, then we can determine x using the B-cycles.

In a similar way, we construct k more cycles C_1, C_2, ..., C_k for determining the y-coordinate of v. For all $i = 1, 2, \ldots, k$, the cycle C_i starts from $(1, p-2) \in F$ and is built using the following rules (cf. Figure 3):

Assume that we currently lie in $(j, p - j - 1) \in F$.
 – If the $(p - j - 1)$-st bit of A_i is 0, we move one step to the right and one down, to the vertex $(j + 1, p - j - 2)$.
 – If the $(p - j - 1)$-st bit of A_i is 1, we move to the left along the $(p - j - 1)$-st row and circle round until we reach the vertex $(j + 1, p - j - 1)$, and move one step down to $(j + 1, p - j - 2)$.

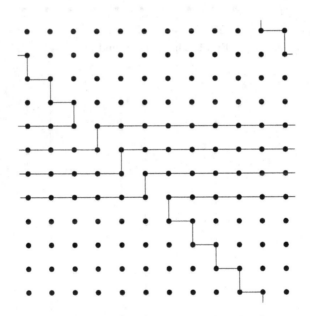

Fig. 3. The cycle C_2.

If p is a power of two, we also take F in our collection of cycles.

We claim that the cycles B_1, B_2, ..., B_k, C_1, C_2, ..., C_k and E, together with D and F if p is a power of two, have the required property. Clearly, we have the right number of cycles.

All vertices are in at least one of our cycles. The idenfication of the empty set is therefore clear. It suffices to show that we can identify an unknown vertex $v = (x, y)$ based on the information, which of our cycles it belongs to.

We first determine whether or not $v \in D$, and similarly whether or not $v \in F$. If neither, then the B-cycles tell us the x-coordinate and the C-cycles the y-coordinate of v, and we are done. Assume that $v \in D$. Since D and F have an empty intersection, we can use the C-cycles to determine the y-coordinate of v. There are only two vertices in D with a given y-coordinate, and exactly one of them belongs to E, so we can identify v. In the same way, if $v \in F$, then the B-cycles tell us the x-coordinate of v. Again the cycle E tells us, which one of the two remaining vertices in F with the same x-coordinate v is. $\square$

To identify the edges using closed walks, we have the following result from [11].

Theorem 6. $\lceil 2 \log_2 p \rceil + 1 \leq E^*(p, 2) \leq 2 \lceil \log_2 p \rceil + 2.$

References

1. U. Blass, I. Honkala, M. G. Karpovsky, S. Litsyn, Short dominating paths and cycles in binary hypercubes, Annals of Combinatorics, to appear.
2. U. Blass, I. Honkala, S. Litsyn, Bounds on identifying codes, Discrete Mathematics, to appear.
3. U. Blass, I. Honkala, S. Litsyn, On binary codes for identification, Journal of Combinatorial Designs 8 (2000), 151–156.
4. K. Chakrabarty, M. G. Karpovsky, L. B. Levitin, "Fault isolation and diagnosis in multiprocessor systems with point-to-point connections" Fault tolerant parallel and distributed systems, Kluwer (1998), pp. 285–301.
5. I. Charon, O. Hudry, A. Lobstein, Identifying codes with small radius in some infinite regular graphs, Electronic Journal of Combinatorics, submitted.
6. I. Charon, I. Honkala, O. Hudry, A. Lobstein, General bounds for identifying codes in some infinite regular graphs, Electronic Journal of Combinatorics, submitted.
7. I. Charon, I. Honkala, O. Hudry, A. Lobstein, The minimum density of an identifying code in the king lattice, Discrete Mathematics, submitted.
8. G. Cohen, I. Honkala, A. Lobstein, G. Zémor, On codes identifying vertices in the two-dimensional square lattice with diagonals, IEEE Transactions on Computers 50 (2001), 174–176.
9. G. Cohen, I. Honkala, A. Lobstein, G. Zémor, Bounds for codes identifying vertices in the hexagonal grid, SIAM Journal on Discrete Mathematics 13 (2000), 492–504.
10. G. Cohen, I. Honkala, A. Lobstein, G. Zémor, "On identifying codes," DIMACS Series in Discrete Mathematics and Theoretical Computer Science, Proceedings of the DIMACS Workshop on Codes and Association Schemes, November 9–12, 1999, pp. 97–109.
11. I. Honkala, M. Karpovsky, S. Litsyn: Cycles for identifying vertices and edges in graphs, submitted.
12. I. Honkala, T. Laihonen, S. Ranto, On codes identifying sets of vertices in Hamming spaces, Designs, Codes and Cryptography, to appear.

13. I. Honkala, T. Laihonen, S. Ranto, On strongly identifying codes, *Discrete Mathematics*, to appear.
14. I. Honkala, A. Lobstein: On the density of identifying codes in the square lattice, submitted.
15. M. G. Karpovsky, K. Chakrabarty, L. B. Levitin, On a new class of codes for identifying vertices in graphs, IEEE Transactions on Information Theory 44 (1998), 599–611.
16. M. G. Karpovsky, S. M. Chaudhry, L. B. Levitin, Detection and location of given sets of errors by nonbinary linear codes, preprint.
17. M. G. Karpovsky, L. B. Levitin, F. S. Vainstein, Diagnosis by signature analysis of test responses, IEEE Transactions on Computers 43 (1994), 1–12.
18. M. G. Karpovsky, S. M. Chaudhry, L. B. Levitin, "Multiple signature analysis: a framework for built-in self-diagnostic," Proceedings of the 22nd International Symposium on Fault-Tolerant Computing, July 8–10, 1992, Boston, Massachusetts (1992), pp. 112–119.
19. T. Laihonen, Sequences of optimal identifying codes, IEEE Transactions on Information Theory, submitted.
20. T. Laihonen, Optimal codes for strong identification, European Journal of Combinatorics, submitted.
21. T. Laihonen, S. Ranto, Families of optimal codes for strong identification, Discrete Applied Mathematics, to appear.
22. S. Ranto, I. Honkala, T. Laihonen, Two families of optimal identifying codes in binary Hamming spaces, IEEE Transactions on Information Theory, submitted.
23. Rosen, K. H. (ed.), Handbook of discrete and combinatorial mathematics, CRC Press, Boca Raton, 2000.
24. L. Zakrevski, M. G. Karpovsky, "Fault-tolerant message routing in computer networks," Proc. Int. Conf. on Parallel and Distributed Processing Techniques and Applications (1999), pp. 2279–2287.
25. L. Zakrevski, M. G. Karpovsky, "Fault-tolerant message routing for multiprocessors," Parallel and Distributed Processing J. Rolim (ed.), Springer (1998), pp. 714–731.

On Algebraic Soft Decision Decoding
of Cyclic Binary Codes

Vladimir B. Balakirsky

Euler Institute for Discrete Mathematics
and its Applications (EIDMA)
c/o Eindhoven University of Technology
P.O.Box 513, 5600 MB Eindhoven, The Netherlands

Abstract. We address a decoding problem for symmetric binary–input memoryless channels when data are transmitted using a cyclic binary code and show that some algebraic methods, like Berlekamp–Massey algorithm, developed for cyclic codes and binary symmetric channels can be effectively used to solve the problem.

1 Possible Extension of Operations in the Galois Field

Let us consider the Galois field $GF(2^m)$ introduced by the primitive polynomial $F(z) \overset{\triangle}{=} \sum_{j=0}^{m} F_j z^j$, where $F_0 = F_m = 1$ and $F_1, \ldots, F_{m-1} \in \{0, 1\}$. Let α denote the primitive element and let $0, \alpha^0, \ldots, \alpha^{2^m-1}$ be the elements of the field, which are represented as binary column–vectors of length m. Each non–zero element can be uniquely expressed as a linear combination of the basis $\alpha^0, \ldots, \alpha^{m-1}$.

Definition 1. *Let ω denote a formal variable. Introduce the infinite set constructed from the elements $0, \omega^\delta \alpha^0, \ldots, \omega^\delta \alpha^{m-1}$ for all $\delta \geq 0$ in such a way that*

$$(\omega^\delta \alpha^j) + 0 \overset{\triangle}{=} \omega^\delta \alpha^j$$

$$(\omega^\delta \alpha^j) 0 \overset{\triangle}{=} 0$$

$$\omega^\delta \alpha^j + \omega^\varepsilon \alpha^j \overset{\triangle}{=} 0$$

$$\omega^0 \alpha^j \overset{\triangle}{=} \alpha^j$$

$$\omega^\delta \alpha^j + \omega^\delta \alpha^{j'} \overset{\triangle}{=} \omega^\delta (\alpha^j + \alpha^{j'})$$

$$(\omega^\delta \alpha^j)^i \overset{\triangle}{=} \omega^{i\delta} \alpha^{ij}$$

$$(\omega^\delta \alpha^j)(\omega^\varepsilon \alpha^{j'}) \overset{\triangle}{=} \omega^{\delta+\varepsilon} \alpha^{j+j'}$$

$$\Big(\sum_{j \in \mathcal{J}} \omega^\delta \alpha^j \Big) \Big(\sum_{j' \in \mathcal{J}'} \omega^\varepsilon \alpha^{j'} \Big) \overset{\triangle}{=} \sum_{j \in \mathcal{J}} \sum_{j' \in \mathcal{J}'} (\omega^\delta \alpha^j)(\omega^\varepsilon \alpha^{j'})$$

and

$$\omega^\delta \alpha^j + \omega^\varepsilon \alpha^{j'} = \omega^\varepsilon \alpha^{j'} + \omega^\delta \alpha^j$$

where $\delta, \varepsilon \geq 0$; $j, j' \in \{0, \ldots, m-1\}$; $\mathcal{J}, \mathcal{J}' \subseteq \{0, \ldots, m-1\}$.

S. Boztaş and I.E. Shparlinski (Eds.): AAECC-14, LNCS 2227, pp. 315–322, 2001.
© Springer-Verlag Berlin Heidelberg 2001

Example 1. Let $m = 4, F(z) = 1 + z + z^4$. Then

$$\alpha^9 = \alpha^1 + \alpha^3, \quad \alpha^{13} = \alpha^0 + \alpha^2 + \alpha^3$$

and

$$w^\delta \alpha^9 = (w^\delta \alpha^1) + (w^\delta \alpha^3)$$
$$w^\varepsilon \alpha^{13} = (w^\varepsilon \alpha^0) + (w^\varepsilon \alpha^2) + (w^\varepsilon \alpha^3).$$

Therefore

$$(w^\delta \alpha^9) + (w^\varepsilon \alpha^{13}) = (w^\varepsilon \alpha^0) + (w^\delta \alpha^1) + (w^\varepsilon \alpha^2)$$
$$= (w^\delta \alpha^1) + (w^\varepsilon \alpha^8)$$
$$\left(w^\delta \alpha^9\right)\left(w^\varepsilon \alpha^{13}\right) = w^{\delta+\varepsilon} \alpha^7.$$

2 Statement of the Decoding Problem

Suppose that there is a discrete memoryless channel defined by the crossover probabilities $W_x(y), x \in \{0,1\}, y \in \mathcal{Y} \overset{\triangle}{=} \{0, \ldots, |\mathcal{Y}| - 1\}$ in a sense that

$$W(\mathbf{y}|\mathbf{x}) \overset{\triangle}{=} \prod_{j=0}^{n-1} W_{x_j}(y_j)$$

is the conditional probability to receive the vector $\mathbf{y} = (y_0, \ldots, y_{n-1}) \in \mathcal{Y}^n$ when the vector $\mathbf{x} = (x_0, \ldots, x_{n-1}) \in \{0,1\}^n$ was sent. We will assume that the channel is symmetric, i.e.,

$$W_0(y) = W_1(|\mathcal{Y}| - 1 - y), \quad \text{for all } y \in \mathcal{Y}. \tag{1}$$

Let a linear block code $\mathcal{C}$ having length n be used for data transmission. Suppose also that the decoder has to construct a codeword having the smallest distortion defined as

$$\delta(\mathbf{x}, \mathbf{y}) \overset{\triangle}{=} \sum_{j=0}^{n-1} \delta(x_j, y_j)$$

where $\delta(x, y) \geq 0$, $x \in \{0,1\}$, $y \in \mathcal{Y}$, are given by the $2 \times |\mathcal{Y}|$ distortion matrix. We will assume that the y-th column of this matrix has exactly one positive entry and denote it by $\delta(y)$ for all $y \in \mathcal{Y}$. For example, if $\mathcal{Y} = \{0, 1, 2, 3\}$, then the distortion matrix can be specified as

$$\begin{bmatrix} \delta(0,0) & .. & \delta(0,3) \\ \delta(1,0) & .. & \delta(1,3) \end{bmatrix} = \begin{bmatrix} 0 & 0 & \delta(2) & \delta(3) \\ \delta(0) & \delta(1) & 0 & 0 \end{bmatrix} = \begin{bmatrix} 0 & 0 & 1 & 4 \\ 4 & 1 & 0 & 0 \end{bmatrix} \tag{2}$$

Given $\mathbf{y} \in \mathcal{Y}^n$, let

$$y_j^* = \begin{cases} 0, & \text{if } \delta(0, y_j) = 0 \\ 1, & \text{if } \delta(1, y_j) = 0 \end{cases} \tag{3}$$

Then, as it is easy to see,

$$\delta(\mathbf{x}, \mathbf{y}) = \sum_{j \in \mathcal{J}(\mathbf{x}, \mathbf{y}^*)} \delta(y_j) \tag{4}$$

where

$$\mathcal{J}(\mathbf{x}, \mathbf{y}^*) = \left\{ j \in \{0, \ldots, n-1\} : x_j \neq y_j^* \right\}. \tag{5}$$

Constructing of the vector $\mathbf{y}^*$ corresponds to quantization of the components of the received vector $\mathbf{y}$. However, the maximum likelihood decoding algorithm applied to the vector $\mathbf{y}^*$ does not find the codeword minimizing the distortion for $\mathbf{y}$ in general case.

Example 2. Suppose that $\mathcal{C}$ is the (15,7) BCH code generated by the polynomial $g(z) = 1 + z^4 + z^6 + z^7 + z^8$ and having the parity check matrix

$$\mathbf{H} = \left[\begin{array}{ccccccccccccccc} 1 & 0 & 0 & 0 & 1 & 0 & 0 & 1 & 1 & 0 & 1 & 0 & 1 & 1 & 1 \\ 0 & 1 & 0 & 0 & 1 & 1 & 0 & 1 & 0 & 1 & 1 & 1 & 1 & 0 & 0 \\ 0 & 0 & 1 & 0 & 0 & 1 & 1 & 0 & 1 & 0 & 1 & 1 & 1 & 1 & 0 \\ 0 & 0 & 0 & 1 & 0 & 0 & 1 & 1 & 0 & 1 & 0 & 1 & 1 & 1 & 1 \\ \hline 1 & 0 & 0 & 0 & 1 & 1 & 0 & 0 & 0 & 1 & 1 & 0 & 0 & 0 & 1 \\ 0 & 0 & 0 & 1 & 1 & 0 & 0 & 0 & 1 & 1 & 0 & 0 & 0 & 1 & 1 \\ 0 & 0 & 1 & 0 & 1 & 0 & 0 & 1 & 0 & 1 & 0 & 0 & 1 & 0 & 1 \\ 0 & 1 & 1 & 1 & 1 & 0 & 1 & 1 & 1 & 1 & 0 & 1 & 1 & 1 & 1 \end{array} \right] \tag{6}$$

Note that the code contains the all–zero codeword $\mathbf{0}$ and the codeword $\mathbf{x}^*$ having components

$$x_j^* = \begin{cases} 0, \text{ if } j \notin \{0, 1, 2, 9, 13\} \\ 1, \text{ if } j \in \{0, 1, 2, 9, 13\}. \end{cases} \tag{7}$$

Let $\mathcal{Y} = \{0, 1, 2, 3\}$ and let the distortion matrix be defined by (2). If

$$y_j = \begin{cases} 0, \text{ if } j \notin \{0, 1, 2, 9, 13\} \\ 1, \text{ if } j \in \{0, 1, 2\} \\ 2, \text{ if } j = 9 \\ 3, \text{ if } j = 13 \end{cases} \tag{8}$$

and the decoder uses the maximum likelihood algorithm for the vector $\mathbf{y}^*$, then the all–zero codeword is decoded. However,

$$\delta(\mathbf{0}, \mathbf{y}) = \delta(y_9) + \delta(y_{13}) = 5$$
$$\delta(\mathbf{x}^*, \mathbf{y}) = \delta(y_0) + \delta(y_1) + \delta(y_2) = 3.$$

3   Possible Algebraic Approach to the Decoding Problem

Lemma. *Given a vector* $\mathbf{y} \in \mathcal{Y}^n$, *let*

$$(\delta_0, \ldots, \delta_{n-1}) \triangleq (\delta(y_0), \ldots, \delta(y_{n-1}))$$

and let

$$S_{i,\mathcal{J}}(\omega) \triangleq \sum_{j \in \mathcal{J}} (\omega^{\delta_j} \alpha^i)^j, \quad \text{for all } \mathcal{J} \subseteq \{0, \ldots, n-1\}. \tag{9}$$

If α^i is the root of the generator polynomial of the code $\mathcal{C}$, then

$$S_{i,\mathcal{J}(\mathbf{x},\mathbf{y}^*)}(\omega) = \sum_{j:\, y_j^* = 1} (\omega^{\delta_j} \alpha^j)^i \tag{10}$$

for all $\mathbf{x} \in \mathcal{C}$, where the set $\mathcal{J}(\mathbf{x}, \mathbf{y}^*)$ is defined in (5).

Proof : Let $\beta \triangleq \alpha^i$. Since $\sum_{j:\, x_j = 1} \beta^j = 0$, we obtain

$$\sum_{j:\, x_j = 1} \omega^{i\delta_j} \beta^j = \left(\sum_{j:\, x_j = 1} \omega^{i\delta_j} \beta^j \right) + \left(\sum_{j:\, x_j = 1} \beta^j \right)$$

$$= \left(\sum_{j:\, x_j = 1} \omega^{i\delta_j} \beta^j \right) + \left(\sum_{j:\, x_j = 1} \omega^0 \beta^j \right)$$

$$= \sum_{j:\, x_j = 1} \left[(\omega^{i\delta_j} \beta^j) + (\omega^0 \beta^j) \right] = \sum_{j:\, x_j = 1} 0 = 0.$$

Therefore

$$\sum_{j:\, y_j^* = 1} \omega^{i\delta_j} \beta^j = \left(\sum_{j:\, y_j^* = 1} \omega^{i\delta_j} \beta^j \right) + \left(\sum_{j:\, x_j = 1} \omega^{i\delta_j} \beta^j \right)$$

$$= \left(\sum_{j:\, x_j \neq y_j^*} \omega^{i\delta_j} \beta^j \right) + \left(\sum_{j:\, x_j = y_j^* = 1} \left[(\omega^{i\delta_j} \beta^j) + (\omega^{i\delta_j} \beta^j) \right] \right)$$

$$= \left(\sum_{j:\, x_j \neq y_j^*} \omega^{i\delta_j} \beta^j \right) + \left(\sum_{j:\, x_j = y_j^* = 1} 0 \right) = \sum_{j:\, x_j \neq y_j^*} \omega^{i\delta_j} \beta^j$$

and (10) follows.

Let us fix $\mathbf{x} \in \mathcal{C}$, and denote

$$\mathcal{J} = \mathcal{J}(\mathbf{x}, \mathbf{y}^*), \quad L = |\mathcal{J}|$$

$$f_{\mathcal{J}}(\omega, z) \triangleq \prod_{j \in \mathcal{J}} \left(\alpha^0 + (\omega^{\delta_j} \alpha^j) z \right) = \sum_{l=0}^{L} \sigma_{l,\mathcal{J}}(\omega) z^l$$

where

$$f_{0,\mathcal{J}}(\omega) \triangleq \alpha^0$$

$$f_{1,\mathcal{J}}(\omega) = \sum_{j \in \mathcal{J}} \omega^{\delta_j} \alpha^j$$

$$f_{2,\mathcal{J}}(\omega) \triangleq \sum_{\substack{j_1, j_2 \in \mathcal{J} \\ j_1 \neq j_2}} \omega^{(\delta_{j_1} + \delta_{j_2})} \alpha^{j_1 + j_2}$$

$$\cdot$$

$$f_{L,\mathcal{J}}(\omega) \triangleq \omega^{\sum_{j \in \mathcal{J}} \delta_j} \alpha^{\sum_{j \in \mathcal{J}} j}$$

It can be easily checked that the introduction of the parameter ω and the arithmetic rules given in Definition 1 do not change the equations for the binary symmetric channels that follow from the Newton's identities [1],

$$S_{i,\mathcal{J}}(\omega) + \sum_{l=1}^{i-1} f_{l,\mathcal{J}}(\omega)S_{i-l,\mathcal{J}}(\omega) = \begin{cases} 0, & \text{if } i \text{ is even} \\ \sigma_{i,\mathcal{J}}(\omega), & \text{if } i \text{ is odd} \end{cases}$$

for all $i = 1, \ldots, L-1$ and

$$\sum_{l=1}^{L} f_{l,\mathcal{J}}(\omega)S_{i-l,\mathcal{J}}(\omega) = 0$$

for all $i = L+1, \ldots, n-1$.

For example, if the parity check matrix of the code is defined in (6), then

$$\sigma_{\{0,1,2\}}(\omega, z) = \left(\alpha^0 + (\omega^{\delta_0}\alpha^0)z\right)\left(\alpha^0 + (\omega^{\delta_1}\alpha^1)z\right)\left(\alpha^0 + (\omega^{\delta_2}\alpha^2)z\right)$$

$$= \alpha^0 + \left(\omega^1\alpha^{10}\right)z + \left(\omega^2\alpha^{11}\right)z^2 + \left(\omega^3\alpha^3\right)z^3$$

$$\sigma_{\{9,13\}}(\omega, z) = \left(\alpha^0 + (\omega^{\delta_9}\alpha^9)z\right)\left(\alpha^0 + (\omega^{\delta_{13}}\alpha^{13})z\right)$$

$$= \alpha^0 + \left((\omega^1\alpha^9) + (\omega^4\alpha^{13})\right)z + \left(\omega^5\alpha^7\right)z^2.$$

Suppose that components of the received vector are defined by (8). Then the vector $\mathbf{y}^*$ has 1's only at positions 9,13, and

$$S_{i,\{9,13\}}(\omega|\mathbf{y}) = (\omega^1\alpha^9)^i + (\omega^4\alpha^{13})^i, \quad i = 0, 1, \ldots$$

Thus,

$$\begin{bmatrix} S_{1,\{9,13\}}(\omega|\mathbf{y}) \\ S_{2,\{9,13\}}(\omega|\mathbf{y}) \\ S_{3,\{9,13\}}(\omega|\mathbf{y}) \\ S_{4,\{9,13\}}(\omega|\mathbf{y}) \end{bmatrix} = \begin{bmatrix} (\omega^1\alpha^9) + (\omega^4\alpha^{13}) \\ (\omega^2\alpha^3) + (\omega^8\alpha^{11}) \\ (\omega^3\alpha^{12}) + (\omega^{12}\alpha^9) \\ (\omega^4\alpha^6) + (\omega^{16}\alpha^7) \end{bmatrix}. \tag{11}$$

Since $\alpha^1, \ldots, \alpha^4$ are roots of the generator polynomial, these functions satisfy (10) for any codeword $\mathbf{x}$.

To describe a variant of the known Berlekamp–Massey algorithm [1] for searching for the coefficients $f_{l,\mathcal{J}}(\omega)$ we need inversions of the descripancies, which are functions of ω.

Definition 2. *Let $K \geq 1$ and let the functions*

$$D(\omega) = \sum_{k=1}^{K} (\omega^{i_k}\alpha^{j_k}), \quad D'(\omega) = \sum_{k=1}^{K} (\omega^{i'_k}\alpha^{j'_k})$$

be given in such a way that $i_k \geq i'_k$ for all $k = 1, \ldots, K$. Introduce the function

$$\left(D_1(\omega) : D_0(\omega)\right) \triangleq \sum_{k=1}^{K} (\omega^{i_k - i'_k}\beta_k)$$

where

$$\beta_k \stackrel{\triangle}{=} \alpha^{j_k} \Big[\sum_{k=1}^{K} \alpha^{j_k} \Big]^{-1}.$$

The algorithm is given below.

1. Set $i = 1$, $h = 0$, $D_i'(\omega) = \omega^0 \alpha^0$, and $(L, f(\omega, z)) = (\lambda, \varphi(\omega, z)) = (0, \omega^0 \alpha^0)$.
2. Increase h by 1 and set

$$D_i(\omega) = S_i(\omega|\mathbf{y}) + f_1(\omega) S_{i-1}(\omega|\mathbf{y}) + \ldots + f_L(\omega) S_{i-L}(\omega|\mathbf{y})$$
$$h_i = h$$
$$(\lambda_i, \varphi_i(\omega, z)) = (\lambda, \varphi(\omega, z))$$
$$(L_i, f_i(\omega, z)) = \begin{cases} (L, f(\omega, z)), & \text{if } D_i(\omega) = 0 \\ \Big(\max\{L, \lambda + h\}, \ f'(\omega, z) \Big), & \text{if } D_i(\omega) \neq 0 \end{cases}$$

where

$$f'(\omega, z) = f(\omega, z) + \Big(D_i(\omega) : D_i'(\omega) \Big) z^h \varphi(\omega, z).$$

3. If $L_i > L$, then set $(\lambda, \varphi(\omega, z)) = (L, f(\omega, z))$, $D_i'(\omega) = D_i(\omega)$, and $h = 0$.
4. Set $(L, f((\omega, z)) = (L_i, f_i(\omega, z))$.
5. If $i < t$, where we assume that $\alpha^1, \ldots, \alpha^t$ are roots of the generator polynomial of the code, then increase i by 1 and go to 2.
6. End.

If $(S_1(\omega), \ldots, S_4(\omega))$ are defined by the matrix on the right–hand side of (11), then the current results of the algorithm are as follows.

$i = 0$:

$$D_0(\omega) = \omega^0 \alpha^0$$
$$f_0(\omega, z) = \omega^0 \alpha^0$$
$$\varphi_0(\omega, z) = \omega^0 \alpha^0$$

$i = 1$:

$$D_1(\omega) = (\omega^1 \alpha^9) + (\omega^4 \alpha^{13})$$
$$\Big(D_1(\omega) : D_1'(\omega) \Big) = (\omega^1 \alpha^9) + (\omega^4 \alpha^{13})$$
$$f_1(\omega, z) = \alpha^0 + \Big((\omega^1 \alpha^9) + (\omega^4 \alpha^{13}) \Big) z$$
$$\varphi_1(\omega, z) = \alpha^0$$

$i = 2$:

$$D_2(\omega) = (\omega^2 \alpha^3) + (\omega^8 \alpha^{11}) + \Big((\omega^1 \alpha^9) + (\omega^4 \alpha^{13}) \Big) \Big((\omega^1 \alpha^9) + (\omega^4 \alpha^{13}) \Big)$$
$$= 0$$
$$f_2(\omega, z) = \alpha^0 + \Big((\omega^1 \alpha^9) + (\omega^4 \alpha^{13}) \Big) z$$
$$\varphi_2(\omega, z) = \alpha^0$$

$i = 3$:

$$D_3(\omega) = (\omega^3 \alpha^{12}) + (\omega^{12} \alpha^9)$$
$$+ \Big((\omega^1 \alpha^9) + (\omega^4 \alpha^{13}) \Big) \Big((\omega^2 \alpha^3) + (\omega^8 \alpha^{11}) \Big)$$
$$= (\omega^6 \alpha^1) + (\omega^9 \alpha^5)$$

$$\Big(D_3(\omega) : D_3'(\omega) \Big) = \omega^5 (\alpha^1 + \alpha^5)(\alpha^9 + \alpha^{13})^{-1}$$
$$= \omega^5 \alpha^7$$

$$f_3(\omega, z) = \alpha^0 + \Big((\omega^1 \alpha^9) + (\omega^4 \alpha^{13}) \Big) z + (\omega^5 \alpha^7) z^2$$
$$\varphi_3(\omega, z) = \alpha^0 + \Big((\omega^1 \alpha^9) + (\omega^4 \alpha^{13}) \Big) z$$

$i = 4$:

$$D_4(\omega) = (\omega^4 \alpha^6) + (\omega^{16} \alpha^7) + \Big((\omega^1 \alpha^9) + (\omega^4 \alpha^{13}) \Big) \Big((\omega^3 \alpha^{12}) + (\omega^{12} \alpha^9) \Big)$$
$$+ (\omega^5 \alpha^7) \Big((\omega^2 \alpha^3) + (\omega^8 \alpha^{11}) \Big)$$
$$= 0$$

$$f_4(\omega, z) = \alpha^0 + \Big((\omega^1 \alpha^9) + (\omega^4 \alpha^{13}) \Big) z + (\omega^5 \alpha^7) z^2$$
$$\varphi_4(\omega, z) = \alpha^0 + \Big((\omega^1 \alpha^9) + (\omega^4 \alpha^{13}) \Big) z$$

Thus, the algorithm sequentially constructs the polynomials

$$f^{(0)}(\omega, z) = \alpha^0$$
$$f^{(1)}(\omega, z) = \alpha^0 + \Big((\omega^1 \alpha^9) + (\omega^4 \alpha^{13}) \Big) z$$
$$f^{(2)}(\omega, z) = \alpha^0 + \Big((\omega^1 \alpha^9) + (\omega^4 \alpha^{13}) \Big) z + \Big(\omega^5 \alpha^7 \Big) z^2.$$

As a result, we know that decoding of the vector $\mathbf{y}^*$ leads to a codeword having the distortion 5 relative to the received vector $\mathbf{y}$. The same conclusion could be received if we run the conventional Berlekamp–Massey procedure, construct the polynomial $f^{(2)}(1, z)$, find its roots, $\alpha^{-9}, \alpha^{-13}$, and compute $\delta_9 + \delta_{13}$. The introduction of the "distortion enumerator" in the definition of the syndrome also allows us to simplify searching for roots (this is not a difficult problem for our example, but we discuss these points in a more general context) : we know that the total number of roots is 2 and that the total distortion is 5; hence, one of the roots, α^{-j_1}, is such that j_1 belongs to the set of positions where $\mathbf{y}$ has symbols 1 or 2, and another root, α^{-j_2}, is such that j_2 belongs to the set of positions where $\mathbf{y}$ has symbols 0 or 3. Furthermore, we know that 3 components, j_1', j_2', j_3', of the vector $\mathbf{y}^*$ should be corrected only if there $\delta_{j_1'} + \delta_{j_2'} + \delta_{j_3'} < 5$, which means that all these components belong to the set of positions where $\mathbf{y}$ has symbols 1 or 2. We also know that $\alpha^{-j_1'}, \alpha^{-j_2'}, \alpha^{-j_3'}$ must be roots of the polynomial

$$f^{(2)}(1, z) + f^{(1)}(1, z)\alpha^j z^2 = \alpha^0 + \alpha^{10} z + (\alpha^0 + \alpha^7) z^2 + \alpha^{10+j} z^3$$

for some $j \in \{0, \ldots, 14\}$. The solution in this case is $j'_1 = 0, j'_2 = 1, j'_3 = 2$ and $j = 8$.

The basic idea of the approach described above is the note that the known computations with syndromes are also possible when we simultaneously process the current distortions of the codewords. This possibility can be useful for algebraic soft decision decoding. We tried to demonstrate them for the Berlekamp–Massey algorithm.

References

1. S. Lin, D. J. Costello, Jr., *Error Control Coding. Fundamentals and Applications.* Prentice–Hall, 1983.

Lifting Decoding Schemes over a Galois Ring

Eimear Byrne

Department of Mathematics, University College Cork,
National University of Ireland,
Cork, Ireland
eimear.byrne@ucc.ie

Abstract. Any vector with components in a Galois ring $R = GR(p^m, n)$ has a unique p−adic representation, given a tranversal on the cosets of $\langle p \rangle$ in R. We exploit this representation to lift a decoding algorithm for an associated code over the residue field of R to a decoding scheme for the original code. The lifted algorithm involves n consecutive applications of the given procedure. We apply these techniques to the decoding of an alternant code over a Galois ring.

1 Introduction

The notion of lifting a decoding scheme for a linear code over $\mathbf{Z}_p$ to a decoding scheme for a code over $\mathbf{Z}_{p^n}$ was introduced in [10], where the authors considered the class of $\mathbf{Z}_{p^n}$− splitting codes, which are free as $\mathbf{Z}_{p^n}$−modules.

In what follows we present techniques for lifting a given decoding algorithm for a linear code of length N over a finite field to a decoding scheme for a linear code of length N over a Galois ring. Specifically, if $\mathbf{H}$ is a parity check matrix for a code C, defined over a Galois ring of characteristic p^n, and $\bar{C}$ is the code with parity check matrix $\bar{\mathbf{H}}$, where $\bar{\mathbf{H}}$ is the image of $\mathbf{H}$ modulo p, then we construct a decoding scheme for C by lifting an algorithm for $\bar{C}$. The lifted algorithm involves n consecutive applications of the decoding procedure for $\bar{C}$ and depends upon the existence of a unique representation of an error vector in the form $\mathbf{e}^0 + p\mathbf{e}^1 + \cdots + p^{n-1}\mathbf{e}^{n-1}$. We also place a constraint on the type of error that can be corrected by such an algorithm, in particular, we require that for each $i \in \{0, ..., n-1\}$, the modulo p image of $\mathbf{e}^i$ be correctable with respect to the algorithm for $\bar{C}$. We implement this technique to derive decoding schemes for alternant codes over Galois rings by lifting decoding schemes for both the Hamming and the Lee distance.

We introduce some notation and definitions. Assume throughout that all rings R and T are finite, local, commutative, rings with unity. Let R have unique maximal ideal $\langle p \rangle$ for some prime p. The polynomial $f \in R[x]$ is called *basic irreducible* if it is irreducible modulo p. We construct a Galois ring as a quotient ring of $\mathbf{Z}_{p^n}[x]$ in the following way.

Definition 1. *Let p be a prime number and let m, n be positive integers. Let $f \in \mathbf{Z}_{p^n}[x]$ be a monic basic irreducible polynomial of degree m. The quotient*

S. Boztaş and I.E. Shparlinski (Eds.): AAECC-14, LNCS 2227, pp. 323–332, 2001.
© Springer-Verlag Berlin Heidelberg 2001

ring $\mathbf{Z}_{p^n}[x]/\langle f \rangle$, denoted $GR(p^m, n)$, is called the Galois ring *of order p^{mn} and characteristic p^n*.

The reader is referred to [11] and [13] for a review of the theory of Galois rings. The integers p, m and n determine uniquely, up to isomorphism, the Galois ring $GR(p^m, n)$. If R is a Galois ring we let k_R denote its unique residue field and R^* its multiplicative group of units. For the remainder, let the symbol R denote the Galois ring $GR(p^m, n)$ and let μ be the natural epimorphism from R onto k_R, defined by $\mu a = a + \langle p \rangle$ for each $a \in R$. Let $\mathcal{T}$ be a transversal on the cosets of $\langle p \rangle$ in R, so that if $v, \rho \in \mathcal{T}$ then $v - \rho \in \langle p \rangle$ if and only if $v = \rho$. An arbitrary element $\theta \in R$ can be represented uniquely by the sum $\theta = \sum_{j=0}^{n-1} p^j \theta_j$, where $\theta_j \in \mathcal{T}$ for each $j \in 0, ..., n-1$, which we call the *$p-$adic representation* of θ for an arbitrary fixed transversal $\mathcal{T}$. For a given $\theta \in R$, the element θ_j (or $(\theta)_j$ if parentheses are required to avoid ambiguity) is the uniquely determined j^{th} component of θ in $\mathcal{T}$. We can extend this notation for vectors in R^N. Let $\mathbf{v} \in R^N$, then $\mathbf{v}$ has the unique $p-$adic representation $\mathbf{v} = \mathbf{v}^0 + p\mathbf{v}^1 + \cdots + p^{n-1}\mathbf{v}^{n-1}$ where each $\mathbf{v}^i$ has components in $\mathcal{T}$. We denote by the symbol $\mathbf{v}^{(i)}$ the truncation of this sum modulo p^{i+1}, so $\mathbf{v}^{(i)} = \mathbf{v}^0 + p\mathbf{v}^1 + \cdots + p^i\mathbf{v}^i$. For $R = \mathbf{Z}_{p^n}$, the Lee distance, is defined as follows. Given $\theta \in \mathbf{Z}_{p^n}$, the *Lee value* of θ, denoted $|\theta|_L$, is defined by

$$|\theta|_L = \left\{ \begin{array}{ll} \theta & \text{if } 0 \leq \theta \leq \frac{p^n}{2} \\ p^n - \theta & \text{if } \frac{p^n}{2} < \theta \leq p^n - 1 \end{array} \right.$$

where the symbol representing θ is an element of $\{0, ..., p^n - 1\}$.

Definition 2. *Let N be a positive integer and let $\mathbf{u}, \mathbf{v}$ be arbitrary vectors in $\mathbf{Z}_{p^n}^N$. The Lee distance, denoted $\mathrm{d}_L(\mathbf{u}, \mathbf{v})$, is defined by*

$$\mathrm{d}_L(\mathbf{u}, \mathbf{v}) = \sum_{j=1}^{N} |u_j - v_j|_L,$$

evaluated over the integers. We denote by $\mathrm{wt}_L(\mathbf{u})$ the Lee weight of a vector $\mathbf{u} \in \mathbf{Z}_{p^n}^N$, where $\mathrm{wt}_L(\mathbf{u}) = \mathrm{d}_L(\mathbf{u}, \mathbf{0})$.

Given an arbitrary vector $\mathbf{v} \in \mathbf{Z}_{p^n}^N$ we denote by $\mathbf{v}^+ = [v_0^+, ..., v_{N-1}^+]$ and $\mathbf{v}^- = [v_0^-, ..., v_{N-1}^-]$ the vectors in $\mathbf{Z}_{p^n}^N$ defined by

$$v_j^+ = \left\{ \begin{array}{ll} v_j & \text{if } v_j = |v_j|_L \\ 0 & \text{otherwise} \end{array} \right. \quad \text{and} \quad v_j^- = \left\{ \begin{array}{ll} v_j & \text{if } p^n - v_j = |v_j|_L \\ 0 & \text{otherwise} \end{array} \right.$$

which gives the decomposition $\mathbf{v} = \mathbf{v}^+ - \mathbf{v}^-$.

Gröbner bases are structures which provide powerful tools for the study of multivariate polynomial ideals. The particular algorithms given in subsequent sections are devised as lifts of decoding algorithms which use Gröbner bases techniques. We mention any relevant details of the theory in its application to decoding alternant codes. The interested reader is referred to [2], [3], [6], [7] and [8] for a review of the general theory of Gröbner bases.

Consider the $R[x]$−module $R[x]^2$. A *term* in $R[x]^2$ is an element of the form $(x^i, 0)$ or $(0, x^j)$ for some integers $i, j \geq 0$. A *monomial* in $R[x]^2$ is a non-zero constant multiple of a term in $R[x]^2$. A term X is *divisible* by a term Y if there exists a nonnegative integer ℓ such that $X = x^\ell Y$.

Definition 3. *A term order $<$ on $R[x]^2$ is defined by the following properties.*

(i) *$<$ is a linear order on the set of terms of $R[x]^2$*
(ii) *if X, Y are terms in $R[x]^2$ such that $X < Y$, then $x^\ell X < x^\ell Y$ for any nonnegative integer ℓ*
(iii) *every strictly descending sequence of terms in $R[x]^2$ terminates*

We define a term order for each integer ℓ as follows.

(i) $(x^{i_1}, 0) <_\ell (x^{i_2}, 0)$ if and only if $i_1 < i_2$ and $(0, x^{j_1}) <_\ell (0, x^{j_2})$ if and only if $j_1 < j_2$
(ii) $(x^i, 0) <_\ell (0, x^j)$ if and only if $i + \ell < j$

Let $(a, b) \in R[x]^2$. The *leading term* of (a, b), denoted $\mathrm{lt}(a, b)$, is identified as the greatest term occurring in an expansion of (a, b) as an R−linear combination of terms. The coefficient attached to $\mathrm{lt}(a, b)$ is the *leading coefficient* of (a, b) and is denoted by $\mathrm{lc}(a, b)$. The *leading monomial* of (a, b) is given by $\mathrm{lm}(a, b) = \mathrm{lc}(a, b)\mathrm{lt}(a, b)$.

Definition 4. *Let A be an $R[x]$−submodule of $R[x]^2$. A set $\mathcal{G} = \{(g_i, h_i)\}_{i=1}^\ell \subseteq A$ of non-zero elements is called a Gröbner basis of A if, for each $(a, b) \in A$, there exists an $i \in \{1, \ldots, \ell\}$ such that $\mathrm{lm}(a, b)$ is divisible by $\mathrm{lm}(g_i, h_i)$.*

Note that, given an arbitrary term order $<$, every Gröbner basis $\mathcal{G}$ of A must contain an element with minimal leading term in A, otherwise some element in the module would have leading monomial divisible by none of the leading monomials of elements in $\mathcal{G}$. The algorithms discussed in Sections 3 and 4 involve finding a minimal element of a module by generating a Gröbner basis.

2   The General Decoding Scheme

Let T be a subring of R and let C be the T−linear code of length N defined by

$$C = \{\mathbf{c} \in T^N : \mathbf{Hc} = \mathbf{0}\}$$

for some $r \times N$ parity check matrix $\mathbf{H}$ with symbols in R and let d be a metric on T^N. Let $\mathbf{c} \in C$ be a transmitted codeword and $\mathbf{v} \in T^N$ the corresponding received word. We assume that $\mathbf{v} = \mathbf{c} + \mathbf{e}$ for some error vector $\mathbf{e} \in T^N$ of minimal weight in $\mathbf{e} + C$. We denote by $\bar{C}$ the k_T−linear code defined by

$$\bar{C} = \{\mathbf{c} \in k_T^N : \bar{\mathbf{H}}\mathbf{c} = \mathbf{0}\}$$

where $\mu\mathbf{H} = \bar{\mathbf{H}}$. Note that $\bar{C} = \mu C$ if and only if C is splitting. Let $\bar{\mathrm{d}}$ be a metric on k_T^N and suppose there exists a decoding algorithm $(\bar{A}, \bar{\mathrm{d}})$ for the code $\bar{C}$, which determines an error vector $\mathbf{e}' \in k_T^N$ of minimal weight given a received word $\mathbf{v}' \in k_T^N$. Explicitly, we suppose that $(\bar{A}, \bar{\mathrm{d}})$ recovers $\mathbf{e}' \in k_T^N$ where $\bar{\mathbf{H}}\mathbf{v}' = \bar{\mathbf{H}}(\mathbf{c}' + \mathbf{e}') = \bar{\mathbf{H}}\mathbf{e}' = \mathbf{s}'$ for some codeword $\mathbf{c}' \in \bar{C}$. Let $\mathcal{T}$ be a transversal on the cosets of $\langle p \rangle$ in R. Express $\mathbf{e} \in T^N$ uniquely in the form $\mathbf{e} = \mathbf{e}^0 + p\mathbf{e}^1 + \cdots + p^{n-1}\mathbf{e}^{n-1}$ where $e_j^i \in \mathcal{T}$ for all $i \in \{0, ..., n-1\}$ and $j \in \{0, ..., N-1\}$. Suppose that $\mu\mathbf{e}^i$ has minimal weight in $\mu\mathbf{e}^i + \bar{C}$ for each $i \in \{0, .., n-1\}$. Then

$$\mathbf{Hv} = \mathbf{H}(\mathbf{c} + \mathbf{e}) = \mathbf{He} = \mathbf{s}$$
$$= \mathbf{He}^0 + p\mathbf{He}^1 + \cdots + p^{n-1}\mathbf{He}^{n-1}$$
$$= \mathbf{s}^{[0]} + p\mathbf{s}^{[1]} + \cdots + p^{n-1}\mathbf{s}^{[n-1]}$$
$$= \mathbf{s}^0 + p\mathbf{s}^1 + \cdots + p^{n-1}\mathbf{s}^{n-1}$$

where $\mathbf{s}^{[i]} = \mathbf{He}^i$ for each $i \in \{0, .., n-1\}$. Note that given any $j \in \{0, ..., n-1\}$, the element $(\mathbf{He}^i)_j$ may not be contained in $\mathcal{T}$, so the vectors $\mathbf{s}^{[i]}$ and $\mathbf{s}^i$ may be distinct. Applying the natural epimorphism to the above yields

$$\mu\mathbf{s}^0 = \mu\mathbf{s} = \bar{\mathbf{H}}\mu\mathbf{v} = \bar{\mathbf{H}}(\mu\mathbf{c} + \mu\mathbf{e}) = \bar{\mathbf{H}}\mu\mathbf{e} = \bar{\mathbf{H}}\mu\mathbf{e}^0.$$

By hypothesis, we may implement the scheme $(\bar{A}, \bar{\mathrm{d}})$ to recover $\mu\mathbf{e}^0$, of minimal weight in $\mu\mathbf{e}^0 + \bar{C}$, from the syndrome $\mu\mathbf{s}^0$ and hence determine $\mathbf{e}^0$, the unique preimage with components in $\mathcal{T}$, of the vector $\mu\mathbf{e}^0$.

We continue iteratively, solving for each $\mathbf{e}^i$ in turn, to extend the decoding scheme $(\bar{A}, \bar{\mathrm{d}})$ to a scheme $(A, \mathrm{d}, \bar{\mathrm{d}})$ for C. At the $(i+1)^{th}$ step let $\mathbf{v}^{\{i\}} = \mathbf{v} - \mathbf{e}^{(i-1)}$, then

$$\mathbf{s}^{\{i\}} = \mathbf{Hv}^{\{i\}} = \mathbf{H}(\mathbf{v} - \mathbf{e}^{(i-1)}) = \mathbf{H}(\mathbf{e} - \mathbf{e}^{(i-1)})$$
$$= p^i\mathbf{He}^i + \cdots + p^{n-1}\mathbf{He}^{n-1}$$
$$= p^i\mathbf{s}^{[i]} + \cdots + p^{n-1}\mathbf{s}^{[n-1]}$$
$$= (\mathbf{s}^{\{i\}})^0 + p(\mathbf{s}^{\{i\}})^1 + \cdots + p^{n-1}(\mathbf{s}^{\{i\}})^{n-1}.$$

It follows that $\mathbf{s}^{\{i\}} \in \langle p^i \rangle$ and $(\mathbf{s}^{\{i\}})^j = \mathbf{0}$ for each $j \in \{0, ..., i-1\}$. Thus

$$p^i((\mathbf{s}^{\{i\}})^i - \mathbf{s}^{[i]}) \in \langle p^{i+1} \rangle$$

and hence

$$\bar{\mathbf{H}}\mu\mathbf{e}^i = \mu\mathbf{s}^{[i]} = \mu(\mathbf{s}^{\{i\}})^i.$$

We invoke $(\bar{A}, \bar{\mathrm{d}})$ to $\mu(\mathbf{s}^{\{i\}})^i$ to recover $\mu\mathbf{e}^i$ and hence the vector $\mathbf{e}^{(i)} = \mathbf{e}^{(i-1)} + p^i\mathbf{e}^i$. After the nth iteration the error vector $\mathbf{e}$ is determined and the received word $\mathbf{v}$ is decoded to $\mathbf{c} = \mathbf{v} - \mathbf{e}$. We summarize these results in the following theorem.

Theorem 1. *Let C be a $R-linear$ code of length N. Suppose that for some distance function $\bar{\mathrm{d}}$ on k_R^N there exists a decoding scheme $(\bar{A}, \bar{\mathrm{d}})$ for $\bar{C}$ which corrects any error pattern $\mathbf{e}' \in k_R^N$ of minimal weight in $\mathbf{e}' + \bar{C}$. Then the decoding scheme $(\bar{A}, \bar{\mathrm{d}})$ can be lifted to a decoding scheme $(A, \mathrm{d}, \bar{\mathrm{d}})$ for C which corrects any error pattern $\mathbf{e} \in R^N$ satisfying*

(i) $\mathbf{e}$ *has minimal weight in* $\mathbf{e} + C$

(ii) $\mu\mathbf{e}^i$ *has minimal weight in* $\mu\mathbf{e}^i + \bar{C}$ *for each* $i \in \{0, ..., n-1\}$.

With $T, R, C, \bar{C}, \mathrm{d}$ and $\bar{\mathrm{d}}$ as before, let $\mathrm{wt}(\mathbf{u}) = \mathrm{d}(\mathbf{u}, \mathbf{0})$ and $\bar{\mathrm{wt}}(\mathbf{v}) = \bar{\mathrm{d}}(\mathbf{v}, \mathbf{0})$ for each $\mathbf{u} \in T^N$ and $\mathbf{v} \in k_T^N$. Let

$$\mathcal{B}_t(\mathbf{u}, \mathrm{d}) = \{\mathbf{w} \in T^N : \mathrm{d}(\mathbf{u}, \mathbf{w}) \le t\}$$

denote the sphere of radius t about $\mathbf{u}$, with respect to d, and let

$$\bar{\mathcal{B}}_t(\mathbf{u}, \bar{\mathrm{d}}) = \{\mathbf{w} \in T^N : \bar{\mathrm{d}}(\mu\mathbf{u}^i, \mu\mathbf{w}^i) \le t, i \in \{0, ..., n-1\}\}.$$

Now let C and $\bar{C}$ have minimum distances d and $\bar{d}$ for the distance functions d and $\bar{\mathrm{d}}$, respectively. Let $t = \lfloor \frac{d-1}{2} \rfloor$ and let $\bar{t} = \lfloor \frac{\bar{d}-1}{2} \rfloor$. Suppose there exists a decoding algorithm $(\bar{A}, \bar{\mathrm{d}}, \bar{t})$ for $\bar{C}$ which corrects any error pattern of weight at most $\bar{t}$ in k_T^N. Then by Theorem 1, we can lift $(\bar{A}, \bar{\mathrm{d}}, \bar{t})$ to construct a decoding algorithm $(A, \mathrm{d}, t, \bar{t})$, which corrects any error pattern $\mathbf{e}$ in T^N satisfying $\mathrm{wt}(\mathbf{e}) \le t$ and $\bar{\mathrm{wt}}(\mu\mathbf{e}^i) \le \bar{t}$ for each $i \in \{0, ..., n-1\}$. So if $\mathbf{v}$ is a received word in T^N, we can implement the lifted algorithm $(A, \mathrm{d}, t, \bar{t})$ to decode $\mathbf{v}$ to a unique codeword $\mathbf{c}$, provided that $\mathbf{v} \in \mathcal{B}_t(\mathbf{c}, \mathrm{d}) \cap \bar{\mathcal{B}}_{\bar{t}}(\mathbf{c}, \bar{\mathrm{d}})$.

Suppose that $\mathbf{v} \in \bar{\mathcal{B}}_{\bar{t}}(\mathbf{c}_1, \bar{\mathrm{d}}) \cap \bar{\mathcal{B}}_{\bar{t}}(\mathbf{c}_2, \bar{\mathrm{d}})$ for codewords $\mathbf{c}_1$ and $\mathbf{c}_2$ in C. Then $\mathbf{v} = \mathbf{c}_1 + \mathbf{e}_1 = \mathbf{c}_2 + \mathbf{e}_2$ for some error vectors $\mathbf{e}_1$ and $\mathbf{e}_2$ in T^N, satisfying $\bar{\mathrm{wt}}(\mu\mathbf{e}_1^i) \le \bar{t}$ and $\bar{\mathrm{wt}}(\mu\mathbf{e}_2^i) \le \bar{t}$ for each $i \in \{0, .., n-1\}$. Then $\mathbf{s} = \mathbf{H}\mathbf{v} = \mathbf{H}\mathbf{e}_1 = \mathbf{H}\mathbf{e}_2$ and $\mu\mathbf{s} = \bar{\mathbf{H}}\mu\mathbf{e}_1 = \bar{\mathbf{H}}\mu\mathbf{e}_2$. Since the algorithm $(\bar{A}, \bar{\mathrm{d}}, \bar{t})$ computes a unique error pattern of weight at most $\bar{t}$, then $\mu\mathbf{e}_1 = \mu\mathbf{e}_2$ and hence $\mathbf{e}_1^0 = \mathbf{e}_2^0$. Repeated applications of the argument show that $\mathbf{e}_1 = \mathbf{e}_2$, indeed if $\mathbf{e}_1^{(j)} = \mathbf{e}_2^{(j)}$ for each $j \in \{0, ..., i-1\}$ then

$$\mathbf{H}(\mathbf{v} - \mathbf{e}_1^{(i-1)}) = \mathbf{H}(\mathbf{v} - \mathbf{e}_2^{(i-1)}),$$

so that $\bar{\mathbf{H}}\mu\mathbf{e}_1^i = \bar{\mathbf{H}}\mu\mathbf{e}_2^{(i)}$ and, since $\bar{\mathrm{wt}}(\mu\mathbf{e}_j^i) \le \bar{t}$ for $j \in \{1, 2\}$, we deduce that $\mu\mathbf{e}_1^{(i)} = \mu\mathbf{e}_2^{(i)}$. At the nth iteration we find that $\mathbf{e}_1 = \mathbf{e}_2$. Thus the $\bar{\mathcal{B}}_{\bar{t}}(\mathbf{c}, \bar{\mathrm{d}})$ are disjoint for distinct $\mathbf{c} \in C$.

Definition 5. *Let* d *be a metric on* k_R^N. *We denote by* $\mathrm{d}_{\max}$ *the metric on* R^N *induced by* d *and defined by*

$$\mathrm{d}_{\max}(\mathbf{u}, \mathbf{v}) = \max\{\mathrm{d}(\mu\mathbf{u}^i, \mu\mathbf{v}^i) : i \in \{0, ..., n-1\}\}.$$

Given an integer ℓ, $\mathrm{d}_{\max}(\mathbf{u}, \mathbf{v}) \le \ell$ if and only if $\mathrm{d}(\mu\mathbf{u}^i, \mu\mathbf{v}^i) \le \ell$ for each $i \in \{0, ..., n-1\}$, so that $\bar{\mathcal{B}}_\ell(\mathbf{u}, \bar{\mathrm{d}}) = \mathcal{B}_\ell(\mathbf{u}, \bar{\mathrm{d}}_{\max})$. Then we decode a received word $\mathbf{v} \in T^N$ to a unique codeword $\mathbf{c} \in C$ if $\mathbf{v} \in \mathcal{B}_{\bar{t}}(\mathbf{c}, \bar{\mathrm{d}}_{\max})$ and hence a bounded-distance decoder for $\bar{C}$ can be lifted to a bounded-distance decoder for C. We have now proved the following result.

Theorem 2. *Let* C *be an* $R-$*linear code of length* N. *Suppose that for some distance function* d *on* k_R^N *there exists a decoding scheme* $(\bar{A}, \mathrm{d}, t)$ *for* $\bar{C}$ *which corrects any error vector* $\mathbf{e}' \in k_R^N$ *such that* $\mathrm{wt}(\mathbf{e}') \le t$. *Then the decoding scheme* $(\bar{A}, \mathrm{d}, t)$ *can be lifted to a decoding scheme* $(A, \mathrm{d}_{\max}, t)$ *for* C *which corrects any error pattern* $\mathbf{e} \in R^N$ *satisfying* $\mathrm{wt}_{\max}(\mathbf{e}) \le t$.

Remark 1. Let C be an R–linear code of length N and minimum Hamming distance d. Suppose the code $\bar{C}$ has minimum Hamming distance $\bar{d} \leq d$ and let $t = \lfloor \frac{\bar{d}-1}{2} \rfloor$. Then the spheres $\mathcal{B}_t(\mathbf{c}, \mathrm{d}_{H\,\mathrm{max}})$ are disjoint for distinct $\mathbf{c} \in C$ and, since for each $i \in \{0, ..., n-1\}$ $\mathrm{wt}_H(\mu \mathbf{v}^i) \leq \mathrm{wt}_H(\mathbf{v})$ for any $\mathbf{v} \in R^N$, it follows that $\mathcal{B}_t(\mathbf{c}, \mathrm{d}_H) \subseteq \mathcal{B}_t(\mathbf{c}, \mathrm{d}_{H\,\mathrm{max}})$ for each $\mathbf{c} \in C$.

Suppose now that C is a $\mathbf{Z}_{p^n}$–linear code with minimum Lee distance d and that the $\mathbf{Z}_p$–linear code $\bar{C}$ has minimum Lee distance $\bar{d}$. Let $t = \lfloor \frac{\bar{d}-1}{2} \rfloor$ and let $\theta \in \mathbf{Z}_{p^n}$. It is not hard to see that $|\mu \theta_i|_L \leq |\theta|_L$ for each $i \in \{0, ..., n-1\}$ and hence for any $\mathbf{v} \in \mathbf{Z}_{p^n}^N$, $\mathrm{wt}_L(\mu \mathbf{v}^i)_L \leq \mathrm{wt}_L(\mathbf{v})$ for each $i \in \{0, ..., n-1\}$. In particular, if $\mathbf{v} \in \mathbf{Z}_{p^n}^N$ and $t = \lfloor \frac{d-1}{2} \rfloor$ then again the spheres $\mathcal{B}_t(\mathbf{c}, d_{L\,\mathrm{max}})$ are disjoint for distinct $\mathbf{c} \in C$ and $\mathcal{B}_t(\mathbf{c}, d_L) \subseteq \mathcal{B}_t(\mathbf{c}, \mathrm{d}_{L\,\mathrm{max}})$ for all $\mathbf{c} \in C$.

Thus any bounded-distance decoding algorithms $(\bar{A}, t, \mathrm{d}_H)$ and $(\bar{A}, t, \mathrm{d}_L)$ for $\bar{C}$ can be lifted to bounded-distance decoding schemes (A, t, d_H) and (A, t, d_L), respectively, for C.

3   Decoding Alternant Codes for the Hamming Distance

We implement the scheme outlined in Section 2, applying Theorem 1 to construct a decoding scheme for an alternant code defined over a Galois ring, by lifting a decoding algorithm for an alternant code defined over the corresponding residue field. Both the given algorithm and the lifted version are implemented with respect to the Hamming distance, and both decoding algorithms are bounded-distance.

Definition 6. *Let R and T be the Galois rings $GR(p^m, n)$ and $GR(p^{m'}, n)$, respectively, where m' divides m. Let N be a nonnegative integer less than p^m. We define $C(N, r, \alpha, \gamma, T)$, the alternant code of length N with symbols in T, by the parity check matrix*

$$H(N, r, \alpha, \gamma, T) = \begin{bmatrix} \gamma_0 & \gamma_1 & \cdots & \gamma_{N-1} \\ \gamma_0 \alpha_0 & \gamma_1 \alpha_1 & \cdots & \gamma_{N-1}\alpha_{N-1} \\ \vdots & \vdots & \vdots & \vdots \\ \gamma_0 \alpha_0^{r-1} & \gamma_1 \alpha_1^{r-1} & \cdots & \gamma_{N-1}\alpha_{N-1}^{r-1} \end{bmatrix}$$

where $\gamma = [\gamma_0, ..., \gamma_{N-1}]$ has its components in R^* and the locator vector, $\alpha = [\alpha_0, ..., \alpha_{N-1}] \in R^N$, satisfies $\alpha_i - \alpha_j \in R^*$ for all distinct i and j in $\{0, ..., N-1\}$

Let C be the code $C(N, r, \alpha, \gamma, T)$, defined as above, with parity check matrix $\mathbf{H} = H(N, r, \alpha, \gamma, T)$. Let $t = \lfloor \frac{r}{2} \rfloor$. The code C has minimum Hamming distance greater than r and corrects all error patterns of Hamming weight at most t (see, for example, [1] or [12]). Then $\bar{C} = C(N, r, \bar{\alpha}, \bar{\gamma}, k_T)$, where $\mu \alpha = \bar{\alpha}$ and $\mu \gamma = \bar{\gamma}$, and $\bar{C}$ is also a t–Hamming error correcting code.

Consider the following decoding algorithm $(\bar{A}, t, \mathrm{d}_H)$ for $\bar{C}$, described in [8], which computes an error vector $\mathbf{e}$ of Hamming weight at most t, given a received word $\mathbf{v} = \mathbf{c} + \mathbf{e}$ for some codeword $\mathbf{c} \in \bar{C}$. Let $\mathcal{J} = \{j \in \{0, .., N-1\} : e_j \neq 0\}$. Determining the error locations $\mathcal{J}$ amounts to solving the key equation

$$\Sigma S \equiv \Omega \bmod x^r \tag{1}$$

where S is the syndrome polynomial associated with $\mathbf{s} = \bar{\mathbf{H}}\mathbf{v}$, the error locator polynomial is $\Sigma = \prod_{j \in \mathcal{J}} (1 - \bar{\alpha}_j x)$ and $\Omega = \sum_{j \in \mathcal{J}} e_j \bar{\gamma}_j \prod_{k \in \mathcal{J}, k \neq j} (1 - \bar{\alpha}_k x)$ is the error evaluator polynomial. It has been shown that the required solution (Σ, Ω) has minimal leading term under $<_{-1}$ of the elements in the module

$$M = \{(a, b) \in k_R[x]^2 : aS \equiv b \bmod x^r\}$$

of all solutions to the key equation (1). In particular, a unit multiple of (Σ, Ω) is contained in any Gröbner basis of M under this term order. Given S, computation of the appropriate basis may be performed by invoking any of the algorithms established in [8]. Then the coefficients of $\mathbf{e}$ may be computed using a method such as a Forney procedure [9], adapted in the obvious way for an alternant code.

Now let $\mathbf{v} \in T^N$ be a received word with corresponding error vector $\mathbf{e} \in T^N$ satisfying $\mathrm{wt}_H(\mathbf{e}) \leq t$. Then for each $i \in \{0, .., n-1\}$,

$$\mathrm{wt}_H(\mu \mathbf{e}^i) = \mathrm{wt}_H(\mathbf{e}^i) \leq \mathrm{wt}_H(\mathbf{e}) \leq t$$

and we define $\mathcal{J}_i = \{j \in \{0, ..., N-1\} : e_j^i \neq 0\}$, an ith error locator polynomial $\Sigma^{[i]} = \prod_{j \in \mathcal{J}_i} (1 - \alpha_j x)$, an ith syndrome polynomial $S^{[i]} = \sum_{k=0}^{r-1} \sum_{j \in \mathcal{J}_i} e_j^i \gamma_j \alpha_j^k x^k$ and an ith key equation $\Sigma^{[i]} S^{[i]} \equiv \Omega^{[i]} \bmod x^r$, where $\Omega^{[i]} = \sum_{j \in \mathcal{J}_i} e_j^i \gamma_j \prod_{k \in \mathcal{J}_i, k \neq j} (1 - \alpha_k x)$. Following the scheme presented in Section 2, at the ith iteration we assume that $\mathbf{e}^{(i-1)}$ is known and set $\mathbf{v}^{\{i\}} = \mathbf{v} - \mathbf{e}^{(i-1)}$. Then $\bar{\mathbf{H}}\mu \mathbf{e}^i = \mu \mathbf{s}^{[i]} = \mu(\mathbf{s}^{\{i\}})^i$ where for each $j, k \in \{0, .., n-1\}$, $\mathbf{s}^{[j]} = \mathbf{H}\mathbf{e}^j$ and $\mathbf{s}^{\{k\}} = \mathbf{H}\mathbf{v}^{\{k\}}$. Consider the equation

$$\mu \Sigma^{[i]} \mu S^{[i]} \equiv \mu \Omega^{[i]} \bmod x^r.$$

Since $\mathrm{wt}_H(\mu \mathbf{e}^i) \leq t$, we may apply the algorithm $(\bar{A}, t, d_H)$ to $\bar{\mathbf{H}}$ and $\mu(\mathbf{s}^{\{i\}})^i$. The ith required solution $(\mu \Sigma^{[i]}, \mu \Omega^{[i]})$ is the unique (up to multiplication by a unit) element of the module

$$\bar{M}^{[i]} = \{(a, b) \in k_R[x]^2 : a\mu S^{[i]} \equiv b \bmod x^r\}$$

with minimal leading term with respect to the term order $<_{-1}$ and hence is contained in a Gröbner basis of $\bar{M}^{[i]}$. Having found $\mu \mathbf{e}^i$, the vector $\mathbf{e}^{(i)} = \mathbf{e}^{(i-1)} + p^i \mathbf{e}^i$ is uniquely determined. After the nth step the error vector is recovered.

Example 1. Let $R = GR(3^2, 3) \simeq \mathbf{Z}_{27}[\xi]$ where ξ is a root of $f = x^2 - 5x - 1 \bmod 27$ and let $\mathcal{T} = \{0, 1, \xi, ..., \xi^7\}$. Let $C = C(8, 4, \alpha, \gamma, \mathbf{Z}_{27})$ and $\mathbf{H} = H(8, 4, \alpha, \gamma, \mathbf{Z}_{27})$ where $\alpha = [1, \xi, \xi^2, ..., \xi^7]$ and $\gamma = [\xi + 3, 2, 6\xi + 1, 1, 9\xi + 1, \xi, 18\xi + 2, 1]$. Then $\bar{\mathbf{H}} = H(8, 4, \bar{\alpha}, \bar{\gamma}, \mathbf{Z}_3)$ and both C and $\bar{C}$ are double-error correcting codes. Let $\mathbf{c}$ be a transmitted codeword, $\mathbf{v}$ the received word and let $\mathbf{e} = \mathbf{v} - \mathbf{c} = [0,1,0,0,0,18,0,0]$ be the error vector. Then $\mu S^{[0]} = \mu S = 2 + 2\bar{\xi}x + \bar{\xi}^3 x^2 + \bar{\xi}^7 x^3$ and, invoking an analogue of the Euclidean Algorithm as outlined in [8], we generate a Gröbner basis for $M^{[0]}$ from the generating set $\{(1, \mu S), (0, x^4)\}$.

(a_1, b_1)	(a_2, b_2)	g
$(0, x^4)$	$(1, \mu S)$	ξx
$(1, \mu S)$	$(\xi^5 x, \xi x + \xi^2 x^2 + \xi^3 x^3)$	ξ^4
$(\xi^5 x, \xi x + \xi^2 x^2 + \xi^3 x^3)$	$(\xi^5 x + 1, \xi)$	

The element $(\bar{\xi}^5 x + 1, \bar{\xi})$ has minimal leading term in $M^{[0]}$ and the polynomial $\bar{\xi}^5 x + 1$ has the root $\bar{\xi}^7$, whose inverse is $\bar{\xi}$ and corresponds to an error occurring at μe_1^0. We compute the error magnitude $\mu e_1^0 = 1$ so $\mu \mathbf{e}^0 = [0,1,0,0,0,0,0,0] = \mathbf{e}^0$. Then $S^{\{1\}} = 18\xi + (9 + 18\xi)x + (9\xi + 9)x^2 + 18x^3 = 9(\xi^5 + \xi^2 x + \xi^7 x^2 + \xi^4 x^3) = 9(S^{\{1\}})^2$ and $\bar{\mathbf{H}}\mu\mathbf{e}^1 = \mu(\mathbf{s}^{\{1\}})^1 = \mu\mathbf{s}^1 = \mathbf{0}$, so that $\mu\mathbf{e}^1 = \mathbf{e}^1 = \mathbf{0}$. Continuing, we find $\mu S^{[2]} = \mu(S^{\{2\}})^2 = \bar{\xi}^5 + \bar{\xi}^2 x + \bar{\xi}^7 x^2 + \bar{\xi}^4 x^3$ and, as before, we compute a Gröbner basis of $M^{[2]}$.

(a_1, b_1)	(a_2, b_2)	g
$(0, x^4)$	$(1, \mu S^{[2]})$	$\xi^4 x$
$(1, \mu S^{[2]})$	$(x, \xi^5 x + \xi^2 x^2 + \xi^7 x^3)$	ξ^5
$(x, \xi^5 x + \xi^2 x^2 + \xi^7 x^3)$	$(\xi x + 1, \xi^5)$	

The element with minimal leading term is $(\bar{\xi}x + 1, \bar{\xi})$, and the inverse of the root of the polynomial $\bar{\xi}x + 1$ is $\bar{\xi}^5$, indicating that an error has occurred at μe_5^2. Computing the error magnitude, we find that $\mu\mathbf{e}^2 = [0,0,0,0,0,2,0,0] = \mathbf{e}^2$. The actual error vector is determined to be $\mathbf{e} = \mathbf{e}^0 + 3\mathbf{e}^1 + 9\mathbf{e}^2 = [0,1,0,0,0,18,0,0]$.

4   Lifting a Decoder for Lee Metric Alternant Codes

We consider the class of alternant codes $C(N, r, \alpha, \mathbf{Z}_{p^n}) = C(N, r, \alpha, \mathbf{1}, \mathbf{Z}_{p^n})$ where $\mathbf{1} = [1, 1, ..., 1]$. We apply Theorem 2 to lift a decoding algorithm $(\bar{A}, t, \mathrm{d}_L)$ for $C(N, r, \bar{\alpha}, \mathbf{Z}_p)$ to a decoder $(A, t, \mathrm{d}_{L\,\mathrm{max}})$ for $C(N, r, \alpha, \mathbf{Z}_{p^n})$, where $\mu\alpha = \bar{\alpha}$. Let $0 \le r < p$, let $C(N, r, \alpha, \mathbf{Z}_{p^n})$ have minimum Lee distance d. If $n > 1$ then $d \ge 2r$ [5]. If $n = 1$ then

$$d \ge \begin{cases} 2r & \text{if } 0 \le r \le \frac{p-1}{2} \\ p & \text{if } \frac{p+1}{2} \le r \le p - 1 \end{cases}$$

[14, Theorem 1]. Decoding schemes for $C(N, r, \alpha, \mathbf{Z}_p)$ have been given in [5] and [14]. We outline an algorithm, $(\bar{A}, t, \mathrm{d}_L)$, which corrects all errors of Lee weight at most t where

$$t = \begin{cases} r - 1 & \text{if } 0 \le r \le \frac{p-1}{2} \\ \frac{p-1}{2} & \text{if } \frac{p+1}{2} \le r \le p - 1. \end{cases}$$

We introduce notation as defined in [14]. Let $\mathbf{c} \in C(N, r, \alpha, \mathbf{Z}_p)$ be a transmitted codeword and let $\mathbf{v}$ be the received word with associated error vector $\mathbf{e} = \mathbf{v} - \mathbf{c}$, satisfying $\mathrm{wt}_L(\mathbf{e}) \leq t$. The vectors $\mathbf{e}^+$ and $\mathbf{e}^-$ are called the positive and negative error vectors. The syndrome values are defined by $S_\ell = \sum_{j=0}^{N-1} e_j \alpha_j^\ell$ for each $\ell \geq 0$,

$$\Sigma^+ = \prod_{j=0}^{N-1} (1 - \alpha_j x)^{e_j^+} \text{ and } \Sigma^- = \prod_{j=0}^{N-1} (1 - \alpha_j x)^{e_j^-} \text{ are the positive and negative}$$

error locator polynomials and the error-locator ratio $\rho = \frac{\Sigma^+}{\Sigma^-} \in R[[x]]$ satisfies

$$S_j + \sum_{i=1}^{j-1} \rho_i S_{j-i} + j\rho_j = 0 \tag{2}$$

for each $j \geq 1$ [14]. Let Φ be the unique polynomial of degree less than r such that $\Phi \equiv \rho \bmod x^r$. Clearly $\Sigma^- \Phi \equiv \Sigma^+ \bmod x^r$, which gives a key equation. Let M be the module of all solutions to the key equation. From [8, Theorem 3.2] the element (Σ^-, Σ^+) has minimal leading term in M with respect to the term order $<_D$, where $D = \partial \Sigma^+ - \partial \Sigma^-$ and

$$D = \left\{ \begin{array}{l} \bar{S}_0 \quad \text{if } 0 \leq \bar{S}_0 \leq t \\ \bar{S}_0 - p \text{ if } p - t < \bar{S}_0 \leq p - 1 \end{array} \right.$$

[14, Theorem 7]. Since $r < p$, for each $i \in \{1, ..., r - 1\}$ Equation 2 can be solved iteratively for unique $\Phi_i = \rho_i$, where we initialise the sequence by setting $\Phi_0 = 1$. Once Φ has been determined, we compute a Gröbner basis of M, again applying algorithms given in [8]. For each $j \in \{0, ..., N - 1\}$, the error magnitudes e_j^+ and e_j^- correspond to the multiplicity of the linear factor $x - \alpha_j$ in Σ^+ and Σ^-, respectively, and may be determined by adopting a modified Chien search [14]. The lifted version of this algorithm proceeds as follows. Let $C = C(N, r, \alpha, \mathbf{Z}_{p^n})$ and suppose the codeword $\mathbf{c} \in C$ is sent and $\mathbf{v} = \mathbf{c} + \mathbf{e}$ is received, where $\mathrm{wt}_L(\mu \mathbf{e}^i) \leq t$ for each $i \in \{0, ..., n - 1\}$. At the ith iteration assume that $\mathbf{e}^{(i-1)}$ is known and set $\mathbf{v}^{\{i\}} = \mathbf{v} - \mathbf{e}^{(i-1)}$. Then $\bar{\mathbf{H}}(\mu \mathbf{e}^i)^+ = (\mu \mathbf{s}^{[i]})^+ = (\mu (\mathbf{s}^{\{i\}})^i)^+$ and $\bar{\mathbf{H}}(\mu \mathbf{e}^i)^- = (\mu \mathbf{s}^{[i]})^- = (\mu (\mathbf{s}^{\{i\}})^i)^-$. We associate with the ith positive (negative) error vector, $(\mathbf{e}^i)^+$ $((\mathbf{e}^i)^-)$, an ith positive (negative) error locator polynomial,

$$(\bar{\Sigma}^{[i]})^+ = \prod_{j=0}^{N-1} (1 - \bar{\alpha}_j x)^{(\mu e_j^i)^+} \, ((\bar{\Sigma}^{[i]})^- = \prod_{j=0}^{N-1} (1 - \bar{\alpha}_j x)^{(\mu e_j^i)^-}) \text{ in } k_R[x] \text{ and an } i^{th}$$

key equation $(\bar{\Sigma}^{[\ell]})^- \bar{\Phi}^{[\ell]} \equiv (\bar{\Sigma}^{[\ell]})^+ \bmod x^r$. Let $D^{[i]} = \partial (\bar{\Sigma}^{[i]})^+ - \partial (\bar{\Sigma}^{[i]})^-$ and let $\bar{M}^{[i]}$ be the ith solution module. Then the ith required solution $((\bar{\Sigma}^{[i]})^+, (\bar{\Sigma}^{[i]})^-)$ has minimal leading term in $\bar{M}^{[i]}$ with respect to the term order $<_{D^{[i]}}$ and hence is contained in a Gröbner basis of $\bar{M}^{[i]}$. We find the polynomial $\bar{\Phi}^{[i]}$ and then compute the required basis. Once $\mu \mathbf{e}^i$ is known, we can determine the vector $\mathbf{e}^{(i)} = \mathbf{e}^{(i-1)} + p^i \mathbf{e}^i$.

Example 2. Let $R = \mathbf{Z}_{49}$, and let $\alpha = [1, 19, 18, 48, 30, 31]$. Let $C = C(6, 3, \alpha, \mathbf{Z}_{49})$ so that $\bar{C} = C(6, 3, \bar{\alpha}, \mathbf{Z}_7)$ and both C and $\bar{C}$ have minimum Lee distance at least 6. Suppose the all-zero codeword is sent, and the error pattern $\mathbf{e} = [1, 0, 0, 8, 0, 42]$

is received. Let $\mathcal{T} = \{0, 1, ..., 6\}$. Then $\mathbf{e} = [1, 0, 0, 1, 0, 0] + 7[0, 0, 0, 1, 0, 6]$ with respect to $\mathcal{T}$ and $\mathrm{wt}_L(\mu \mathbf{e}^i) = 2$ for $i = 0, 1$. We perform two consecutive implementations of an algorithm $(\bar{A}, 2, d_L)$ for $\bar{C}$. Now $\mu \mathbf{s} = \bar{\mathbf{H}} \mu \mathbf{e} = [2, 0, 2]$ so that $\mu S_0^{[0]} = 2, \mu S^{[0]} = \mu S = 2x^2$, and $D^{[0]} = 2$. We solve for $\bar{\Phi} = 1 + 6x^2$. Then $\{(1, 1 + 6x^2), (0, x^3)\}$ is a Gröbner basis of $\bar{M}^{[0]}$ with respect to the term order $<_2$. The minimal element is $(1, 1 + 6x^2)$ and the polynomial $1 + 6x^2$ has roots $6 = 5^3$ and 1, with corresponding inverses 5^3 and 1, indicating that $(\mu \mathbf{e}^0)^+ = [1, 0, 0, 1, 0, 0]$ and $(\mu \mathbf{e}^0)^- = \mathbf{0}$, so $\mu \mathbf{e}^0 = [1, 0, 0, 1, 0, 0]$. Then $\mu \mathbf{s}^{[1]} = \bar{\mathbf{H}}(\mu \mathbf{v} - \mu \mathbf{e}^0) = [0, 3, 6], \mu S_0^{[1]} = 0, \mu S^{[1]} = 3x + 6x^2$, and $D^{[1]} = 0$. The polynomial $\bar{\Phi}^{[1]}$ is given by $1 + 4x + 5x^2$ and we compute $\{(2x, 2x + x^2), (2 + x, 2 + 2x)\}$, a Gröbner basis of $\bar{M}^{[1]}$ with respect to $<_0$. Then $((\bar{\Sigma}^{[1]})^-, (\bar{\Sigma}^{[1]})^+) = (2 + x, 2 + 2x)$ and the inverses of the roots of $(\bar{\Sigma}^{[1]})^-$ and $(\bar{\Sigma}^{[1]})^+$ are given by $3 = 5^5$ and $6 = 5^3$. It follows that $(\mu \mathbf{e}^1)^+ = [0, 0, 0, 1, 0, 0], (\mu \mathbf{e}^1)^- = [0, 0, 0, 0, 0, 1]$ and thus $\mu \mathbf{e}^1 = [0, 0, 0, 1, 0, 6]$. The error vector is then calculated as $\mathbf{e} = [1, 0, 0, 1, 0, 0] + 7[0, 0, 0, 1, 0, 6] = [1, 0, 0, 8, 0, 42]$.

References

1. A. de Andrade, J. C. Interlando, and R. Palazzo Jnr., "On Alternant Codes Over Commutative Rings", *IEEE Int. Symposium on Information Theory and its Applications*, 1, pp. 231-236, Mexico, 1998.
2. W. W. Adams and P. Loustaunau, "An Introduction to Gröbner Bases", *Graduate Studies in Mathematics*, vol. 3, American Mathematical Society, 1994.
3. E. Byrne and P. Fitzpatrick, "Gröbner Bases over Galois Rings with an Application to Decoding Alternant Codes", *Journal of Symbolic Computation*, to appear.
4. ———, "Hamming Metric Decoding of Alternant Codes Over Galois Rings", preprint.
5. E. Byrne, "Decoding a Class of Lee Metric Codes Over a Galois Ring", preprint.
6. T. Becker, V. Weispfenning, "Gröbner Bases, a Computational Approach to Commutative Algebra", *Graduate Texts in Mathematics*, 141, Springer-Verlag, New York, 1993.
7. D. Cox, J. Little, and D. O'Shea, *Ideals, Varieties, and Algorithms*. New York: Springer-Verlag, 1992.
8. P. Fitzpatrick, "On the Key Equation", *IEEE Trans. Inform. Theory*, vol. 41, pp. 1290-1302, 1995.
9. G. D. Forney, Jr., "On Decoding BCH Codes," *IEEE Trans. Inform. Theory*, vol. 11, pp. 549-557, 1965.
10. M. Greferath, U. Vellbinger, "Efficient Decoding of $\mathbf{Z}_{p^k}$ –Linear Codes", *IEEE Trans. Inform. Theory*, vol. 44, pp. 1288-1291, 1998.
11. B. R. McDonald, *Finite Rings with Identity*, New York: Marcel Dekker, 1974.
12. G. H. Norton and A. Salagean-Mandache, "On the Key Equation Over a Commutative Ring", *Designs, Codes, and Cryptography*, 20(2): pp.125-141, 2000.
13. R. Raghavendran, "Finite Associative Rings", *Compositio Mathematica*, vol. 21, pp.195-229, 1969.
14. R. Roth and P. H. Siegel, "Lee-Metric BCH Codes and Their Application to Constrained and Partial-Response Channels", *IEEE Trans. Inform. Theory*, vol. 40, pp. 1083-1095, 1994.

Sufficient Conditions on Most Likely Local Sub-codewords in Recursive Maximum Likelihood Decoding Algorithms

Tadao Kasami[1], Hitoshi Tokushige[1], and Yuichi Kaji[2]

[1] Faculty of Information Sciences, Hiroshima City University
4–1, Ozuka-Higashi 3, Asaminami, Hiroshima 731–3194, Japan
kasami@cs.hiroshima-cu.ac.jp, tokusige@cs.hiroshima-cu.ac.jp
[2] Graduate School of Information Science, Nara Institute of Science and Technology
8916–5 Takayama, Ikoma, Nara, 630–0101 Japan
kaji@is.aist-nara.ac.jp

Abstract. First, top-down RMLD(recursive maximum likelihood decoding) algorithms are reviewed. Then, in connection with adjacent sub-codewords, a concept of conditional syndrome is introduced. Based on this, sufficient conditions of most likely local sub-codewords in top-down RMLD algorithms are presented. These conditions lead to efficient implementations of top-down RMLD algorithms.

1 Introduction

For $i \leq j$, $[i, j]$ denotes the set of integers from i to j, called a section. For a positive integer n, V^n denotes the set of binary n-tuples. For $\boldsymbol{u} = (u_1, u_2, \ldots, u_n) \in V^n$ and a subset $I = \{i_1, i_2, \ldots, i_m\}$ of $[1, n]$, $p_I \boldsymbol{u} \triangleq (u_{i1}, u_{i2}, \ldots, u_{im})$. For $U \subseteq V^n$, $p_I U \triangleq \{p_I \boldsymbol{u} : \boldsymbol{u} \in U\}$ and $U_I \triangleq p_I \{\boldsymbol{u} \in U : \sup(\boldsymbol{u}) \subseteq I\}$, where $\sup(\boldsymbol{u})$ denotes the support of $\boldsymbol{u}$. For a matrix M with n columns, $p_I M$ denotes the submatrix of M consisting of the i_1-th, the i_2-th,..., the i_m-th columns in this order.

We assume that a binary (N, K) linear block code C is used over an AWGN channel with BPSK signaling and each codeword is equally transmitted. For a received sequence $\boldsymbol{r} = (r_1, r_2, \ldots, r_N)$, let $\boldsymbol{z} = (z_1, z_2, \ldots, z_N)$ denote the binary hard-decision sequence for $\boldsymbol{r}$. For $I \subseteq [1, N]$ and $\boldsymbol{u} \in p_I V^N$, define

$$L(\boldsymbol{u}) = \sum_{\{i \in I \,:\, u_i \neq z_i\}} |r_i|. \tag{1}$$

$L(\boldsymbol{u})$ is called the correlation discrepancy of $\boldsymbol{u}$. By definition, $L(\boldsymbol{u}) \geq L(\boldsymbol{z}) = 0$. For $U \subseteq p_I V^N$, define $L[U] \triangleq \min_{\boldsymbol{u} \in U} L(\boldsymbol{u})$ and for $\boldsymbol{u} \in U$ such that $L(\boldsymbol{u}) = L[U]$, we write $\boldsymbol{u} = v[U]$ and call it the best (or the most likely) in U. For convenience, define $L[\emptyset] = \infty$ for the empty set $\emptyset$. For the most likely codeword $\boldsymbol{c}_{\mathrm{ML}}$ of C, $\boldsymbol{c}_{\mathrm{ML}} = v[C]$.

S. Boztaş and I.E. Shparlinski (Eds.): AAECC-14, LNCS 2227, pp. 333–342, 2001.

We briefly review RMLD(recursive maximum likelihood decoding [1]) and introduce top-down RMLD based on a "call by need" approach [2]-[5]. For a binary linear block code A and its linear subcode B, let A/B denote the set of cosets of B in A. Let $I \subseteq [1, N]$ and D be a coset in $p_I C / C_I$.

Local Optimum [1] : For any codeword $u' \in C$ such that $p_I u' \in D$, there is $u \in C$ such that

$$p_I u = v[D] \text{ and } L(u) \leq L(u'). \tag{2}$$

$v[D]$ is called **the most likely local (MLL) sub-codeword** in D.

Let I and J be disjoint subsets in $[1, N]$. For $u \in p_I V^N$ and $v \in p_J V^N$, $u \circ v$ denotes a binary $(|I| + |J|)$-tuple w such that $p_I w = u$ and $p_J w = v$. Note that $u \circ v = v \circ u$, by definition. For $u \in p_I C$ and $v \in p_J C$, u and v are said to be **adjacent**, if and only if $u \circ v \in p_{I \cup J} C$. For $U_I \subseteq p_I V^N$ and $U_J \subseteq p_J V^N$, define $U_I \circ U_J$ as $\{u \circ v : u \in U_I \text{ and } v \in U_J\}$. Iff u and v are adjacent, $\{u + C_I\} \circ \{v + C_J\} \subseteq p_{I \cup J} C$ and therefore, cosets $\{u + C_I\}$ and $\{v + C_J\}$ are said to be adjacent. The following lemma holds.

Decomposition Lemma [1] : Let I and J be disjoint nonempty subsets of $[1, N]$. Let D be a coset in $p_{I \cup J} C / C_{I \cup J}$. Then there is a unique pair (D_I, D_J) such that $D_I \in p_I C / C_I, D_J \in p_J C / C_J$ and $v[D] = v[D_I] \circ v[D_J]$.

Section Tree : A binary tree, called a section tree, is used to show the partition of local decoding sections in RMLD. A section tree ST is chosen independently of received signal sequences. Each node of ST represents a section and is labeled I_α, where α is a binary sequence. The level of node I_α is defined as $|\alpha|$, the length of α. The root node represents $[1, N]$ and is labeled I_λ. Nonleaf node I_α has two successor nodes denoted $I_{\alpha 0}$ and $I_{\alpha 1}$, called a brother to each other. A complete uniform binary section tree with $N = 2^m, |I_\alpha|$(the length of section I_α) = $2^{m-|\alpha|}$ and $0 \leq |\alpha| \leq m$ represents a uniform binary sectionalization.

We abbreviate $p_I C / C_I$ as T_I, I_α as α and T_{I_α} as T_α. For any index α and an integer l with $1 \leq l \leq |T_\alpha|$, let $v_\alpha(l)$ denote the MLL sub-codeword with the l-th smallest discrepancy in T_α, that is, with the smallest discrepancy in $T_\alpha \backslash \bigcup_{h=1}^{l-1} \{v_\alpha(h) + C_\alpha\}$. From Decomposition Lemma, there is a unique pair $i_\alpha(l)$ and $j_\alpha(l)$, such that $1 \leq i_\alpha(l) \leq |T_{\alpha 0}|, 1 \leq j_\alpha(l) \leq |T_{\alpha 1}|$ and

$$v_\alpha(l) = v_{\alpha 0}(i_\alpha(l)) \circ v_{\alpha 1}(j_\alpha(l)). \tag{3}$$

The most likely codeword $v_\lambda(1)$ is derived from $v_0(i_\lambda(1))$ and $v_1(j_\lambda(1))$ which can be obtained in turn recursively by (3). Simulation results [2]-[4] show $i_\alpha(l) \ll |T_\alpha|$ or $j_\alpha(l) \ll |T_\alpha|$ for almost all cases of relatively small $|\alpha|$. To make effective use of this fact, top-down RMLD is designed.

Suppose $v_{\alpha 0}(i)$ (or $v_{\alpha 1}(j)$) has been found. Then how can we find the best v which is adjacent to $v_{\alpha 0}(i)$ (or $v_{\alpha 1}(j)$)? As is shown in Sections 2 and 3, v is in a small block of $T_{\alpha 1}$ (or $T_{\alpha 0}$) by analyzing the adjacent structure. For example, if $\lambda = 0$, then the block consists of only a single coset. The block can be specified by conditional syndromes introduced in Section 3. In Sections 4 and 5, a procedure for finding $v_\alpha(l)$ and two different types of sufficient conditions that $v_{\alpha 0}(i) \circ v_{\alpha 1}(j) = v_\alpha(l)$ are presented.

2 Adjacency

For $I \subseteq [1, N]$, define $\overline{I} = [1, N] \setminus I$. Let J be a subset of $[1, N]$ disjoint from I, called a conditional subset. For $\boldsymbol{w} \in p_J C$, called a condition vectors, define

$$A_{I|J}(\boldsymbol{w}) \triangleq \{\boldsymbol{u} \in p_I C : \boldsymbol{u} \circ \boldsymbol{w} \in p_{I \cup J} C\}. \tag{4}$$

For $\boldsymbol{u} \in A_{I|J}(\boldsymbol{w})$ and $\boldsymbol{u}' \in p_I C$, $\boldsymbol{u}' \in A_{I|J}(\boldsymbol{w})$ iff

$$\boldsymbol{u} + \boldsymbol{u}' \in p_I C_{\overline{J}}. \tag{5}$$

That is, $A_{I|J}(\boldsymbol{w})$ is a coset of $p_I C / p_I C_{\overline{J}}$, abbreviated as $A_{I|J}$.

Let B_1, B_2, B_3 be linear block codes such that $B_1 \supseteq B_2 \supseteq B_3$. A coset $D_2 \in B_1/B_2$ consists of $|B_2/B_3|$ cosets in B_1/B_3. Let D_2/B_3 denote $\{D_3 \in B_1/B_3 : D_3 \subseteq D_2\}$ and $(B_1/B_2)/B_3$ denote the family of cosets $\{D_3 \in D_2/B_3 : D_2 \in B_1/B_2\}$. Each $D_3 \in D_2/B_3$ is called a B_1/B_2 block.

Since $p_I C \supseteq p_I C_{\overline{J}} \supseteq C_I$, $T_I (= p_I C / C_I)$ is partitioned by $(p_I C / p_I C_{\overline{J}})/C_I$. Each block is called an $A_{I|J}$ block of the same size $B_{I|J} \triangleq |p_I C_{\overline{J}}/C_I|$. If $J = \overline{I}$, then $p_I C_{\overline{J}} = C_I$ and $A_{I|\overline{I}} = T_I$, that is, an $A_{I|\overline{I}}$ block consists of a single coset of T_I.

Let $\{I_1, I_2\}$ be a partition of I. J may be empty. The following lemma holds.

Lemma 1. (Adjacent Structure)

(i) For $\boldsymbol{u}$ and $\boldsymbol{u}'$ in $A_{I|J}(\boldsymbol{w})$, suppose

$$p_{I_1} \boldsymbol{u} \circ p_{I_2} \boldsymbol{u}' \in A_{I|J}(\boldsymbol{w}). \tag{6}$$

Then,

$$p_{I_1} \boldsymbol{u}' \circ p_{I_2} \boldsymbol{u} \in A_{I|J}(\boldsymbol{w}), \tag{7}$$

$$\boldsymbol{u} + \boldsymbol{u}' \in p_{I_1} C_{\overline{I_2 \cup J}} \circ p_{I_2} C_{\overline{I_1 \cup J}}. \tag{8}$$

(ii) Conversely, if $\boldsymbol{u} \in A_{I|J}(\boldsymbol{w})$, $\boldsymbol{u}' \in p_I C$ and (8) holds, then $\boldsymbol{u}' \in A_{I|J}(\boldsymbol{w})$ and (6) and (7) hold.

(Proof) Define $p_{I_i} \boldsymbol{u} \triangleq \boldsymbol{u}_i$ and $p_{I_i} \boldsymbol{u}' \triangleq \boldsymbol{u}'_i$ for $i \in \{1, 2\}$.

(i) From (4), $\boldsymbol{u}_1 \circ \boldsymbol{u}_2 \circ \boldsymbol{w}$ and $\boldsymbol{u}'_1 \circ \boldsymbol{u}'_2 \circ \boldsymbol{w}$ are in $p_{I \cup J} C$. If $\boldsymbol{u}_1 \circ \boldsymbol{u}'_2 \in A_{I|J}(\boldsymbol{w})$, then $\boldsymbol{u}_1 \circ \boldsymbol{u}'_2 \circ \boldsymbol{w}$ is also in $p_{I \cup J} C$. Hence, $\boldsymbol{u}_1 \circ \boldsymbol{u}_2 \circ \boldsymbol{w} + \boldsymbol{u}'_1 \circ \boldsymbol{u}'_2 \circ \boldsymbol{w} + \boldsymbol{u}_1 \circ \boldsymbol{u}'_2 \circ \boldsymbol{w} = \boldsymbol{u}'_1 \circ \boldsymbol{u}_2 \circ \boldsymbol{w} \in p_{I \cup J} C$. That is, $\boldsymbol{u}'_1 \circ \boldsymbol{u}_2 \in A_{I|J}(\boldsymbol{w})$. Since $\boldsymbol{u}_1 \circ \boldsymbol{u}_2$ and $\boldsymbol{u}'_1 \circ \boldsymbol{u}'_2 \in A_{I|J}(\boldsymbol{w})$, $\boldsymbol{u}_1 + \boldsymbol{u}'_1 \in p_{I_1} C_{\overline{I_2 \cup J}}$ and $\boldsymbol{u}_2 + \boldsymbol{u}'_2 \in p_{I_2} C_{\overline{I_1 \cup J}}$. Since $\boldsymbol{u} + \boldsymbol{u}' = (\boldsymbol{u}_1 + \boldsymbol{u}'_1) \circ (\boldsymbol{u}_2 + \boldsymbol{u}'_2)$, (8) holds.

(ii) Since $\boldsymbol{u} \in A_{I|J}(\boldsymbol{w})$ and $\boldsymbol{u} + \boldsymbol{u}' \in p_{I_1} C_{\overline{I_2 \cup J}} \circ p_{I_2} C_{\overline{I_1 \cup J}} \subseteq p_I C_{\overline{J}}$, from (5) $\boldsymbol{u}' \in A_{I|J}(\boldsymbol{w})$. From (8), $\boldsymbol{u}_2 + \boldsymbol{u}'_2 \in p_{I_2} C_{\overline{I_1 \cup J}}$, that is, $\boldsymbol{0}_{I_1} \circ (\boldsymbol{u}_2 + \boldsymbol{u}'_2) \circ \boldsymbol{0}_J \in p_{I \cup J} C$, where $\boldsymbol{0}_{I_1}$ and $\boldsymbol{0}_J$ denote zero vectors over I_1 and J, respectively. Since $\boldsymbol{u}_1 \circ \boldsymbol{u}_2 \circ \boldsymbol{w} \in p_{I \cup J} C$, $\boldsymbol{u}_1 \circ \boldsymbol{u}'_2 \circ \boldsymbol{w} \in p_{I \cup J} C$, that is, $\boldsymbol{u}_1 \circ \boldsymbol{u}'_2 \in A_{I|J}(\boldsymbol{w})$. From (i), (7) also holds. $\triangle$

Define
$$D_i(\boldsymbol{w}) \triangleq A_{I_i|J}(\boldsymbol{w})/C_{I_i}, i \in \{1,2\}. \tag{9}$$
The following graph provides a good insight into the adjacency relation between cosets in $D_1(\boldsymbol{w})$ and $D_2(\boldsymbol{w})$ in $A_{I|J}(\boldsymbol{w})$.

Adjacency Graph : Define G_A as a bipartite graph such that (i) the two sets N_1 and N_2 of nodes labeled with cosets in $D_1(\boldsymbol{w})$ and $D_2(\boldsymbol{w})$, respectively, in a one-to-one way and (ii) two nodes labeled with $\delta_1 \in D_1(\boldsymbol{w})$ and with $\delta_2 \in D_2(\boldsymbol{w})$ are connected by a branch with label $\delta_1 \circ \delta_2$ iff $\delta_1 \circ \delta_2 \subseteq A_{I|J}(\boldsymbol{w})$. $\qquad\triangle$

It follows from (4) that for $i \in \{1,2\}$,

$$p_{I_i} A_{I|J}(\boldsymbol{w}) = A_{I_i|J}(\boldsymbol{w}). \tag{10}$$

Hence the set of branch labels in G_A, denoted $B(G_A)$, is

$$A_{I|J}(\boldsymbol{w})/(C_{I_1} \circ C_{I_2}). \tag{11}$$

Note that
$$p_I C \supseteq p_I C_{\overline{J}} \supseteq p_{I_1} C_{\overline{I_2 \cup J}} \circ p_{I_2} C_{\overline{I_1 \cup J}} \supseteq C_{I_1} \circ C_{I_2}. \tag{12}$$
Then, $A_{I|J}(\boldsymbol{w}) \in p_I C/p_I C_{\overline{J}}$ is partitioned into $A_{I|J}(\boldsymbol{w})/(p_I C_{\overline{I_2 \cup J}} \circ p_{I_2} C_{\overline{I_1 \cup J}})$, whose each block (a coset of $p_I C/(p_{I_1} C_{\overline{I_2 \cup J}} \circ p_{I_2} C_{\overline{I_1 \cup J}})$) consists of $|p_{I_1} C_{\overline{I_2 \cup J}} \circ p_{I_2} C_{\overline{I_1 \cup J}}|/|C_{I_1} \circ C_{I_2}|$ cosets of $p_I C/(C_{I_1} \circ C_{I_2})$. The resulting set of blocks is a partition of $B(G_A)$. Therefore, (i) and (ii) of the following lemma hold from Lemma 1.

Lemma 2. (i) G_A consists of $|p_I C_{\overline{J}}|/|p_{I_1} C_{\overline{I_2 \cup J}} \circ p_{I_2} C_{\overline{I_1 \cup J}}|$ isomorphic complete bipartite subgraphs, called parallel components, and there are no cross connection between them.

(ii) For a parallel component P, its label set of branches, denoted $B(P)$, is $D/(C_{I_1} \circ C_{I_2})$, where D is a coset in $A_{I|J}(\boldsymbol{w})/(p_{I_1} C_{\overline{I_2 \cup J}} \circ p_{I_2} C_{\overline{I_1 \cup J}})$. The two label sets of nodes, denoted $N_i(P)$ with $\{1,2\}$, are $p_{I_i} B(P)$. For any $\boldsymbol{u}_1 \circ \boldsymbol{u}_2 \subseteq \delta_1 \circ \delta_2 \in B(P)$, $N_i(P) = A_{I_i|(I_{i'} \cup J)}(\boldsymbol{u}_{i'} \circ \boldsymbol{w})/C_{I_i} \in (p_{I_i} C/p_{I_i} C_{\overline{I_{i'} \cup J}})/C_{Ii}$, where $i' = 2,1$ for $i = 1,2$, respectively.

(iii) For different $\delta_1 \circ \delta_2 \in B(P)$ and $\delta_1' \circ \delta_2' \in B(P')$, they are included in the same coset of $p_I C/C_I$ only if $P \neq P'$.

(Proof) (iii) For different $\delta_1 \circ \delta_2 \in B(P)$ and $\delta_1' \circ \delta_2' \in B(P')$, suppose $\delta_1 \circ \delta_2 + \delta_1' \circ \delta_2' = (\delta_1 + \delta_1') \circ (\delta_2 + \delta_2') \in C_I$. Since $(p_{I_1} C_{\overline{I_2 \cup J}} \circ p_{I_2} C_{\overline{I_1 \cup J}}) \cap C_I = C_{I_1} \circ C_{I_2}$, if $P = P'$, that is, $\delta_1 \circ \delta_2 + \delta_1' \circ \delta_2' = (\delta_1 + \delta_1') \circ (\delta_2 + \delta_2') \subseteq p_{I_1} C_{\overline{I_2 \cup J}} \circ p_{I_2} C_{\overline{I_1 \cup J}}$, then $\delta_i + \delta_i' \in C_{I_i}$ for $i \in \{1,2\}$. Since δ_i and $\delta_i' \in p_{I_i} C/C_{I_i}$, $\delta_i = \delta_i'$, a contradiction.

Property (iii) of Lemma 2 is refined as follows: For a parallel component P, $T_I(P) \triangleq \{D \in T_I : \text{there is } \delta_1 \circ \delta_2 \in B(P) \text{ such that } \delta_1 \circ \delta_2 \subseteq D\}$.

Further property of G_A: The set of parallel components is partitioned into $|p_I C_{\overline{J}}|/|C_I + (p_{I_1} C_{\overline{I_2 \cup J}} \circ p_{I_2} C_{\overline{I_1 \cup J}})|$ blocks of the same size $|C_I + (p_{I_1} C_{\overline{I_2 \cup J}} \circ p_{I_2} C_{\overline{I_1 \cup J}})|/|p_{I_1} C_{\overline{I_2 \cup J}} \circ p_{I_2} C_{\overline{I_1 \cup J}}|$ in such a way that for parallel components P and P', if P and P' are in the same block, then $T_I(P) = T_I(P')$ and otherwise, $T_I(P) \cap T_I(P') = \emptyset$.

3   Conditional Syndromes

For a linear code A, $A^\perp$, $H(A)$ and $\dim(A)$ denote the dual code of A, a parity check matrix of A and the number of information symbols of A, respectively. For a matrix M, M^T, $r(M)$ and $\mathrm{rank}(M)$ denote the transposition, the number of rows and the rank of M, respectively. For a linear subcode B of A, define $\dim A/B \triangleq \dim A - \dim B$. We can construct a parity check matrix of B whose submatrix of the last $\dim(A^\perp)$ rows is a parity matrix of A. The remaining submatrix of the first $\dim A/B$ rows is called a syndrome matrix of A/B, denoted $H(A/B)$. For an appropriately chosen $H(B)$,

$$H(B) = \begin{bmatrix} H(A/B) \\ H(A) \end{bmatrix}. \tag{13}$$

For a linear subcode E such that $A \supseteq E \supseteq B$, we can construct a syndrome matrix A/B whose submatrix of the last $\dim A/E$ rows is a syndrome matrix of A/E. The remaining submatrix is called a syndrome matrix of $(A/E)/B$.

Let I be a section. For convenience, define the following I-bit order " $\underset{I}{<}$ ". For i and j in $[I, N]$,

$$i \underset{I}{<} j \begin{cases} \text{for } i \text{ and } j \text{ in } I \text{ such that } i < j, \\ \text{for } i \text{ and } j \text{ in } \overline{I} \text{ such that } i < j \text{ in any given ordering of } \overline{I}, \\ \text{for } i \in I \text{ and } j \in \overline{I}. \end{cases}$$

For a section I of main concern, called an m-section, we use the following matrix $H^{(I)}$ as a parity check matrix of C. $H^{(I)}$ is a trellis oriented generator matrix (TOGM [6]) for $C^\perp$ with respect to the I-bit order. For $1 \leq i \leq N - K$, let $ld(i)$ and $tr(i)$ denote the column numbers of the leading '1' and the trailing '1' of the i-th row, denoted $\boldsymbol{h}^{(I)}(i)$, of $H^{(I)}$, respectively. Then by following the definition of TOGM,

$$ld(i) \underset{I}{<} ld(i') \text{ and } tr(i) \neq tr(i'), \text{ for } 1 \leq i < i' \leq N - K. \tag{14}$$

For an m-section, we omit the super index I in $H^{(I)}$ and $\boldsymbol{h}^{(I)}(i)$.

It holds [7] that for $J \subseteq [1, N]$,

$$(C_J)^\perp = p_J(C^\perp), \quad (p_J C)^\perp = (C^\perp)_J. \tag{15}$$

We will choose the following matrices $H_{C,J}$ and H_J as $H(C_J)$ and $H(p_J C)$, respectively.

(H1) $H_{C,J}$: a submatrix of $p_J H$ whose row set is a maximal subset of linearly independent rows of $p_J H$.

$\qquad H_{C,I}$: the submatrix of $p_I H$ consisting of nonzero rows of $p_I H$ from (14).

(H2) H_J : the matrix whose set of rows is $p_I R_J$, where

$$R_J \triangleq \{\boldsymbol{h}(i) : \sup(\boldsymbol{h}(i)) \subseteq J \text{ for } 1 \leq i \leq N - K\}. \tag{16}$$

It follows from (14) that $\operatorname{rank}(H_{C,J}) = \operatorname{r}(H_{C,J})$ and $\operatorname{rank}(H_J) = \operatorname{r}(H_J)$.

A syndrome matrix of $T_I (= p_I C / C_I)$ can be obtained as the submatrix derived from $H_{C,I}$ by deleting the rows of its submatrix H_I. This specific form of $H(T_I)$ is denoted by $H_{S,I}$. For $\boldsymbol{u} \in p_I C$, define the syndrome of $\boldsymbol{u}$, denoted $s_I(\boldsymbol{u})$, as

$$s_I(\boldsymbol{u}) = \boldsymbol{u} H_{S,I}^T. \tag{17}$$

s_I is a linear one-to-one mapping from T_I to $V^{\dim T_I}$ such that for $\boldsymbol{u}$ and $\boldsymbol{u}' \in p_I C$,

$$s_I(\boldsymbol{u}) = s_I(\boldsymbol{u}') \iff \boldsymbol{u} + \boldsymbol{u}' \in C_I. \tag{18}$$

For a nonempty subset D of a coset in $p_I C / C_I$, define $s_I(D) \triangleq s_I(\boldsymbol{u})$ for $\boldsymbol{u} \in D$.

Next we introduce conditional syndromes. Let I be an m-section and J be a set of $[1, N]$ disjoint from I. J is called a conditional subset. From (H2), the rows of $H_{I \cup J}$ can be partitioned into $p_{I \cup J} R_{I,J}$, $p_{I \cup J} R_I$ and $p_{I \cup J} R_J$, where

$$R_{I,J} \triangleq \{ \boldsymbol{h}(i) : \sup(\boldsymbol{h}(i)) \subseteq I \cup J, \sup(\boldsymbol{h}(i)) \cap I \neq \emptyset, \sup(\boldsymbol{h}(i)) \cap J \neq \emptyset, 1 \le i \le N-K \}. \tag{19}$$

The submatrix consisting of the rows of $p_{I \cup J}(R_I \cup R_J)$ is an $H(p_I C \circ p_J C)$ from (H2). Since $p_I C \circ p_J C \supseteq p_{I \cup J} C$, the submatrix consisting of the rows of $p_{I \cup J} R_{I,J}$ is a syndrome matrix of $(p_I C \circ p_J C)/p_{I \cup J} C$. From (14), $p_I R_{I,J}, p_J R_{I,J}$ are linearly independent row sets. Let $H_{I|J}$ and $H_{J|I}$ denote the matrices whose rows sets are given by $p_I R_{I,J}$ and $p_J R_{I,J}$, respectively. From (H1) and (H2), the following lemma holds.

Lemma 3. (Adjacency Lemma)

(i) $\begin{bmatrix} H_{I|J} \\ H_I \end{bmatrix}$ and $\begin{bmatrix} H_{J|I} \\ H_J \end{bmatrix}$ are parity check matrices of $p_I C_{\overline{J}}$ and $p_J C_{\overline{I}}$, respectively.

(ii) For $\boldsymbol{u} \in p_I C$ and $\boldsymbol{w} \in p_J C$, $\boldsymbol{u} \circ \boldsymbol{w} \in p_{I \cup J} C$ iff $\boldsymbol{u} H_{I|J}^T = \boldsymbol{w} H_{J|I}^T$.

$\triangle$

Note that $H_{I|J}$ and $H_{J|I}$ are syndrome matrices $H(p_I C / p_I C_{\overline{J}})$ and $H(p_J C / p_J C_{\overline{I}})$, respectively.

Recall that the row set of $H_{S,I}$ is $p_I R_{I,\overline{I}}$. Hence $H_{S,I}$ can be partitioned into two submatrices $H_{I|J}$ and the remaining submatrix, denoted $H_{I|R}$, whose row set is $p_I \{ \boldsymbol{h}(i) : \sup(\boldsymbol{h}(i)) \cap I \neq \emptyset, \sup(\boldsymbol{h}(i)) \cap (\overline{I} \setminus J) \neq \emptyset, 1 \le i \le N - K \}$.

Then we have the following form of $H_{S,I}$:

$$H_{S,I} = \begin{bmatrix} H_{I|R} \\ H_{I|J} \end{bmatrix}, \tag{20}$$

where $H_{I|R}$ is a syndrome matrix of $A_{I|J}/C_I$.

For $\boldsymbol{u} \in p_I C$ and $\boldsymbol{w} \in p_J C$, $s_{I|R}$ and $s_{I|J}$ syndromes of $\boldsymbol{u}$ and $s_{J|I}$ syndrome of $\boldsymbol{w}$ are defined as

$$s_{I|R}(\boldsymbol{u}) \triangleq \boldsymbol{u} H_{I|R}^T, \quad s_{I|J}(\boldsymbol{u}) \triangleq \boldsymbol{u} H_{I|J}^T, \quad s_{J|I}(\boldsymbol{w}) \triangleq \boldsymbol{w} H_{J|I}^T. \tag{21}$$

It follows from (17), (20) and (21) that

$$s_I(\boldsymbol{u}) = s_{I|R}(\boldsymbol{u}) \circ s_{I|J}(\boldsymbol{u}), \tag{22}$$

and if $\boldsymbol{u} \circ \boldsymbol{w} \in p_{I \cup J}C$,

$$s_I(\boldsymbol{u}) = s_{I|R}(\boldsymbol{u}) \circ s_{J|I}(\boldsymbol{w}). \tag{23}$$

For $\boldsymbol{u}$ and $\boldsymbol{u}' \in p_I C$, $s_{I|J}(\boldsymbol{u}) = s_{I|J}(\boldsymbol{u}')$ iff cosets $\boldsymbol{u} + C_I$ and $\boldsymbol{u}' + C_I$ belong to the same $A_{I|J}$ block. That is, $s_{I|J}(\boldsymbol{u})$ identifies the $A_{I|J}$ block containing $\boldsymbol{u} + C_I$, denoted $A_{I|J}[s_{I|J}(\boldsymbol{u})]$. If $\boldsymbol{u} \circ \boldsymbol{w} \in p_{I \cup J}C$, this block is the same as $A_{I|J}(\boldsymbol{w})$ which is also represented as $A_{I|J}[s_{J|I}(\boldsymbol{w})]$ for convenience. On the other hand, $s_{I|R}(\boldsymbol{u})$ identifies the coset $\boldsymbol{u} + C_I$ in the $A_{I|J}$ block. The above conditional syndrome can be generalized to multiple conditional subsets [5].

4   Search Procedure

For a nonleaf section I_α with $\alpha = a_1 a_2 \cdots a_h \in \{0,1\}^h$, we will present an outline of a recursive procedure for finding $\boldsymbol{v}_\alpha(l)$ based on the decomposition (3). The brother section of an ancestor section $I_{\alpha'}$ of I_α can be a conditional section, where α' is a nonnull prefix of α. Then a conditional set is a union of sections whose index set is a subset of (24):

$$\{\alpha^{(i)} = a_1 a_2 \cdots a_{i-1} \bar{a}_i, 1 \le i \le h\}, \tag{24}$$

where $\bar{0} \triangleq 1$ and $\bar{1} \triangleq 0$. Empty J means no conditional set.

Given a condition vector $\boldsymbol{w} \in p_J C$, our problem is to find the l-th best MLL sub-codeword in $A_{I_\alpha|J}[s]$ with $s \triangleq s_{J|I_\alpha}(\boldsymbol{w})$, denoted by $\boldsymbol{v}_\alpha(J,s;l)$.

Let $I_{\alpha 0}$ and $I_{\alpha 1}$ be a partition of I_α. Either J or $J_b \triangleq J \cup I_{\alpha \bar{b}}$ is a conditional set for $I_{\alpha b}$ with $b \in \{0,1\}$. The following abbreviations will be used:

$$s_b \triangleq s_{J|I_{\alpha b}}(\boldsymbol{w}), \tag{25}$$

$$s_{\bar{b}}(\boldsymbol{u}) \triangleq s_{J_b|I_{\alpha \bar{b}}}(\boldsymbol{w} \circ \boldsymbol{u}), \text{ for } \boldsymbol{u} \in p_{\alpha b}C, \tag{26}$$

$$\boldsymbol{u}_b(i) \triangleq \boldsymbol{v}_{\alpha b}(J, s_b; i), \text{ for } 1 \le i \le B_{I_{\alpha b}|J} = |p_{\alpha b} C_{\overline{J}}/C_{\alpha b}|, \tag{27}$$

$$S_{\alpha,J,s,l} \triangleq \{s_{I_\alpha}(\boldsymbol{v}_\alpha(J,s;i)) : 1 \le i < l\}, \text{ for } 1 \le l \le B_{I_\alpha|J} = |p_\alpha C_{\overline{J}}/C_\alpha|. \tag{28}$$

Let $\boldsymbol{u}'_{\bar{b}}(i)$ denote the best MLL sub-codeword in those cosets $D \in p_{\alpha \bar{b}}C/C_{\alpha \bar{b}}$ which belong to $A_{I_{\alpha \bar{b}}|J_b}[s_{\bar{b}}(\boldsymbol{u}_b(i))]$ block such that $s_{I_\alpha}(\{\boldsymbol{u}_b(i)\} \circ D) \notin S_{\alpha,J,s,l}$.

In the adjacency graph G_A where $I = I_\alpha, I_1 = I_{\alpha 0}$ and $I_2 = I_{\alpha 1}$, $N_1(\boldsymbol{w}) = A_{I_{\alpha 0}|J}[s_0]$, $N_2(\boldsymbol{w}) = A_{I_{\alpha 1}|J}[s_1]$ and $B(G_A) = A_{I|J}[s]/(C_{\alpha 0} \circ C_{\alpha 1})$. For a parallel component P in G_A such that $\{\boldsymbol{u}_0(i) + C_{\alpha 0}\} \circ \{\boldsymbol{u}'_1(i) + C_{\alpha 1}\} \in B(P)$, $N_1(P) = A_{I_{\alpha 0}|J_1}[s_0(\boldsymbol{u}'_1(i))]/C_{\alpha 0}$ and $N_2(P) = A_{I_{\alpha 1}|J_0}[s_1(\boldsymbol{u}_0(i))]/C_{\alpha 1}$. Since P is a complete bipartite graph and the s_{I_α} syndromes of branch labels in P are all different, $\boldsymbol{u}'_1(i)$ is the best MLL sub-codeword in $N_2(P)$ such that $s_{I_\alpha}(\boldsymbol{u}_0(i) \circ \boldsymbol{u}'_1(i)) \notin S_{\alpha,J,s,l}$. Such $\boldsymbol{u}'_1(i)$ exists, if $l \le B_{I_{\alpha 1}|J_0}$.

From Decomposition Lemma, there exist integers j_0 and j_1 such that $1 \leq j_b \leq B_{I_{\alpha b}|J}$ with $b \in \{0, 1\}$,

$$v_\alpha(J, s; l) = v_{\alpha 0}(J, s_0; j_0) \circ v_{\alpha 1}(J, s_1; j_1) \tag{29}$$

$$= u_0(j_0) \circ u_1(j_1), \tag{30}$$

If follows from (30) and the definition of $u'_{\bar{b}}(i)$ that

$$v_\alpha(J, s; l) = u_0(j_0) \circ u'_1(j_0) = u'_1(j_1) \circ u_1(j_1). \tag{31}$$

For $1 \leq \bar{i}_b \leq B_{I_{\alpha b}|J}$ with $b \in \{0, 1\}$,

$$U_f \triangleq \{u_b(i) \circ u'_{\bar{b}}(i) : 1 \leq i \leq \bar{i}_b, b \in \{0, 1\}\}. \tag{32}$$

The next lemma provides a sufficient condition that $v[U_f] = v_\alpha(J, s; l)$. Define

$$L_b(i) = L(u_b(i)), \quad L'_{\bar{b}}(i) = L(u'_{\bar{b}}(i)). \tag{33}$$

Minimality Lemma : Suppose that

$$L(v[U_f]) \leq L_0(\bar{i}_0) + L_1(\bar{i}_1). \tag{34}$$

Then,

$$v[U_f] = v_\alpha(J, s; l). \tag{35}$$

(Proof) If $j_0 \leq \bar{i}_0$ or $j_1 \leq \bar{i}_1$ in (30), then $v_\alpha(J, s; l) = v[U_f]$ from (31) and (32). Suppose $j_0 > \bar{i}_0$ and $j_1 > \bar{i}_1$. Then $L(v_\alpha(J, s; l)) = L_0(j_0) + L_1(j_1) \geq L_0(\bar{i}_0) + L_1(\bar{i}_1) \geq L(v[U_f])$. Hence (35) holds.

5 Sufficient Conditions for Early Termination

For $I \subseteq [1, N]$, let $d_I(C)$ and $d_I(x, y)$ denote the minimum distance of $p_I(C)$ and the Hamming distance between binary $|I|$-tuples x and y, respectively. Let J be a subset disjoint from I.

For $w \in p_J C$ and different x and $y \in A_{I|J}(w), x + y \in p_I C_{\bar{J}}$ and therefore,

$$d_I(x, y) \geq d_I(C_{\bar{J}}) \geq d_{I \cup J} C_{\bar{J}} \geq d_{I \cup J} C. \tag{36}$$

For a subset $B \subseteq A_{I|J}(w)$,

$$A_{I|J}(w) \backslash B \subseteq \{x \in V^{|I|} : d_I(x, u) \geq d_I(C_{\bar{J}}) \text{ for } u \in B\}. \tag{37}$$

Hence we have the following lower bound on $L[A_{I|J}(w) \backslash B]$:

$$L[A_{I|J}(w) \backslash B] \geq \min_{\bigcap_{u \in B} \{x \in V^{|I|} : d_I(x, u) \geq d_I(C_{\bar{J}})\}} L(\bar{x}). \tag{38}$$

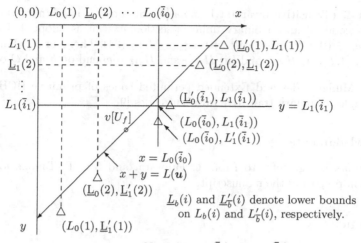

$$\underline{L}_b(i) \text{ and } \underline{L}'_{\bar{b}}(i) \text{ denote lower bounds} \\ \text{on } L_b(i) \text{ and } L'_{\bar{b}}(i), \text{ respectively.}$$

Note that $\boldsymbol{u}_0(\bar{i}_0)$ and $\boldsymbol{u}_1(\bar{i}_1)$
are not necessarily adjacent.

Fig. 1. Illustration for Minimality Lemma

The right-hand side expression can be evaluated by an integer programming approach [8].

Example 1 : A lower bound on $L(\boldsymbol{v}_\alpha(J, s; l))$, where $s = s_{J|I_\alpha}(\boldsymbol{w})$ for $\boldsymbol{w} \in p_J C$:

Let $I = I_\alpha$ and $B = \{\boldsymbol{v}_\alpha(J, s; j) : 1 \le j < l\} \cup C_g$, where $C_g \subseteq U_f$. It follows from (39) that since $L[C_g] \ge L(\boldsymbol{v}_\alpha(J, s; l))$, either $L[C_g] = L(\boldsymbol{v}_\alpha(J, s; l))$ or

$$L(\boldsymbol{v}_\alpha(J, s; l)) \ge \bigcap_{\boldsymbol{u} \in B} \min_{\{\boldsymbol{x} \in V^{|I_\alpha|} : d_{I_\alpha}(\boldsymbol{x}, \boldsymbol{u}) \ge d_{I_\alpha}(C_{\overline{J}})\}} L(\boldsymbol{x}). \tag{39}$$

That is, a sufficient condition that $v[C_g] = \boldsymbol{v}_\alpha(J, s; l)$ is given by

$$L(v[C_g]) \le \text{the right-hand side of (39).} \tag{40}$$

Similarly, lower bounds on $L(\boldsymbol{v}_{\alpha b}(J, s_b; j))$ and $L(\boldsymbol{v}_{\alpha\bar{b}}(J_b, s_{\bar{b}}(\boldsymbol{u}); j))$ can be derived.

6   On Implementation

An effective implementation of the search procedure presented in Section 4 is under study [2]-[4]. Besides the results shown in Sections 2, 3 and 5, the following fact is also of use.

Lemma 4. (Transitive Invariant Lemma [5]) Suppose C is a binary transitive invariant code [9] and a binary uniform sectionalization is adopted. For $\alpha = a_1 a_2 \cdots a_h \in \{0,1\}^h$, define $\alpha^{(j)}$ with $1 \le j \le h$ by (24). Then, for $J = \cup_{j \in Q} I_{\alpha^{(j)}}$ with $Q \subseteq [1, h]$, (i) H_{S, I_α} and (ii) $H_{I_\alpha | J}$ and $H_{J | I_\alpha}$ depend on h and Q only.

Reed-Muller codes and Extended permuted codes of primitive BCH codes are examples of binary transitive invariant codes [9].

Acknowledgments

The authors are grateful to Prof. T. Fujiwara, and Mr. T. Tanoue for their support in preparing the manuscript.

References

1. T. Fujiwara, H. Yamamoto, T. Kasami and S. Lin, "A trellis-based recursive maximum likelihood decoding algorithm for linear codes," *IEEE Trans. Inform. Theory*, vol. 44, pp. 714–729, Mar. 1998.
2. Y. Kaji, T. Fujiwara and T. Kasami, "An efficient call-by-need algorithm for the maximum likelihood decoding of a linear code," *Proc. of ISITA2000*, Honolulu, HA, USA, pp. 335–338, Nov., 2000.
3. T. Kasami, H. Tokushige and Y. Kaji, "Search procedures in top-down recursive maximum likelihood decoding algorithm," *Proc. of SITA2000*, Aso, Kumamoto, Japan, pp. 535–538, Oct. 2000.
4. Y. Kaji, H. Tokushige and T. Kasami, "An improved search algorithm for the adaptive and recursive MLD," *Proc. of 2001 ISIT*, Washington, D.C., Jun. 2001.
5. T. Kasami, "On Recursive Maximum Likelihood Decoding Algorithms for Linear Codes," *Technical Report of IEICE*, IT2000-58, pp. 75–83, Mar. 2001.
6. S. Lin, T. Kasami, T. Fujiwara and M. Fossorier, Trellises and Trellis-based Decoding Algorithms for Linear Block Codes, Kluwer Academic Publishers, Boston, MA, 1998.
7. G. D. Forney Jr., "Coset codes II: Binary lattice and related codes," *IEEE Trans. Inform. Theory*, vol. 34, pp. 1152–1187, Sept. 1988.
8. T. Kasami, "On integer programming problems related to soft-decision iterative decoding algorithms," *Proc. of AAECC-13*, Honolulu, Hawaii, USA. Lecture Notes in Computer Science, vol. 1719, pp. 43–54, Spring Verlag, Nov. 1999.
9. T. Kasami, H. Tokushige, T. Fujiwara, H. Yamamoto and S.Lin, "A recursive maximum likelihood decoding algorithm for some transitive invariant binary block codes," *IEICE Trans. Fundamentals*, vol. E81-A, pp. 1916–1924, Sept. 1998.

A Unifying System-Theoretic Framework for Errors-and-Erasures Reed-Solomon Decoding

Margreta Kuijper[1], Marten van Dijk[2], Henk Hollmann[2] and Job Oostveen[2]

[1] Dept. of EE Engineering, University of Melbourne, VIC 3010, Australia,
m.kuijper@ee.mu.oz.au
[2] Philips Research, 5656 AA Eindhoven, The Netherlands,
{marten.van.dijk,henk.d.l.hollmann,job.oostveen}@philips.com

Abstract. In the literature there exist several methods for errors-and-erasures decoding of RS codes. In this paper we present a unified approach that makes use of behavioral systems theory. We show how different classes of existing algorithms (e.g., syndrome based or interpolation based, non-iterative, erasure adding or erasure deleting) fit into this framework. In doing this, we introduce a slightly more general WB key equation and show how this allows for the handling of erasure locations in a natural way.

1 Introduction

Reed-Solomon (RS) codes find applications in storage and communication systems. Their algebraic structure has given rise to several low-complexity algorithms for error correction. The most well known are the Berlekamp-Massey (BM), the Euclidean and the Welch-Berlekamp (WB) algorithm.

The importance of having an errors-and-erasures correcting algorithm became truly apparent in the seminal paper [9] of G.D. Forney Jr., which presents a generalized minimum distance (GMD) decoding method which repeatedly employs errors-and-erasures decoding. In particular, efficient GMD decoding needs a fast iterative processing of Erasures, i.e., a fast way to obtain the solution for f erasures from the solution with either $f + 1$ or $f - 1$ erasures (named *erasure deletion* and *erasure addition*, respectively).

The decoding of corrupted RS code words boils down to solving a key equation. Classical key equations are the BM key equation and the WB key equation. Araki et al.[1] introduced the generalized key equation of which the classical ones are particular examples. As is shown in Section 2, these key equations can be reformulated in terms of behavioral modeling. Behavioral modeling has already been used to provide a good understanding of errors-only decoding [15,16,17,18].

The main contribution of this paper is in Section 3, where our approach straightforwardly gives rise to a range of errors-and-erasures decoding algorithms, and where we make connections with the existing literature. We unify several presently known iterative errors-and-erasures decoding algorithms in one conceptually clear framework. We explain these algorithms and also give a new

S. Boztaş and I.E. Shparlinski (Eds.): AAECC-14, LNCS 2227, pp. 343–352, 2001.
© Springer-Verlag Berlin Heidelberg 2001

proof of the correctness of classical noniterative errors-and-erasures decoding in terms of behavioral modeling. Further, we generalize the WB key equation, which gives rise to a variant of the WB algorithm with which we can handle erasures.

2   General Framework

2.1   Preliminaries on RS Codes and Key Equations for Errors-Only Decoding

Let $\{x_1, \cdots, x_n\}$ be a subset of a finite field $\mathbb{F}$ with all x_i's distinct. We define aRS code as a set of codewords of the form $c = (M(x_1), \ldots, M(x_n))$, where $M(x)$ is a polynomial of degree $< k$. RS codes are maximum distance separable (MDS), i.e. the minimum Hamming distance d of a (n, k) RS code equals $n - k + 1$. As a result, t errors and f erasures can be corrected if $2t + f \leq d - 1 = n - k$.

For decoding, the above definition naturally leads to the key equation

$$D(x_i)y_i = N(x_i)\eta_i \tag{1}$$

for $i = 0, \ldots, n - k$. Here y_i and η_i are data derived form the received word–in this decoding context all η_i's are nonzero and $y_{n-k} = 0$. The aim of errors-only decoding is to find polynomials $D(x)$ and $N(x)$ that satisfy (1) and for which deg $N \leq$ deg D and deg D is minimal. The error locations are then computed as the zeros of $D(x)$. A well known algorithm for solving this problem is the WB algorithm, which processes the interpolation data (x_i, y_i, η_i) iteratively for $i = 0, \ldots, n - k$. In the literature η_i usually equals 1. In this paper, however, we prefer to leave η_i unspecified (possibly zero) as this allows us to incorporate erasure decoding in Section 3.

Alternatively, a RS code is defined as a set of codewords which have zeros at zero locations $z_1, \ldots, z_{n-k}$. Here the zero locations are prespecified consecutive powers of a primitive element in $\mathbb{F}$. Decoding methods are then derived on the basis of the syndrome sequence $(S_1, \ldots, S_{n-k}) := (r(z_1), \ldots, r(z_{n-k}))$, where $r(x)$ denotes the received polynomial. A relevant equation is Berlekamp's classical key equation

$$\Lambda(x)S(x) = \Omega(x) \mod x^{n-k+1} . \tag{2}$$

Here $S(x) := S_1 x + \cdots + S_{n-k}x^{n-k}$ is the syndrome polynomial. The aim of errors-only decoding is to find polynomials $\Lambda(x)$ and $\Omega(x)$ that satisfy (2) and for which $\Lambda(0) \neq 0$ and max $\{$ deg Λ, deg $\Omega\}$ is minimal. The error locations are then computed as the reciprocals of the zeros of $\Lambda(x)$. This problem is solved by the BM algorithm which iteratively processes the syndrome components. Note that $\Lambda(x)$ corresponds to a shortest LFSR for the syndrome components $S_1, \ldots, S_{n-k}$.

Both of the above described decoding methods are instances of polynomial interpolation. In the first method the interpolation points $x_0, \ldots, x_{n-k}$ are all distinct whereas the second method performs repeated interpolation at one single point $x = 0$ (these originate from interpolation requirements on derivatives of the key equation). We denote the latter as interpolation at $(0, \{0, S_1, \ldots, S_{n-k}\})$. This common interpolation aspect is exploited in recent work [5] by Blackburn who presents a generalized interpolation method that incorporates both types of interpolation.

2.2   Errors-Only Decoding of RS Codes in a Behavioral Framework

Formulation in Terms of Behavioral Modeling. Here we recall how decoding in terms of the above key equations is reformulated as behavioral modeling of certain trajectories of time. Let us start with Berlekamp's classical key equation (2). From the syndromes $S_1, \ldots, S_d$ we define the trajectory $b : \mathbb{Z}_+ \mapsto \mathbb{F}^2$ given by

$$b = \left(\begin{bmatrix} S_{d-1} \\ 0 \end{bmatrix}, \ldots, \begin{bmatrix} S_1 \\ 0 \end{bmatrix}, \begin{bmatrix} 0 \\ 1 \end{bmatrix}, \begin{bmatrix} 0 \\ 0 \end{bmatrix}, \begin{bmatrix} 0 \\ 0 \end{bmatrix} \cdots \right) . \tag{3}$$

It can now be easily verified that $\Lambda(x)$ and $\Omega(x)$ are solutions of (2) if and only if the trajectory b is a solution of the difference equation

$$[\Lambda(\sigma) \quad - \Omega(\sigma)] \begin{bmatrix} w_1 \\ w_2 \end{bmatrix} = 0 \tag{4}$$

in the variable $\begin{bmatrix} w_1 \\ w_2 \end{bmatrix} : \mathbb{Z}_+ \mapsto \mathbb{F}^2$. Here σ stands for the backward shift operator.

Let us now consider the WB key equation (1). From the interpolation data we define d trajectories $b_i : \mathbb{Z}_+ \mapsto \mathbb{F}^2$ given by

$$b_i = \begin{bmatrix} y_i \\ \eta_i \end{bmatrix} (1, x_i, x_i{}^2, \ldots) \qquad \text{for } i = 0, \ldots, d-1. \tag{5}$$

Clearly, the polynomials $D(x)$ and $N(x)$ are solutions of (1) if and only if all trajectories b_i $(i = 0, \ldots, d-1)$ are solutions of the difference equation

$$[D(\sigma) \quad - N(\sigma)] \begin{bmatrix} w_1 \\ w_2 \end{bmatrix} = 0 . \tag{6}$$

For decoding we require in addition that the row degrees of $[\Lambda(x) \quad - \Omega(x)]$ and $[D(x) \quad - N(x)]$, respectively, are minimal. Here the row degree of a polynomial row vector is defined as the maximum degree of its entries. Furthermore, for decoding, we require $\Lambda(0) \neq 0$ for the solution of (4) and $\deg N \leq \deg D$ for the solution of (6). The fact that these requirements differ is solely due to the fact that Berlekamp's key equation (2) aims at reciprocals of error locations rather than at the locations themselves.

Remark 1. Note that, if we process the syndromes in a reversed order then the requirement that $\Lambda(0) \neq 0$ is to be replaced by the requirement that $\deg \Omega \leq \deg \Lambda$.

Having reformulated the two decoding problem statements in a behavioral setting, how do we go about solving it? A model of the form (4)

$$[\Lambda(\sigma) \quad - \Omega(\sigma)] \begin{bmatrix} w_1 \\ w_2 \end{bmatrix} = 0$$

clearly gives rise to a linear σ-invariant solution space ("behavior") spanned by infinitely many trajectories from $\mathbb{Z}_+$ to $\mathbb{F}^2$. For our decoding we require that this behavior contains the given trajectory b, defined by (3). The smallest σ-invariant behavior $\mathcal{B}^*$ that contains b is clearly finite dimensional and given by the span of $b, \sigma b, \ldots, \sigma^d b$. This behavior $\mathcal{B}^*$ is called the *Most Powerful Unfalsified Model (MPUM)* for the data set $\{b\}$, see [23]. For $\mathcal{B}^*$ we can immediately write down a representation, namely

$$\begin{bmatrix} 1 & -(S_1\sigma + \cdots + S_{d-1}\sigma^{d-1}) \\ 0 & \sigma^d \end{bmatrix} w = 0 \ . \tag{7}$$

The above representation is not unique–in fact, all other representations of $\mathcal{B}^*$ can be obtained by left multiplying the matrix in (7) by a unimodular polynomial matrix, i.e. a polynomial matrix whose determinant is a nonzero constant. Note that it follows that the degree of the determinant of any matrix that represents $\mathcal{B}^*$ equals $d = \dim \mathcal{B}^*$, whereas the sum of the row degrees of any such matrix is larger than or equal to d. It can be proven [23] that there exists a representation of $\mathcal{B}^*$ for which equality holds. This representation has minimal row degrees and is called "row reduced". A solution $[\Lambda(x) \quad - \Omega(x)]$ of the decoding problem is simply found by selecting from the two rows in a row reduced representation of $\mathcal{B}^*$ the row of minimal degree that satisfies the additional requirement (here: $\Lambda(0) \neq 0$).

In the case of the WB key equation (1) the approach is completely analogous: simply replace $\{b\}$ by $\{b_0, \ldots, b_{d-1}\}$, defined in (5) and find a row reduced representation for its MPUM accordingly, see [18,19]. In this case we choose the row of minimal degree that satisfies the additional requirement that $\deg N \leq \deg D$.

Algorithms. A well known noniterative algorithm for solving the above decoding problems is the Euclidean algorithm. In [18] it has been explained that the Euclidean algorithm simply brings the matrix in (7) in row reduced form.

Alternatively, the general iterative behavioral modeling procedure of [23, p. 289] can be used. For key equation (2) it is explained in detail in [15] how the BM algorithm can be interpreted as an instance of this procedure.

It has been shown in [18,19] how the same general iterative behavioral modeling procedure of [23] can also be put to work to produce an iterative algorithm for solving key equation (1). The resulting algorithm closely resembles the WB algorithm but involves a different update parameter [18, Sect. 4.4]. It plays a key role in the sequel of this paper. We believe that the behavioral set-up enables a particularly transparent explanation. For this reason we now explain the algorithm as clearly as possible, see also [18, Thm. 4.2].

Algorithm 1. As a first step we initialize

$$R_{-1}(x) := \begin{bmatrix} 1 & 0 \\ 0 & 1 \end{bmatrix}.$$

Note that the row degrees L^1_{-1} and L^2_{-1} of this matrix both equal 0. The behavior represented by $R_{-1}(\sigma)\boldsymbol{w} = 0$ equals $\{0\}$. We now proceed by processing the data (x_i, y_i, η_i) step by step. At step i $(i = 0, \ldots, d-1)$ we process the corresponding trajectory $\boldsymbol{b}_i$ given by (5). For this, we first compute the error trajectory $\boldsymbol{e}_i := R_{i-1}(\sigma)\boldsymbol{b}_i$, which is easily shown to be of the form

$$\boldsymbol{e}_i = \begin{bmatrix} \Delta_i \\ \Gamma_i \end{bmatrix} (1, x_i, x_i^2, \ldots).$$

In fact, Δ_i and Γ_i are computed as

$$\begin{bmatrix} \Delta_i \\ \Gamma_i \end{bmatrix} := R_{i-1}(x_i) \begin{bmatrix} y_i \\ \eta_i \end{bmatrix}.$$

We then choose an update matrix $V_i(x)$ such that $V_i(\sigma)\boldsymbol{w} = 0$ represents the MPUM for $\{\boldsymbol{e}_i\}$. Defining $R_i(x) := V_i(x)R_{i-1}(x)$, we then have that $R_i(\sigma)\boldsymbol{w} = 0$ is a representation that models all data $\boldsymbol{b}_0, \ldots, \boldsymbol{b}_i$ processed so far. We need to choose $V_i(x)$ carefully, so as to produce a row reduced matrix $R_i(x)$. Recall that this means that the sum of the first row degree L^1_i and the second row degree L^2_i of $R_i(x)$ equals the degree of the determinant of $R_i(x)$. This is achieved by making sure that only one of the row degrees of $R_{i-1}(x)$ is increased by one when left multiplied by $V_i(x)$. The following specification satisfies this requirement: if $(\Gamma_i \neq 0$ and $L^1_{i-1} \geq L^2_{i-1})$ or $\Delta_i = 0$ then

$$V_i(x) := \begin{bmatrix} \Gamma_i & -\Delta_i \\ 0 & x - x_i \end{bmatrix}; \qquad L^1_i := L^1_{i-1} \quad \text{and} \quad L^2_i := L^2_{i-1} + 1$$

and, if otherwise,

$$V_i(x) := \begin{bmatrix} x - x_i & 0 \\ \Gamma_i & -\Delta_i \end{bmatrix} \qquad L^1_i := L^1_{i-1} + 1 \quad \text{and} \quad L^2_i := L^2_{i-1}.$$

Note that for efficient implementation it is sufficient to update only L^1_i since $L^1_i + L^2_i = i + 1$ at each step i. After processing all data (x_i, y_i, η_i) for $i = 0, \ldots, d-1$, the matrix $R_d(x)$ is a row reduced representation of the MPUM $\mathcal{B}^*$ of $\{\boldsymbol{b}_0, \ldots, \boldsymbol{b}_{d-1}\}$. It can be proven that $R_{d-1}(x)$ also has the property that the degree of its lower left entry is strictly smaller than the degree of the lower right entry. From the row reducedness of $R_d(x)$ it then follows that the upper left entry $D(x)$ and the upper right entry $N(x)$ are a solution of key equation (1) for which $\deg N \leq \deg D$ and $\deg D$ is minimal.

Remark 2. The above algorithm can be easily adapted (see [5]) so as to process repeated interpolations at $x = 0$, say involving the reversed syndrome polynomial $S_{d-1}x + \cdots + S_1 x^{d-1}$. This straightforwardly gives rise to the algorithm of [18, sect. 4.2] which computes a polynomial whose zeros are the error locations rather than the reciprocals of the error locations, see Remark 1. In this case the discrepancies Δ_i and Γ_i are computed as

$$\begin{bmatrix} \Delta_i \\ \Gamma_i \end{bmatrix} := \text{coeff of } x^i \text{ in } R_{i-1}(x) \begin{bmatrix} S_{d-1}x + \ldots + S_1 x^{d-1} \\ 1 \end{bmatrix} .$$

3   Errors-and-Erasures RS Decoding

In this section we present various methods for errors-and-erasures decoding, most of which can be found in the literature. The main aim of this section is to cast all methods into one conceptually clear framework by reformulation in behavioral modeling terms.

3.1   Noniterative Processing of Erasures

Here we deal with a situation where $f < d$ erasure locations $\alpha_1, \alpha_2, \ldots, \alpha_f$ are a priori specified, for example through the erasure locator polynomial $\Gamma(x) := \prod_{j=1}^{f}(1 - \alpha_j x)$. We seek to find the corresponding error values (possibly zero) as well as additional errors in the non-erased locations. Substituting zeros in the erased positions we first derive syndrome values $S_1, \ldots, S_{d-1}$. Errors-and-erasures decoding amounts to finding the shortest LFSR $\Lambda(x)$ for $S_1, \ldots, S_{d-1}$ that contains $\Gamma(x)$ as a factor. Methods for solving this problem are well known and can be found in e.g. [6,22]. In this subsection we first seek to reformulate the problem in behavioral modeling terms. We then outline how a range of different classical solution methods fits into our framework.

In terms of trajectories, the above requirement that $\Gamma(x)$ is a factor of the errors-and-erasures locator polynomial $\Lambda(x)$ is easily reformulated as the requirement to model not only the trajectory $\boldsymbol{b} : \mathbb{Z}_+ \mapsto \mathbb{F}^2$ given by (3), but also, for $j = 1, \ldots, f$, the trajectories

$$\boldsymbol{b}_j := \begin{bmatrix} 1 \\ 0 \end{bmatrix} (1, \alpha_j^{-1}, \alpha_j^{-2}, \ldots) . \tag{8}$$

With $S(x) := S_1 x + \cdots + S_{d-1} x^{d-1}$, a representation for the MPUM for the set of trajectories $\{\boldsymbol{b}, \boldsymbol{b}_1, \ldots, \boldsymbol{b}_f\}$ is readily obtained as

$$\begin{bmatrix} \Gamma(\sigma) & -\bar{S}(\sigma) \\ 0 & \sigma^d \end{bmatrix} \boldsymbol{w} = 0,$$

where $\bar{S}(x) := \Gamma(x)S(x) \mod x^d$ is the modified syndrome, see e.g. [22]. The task at hand is now simply to bring the matrix in the above equation in row reduced form. As described below, this can be done in a convenient way by making use of the next lemma (whose proof is straightforward) and the decomposition $\bar{S}(x) = \bar{S}_1(x) + x^f \bar{S}_2(x)$, where $\bar{S}_1(x) := \bar{S}_1 x + \bar{S}_2 x^2 + \cdots + \bar{S}_f x^f$ and $\bar{S}_2(x) := \bar{S}_{f+1} x + \cdots + \bar{S}_{d-1} x^{d-1-f}$.

Lemma 1. *Let $a(x)$ and $b(x)$ be polynomials of degree f and let $c(x)$ be a polynomial of degree $\leq f$. Let $F(x)$ be a 2×2 polynomial matrix that is row reduced. Then*

$$F(x) \begin{bmatrix} a(x) & c(x) \\ 0 & b(x) \end{bmatrix}$$

is row reduced.

Theorem 1. *Let*

$$\begin{bmatrix} \Lambda(\sigma) & -\Omega(\sigma) \\ \lambda(\sigma) & -\omega(\sigma) \end{bmatrix} w = 0$$

be a row reduced representation of the MPUM of the trajectory

$$\left(\begin{bmatrix} \bar{S}_{d-1} \\ 0 \end{bmatrix}, \ldots, \begin{bmatrix} \bar{S}_{f+1} \\ 0 \end{bmatrix}, \begin{bmatrix} 0 \\ 1 \end{bmatrix}, \begin{bmatrix} 0 \\ 0 \end{bmatrix}, \begin{bmatrix} 0 \\ 0 \end{bmatrix}, \ldots \right).$$

Define

$$R(x) = \begin{bmatrix} \bar{\Lambda}(x) & -\bar{\Omega}(x) \\ \bar{\lambda}(x) & -\bar{\omega}(x) \end{bmatrix} := \begin{bmatrix} \Lambda(x) & -\Omega(x) \\ \lambda(x) & -\omega(x) \end{bmatrix} \begin{bmatrix} \Gamma(x) & -\bar{S}_1(x) \\ 0 & x^f \end{bmatrix}.$$

Then $R(\sigma)w = 0$ is a row reduced representation of the MPUM of $\{b, b_1, \ldots, b_f\}$, as defined in (3) and (8).

Proof. Applying the above lemma for $a(x) = \Gamma(x)$, $b(x) = x^f$ and $c(x) = -\bar{S}_1(x)$, it follows that $R(x)$ is row reduced. It can also be easily seen that $R(\sigma)w = 0$ represents the MPUM of $\{b, b_1, \ldots, b_f\}$.   □

Because of the above theorem we can perform errors-and-erasures decoding by computing the modified syndrome values $\bar{S}_{f+1}, \ldots \bar{S}_{d-1}$ and constructing a shortest LFSR for them. The latter can be done either noniteratively, by applying the Euclidean algorithm on the polynomials x^{d-f} and $\bar{S}_2(x)$ or iteratively by applying the BM algorithm on $\bar{S}_{f+1}, \ldots \bar{S}_{d-1}$. Both methods are classical and can be found in e.g. [6], see also [8]. The BM type method is essentially equivalent to the method recounted in [21, Sect. II-A] and [6,13]: it can be easily verified that applying BM on $\bar{S}_{f+1}, \ldots \bar{S}_{d-1}$ is the same as applying BM on $S_1, \ldots, S_N$ and initializing with

$$\begin{bmatrix} \Gamma(x) & 0 \\ 0 & x \end{bmatrix}.$$

3.2   Iterative Processing of Erasures

Erasure Deletion through Interpolation at Distinct Points. The most natural way [2,3,4,20] to deal with erasures is to employ an approach based on interpolation at the code locations. Indeed, in this approach the interpolation points can be chosen complementary to the erasure locations, which are thus ignored ("erased"). In fact, we can regard the preliminary step of the WB algorithm, in which k entries are re-encoded, as a case of erasures-only decoding in which $n - k = d - 1$ code locations are erased. In each subsequent step of the WB algorithm one erasure is deleted from the full set of $d - 1$ erasures, until at the last $(d - 1)$st step all erasures have been deleted and errors-only decoding is completed. Thus the WB algorithm and the closely related Algorithm 1 can be regarded as instances of an iterative errors-and-erasures decoding method in which erasures are successively deleted.

Syndrome-Based Erasure Addition. Alternatively, it is possible to formulate a syndrome-based errors-and-erasures decoding method that processes the erasures iteratively, as presented by Kötter in [14]. Indeed, the exposition in Section 3.1 is easily modified to reformulate decoding as the construction of a row reduced representation for the MPUM of the data set $\{\tilde{b}, \tilde{b}_1, \ldots, \tilde{b}_f\}$, where

$$\tilde{b} := \left(\begin{bmatrix} S_1 \\ 0 \end{bmatrix}, \ldots, \begin{bmatrix} S_{d-1} \\ 0 \end{bmatrix}, \begin{bmatrix} 0 \\ 1 \end{bmatrix}, \begin{bmatrix} 0 \\ 0 \end{bmatrix}, \begin{bmatrix} 0 \\ 0 \end{bmatrix}, \ldots \right), \tag{9}$$

and, for $j = 1, \ldots, f$,

$$\tilde{b}_j := \begin{bmatrix} 1 \\ 0 \end{bmatrix} (1, \alpha_j, \alpha_j^2, \ldots) . \tag{10}$$

In the notation of Section 2.1, the decoding problem is thus an interpolation problem with interpolation data $(0, \{0, S_{d-1}, \ldots, S_1\}), (\alpha_1, 1, 0), \ldots, (\alpha_f, 1, 0)$. This approach is close to the work by Kötter [14] who, in behavioral terms, first constructs a row reduced representation for the syndromes, then takes its reciprocal model and proceeds by performing interpolation at the erasure locations. In our set-up we process the syndrome components in a reversed order so that a reciprocal model needs not be computed. Note that the order in which erasures are added is not important. In fact, erasures can even be added after any intermediate syndrome processing iteration, an observation which was also made in [13], where a similar algorithm is presented.

Syndrome-Based Erasure Deletion. In [21] Taipale and Seo employ an erasure deleting approach that is syndrome-based. Their algorithm produces a polynomial whose zeros are the reciprocals of the error locations. Below we present an algorithm which resembles the algorithm in [21] but produces a polynomial whose zeros are the error locations. We found that setting up the algorithm in this way rather than in the reciprocal domain enhances its insightfulness. Similar algorithms to ours have been presented in [10,11,12].

For our syndrome-based erasure deletion approach, we first consider erasures-only decoding, specifying $d - 1$ erasure locations $\alpha_1, \alpha_2, \ldots, \alpha_{d-1}$ and defining $\tilde{\Gamma}(x) := \prod_{j=1}^{d-1}(x - \alpha_j)$. We initialize our algorithm with

$$R_0(x) = \begin{bmatrix} \tilde{\Gamma}(x) & -\tilde{S}(x) \\ 0 & x^d \end{bmatrix},$$

where $\tilde{S}(x) := \tilde{\Gamma}(x)(S_{d-1}x + \cdots + S_1 x^{d-1}) \mod x^d$. Note that the representation

$$\begin{bmatrix} \tilde{\Gamma}(\sigma) & -\tilde{S}(\sigma) \\ 0 & \sigma^d \end{bmatrix} \boldsymbol{w} = 0$$

models $\{\tilde{\boldsymbol{b}}, \tilde{\boldsymbol{b}}_1, \ldots, \tilde{\boldsymbol{b}}_f\}$, given by (9-10). Erasure deletion comes down to removing, one by one, the erasure trajectories $\tilde{\boldsymbol{b}}_j$ $(j = 1, \ldots, d-1)$. After erasing all $d-1$ erasures, the output of the algorithm achieves errors-only decoding. Not surprisingly, the algorithm operates inversely to Algorithm 1. For the sake of brevity we omit its proof here.

Algorithm 2. Initialize

$$R_0(x) := \begin{bmatrix} \tilde{\Gamma}(x) & -\tilde{S}(x) \\ 0 & x^d \end{bmatrix}; \qquad L_0^1 := d-1 \quad \text{and} \quad L_0^2 := d .$$

At step i, process the erasure α_i $(i = 1, \ldots, d-1)$ by computing

$$\begin{bmatrix} \Delta_i \\ \Gamma_i \end{bmatrix} := R_{i-1}(\alpha_i) \begin{bmatrix} 0 \\ 1 \end{bmatrix} .$$

Then define $R_i(x) := V_i(x) R_{i-1}(x)$ where
if ($L_{i-1}^1 \geq L_{i-1}^2$ and $\Gamma_i \neq 0$)   or $\Delta_i = 0$, then

$$V_i(x) := \begin{bmatrix} \frac{\Gamma_i}{x-\alpha_i} & \frac{-\Delta_i}{x-\alpha_i} \\ 0 & 1 \end{bmatrix} \qquad L_i^1 := L_{i-1}^1 - 1 \quad \text{and} \quad L_i^2 := L_{i-1}^2$$

and, if otherwise,

$$V_i(x) := \begin{bmatrix} 1 & 0 \\ \frac{\Gamma_i}{x-\alpha_i} & \frac{-\Delta_i}{x-\alpha_i} \end{bmatrix} \qquad L_i^1 := L_{i-1}^1 \quad \text{and} \quad L_i^2 := L_{i-1}^2 - 1 .$$

Now, the zeros of the upper left entry of $R_i(x)$ are candidate error locations for errors-and-erasures decoding with $d-1-i$ erasures specified.

Note again that for efficient implementation only L_i^1 needs to be specified since $L_i^1 + L_i^2 = 2d - 1 - i$ at each step i.

4   Conclusions

In this paper we put behavioral systems theory to work to provide a unified explanation of a range of iterative errors-and-erasures decoding algorithms in the literature. In doing this, we introduced a slightly more general version of the WB key equation (by introducing the η_i in equation (1)) to accomodate the handling of erasure trajectories. We classified several known iterative procedures for errors-and-erasures RS decoding and gave an overview of the relationships between our framework and the currently known schemes.

References

1. Araki, K. and I. Fujita (1992). Generalized syndrome polynomials for decoding Reed-Solomon codes. *IEICE Trans. Fundamentals* **E75-A**, 1026-1029.
2. Araki, K., Takada, M. and M. Morii (1992). On the efficient decoding of Reed-Solomon codes based on GMD criterion. *Proc. 22nd Int. Symp. on Multiple Valued Logic*, 138-142.
3. Araki, K., Takada, M. and M. Morii (1993). The efficient GMD decoders for BCH codes. *IEICE Trans. Inform. and Systems* **E76-D**, 594-604.
4. Berlekamp, E.R. (1996). Bounded distance+1 soft-decision Reed-Solomon decoding. *IEEE Trans. Inform. Theory* **42**, 704-721.
5. Blackburn, S.R. (1997). A generalized rational interpolation problem and the solution of the WB algorithm. *Designs, Codes and Cryptography* **11**, 223-234.
6. Blahut, R.E. (1983). *Theory and Practice of Error Control Codes*, Addison-Wesley.
7. Elias, P. (1954). Error-free coding. *IRE Trans. Inform. Theory* **PGIT-4**, 29-37.
8. Fitzpatrick, P. (1995). On the key equation. *IEEE Trans. Inform. Theory* **41**, 1290-1302.
9. Forney, G.D., Jr. (1966). Generalized minimum distance decoding. *IEEE Trans. Inform. Theory* **12**, 125-131.
10. Fujisawa, M. and S. Sakata (1999). On fast generalized minimum distance decoding for algebraic codes. *Preproc. AAECC-13*, 82-83.
11. Kamiya, N. (1995). On multisequence shift register synthesis and generalized-minimum-distance decoding of Reed-Solomon codes. *Finite Fields Appl.* **1**, 440-457.
12. Kamiya, N. (1999). A unified algorithm for solving key equations for decoding alternant codes. *IEICE Trans. Fundamentals*, **E82-A**, 1998-2006.
13. Kobayashi, Y., Fujisawa, M. and S. Sakata (2000). Constrained shiftregister synthesis: fast GMD decoding of 1D algebraic codes. *IEICE Trans. Fundamentals* **83**, 71-80.
14. Kötter, R. (1996). Fast generalized minimum distance decoding of algebraic geometric and Reed-Solomon codes. *IEEE Trans. Inform. Theory* **42**, 721-738 .
15. Kuijper, M. and J.C. Willems (1997). On constructing a shortest linear recurrence relation. *IEEE Trans. Aut. Control* **42**, 1554-1558.
16. Kuijper, M. (1999). The BM algorithm, error-correction, keystreams and modeling. In *Dynamical Systems, Control, Coding, Computer Vision*, G. Picci, D. S. Gilliam (eds.), Birkhäuser
17. Kuijper, M. (1999). Further results on the use of a generalized BM algorithm for BCH decoding beyond the designed error-correcting capability, *Proc. 13th AAECC*, Hawaii, USA (1999), 98-99.
18. Kuijper, M. (2000). Algorithms for Decoding and Interpolation. In *Codes, Systems, and Graphical Models* , IMA Series Vol. 123, B. Marcus and J. Rosenthal (eds.), pp. 265-282, Springer-Verlag.
19. Kuijper, M. (2000). A system-theoretic derivation of the WB algorithm, in *Proc. IEEE International Symposium on Information Theory* (ISIT'00) p. 418.
20. Sorger, U. (1993). A new Reed-Solomon decoding algorithm based on Newton's interpolation. *IEEE Trans. Inform. Theory* **39**, 358-365.
21. Taipale, D.J. and M.J. Seo (1994). An efficient soft-decision Reed-Solomon decoding algorithm. *IEEE Trans. Info. Theory* **40**, 1130-1139.
22. Wicker, S.B. (1995). *Error control systems*, Prentice Hall.
23. Willems, J.C. (1991). Paradigms and puzzles in the theory of dynamical systems. *IEEE Trans. Aut. Control* **36**, 259-294.

An Algorithm
for Computing Rejection Probability
of MLD with Threshold Test over BSC

Tadashi Wadayama

Okayama Prefectural University, Okayama, 719-1197, Japan
wadayama@c.oka-pu.ac.jp
http://vega.c.oka-pu.ac.jp/~wadayama/welcome.html

Abstract. An algorithm for evaluating the decoding performance of maximum likelihood decoding (MLD) with threshold test over a binary symmetric channel(BSC) is presented. The proposed algorithm, which is based on the dynamic programming principle, computes the exact values of correct correction, rejection and undetected-error probabilities. The computational complexity of the algorithm is $O(n2^{(1-r)n})$, where n and r denote length and coding rate of the code.

1 Introduction

Binary linear codes such as BCH codes are widely exploited in practical error control systems. On the receiver side of the such systems (e.g., automatic repeat request(ARQ) systems, concatenated coding systems, product coding system, etc.), maximum likelihood decoding (MLD) with *rejection test* is often used. The rejection test examines whether the output of a ML decoder is reliable enough or not. If the output of the ML decoder is rejected, the received word is treated as an *erasure*. In an ARQ system, the receiver sends a request for retransmission of the erased block to the sender. In a concatenated coding system(or a product coding system), the erasure information of an inner code is also utilized for outer code decoding. The erasure information improves overall decoding performance of a concatenated coding system.

In order to design an error control system including MLD with rejection test, we need to evaluate its decoding performance. There are three probabilities that we would like to evaluate: the correct correction probability P_{cr}, the rejection probability P_{rj} and undetected-error probability P_{ud}. In [1], Hashimoto presented performance analysis of several rejection tests and compared their performances based on their error exponents. The exponents reveal the asymptotic behavior of the rejection tests. Hashimoto also reported simulation results for several short codes such as BCH codes of length 15,31 [2]. Almost all rejection tests (which he has tested) give similar performances. ¿From the observation, he conjectured that there are no (or few) differences on performance among them over a binary symmetric channel(BSC).

S. Boztaş and I.E. Shparlinski (Eds.): AAECC-14, LNCS 2227, pp. 353–362, 2001.
© Springer-Verlag Berlin Heidelberg 2001

In this paper, we discuss exact calculation of the decoding performance of MLD with a rejection test. In particular, we here only deal with *threshold test* as the rejection test. The threshold test accepts an output of the ML decoder if the weight of the estimated error is smaller than or equal to τ, where τ is a predetermined threshold value. Otherwise, the threshold test rejects the output. Although the decoding scheme is easy to describe, it is not easy to analyze it even for a BSC. This is because P_{rj} is closely related to enumeration of the complete coset weight distribution of a binary linear code. The complete coset weight distribution is the set of the weight distributions of all the cosets of C. The computation of the complete coset weight distribution is time-consuming; it takes $O(n2^n)$-time with a brute force algorithm[11], where n denotes code length. Furthermore, there are few codes whose complete coset weight distribution is known. These difficulties prevent us deriving the rejection and the undetected-error probabilities of a non-trivial code.

A new algorithm for computing P_{cr}, P_{rj} and P_{ud} over a BSC is presented in this paper. The computational complexity of the algorithm is $O(n2^{(1-r)n})$, where r denotes coding rate of the code. Thus, the proposed algorithm is applicable to any binary linear code with redundancy up to nearly 25–30 bits with a typical computer. For example, the rejection and the undetected-error probabilities of several BCH codes, such as the (63,39,9) BCH code, have been successfully computed with the proposed algorithm.

The proposed algorithm may be used also for the decoding performance analysis of the Chase algorithm and list decoding algorithms which can correct errors of weight beyond $\lfloor (d-1)/2 \rfloor$, where d denotes minimum distance of the code.

2   Preliminaries

2.1   MLD with Threshold Test

Let C be an (n, k, d) binary linear code, where k denotes dimension of the code. A codeword $\boldsymbol{x} \in C$ is assumed to be transmitted to the BSC with the bit error probability p. The vector $\boldsymbol{y}$ is the received word such that: $\boldsymbol{y} = \boldsymbol{x} \oplus \boldsymbol{e}$; the error vector $\boldsymbol{e}$ occurs with the probability $p^{w_H(\boldsymbol{e})}(1-p)^{n-w_H(\boldsymbol{e})}$. The notation $w_H(\boldsymbol{x})$ denotes the Hamming weight of $\boldsymbol{x}$. The operator $\oplus$ represents the component-wise addition over F_2, where F_2 is the Galois field with two elements.

The following explains the decoding rule of MLD with threshold test considered here.

[MLD with threshold test over BSC] For a given received word $\boldsymbol{y}$, we first perform MLD and obtain an estimated word $\hat{\boldsymbol{x}}$ satisfying

$$\hat{\boldsymbol{x}} = \arg\max\{P(\boldsymbol{y}|\boldsymbol{x}) : \boldsymbol{x} \in C\}, \tag{1}$$

where $P(\boldsymbol{y}|\boldsymbol{x}) = p^{d_H(\boldsymbol{x},\boldsymbol{y})}(1-p)^{n-d_H(\boldsymbol{x},\boldsymbol{y})}$ and $d_H(\cdot, \cdot)$ denotes the Hamming distance function. We should clarify the meaning of $\arg\max$ in (1). Let f be a real-valued function whose domain is a finite set X with a total order. We define

$$\arg\max\{f(x) : x \in X\} \stackrel{\triangle}{=} \min\{x \in X : f(x) = f_{max}\}, \qquad (2)$$

where $f_{max} \stackrel{\triangle}{=} \max\{f(x) : x \in X\}$ and the minimum in the right hand side of (2) means the minimum with respect to the total order of X. In a similar way, $\arg\min$ is also defined. In (1), we implicitly assumed that a total order is defined on C. However, we do not specify the order because the order does not affect the following analysis.

After MLD, $\hat{x}$ is examined by the following threshold test: if the weight of the estimated error $\hat{e} \stackrel{\triangle}{=} y \oplus \hat{x}$ is larger than τ, then $\hat{x}$ is rejected. Otherwise, the decoder accepts $\hat{x}$ and outputs it. The parameter $\tau (0 \le \tau \le \rho)$ is the threshold parameter and ρ is the covering radius of C.   □

The events which occur after decoding are classified into the following three categories: *correct decoding*, *undetected-error* and *rejection*. We assume that every codeword in C is transmitted equally likely. The correct decoding probability P_{cr} is given by $P_{cr} \stackrel{\triangle}{=} (1/2^k) \sum_{x \in C} P(\hat{x} = x, w_H(\hat{e}) \le \tau | x)$. The undetected-error probability P_{ud} is given by $P_{ud} \stackrel{\triangle}{=} (1/2^k) \sum_{x \in C} P(\hat{x} \ne x, w_H(\hat{e}) \le \tau | x)$. The rejection probability P_{rj} is given by $P_{rj} \stackrel{\triangle}{=} (1/2^k) \sum_{x \in C} P(w_H(\hat{e}) > \tau | x)$. ¿From the definition of these probabilities and the decoding rule (1), we have the relation $P_{cr} + P_{ud} + P_{rj} = 1$.

2.2   Coset of a Binary Linear Code

An (n, k, d) binary linear code C gives coset decomposition of the vector space F_2^n. Let the *weight of a coset* be the Hamming weight of the minimal weight vector in the coset. A binary vector whose Hamming weight is minimal in a coset is called the *coset leader* of the coset. The number of the coset leaders with the Hamming weight $i (0 \le i \le \rho)$ is denoted by α_i[3]. The $\rho + 1$-tuple $(\alpha_0, \alpha_1, \alpha_2, \ldots, \alpha_\rho)$ is called *the weight distribution of coset leaders* of C.

Let $L(C) = \{v_1, v_2, \ldots, v_{2^{n-k}}\}$ be a set of the coset leaders of C. A coset including the coset leader $v \in L(C)$ is represented by $C_v \stackrel{\triangle}{=} \{u \oplus v : u \in C\}$. MLD can be achieved by *standard array decoding* (see [4]). Standard array decoding is the following procedure: if a received word y belongs to $C_{v'}$, the decoder outputs $\hat{x} = y \oplus v'$ as the estimated word.

3   Performance of MLD with Threshold Test

3.1   Correct Decoding Probability

Hereafter, we assume that the transmitted word x is the zero vector $\mathbf{0}$ without loss of generality. The notation $\mathbf{0}$ means the n-tuple of zeros. Using the notation on cosets defined above, we can rewrite P_{cr} into the following form:

$$P_{cr} = P(\hat{x} = \mathbf{0}, w_H(y) \le \tau | \mathbf{0}) \qquad (3)$$

$$= \sum_{\boldsymbol{y} \in L(C), w_H(\boldsymbol{y}) \leq \tau} P(\boldsymbol{y}|\boldsymbol{0}). \tag{4}$$

It is evident that the received word $\boldsymbol{y}$ is successfully corrected when $\boldsymbol{y} \in L(C)$ and $w_H(\boldsymbol{y}) \leq \tau$. Thus, the above equation can be further simplified in such a way:

$$P_{cr} = \sum_{i=0}^{\tau} \alpha_i p^i (1-p)^{n-i}. \tag{5}$$

The equation (5) implies that the knowledge on the weight distribution of coset leaders is required for evaluating P_{cr}. When $\tau = \rho$, we have $P_{ML} \overset{\triangle}{=} \sum_{i=0}^{\rho} \alpha_i p^i (1-p)^{n-i}$, which is the correct decoding probability of MLD (without threshold test)[3].

3.2   Calculation of the Probabilities: Case I ($\tau \leq \lfloor (d-1)/2 \rfloor$)

In the case where $\tau \leq \lfloor (d-1)/2 \rfloor$, the correct decoding region $\{\boldsymbol{y} \in L(C) : w_H(\boldsymbol{y}) \leq \tau\}$ coincides with the n-dimensional Hamming sphere $\{\boldsymbol{u} \in F_2^n : w_H(\boldsymbol{u}) \leq \tau\}$. This property greatly simplifies the analysis on P_{cr}, P_{ud} and P_{rj}. For example, it is known that $\alpha_i = \binom{n}{i}$ holds for $0 \leq i \leq \lfloor (d-1)/2 \rfloor$[3]. Hence, we can easily compute P_{cr} by (5).

It is also known that, for the case $\tau \leq \lfloor (d-1)/2 \rfloor$, the probabilities P_{rj} and P_{ud} can be obtained based on the knowledge of the weight distribution of C[5][6].

3.3   Calculation of the Probabilities: Case II($\tau > \lfloor (d-1)/2 \rfloor$)

In the case where $\tau > \lfloor (d-1)/2 \rfloor$, the situation becomes complicated. This is because the acceptance region is no longer a Hamming sphere. We require more detailed feature of C than its weight distribution.

If the received word $\boldsymbol{y}$ is fallen to a coset with weight larger than τ, the standard array ML decoder outputs $\boldsymbol{y} \oplus \hat{\boldsymbol{e}}$, where $w_H(\hat{\boldsymbol{e}}) > \tau$. Therefore, the rejection probability is given by $P_{rj} = \sum_{\boldsymbol{y} \in V(C,\tau)} P(\boldsymbol{y}|\boldsymbol{0})$, where $V(C,\tau)$ is the set of the vectors contained in the cosets of C with weight larger than τ:

$$V(C,\tau) \overset{\triangle}{=} \bigcup_{\boldsymbol{v} \in L(C), w_H(\boldsymbol{v}) > \tau} C_{\boldsymbol{v}}.$$

Let $W_{C_{\boldsymbol{v}}}(x,y)$ be the weight enumerator of the coset $C_{\boldsymbol{v}}$: $W_{C_{\boldsymbol{v}}}(x,y) = \sum_{\boldsymbol{w} \in C_{\boldsymbol{v}}} x^{w_H(\boldsymbol{w})} y^{n-w_H(\boldsymbol{w})}$. Using the weight enumerator of the cosets, the rejection probability P_{rj} is represented by $P_{rj} = \sum_{\boldsymbol{v} \in L(C) \cap V(C,\tau)} W_{C_{\boldsymbol{v}}}(p, 1-p)$. The undetected-error probability P_{ud} is given by $P_{ud} = 1 - P_{cr} - P_{rj}$.

As we have seen above, the knowledge of the complete coset weight distribution enables us to compute P_{rj} and P_{ud} when $\tau \geq \lfloor (d-1)/2 \rfloor$.

4   Algorithm for Computing Rejection Probability

An algorithm for computing the rejection probability for a given triple (C, p, τ) is presented here. As discussed in Section 3.2, there is a simple method for computing these probabilities when $\tau \leq \lfloor (d-1)/2 \rfloor$. The following algorithm makes computation for $\tau > \lfloor (d-1)/2 \rfloor$ feasible.

4.1 Proposed Algorithm

An $(n-k) \times n$ parity check matrix of C is denoted by $H = \{h_1, h_2, \ldots, h_n\}$, where h_i is the i-th binary column vector of length $n-k$. In this paper, we consider that the binary column vector h_i belongs to F_2^{n-k}.

For each $\sigma \in F_2^{n-k}$ and $0 \leq t \leq n$, we define $P(\sigma, t)$ and $W(\sigma, t)$ by the following recursive formulas:

$$P(\sigma, 0) \stackrel{\triangle}{=} \begin{cases} 1, \sigma = 0 \\ 0, \sigma \neq 0 \end{cases} \tag{6}$$

$$P(\sigma, t) \stackrel{\triangle}{=} (1-p)P(\sigma, t-1) + pP(\sigma \oplus h_t, t-1) \tag{7}$$

$$W(\sigma, 0) \stackrel{\triangle}{=} \begin{cases} 0, \sigma = 0 \\ \infty, \sigma \neq 0 \end{cases} \tag{8}$$

$$W(\sigma, t) \stackrel{\triangle}{=} \min\{W(\sigma, t-1), W(\sigma \oplus h_t, t-1) + 1\}. \tag{9}$$

The next lemma plays a key role in the proposed algorithm.

Lemma 1. *The probabilities P_{cr}, P_{rj} and P_{ud} are given by*

$$P_{cr} = \sum_{i=0}^{\tau} \alpha_i p^i (1-p)^{n-i} \tag{10}$$

$$P_{rj} = \sum_{\sigma \in U(C,\tau)} P(\sigma, n) \tag{11}$$

$$P_{ud} = 1 - P_{cr} - P_{rj}, \tag{12}$$

where

$$U(C, \tau) \stackrel{\triangle}{=} \{\sigma \in F_2^n : W(\sigma, n) > \tau\} \tag{13}$$

and α_i's $(0 \leq i \leq \rho)$ are obtained by

$$\alpha_i = |\{\sigma \in F_2^{n-k} : W(\sigma, n) = i\}|. \tag{14}$$

Proof: For $\sigma \in F_2^{n-k}$, $0 \leq t \leq n$, we define $C(\sigma, t)$ by

$$C(\sigma, t) \stackrel{\triangle}{=} \{u \in F_2^t : (u \cdot 0^{n-t})H^T = \sigma\}. \tag{15}$$

$$C(\sigma, 0) \stackrel{\triangle}{=} \emptyset, \quad \sigma \in F_2^{n-k}, \tag{16}$$

where 0^{n-t} represents the zero vector of length $n-t$. The symbol $\cdot$ denotes the concatenation operator. ¿From the recursive formula on $W(\sigma, t)$ (9) and the definition (15), we see that $W(\sigma, t)$ coincides with the weight of the minimal weight vector in $C(\sigma, t)$. It is also obvious that $C(\sigma, n)$ is the coset of C corresponding to the syndrome σ. Since $W(\sigma, n)$ is the weight of the minimal weight vector in $C(\sigma, n)$, $W(\sigma, n)$ can be considered as the weight of the coset corresponding to syndrome σ. This explains the validity of (14).

The recursive formulas on $C(\sigma, t)$ and $P(\sigma, t)$ lead to

$$P(\sigma, t) = \sum_{c \in C(\sigma, t)} p^{w_H(c)} (1 - p)^{t - w_H(c)}. \tag{17}$$

Note that $P(\sigma, n)$ can be rewritten into:

$$P(\sigma, n) = \sum_{c \in C(\sigma, n)} p^{w_H(c)} (1 - p)^{n - w_H(c)} \tag{18}$$

$$= \sum_{y \in C(\sigma, n)} P(y|0). \tag{19}$$

The right hand side of (19) is the probability such that the received word y is fallen into the coset corresponding to syndrome σ under the assumption $x = 0$. The rejection probability is, thus, obtained by

$$P_{rj} = \sum_{y \in V(C, \tau)} P(y|0) \tag{20}$$

$$= \sum_{\sigma \in U(C, \tau)} P(\sigma, n). \tag{21}$$

□

The recursive formulas (7) and (9) naturally give the following algorithm for computing the target probabilities.

[Proposed Algorithm]

Step 1 Set $P(\sigma, 0)$ and $W(\sigma, 0)$ from to (6) and (8).
Step 2 Set $t := 1$.
Step 3 Compute $P(\sigma, t)$ and $W(\sigma, t)$ for each $\sigma \in F_2^{n-k}$ from (7) and (9).
Step 4 If $t < n$, then set $t := t + 1$ and go to Step 3.
Step 5 Compute (P_{cr}, P_{rj}, P_{ud}) from Lemma 1 and output them.   □

The proposed algorithm is based on the dynamic programming principle and similar to the Viterbi algorithm (recursive computation of $W(\sigma, t)$) and forward computation of the BCJR algorithm[13] (recursive computation of $P(\sigma, t)$).

4.2   Syndrome Trellis Representation

The best way to understand the proposed algorithm is to use the *syndrome trellis* representation of F_2^n. The syndrome trellis is closely related to the Wolf-trellis[14] and the BCJR-trellis[13]. The BCJR-trellis is naturally defined from the parity check matrix of a target code and it contains all the codewords of C as its label sequences. For our purpose, we define the syndrome trellis which is also defined based on the parity check matrix of C.

The syndrome trellis of C is a directed graph with edge labels. Each node in the syndrome trellis is associated with a pair (σ, t) for $\sigma \in F_2^{n-k}, 0 \le t \le n$. There exists an edge with label 0 between the nodes $(\sigma, t-1)$ and (σ, t). In a similar way, there exists an edge with label 1 between the nodes $(\sigma, t-1)$ and $(\sigma \oplus h_t, t)$. The set of the label sequences from $(0, 0)$ to $(0, n)$ coincides with C. Furthermore, the syndrome trellis contains all the binary vectors of length n. For example, the set of the label sequences from $(0, 0)$ to (σ, n) coincide with the coset of C corresponding to the syndrome σ. More detailed definition of the syndrome trellis can be found in [9][11].

4.3   Related Algorithms

The computation of the weight distribution of a binary linear code using trellis structure has started from the work by Desaki et al.[8]. They presented an efficient algorithm for computing the weight distribution of a given binary linear code using its minimal trellis. Wadayama et al.[9] proposed an algorithm(WWK-algorithm) for enumerating the weight distribution of coset leaders for a given binary linear code. They computed several weight distributions of coset leaders for primitive BCH codes, extended primitive BCH codes and Reed-Muller codes with $n - k \le 28$ and $n \le 128$. This algorithm is based on the syndrome trellis of a target code and the recursive formula (9). Thus, we can regard the proposed algorithm as a natural extension of the WWK-algorithm. Recently, an improved algorithm has been presented by Maeda et al.[10]. With their algorithm, they disclosed weight distributions of coset leaders for several codes with $n \le 128$ and $n - k \le 42$ such as the (64,22) BCH-code. Fujita et al.[11] extended the WWK-algorithm to an algorithm for computing the complete coset weight distribution. They have computed the complete coset weight distributions of several codes such as the (63,39)-BCH code. Their algorithm (FW-algorithm) is based on the same principle used in the proposed algorithm. Let $E(\sigma, t)$ be the the weight enumerator for $C(\sigma, t)$ defined by $E(\sigma, t) \stackrel{\triangle}{=} \sum_{c \in C(\sigma, t)} z^{w_H(c)}$. There is a recursive relation on the weight enumerators for $\sigma \in F_2^{n-k}, 1 \le t \le n$:

$$E(\sigma, 0) \stackrel{\triangle}{=} \begin{cases} 1, & \sigma = 0^{n-k} \\ 0, & \sigma \neq 0^{n-k} \end{cases} \tag{22}$$

$$E(\sigma, t) \stackrel{\triangle}{=} E(\sigma, t-1) + zE(\sigma \oplus h_t, t-1). \tag{23}$$

The weight enumerator $E(\sigma, n)$ is the weight enumerator of the coset corresponding to the syndrome σ. This recursive formula (23) gives the FW-algorithm. Note that this recursive formula is exactly the same as the one used in the algorithm by Desaki et al[7][8]. The difference of the two algorithms lies in the trellis where the two algorithms work on. Table 1 summarize the algorithms closely related to the proposed algorithm.

Table 1. Summary of related algorithms

Target	Required recursive formula	Reference
covering radius	(9)	[9][10]
weight distribution	(23)	[7][8]
weight distribution of coset leaders	(9)	[9][10]
complete coset weight distribution	(23)	[11]
$P_{cr}, P_{rj}, P_{ud}(0 \leq \tau \leq \rho)$	(7) and (9)	This section
average distortion	(7) and (9)	Appendix

4.4   Time and Space Complexity of the Proposed Algorithm

In order to obtain the probabilities (P_{cr}, P_{ud}, P_{rj}), we need to evaluate $P(\boldsymbol{\sigma}, t)$ and $W(\boldsymbol{\sigma}, t)$ for each $\boldsymbol{\sigma} \in F_2^{n-k}$ and $1 \leq t \leq n$. There are at most $n2^{n-k}$-pairs of $(\boldsymbol{\sigma}, t)$ and it takes constant time to calculate (7) and (9) for each pair. Hence, the time complexity of the proposed algorithm is $O(n2^{(1-r)n})$, where $r \overset{\triangle}{=} k/n$.

In the above discussion, we assumed that the addition $\boldsymbol{a} \oplus \boldsymbol{b}, \boldsymbol{a}, \boldsymbol{b} \in F_2^{n-k}$ which appears in the recursive formulas takes constant time to compute regardless of $n - k$. This assumption seems somewhat artificial but the behavior of the proposed algorithm implemented in a computer is well described with this assumption. This is because most computers can perform addition of several bits(32 or 64) in parallel and it is impossible to treat the code with $n - k \geq 64$.

To compute $W(\boldsymbol{\sigma}, t)$, we only need to keep $W(\boldsymbol{\sigma}, t-1)$ for all $\boldsymbol{\sigma}$. We thus need memory space proportional to $\log_2 n2^{n-k}$ for computing $W(\boldsymbol{\sigma}, t)$. In a similar way, memory space proportional to 2^{n-k} is required for computing $P(\boldsymbol{\sigma}, t)$. As a result, the space complexity of the proposed algorithm is $O(\log_2 n2^{(1-r)n})$.

A straightforward implementation of the proposed algorithm (without any modification) works well when $n - k \leq 25\text{--}30$ with a today's typical personal computer. The technique devised by Maeda et al.[10] could be used to improve this upper limitation.

The known algorithm for computing the complete coset weight distribution is much slower and less space efficient than the proposed algorithm. For example, the FW-algorithm[11] is also slower and less space efficient. Table 2 presents the comparison of the time and space complexities of the proposed algorithm, the FW-algorithm and a brute force(BF) algorithm. The BF algorithm generates all the binary vectors of length n to compute the target probabilities. ¿From Table 2, we see that the proposed algorithm is superior to the FW-algorithm and the BF-algorithm in terms of the time and space complexities.

Appendix

Average Distortion

As an application of the proposed algorithm, we here discuss a performance analysis method for a vector quantizer using a binary linear code C. The following

Table 2. Comparison of time and space complexities

	Time complexity	Space complexity
Proposed	$O(n2^{(1-r)n})$	$O(\log n2^{(1-r)n})$
FW[11]	$O(n^2 2^{(1-r)n})$	$O(n^3 2^{(1-r)n})$
BF	$O(n2^n)$	–

scenario is supposed. ¿From an information source, a binary vector $y \in F_2^n$ is equally likely generated. The vector quantizer finds the nearest codeword to y in such a way: $\hat{x} = \arg\min\{d_H(y, x) : x \in C\}$. ¿From the above scenario, the average distortion of C is defined by

$$D(C) \triangleq E[d_H(y, \hat{x})] \tag{24}$$

$$= \sum_{y \in F_2^n} \frac{1}{2^n} d_H(y, \hat{x}). \tag{25}$$

Assume the BSC with $p = 1/2$. Of course, in this case, the received word y is independent from the transmitted word x. The quantization process is equivalent to MLD over this BSC. Hence, if x belongs to the coset C_v, the distance $d_H(x, \hat{x})$ becomes the weight of the coset, namely $w_H(v)$. ¿From this observation, we can express $D(C)$ by $W(\sigma, n)$ and $P(\sigma, n)$:

$$D(C) = \sum_{v \in L(C)} w_H(v) P(y \in C_v | 0) \tag{26}$$

$$= \sum_{v \in L(C)} w_H(v) \sum_{y \in C_v} P(y | 0) \tag{27}$$

$$= \sum_{\sigma \in F_2^{n-k}} W(\sigma, n) P(\sigma, n). \tag{28}$$

Using the proposed algorithm and the above relation, we can compute $D(C)$ efficiently. In Table 3, the average distortions of several codes computed by the above algorithm are shown.

Table 3. Average distortion of several codes

Code	Average distortion
(24,12,8) Golay code	3.353
(31,16,7) BCH code	4.282
(31,11,11) BCH code	6.069
(63,45,7) BCH code	4.061
(63,39,9) BCH code	5.595

References

1. T. Hashimoto, "Composite scheme LR+Th for decoding with erasures and its effective equivalence to Forney's rule," *IEEE Trans. Inform. Theory*, IT-45, No.1, pp.78–93 (1999).
2. T. Hashimoto, "On the erasure/undetected-error probabilities over BSC," *IEICE Technical Report*, IT-2000-40,pp.23–27, Tokyo (2000).
3. F.J.MacWilliams and N.J.A.Sloane, "The theory of error correcting codes," North-Holland (1986).
4. W.W. Peterson and E.J.Weldon,Jr., "Error-correcting codes,"2nd ed.,MIT press (1972).
5. E.R.Berlekamp, "Algebraic coding theory," Revised 1984 edition, Aegean Park Press (1984).
6. H. Imai, "Fugou riron (coding theory)," IEICE (1990).
7. Y.Desaki, T.Fujiwara and T.Kasami, "The weight distributions of extended binary primitive BCH codes of length 128," *IEEE Trans. Inform. Theory*, vol. IT-43, pp. 1364–1371, July (1997).
8. Y.Desaki, T.Fujiwara, and T.Kasami, "A method for computing the weight distribution of a block code by using its trellis diagram," *IEICE Trans. Fundamentals*, vol. E77-A, No8, pp. 1230–1237 Aug. (1994).
9. T. Wadayama, K. Wakasugi and M. Kasahara, "A trellis-based algorithm for computing weight distribution of coset leaders for binary linear codes," in *Proceedings of International Symposium on Information Theory and Its Application (ISITA 1998)*, Mexico City, Mexico, Oct. (1998).
10. M.Maeda and T.Fujiwara, "An algorithm for computing weight distribution of coset leaders of binary linear block codes," in *Proceedings of International Symposium on Information Theory (ISIT 2000)*, Sorrento, Italy, June (2000).
11. Y. Fujita and T. Wadayama, "An algorithm for computing complete coset weight distribution of binary linear codes," in *Proceedings of 23th Symposium on Information Theory and Its Applications (SITA 2000)*, Aso, pp.647–650, Oct. (2000).
12. A.Kuznetsov, Francis Swarts, A.J.Han Vinck and Hendrik C. Ferreira, "On the undetected error probability of linear block codes on channels with memory," *IEEE Trans. Inform. Theory*, vol.IT-42,pp.303–309 (1996).
13. L.R.Bahl, J.Cocke , F.Jelinek and J.Raviv, "Optimal decoding of linear codes for minimizing symbol error rate," *IEEE Trans. Inform. Theory*, vol.IT-20,pp.284–287 (1974).
14. J. K. Wolf, "Efficient maximum likelihood decoding of linear block codes using a trellis," *IEEE Trans. Inform. Theory*, vol.IT-24, pp.76–80 (1978).

Cartan's Characters
and Stairs of Characteristic Sets

François Boulier and Sylvain Neut

Université Lille I, Laboratoire d'Informatique Fondamentale
59655 Villeneuve d'Ascq, France
{boulier, neut}@lifl.fr

Abstract. Differential geometry and differential algebra are two formalisms which can be used to study systems of partial differential equations. Cartan's characters are numbers which naturally appear in the former case ; stairs of characteristic sets are pictures naturally drawn in the latter. In this paper, we clarify the relationship between these two notions. We prove also some invariant properties of characteristic sets.

1 Introduction

Cartan's characters are numbers (ranks of matrices) associated to analytic systems of exterior differential forms. Exterior differential systems come from differential geometry [4,11]. A solution of an analytic exterior differential system S is a differential manifold (the integral manifold of S) the tangent space of which annihilates the differential system (Cartan–Kähler theorem). The Cartan's characters of an analytic exterior differential system S indicate if S has solutions, provide the dimension of the integral manifold of S and the number of the arbitrary analytic functions of the highest number of variables the integral manifold depends on. Cartan's characters are geometric objects, in the sense that they are invariant under the action of changes of coordinates. Every analytic system Σ of partial differential equations can be transformed as an analytic exterior differential system S. The integral manifold of S is however a solution of the PDE system Σ only if the exterior differential system S satisfies an additional hypothesis: being in involution.

A characteristic set is a set of differential polynomials. This notion is defined in differential algebra [13,6] which aims at solving systems of ordinary or partial differential equations from a purely algebraic point of view. A solution of a system of differential polynomials is a point with coordinates in some differential field extension of the base field of the polynomials. A characteristic set of the differential ideal $\mathfrak{a}$ generated by a system of differential polynomials Σ indicates if this system has solutions and permits to compute formal power series solutions of Σ. These formal power series may not be convergent in any neighborhood of the expansion point whence do not necessarily represent analytic solutions of Σ. The characteristic sets of $\mathfrak{a}$ are not geometric objects in the sense that they depend on some ranking on the sets of the derivatives of the differential

S. Boztaş and I.E. Shparlinski (Eds.): AAECC-14, LNCS 2227, pp. 363–372, 2001.
© Springer-Verlag Berlin Heidelberg 2001

indeterminates and that their images under changes of coordinates are usually not characteristic sets. The stairs generated by a characteristic set C of $\mathfrak{a}$ are pictures which can be drawn by looking at C. They reflect some properties of C and of the solutions of Σ.

In this paper, we clarify the relationship between Cartan's characters and stairs of a particular class of characteristic sets. We show how to read the first Cartan's character of the exterior differential system obtained from a PDE system Σ in the stairs of the orderly characteristic sets of $\mathfrak{a}$ and we give a conjecture for the last Cartan's character. To support this conjecture, we prove that the properties of orderly characteristic sets which give this last Cartan's character are invariant under change of orderly ranking (theorem 2) and under the action of some changes of coordinates (theorem 3). We reformulate also Cauchy–Kovalevskaya theorem (which is the base of the Cartan–Kähler theorem) in terms of characteristic sets and show its relationship with orderly characteristic sets.

We assume for legibility that the differential ideal $\mathfrak{a}$ is prime. The results we give generalize to regular differential ideals using [1,2,8, Lazard's lemma].

2   Differential Exterior Algebra

2.1   Cartan's Characters

Let S be an exterior differential system. Cartan's characters give the maximal dimension n of integral elements of S. They are obtained by constructing successive integral elements of increasing dimension. The Cartan–Kähler [11, Theorem 15.7] theorem asserts that the integral elements of dimension n form the tangent space of some differential manifold of dimension n: the integral manifold of S.

2.2   PDE Systems to Exterior Differential Systems

Every analytic PDE system Σ can be converted to an analytic exterior differential system S and conversely [4, page 88]. The differential manifolds solutions of Σ are the integral manifolds of S (provided that S is in involution).

Let Σ be a system in the partial derivatives of the dependent variables $\{u^\alpha, \alpha = 1, \ldots, n\}$ with respect to the independent variables $\{x^i, i = 1, \ldots, p\}$. Let u^α_J denote the derivatives of the dependent variables u^α, where J denotes the multi–index $(j_1, \ldots, j_t), 1 \le j_k \le p$. With this notation u^α_J represents the derivative of u^α with respect to $(x^{j_1} \ldots x^{j_t})$. The order of J denoted by $\#J$ is equal to t. The derivative of the variable u^α_J with respect to the independent variable x^i is denoted $u^\alpha_{J,i}$. Let $J^q = \{x^i, u^\alpha_J\}$, $1 \le i \le p$, $0 \le \#J \le q$ denote the jet space of order q. Assume that Σ is a system of PDE of J^q and suppose to simplify that Σ has solutions.

$$\Sigma \tag{1}$$

$$du^\alpha_J - u^\alpha_{J,i} dx^i = 0, 0 \le \#J < q \tag{2}$$

$$d\Sigma \tag{3}$$

$$dx^i \wedge du^\alpha_{J,i} = 0 \tag{4}$$

The exterior differential system S corresponding to Σ is displayed above, where (1) is the system of PDE (the 0–forms of the exterior system), (2) is the set of *contact forms* on J^q, (3) and (4) are the exterior derivatives of (1) and (2) respectively. The system S is closed.

2.3   System in Involution

In exterior differential algebra, there is a priori no distinction between independent and dependent variables. Therefore, the solutions of S may imply relations between the independent variables of Σ. This we want to avoid. Roughly speaking, a system S is in *involution* with respect to $x^1, \ldots, x^p$ if its solutions do not imply any relation between these variables (i.e $dx^1 \wedge \cdots \wedge dx^p \neq 0$).

Every exterior differential system can be transformed as a differential system in involution by the *prolongation* process.

Definition 1. *([4, page 88]). An exterior differential system S is said to be* in involution with respect to $x^1, \ldots, x^p$ *if the equations of its integral elements of dimension p, do not involve any relation on $dx^1, \ldots, dx^p$.*

To check if a differential exterior system is in involution, one has to compute the reduced Cartan's characters. They are obtained by a similar computation as the Cartan's characters but imposing the independence of some variables [3].

Theorem 1. *(algorithmic criterion for involution) A differential exterior system is in involution if and only if its reduced Cartan's characters are equal to its Cartan's characters.*

Proof. See [10, pages 467–468]. ∎

2.4   Prolongation of an Exterior Differential System

Consider again the exterior differential system S obtained from Σ. Assume that S lies in J^q and that S is not in involution. By the [11, Cartan–Kuranishi prolongation theorem] the system S can be *prolongated* up to order $q(S)$ to obtain a system S' which is in involution (we denote $q(S)$ the smallest nonnegative integer such that the prolongated system is in involution). In this section, we assume that 0–forms are polynomials since there are decisions problems which are not algorithmic when one considers wider classes of 0–forms. To prolongate S from J^q to $J^{q(S)}$ one performs the following steps:

1. Enlarge S with the contact forms which lie in J^{q+1} but not in J^q and with their exterior derivatives (this amounts to consider the equations defining a plan element of dimension p).
2. Enlarge S with a basis B of the algebraic ideal of the 0–forms which are consequences of S and of the independence of $dx^1, \ldots, dx^p$. This ideal is generated by the coefficients of all the relations (5) lying in the exterior (nondifferential) ideal generated by S.

$$c_i dx^i = 0, \quad c_{i,j} dx^i \wedge dx^j = 0, \quad \ldots . \tag{5}$$

Algorithmically, one can first compute a Gröbner basis G of the exterior ideal generated by S for an elimination ordering such that $du_j^\alpha \gg dx^i, \ldots$ (the dots standing for the indeterminates occuring in the coefficients c_k of the forms). A basis B is given by the set of the coefficients of the forms of G which only involve $dx^i, \ldots$ and not any du_j^α.

3. Repeat steps 1 and 2 as long as the system is not in involution.

2.5   Implementation

The computation of Cartan's characters and of the prolongation process were implemented by the second author in a MAPLE VI package. This package handles polynomial 0–forms which are not necessarily solved w.r.t. some set of indeterminates (this is an improvement w.r.t. [5]). The necessary computations modulo the 0–forms are performed using Gröbner bases methods and a MAPLE implementation by F. Lemaire of triangular sets algorithms [7].

The package involves also specialized algorithmic techniques which replace Gröbner bases in the case of exterior differential systems coming from PDE polynomial systems (these techniques only apply when the exterior system is made with forms of degree at most 2).

2.6   Example

Consider the system $\Sigma : u_{xx} = 1, u_{yy} = 1$. A detailed analysis is given in [3]. The differential exterior system obtained from Σ must be prolonged once in order to be in involution. This done, one finds the Cartan's characters $s_0 = 10$ and $s_1 = 0$. The reduced Cartan's characters are identical. By the Cartan–Kähler theorem, S' admits an integral manifold of dimension 2 with independent variables x and y which does not depend on arbitrary functions (since $s_1 = \sigma_2 = 0$). The manifold depends thus on arbitrary constants and the equations of the integral elements of dimension 2 imply that for one x and one y, only u, u_x, u_y and u_{xy} are undetermined. These are the four constants the manifold of S depends on.

Remark. For systems as simple as this one there are more efficient methods. See in particular [18] for an approach based on Janet's theory.

3   Differential Algebra

Basic notations can be found in [3]. They are very close to that of [6]. Let $\mathfrak{a}$ be a differential ideal of R.

Definition 2. *An autoreduced subset $C \subset \mathfrak{a}$ is said to be a characteristic set of $\mathfrak{a}$ if $\mathfrak{a}$ contains no nonzero differential polynomial reduced w.r.t. $\mathfrak{a}$.*

One can associate a set of diagrams to any characteristic set C: there is one diagram per differential indeterminate. On each diagram, there are as many axes as there are derivations. The leaders of the elements of C are represented as black circles. The area which contains their derivatives is striped. We call *stairs generated by* C the pictures drawn over the diagrams associated to C.

The derivatives which are not derivatives of any leader of C i.e. the derivatives which lie in the nonstriped areas are called *derivatives under the stairs of* C. The nonstriped areas can be represented as a finite, irredundant union of *bands*, either perpendicular or parallel to each axis, called *bands under the stairs of* C. These bands may have different dimensions (e.g. dimension 2 bands are planes, dimension 1 bands are lines, dimension 0 bands are points).

Consider for instance the heat equation $u_t = u_{xx}$. The situation is very simple for there is only one linear equation. This equation already forms a characteristic set of the prime differential ideal it generates in $R = K\{u\}$ endowed with derivations w.r.t. x and t. There are actually two characteristic sets possible: one such that u_t is the leader of the equation (w.r.t. some nonorderly ranking), which generates one stair with two dimension 1 bands and one such that u_{xx} is the leader of the equation (w.r.t. some orderly ranking), with one dimension 1 band.

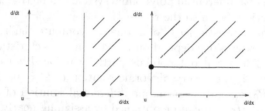

The following example illustrates a more complicated, nonlinear, situation where there are only finitely many derivatives (dimension zero bands) under the stairs. The system C is a characteristic set w.r.t. the orderly ranking

$$\cdots > v_{xx} > v_{xy} > v_{yy} > u_{xx} > u_{xy} > u_{yy} > v_x > v_y > u_x > u_y > v > u.$$

of the prime differential ideal $[C]\colon H_C^\infty$ in the ring $\mathbb{Q}\{u, v\}$ endowed with derivations w.r.t. x and y. Ranks are on the left hand side of the equal signs. Initials of the differential polynomials are denominators of the right hand side.

$$C = \{v_{xx} = u_x,\ v_y = (u_x\, u_y + u_x\, u_y\, u)/(4\, u),\ u_x^2 = 4\, u,\ u_y^2 = 2\, u\}$$

It generates the following set of two stairs:

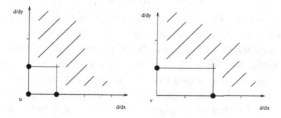

The following system $\overline{C}$ is a characteristic set w.r.t. the nonorderly ranking

$$\cdots > u_x > u_y > u > \cdots > v_{xx} > v_{xy} > v_{yy} > v_x > v_y > v.$$

of the same differential ideal $\mathfrak{a}$.

$$\overline{C} = \{u = v_{yy}^2,\ v_{xx} = 2\,v_{yy},\ v_{xy} = (v_{yy}^3 - v_{yy})/v_y,\ v_{yy}^4 = 2\,v_{yy}^2 + 2\,v_y^2 - 1\}$$

Here are the stairs generated by $\overline{C}$.

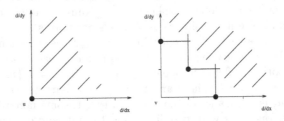

Seidenberg proved [17, page 160] that every abstract solution (taken in some differential field) of a differential polynomial system can be translated as a formal power series solution. See also the recent [15].

Given any characteristic set C of $\mathfrak{a}$, one can compute the Taylor expansions $\tilde{u}_i$. These Taylor expansions depend on the chosen characteristic set. Denote R_0 the ring of the differential polynomials partially reduced w.r.t. C. Seidenberg proved [16, page 52] that every algebraic solution of $C = 0$, $H_C \neq 0$, viewed as a system of R_0, extends to a unique differential solution of $\mathfrak{a}$. The algebraic solution of $C = 0$, $H_C \neq 0$ can be obtained by assigning nearly arbitrary values to the derivatives under the stairs of C (for they are algebraically independent modulo $\mathfrak{a}$) and assigning to the leaders of the elements of C values which are algebraic over them (solving the system as an algebraic dimension zero system).

The bands of dimension k under the stairs of C can be viewed as arbitrary functions of k variables the computed Taylor expansions depend on. Assume that the set of the derivatives under the stairs of C is formed of b_k bands of dimension k, for $0 \leq k \leq m$ and consider one band of dimension k. Renaming the derivations if needed, assume it is parallel to the axes $x_1, \ldots, x_k$ and perpendicular to the hyperplane which contains the axes $x_{k+1}, \ldots, x_m$. Assume it crosses this hyperplane at $u_{x_{k+1}^{e_1} \cdots x_m^{e_m}}$. Then we can say that the values assigned to the derivatives lying on that band are given by an arbitrary function f of k variables $u_{x_{k+1}^{e_1} \cdots x_m^{e_m}}(x_1, \ldots, x_k, 0, \ldots, 0) = f(x_1, \ldots, x_k)$. Some bands may overlap. In such a situation, the values at the origin of the corresponding arbitrary functions and some of their derivatives must be the same.

It is now tempting to claim that the solutions of $\mathfrak{a}$ depend on b_k arbitrary functions of k variables, for $0 \leq k \leq m$. Such a claim is however problematic for the numbers $b_0, \ldots, b_m$ depend on the chosen characteristic set and not on the differential ideal. The heat equation provides the simplest example. Its solution would depend either of one or of two arbitrary functions of one variable, depending on the chosen characteristic set.

This example was already considered by Cartan in [4, page 76] (and formerly by S. Kovalevskaya, to explicit the importance of solving PDEs w.r.t. the highest order derivatives to provide convergence of power series[1]). who chose to consider that the solutions of the heat equation depend on two arbitrary functions $u(0,t) = f_0(t)$ and $u_x(0,t) = f_1(t)$ for the following reason: with such a choice, if the two arbitrary functions are analytic then the formal power series computed from the characteristic set $u_{xx} = u_t$ are analytic because of the Cauchy–Kovalevskaya theorem. If we consider however that the solutions of the heat equation depend on only one arbitrary function $u(x,0) = f(x)$ then the formal power series computed from the characteristic set $u_t = u_{xx}$ may not be analytic, even if $f(x)$ is analytic. An example is $f(x) = 1/(1-x)$ for which the power series only converges for $x = 0$. We have formulated Cartan's analysis in terms of characteristic sets but Cartan does not.

Let's now reformulate Cauchy–Kovalevskaya theorem [12,10] in terms of characteristic sets too. This theorem deals with systems of equations in the partial derivatives of unknown functions $u_1, \ldots, u_n$ w.r.t. independent variables $t, x_1, \ldots, x_m$ (the independent variable t is distinguished).

$$\frac{\partial^{r_i} u_i}{\partial t^{r_i}} = \Phi_i \left(t, x_1, \ldots, x_n, \frac{\partial^k u_j}{\partial t^{k_0} \partial x_1^{k_1} \cdots \partial x_m^{k_m}} \right)$$

The indices must be such that $1 \leq i \leq n$ and $k_0 + \cdots + k_m = k \leq r_j$ and $k_0 < r_j$. We claim therefore that, if the Φ_i functions were differential polynomials then this system would form a characteristic set of a prime differential ideal w.r.t. some orderly ranking. Indeed, it would be *coherent* in the sense of [14] for there is one equation per differential indeterminate. It would be autoreduced. Last, it would be *orthonomic* (i.e. the initials and separants of all differential polynomials would be equal to 1) whence would generate a prime differential ideal. □ This characteristic set would generate stairs of a very simple form. The set of the derivatives under the stairs generated be the system are formed of $r_1 + \cdots + r_n$ bands of dimension m to which correspond $r_1 + \cdots + r_n$ arbitrary function $f(x_1, \ldots, x_m)$. Assume we are looking for a solution in the neighborhood of zero. The Cauchy–Kovalevskaya theorem then states that, if the Φ_i and the f's functions are analytic in the neighborhood of the expansion point then the differential system has a unique analytic solution. One often says that the solutions of the system depend of $r_1 + \cdots + r_n$ arbitrary functions of m variables.

3.1   Some Invariant Properties of Stairs

We do not know if Cauchy–Kovalevskaya holds for general orderly characteristic sets. We have never heard of any such generalization but we believe it is true. This conjecture is strongly enforced by some invariant properties of orderly characteristic sets that we prove below. Let C be an orderly characteristic set of some differential prime ideal $\mathfrak{a}$ in $R = K\{U\}$. Let N be the set of the derivatives under the stairs of C and d be any nonnegative integer.

---

[1] We would like to thank the referee for this remark.

Proposition 1. *The set $N_d = \{w \in N \mid \operatorname{ord} w \leq d\}$ constitutes a transcendence basis of the field extension $G_d = \operatorname{Fr}(R_d/(\mathfrak{a} \cap R_d))$ over K.*

Proof. This is a very easy proposition (though we do not know any reference for it). See [3] for an elementary proof.

Since two different transcendence bases of G_d have the same number of elements, we see that the number of derivatives of order less than or equal to d under the stairs of C is the same for every orderly characteristic set of $\mathfrak{a}$. By some combinatorial argument, we may conclude that

Theorem 2. *Let $\mathfrak{a}$ be any prime differential ideal. The number of bands of highest dimension is the same under the stairs of all orderly characteristic sets of $\mathfrak{a}$.*

The theorem above is strongly related to [6, Theorem 6, page 115] on the differential analogue of Hilbert's characteristic polynomial. This polynomial only depends on the differential ideal $\mathfrak{a}$ and not on any ranking. Denote ω this polynomial. For every sufficiently big d the integer $\omega(d)$ provides the transcendence degree of G_d over K (with the notation of our proposition 1). Moreover, from any orderly characteristic set C of $\mathfrak{a}$ satisfying [6, item (e), Lemma 16, page 51], it is possible do extract an explicit formula for ω, which is invariant by any change of orderly ranking. Therefore, our theorem follows Kolchin's theorem if all characteristic sets of $\mathfrak{a}$ satisfy [6, item (e), Lemma 16, page 51]. Our proof covers all cases and relies on much simpler arguments.

We are going to prove that this number is invariant under the action of *orderly invertible* changes of coordinates. Let $R = K\{u_1, \ldots, u_n\}$ endowed with derivations w.r.t. independent variables $x_1, \ldots, x_m$ and $\overline{R} = K\{\overline{u}_1, \ldots, \overline{u}_n\}$ endowed with derivations w.r.t. independent variables $\overline{x}_1, \ldots, \overline{x}_m$ be two differential polynomial rings. We consider an invertible change of coordinates $\phi : R \to \overline{R}$ i.e. a change of coordinates such that the solutions of any system of differential polynomials $\Sigma \subset R$ can be obtained by first solving $\phi\Sigma$, then applying the inverse change of coordinates ϕ^{-1} over the solutions of $\phi\Sigma$. A simple example is given by $\phi : w = x + y, z = x - y$ over the differential ring $R = K\{u\}$ endowed with derivations w.r.t x and y. The field $K = \mathbb{Q}(x, y)$. The inverse change is $\phi^{-1} : x = (w + z)/2, y = (w - z)/2$. The change of coordinates we consider are K–algebra isomorphisms. They map derivatives to differential polynomials (e.g. $\phi(u_{xx}) = u_{ww} + 2u_{wz} + u_{zz}$). They establish a bijection between the differential ideals of R and that of $\overline{R}$, the following proposition shows.

Proposition 2. *For every differential polynomial $p \in R$ and every system of differential polynomials Σ of R we have $p \in \sqrt{[\Sigma]} \Leftrightarrow \phi p \in \sqrt{[\phi\Sigma]}$.*

Proof. Using the differential theorem of zeros. See [3].

A change of coordinates ϕ is said to be *orderly* if it maps any derivative of order d to a differential polynomial involving only derivatives of order d. This is the case in our example. Let $\phi : R \to \overline{R}$ be an orderly change of coordinates. Let C (resp. $\overline{C}$) be an orderly characteristic set of some prime differential ideal $\mathfrak{a}$ (resp. $\overline{\mathfrak{a}} = \phi\mathfrak{a}$). Denote N_d (resp. $\overline{N}_d$) the set of the derivatives under the stairs of C (resp. $\overline{C}$) with order less than or equal to d.

Proposition 3. *The sets N_d and $\overline{N}_d$ have the same number of elements.*

Proof. A transcendence degree argument. See [3].

By some combinatorial argument, we may conclude that

Theorem 3. *If ϕ is an invertible orderly change of coordinates, $\mathfrak{a}$ is a prime differential ideal and $\overline{\mathfrak{a}} = \phi\mathfrak{a}$ then the number of bands of highest dimension is the same under the stairs of all orderly characteristic sets of $\mathfrak{a}$ and $\overline{\mathfrak{a}}$.*

4   Cartan's Characters and Characteristic Sets

Let C be an orderly characteristic set of a differential prime ideal $\mathfrak{a} = [C] : H_C^\infty$ of some differential polynomial ring R. When one converts C to an exterior differential system S, one does not necessarily get an exterior differential system in involution. The linear system Σ given in section 2.6 provides such an example. However there exists a positive integer $q(S) \geq \operatorname{ord} C$ such that the exterior differential system S' obtained by prolongation of S is in involution.

4.1   Reading the First Cartan's Character

Define $C_{q(S)} = \{r \in \Theta C \mid \operatorname{ord} r \leq q(S)\}$. The set $C_{q(S)}$ may not be triangular for it may contain different differential polynomials having the same leader. Define $C'_{q(S)}$ as any triangular subset of $C_{q(S)}$ having the same set of leaders as $C_{q(S)}$. Denote $c_1, \ldots, c_k$ the coefficients obtained at the step 2 of the prolongation process when computing S' from C (see section 2.4).

Proposition 4. *We have $(C \cup \{c_1, \ldots, c_k\}) : H_C^\infty = (C'_{q(S)}) : H_C^\infty$.*

Proof. See [3].

Corollary 1. *The vector space spanned by $dC \cup \{dc_1, \ldots, dc_k\}$ over the field of fractions G of $R_{q(S)}/(C'_{q(S)}) : H_C^\infty$ is equal to the vector space spanned by $dC'_{q(S)}$.*

Proof. See [3].

Theorem 4. *The first Cartan's character s_0 of S' is equal to $\dim J^{q(S)-1}$ minus the number of independent variables plus the number of derivatives of leaders of $C'_{q(S)}$ which have order $q(S)$.*

Consider again the system $\{u_{xx} = 1, u_{yy} = 1\}$. Here $q(S) = 3$, $\dim J^{q(S)-1} = 8$ and there are two independent variables. The system forms a characteristic set for any orderly ranking. As in section 2.6, the theorem gives $s_0 = 10$.

4.2   Reading the Last Cartan's Character

From the theorem of Cartan–Kähler and theorems 2 and 3, we conjecture

Conjecture 1. The last nonzero Cartan's character s_i is equal to the number of bands of highest dimension under the stairs of all orderly characteristic sets of the differential ideal $\mathfrak{a}$, provided that $i \neq 0$.

Consider for instance the heat equation. The last nonzero Cartan's caracter is $s_1 = 2$ [4, page 76]. There are two bands of dimension 1 under the stairs of the only orderly characteristic set of the differential ideal $[u_{xx} - u_t]$. [5], [18]

Acknowledgements. We would like to thank the referee for his very constructive report.

References

1. François Boulier, Daniel Lazard, François Ollivier, and Michel Petitot. Representation for the radical of a finitely generated differential ideal. In *ISSAC'95*, p. 158–166, Montréal, Canada, 1995.
2. François Boulier, Daniel Lazard, François Ollivier, and Michel Petitot. Computing representations for radicals of finitely generated differential ideals. Tech. rep., Univ. Lille I, LIFL, France, 1997.
3. François Boulier and Sylvain Neut. Cartan's characters and stairs of characteristic sets. Tech. rep., Univ. Lille I, LIFL, France, 2001. (ref. LIFL 2001–02).
4. É. Cartan. *Les systèmes différentiels extérieurs et leurs applications géométriques.* Hermann, 1945.
5. David Hartley and Robin W. Tucker. A Constructive Implementation of the Cartan–Kähler Theory of Exterior Differential Systems. *JSC*, 12:655–667, 1991.
6. Ellis R. Kolchin. *Differential Algebra and Algebraic Groups.* A.P., N.Y., 1973.
7. Marc Moreno Maza. On Triangular Decompositions of Algebraic Varieties. Tech. rep. 2000. (pres. at MEGA2000).
8. Sally Morrison. The Differential Ideal $[P] : M^\infty$. *JSC*, 28:631–656, 1999.
9. Sylvain Neut. Algorithme de complétion en algèbre différentielle extérieure. Mémoire de DEA, LIFL, Univ. Lille I, July 1997.
10. Peter J. Olver. *Applications of Lie groups to differential equations*, volume 107 of *Graduate Texts in Mathematics.* Springer Verlag, second edition, 1993.
11. Peter J. Olver. *Equivalence, Invariants and Symmetry.* Cambridge Univ. Press, New York, 1995.
12. I. G. Petrovsky. *Lectures on Partial Differential Equations.* I.P., 1950.
13. Joseph Fels Ritt. *Differential Algebra.* Dover Publ. Inc., New York, 1950.
14. Azriel Rosenfeld. Specializations in differential algebra. *TAMS*, 90:394–407, 1959.
15. C. J. Rust, Gregory J. Reid, and Allan D. Wittkopf. Existence and Uniqueness Theorems for Formal Power Series Solutions of Analytic Differential Systems. In *ISSAC'99*, Vancouver, Canada, 1999.
16. Abraham Seidenberg. An elimination theory for differential algebra. *Univ. California Publ. Math. (New Series)*, 3:31–65, 1956.
17. Abraham Seidenberg. Abstract differential algebra and the analytic case. *PAMS*, 9:159–164, 1958.
18. Werner Markus Seiler. *Analysis and Application of the Formal Theory of Partial Differential Equations.* PhD thesis, Lancaster Univ., U. K., 1994.

On the Invariants of the Quotients of the Jacobian of a Curve of Genus 2

P. Gaudry[1] and É. Schost[2]

[1] LIX, École polytechnique 91128 Palaiseau, France
[2] Laboratoire GAGE, UMS MEDICIS, École polytechnique 91128 Palaiseau, France

Abstract. Let $\mathcal{C}$ be a curve of genus 2 that admits a non-hyperelliptic involution. We show that there are at most 2 isomorphism classes of elliptic curves that are quotients of degree 2 of the Jacobian of $\mathcal{C}$.
Our proof is constructive, and we present explicit formulae, classified according to the involutions of $\mathcal{C}$, that give the minimal polynomial of the j-invariant of these curves in terms of the moduli of $\mathcal{C}$. The coefficients of these minimal polynomials are given as rational functions of the moduli.

Introduction

Among the curves of genus 2, those with reducible Jacobian have a particular interest. For instance, the present records for rank or torsion are obtained on such curves [3]. Also, it is in this particular setting that Dem'janenko-Manin's method yields all the rational points of a curve [7].

The aim of this paper is to give a constructive proof of the following theorem.

Theorem 1 *Let $\mathcal{C}$ be a curve of genus 2 with (2,2)-reducible Jacobian. Then there are at most 2 elliptic curves that are quotients of degree 2 of its Jacobian, up to isomorphism.*

In this case, we present rational formulae that give the j-invariant of these elliptic curves in terms of the moduli of $\mathcal{C}$.

The moduli of the curves of genus 2 form a 3-dimensional variety that was first described by Igusa in [4]. His construction relies on 4 covariants of the associated sextic, denoted by (A, B, C, D); see also [11]. We use the moduli (j_1, j_2, j_3) proposed in [5], which are ratios of these covariants. If A is not zero, they are given by

$$j_1 = 144\frac{B}{A^2}, \quad j_2 = -1728\frac{AB - 3C}{A^3}, \quad j_3 = 486\frac{D}{A^5}.$$

The special case $A = 0$ is dealt with in Appendix. In the sequel, the characteristic of the basefield will be supposed different from 2, 3 and 5. We will regularly work over an algebraic closure of the initial field of definition of the curves.

S. Boztaş and I.E. Shparlinski (Eds.): AAECC-14, LNCS 2227, pp. 373–386, 2001.

Acknowledgements

The computations necessary to obtain the formulae given here were done on the machines of UMS MEDICIS 658 (CNRS – École polytechnique). We thank Philippe Satgé for his careful reading of this paper, and François Morain for his numerous comments and suggestions.

1   Preliminaries

Definition 2 *The Jacobian of a curve C of genus 2 is (2,2)-reducible if there exists a (2,2)-isogeny between Jac(C) and a product $\mathcal{E}_1 \times \mathcal{E}_2$ of elliptic curves. The curve $\mathcal{E}_1$ is then called a quotient of Jac(C) of degree 2.*

As usual, the prefix $(2,2)$ means that the kernel of the isogeny is isomorphic to $\mathbb{Z}/2\mathbb{Z} \times \mathbb{Z}/2\mathbb{Z}$.

A curve of genus 2 always admits the hyperelliptic involution, denoted ι, which commutes with all other automorphisms. The following lemma, in substance in [4], relates the reducibility to the existence of other involutions.

Lemma 3 *Let C be a curve of genus 2. The set of the non-hyperelliptic involutions of C is mapped onto the isomorphisms classes of elliptic curves which are quotient of degree 2 of the Jacobian of C, via $\tau \mapsto C/\tau$. As a consequence the Jacobian of C is (2,2)-reducible if and only if C admits a non-hyperelliptic involution.*

Proof. Let τ be a non-hyperelliptic involution of C. The quotient of C by τ is a curve $\mathcal{E}$ of genus 1 [4]; this curve is a also quotient of the Jacobian of C. The Jacobian projects onto $\mathcal{E}$, and the kernel of this map is another elliptic curve $\mathcal{E}'$. Consequently, the Jacobian of C splits as $\mathcal{E} \times \mathcal{E}'$.

Let now $\mathcal{E}$ be an elliptic quotient of degree 2 of Jac(C). There exists a morphism φ of degree 2 from C onto $\mathcal{E}$. For a generic point p on C, the fiber $\varphi^{-1}(\varphi(p))$ can be written $\{p, q(p)\}$, where q is a rational function of p. We define τ as the map $p \mapsto q(p)$. Since the curve $\mathcal{E}$ has genus one, τ is not the hyperelliptic involution. $\square$

Bolza [1], Igusa [4] and Lange [8] have classified the curves with automorphisms, in particular the curves with involutions. The moduli of such curves describe a 2-dimensional subvariety of the moduli space; we will denote this set by $\mathcal{H}_2$. In our local coordinates, this hypersurface is described by the following equation $R(j_1, j_2, j_3)$, whose construction is done in [11].

$$
\begin{aligned}
R: \ & 8393900389396594682757122 j_3^2 + 9211413321697223245824000000 j_3^3 \\
& + 3298357634722313011200000 j_1^2 j_3^2 + 182200942574622720 j_3 j_1 j_2^2 \\
& - 374813367582081024 j_3 j_1^2 j_2 + 9995023135522160640000 j_3^2 j_1 j_2 \\
& + 94143178827 j_2^4 - 562220051373121536 j_3 j_2^2 - 562220051373121536 j_3 j_1^3 \\
& + 43381176803481600 j_3 j_2^2 - 71964166575759556608000 j_3^2 j_2 \\
& - 388606499509101605683200 j_3^2 j_1 - 1156831381426176 j_1^5 j_3 \\
& - 31381059609 j_1^7 + 62762119218 j_1^4 j_2^2 + 13947137604 j_1^3 j_2^3 \\
& - 31381059609 j_1 j_2^4 - 188286357654 j_1^3 j_2^2 - 6973568802 j_1^3 j_2 \\
& + 192612425007458304 j_1^4 j_3 + 94143178827 j_1^6 - 6973568802 j_2^5 \\
& + 28920784535654400 j_1^2 j_3 j_2^2 + 164848471853230080 j_1^3 j_3 j_2 = 0.
\end{aligned}
$$

We will call *reduced group of automorphisms* of a curve the quotient of its group of automorphisms by $\{1, \iota\}$. Then the points on $\mathcal{H}_2$ can be classified according to their reduced group of automorphisms $\mathcal{G}$.

- $\mathcal{G}$ is the dihedral group D_6; this is the case for the point on $\mathcal{H}_2$ associated to the curve $y^2 = x^6 + 1$.
- $\mathcal{G}$ is the symmetric group $\mathfrak{S}_4$; this is the case for the point associated to the curve $y^2 = x^5 - x$.
- $\mathcal{G}$ is the dihedral group D_3; the corresponding points describe a curve $\mathcal{D}$ on $\mathcal{H}_2$, excluding the two previous points.
- $\mathcal{G}$ is Klein's group V_4. The corresponding points describe a curve $\mathcal{V}$ on $\mathcal{H}_2$, excluding the two previous points; these 2 points form the intersection of $\mathcal{D}$ and $\mathcal{V}$.
- $\mathcal{G}$ is the group $\mathbb{Z}/2\mathbb{Z}$. This corresponds to the open subset $\mathcal{U} = \mathcal{H}_2 - \mathcal{D} - \mathcal{V}$; this situation will be called the *generic case*.

In the sequel, we characterize all these cases, except the two isolated points, in terms of the moduli of $\mathcal{C}$, describe the involutions of $\mathcal{C}$ and compute the corresponding j-invariants.

In the generic case, we introduce two characteristic invariants of the isomorphism classes. Our explicit formuae give an easy proof of the fact that the curves with moduli on $\mathcal{D}$ admit a real multiplication by $\sqrt{3}$. Finally, the involutions are naturally paired as $(\tau, \tau\iota)$, and these involutions correspond in general to distinct elliptic curves; we show that on the curve $\mathcal{V}$, each pair $(\tau, \tau\iota)$ yields a single elliptic curve.

The proof of Theorem 1 could be achieved through the exhaustive study of all possible automorphism groups, which would require to consider groups of order up to 48. We follow another approach, which relies on the computer algebra of polynomials systems.

This method brings to treat many polynomial systems. While most of them can be easily treated by the Gröbner bases package of the Magma Computer Algebra System [10], the more difficult one in Section 2 requires another approach, which we will briefly describe. The systems we solved cannot be given here, for lack of space; they are available upon request. The study of the group action in Section 2 was partly conducted using the facilities of Magma for computing in finite groups.

2   The Generic Case

In the open set $\mathcal{U}$, the reduced group of automorphisms is $\mathbb{Z}/2\mathbb{Z}$. Consequently, the whole group of automorphisms has the form $\{1, \iota, \tau, \tau\iota\}$, and Lemma 3 implies that there are at most two elliptic quotients. Our goal is to compute a polynomial of degree 2 giving their j-invariants in terms of the moduli (j_1, j_2, j_3).

2.1   The Minimal Polynomial from a Rosenhain Form

As a first step, we obtain the j-invariants from a Rosenhain form. The following result is based on [4], which gives the Rosenhain form of a $(2, 2)$-reducible curve.

Theorem 4 *Let C be a curve of genus 2 whose moduli belong to $\mathcal{H}_2$. On an algebraic closure of its definition field, C is isomorphic to a curve of equation*

$$y^2 = x(x-1)(x-\lambda)(x-\mu)(x-\nu), \text{ where } \mu = \nu\frac{1-\lambda}{1-\nu},$$

and λ, ν, μ are pairwise distinct, different from 0 and 1. The Jacobian of C is $(2,2)$-isogeneous to the product of the elliptic curves of equation $y^2 = x(x-1)(x-\Lambda)$, where Λ is a solution of

$$\nu^2\lambda^2\Lambda^2 + 2\nu\mu(-2\nu+\lambda)\Lambda + \mu^2 = 0. \tag{1}$$

Proof. The curve C has 6 Weierstraß points, and an isomorphism from C to another curve is determined by the images of 3 of these points. Let τ be a non-hyperelliptic involution of C, and P_1, P_2, P_3 be Weierstraß points on C that represent the orbits of τ. The curve C' defined by sending $\{P_1, P_2, P_3\}$ to $\{0, 1, \infty\}$ admits the equation $y^2 = x(x-1)(x-\lambda)(x-\mu)(x-\nu)$. This curve is not singular, so λ, ν, μ are pairwise distinct, and different from 0 and 1.

The image of the involution of C on C' is still denoted by τ. This involution permutes the Weierstraß points of C'; up to a change of names, we have $\tau(0) = \lambda$, $\tau(1) = \mu$ and $\tau(\infty) = \nu$. On another hand, τ can be written

$$\tau(x,y) = \left(\frac{ax+b}{cx+d}, \frac{wy}{(cx+d)^3}\right),$$

and since it has order 2, we have $a = -d$ and $w = \pm(ad-bc)^{3/2}$. The involution τ is determined by $\tau(0) = \lambda$ and $\tau(\infty) = \nu$, which gives

$$\tau(x,y) = \left(\nu\frac{x-\lambda}{x-\nu}, \frac{u^3y}{(x-\nu)^3}\right),$$

where $u = \pm\sqrt{\nu(\nu-\lambda)}$. Changing the sign of u amounts to composing τ with ι. The relation $\tau(1) = \mu$ then yields the first assertion $\mu = \nu(1-\lambda)/(1-\nu)$.

We now look for a curve isomorphic to C', where the involution can be written $(x,y) \mapsto (-x,y)$. This means that we consider a transformation

$$\varphi : x \mapsto \frac{ax+b}{cx+d}$$

such that $\varphi(0) = -\varphi(\lambda)$, $\varphi(1) = -\varphi(\mu)$, $\varphi(\infty) = -\varphi(\nu)$. The map

$$\varphi(x) = \frac{x-\nu-u}{x-\nu+u},$$

is such a transformation. As a result, the curve C is isomorphic to the curve C'' of equation $y^2 = (x^2-x_1^2)(x^2-x_2^2)(x^2-x_3^2)$, where

$$x_1 = \varphi(\infty) = 1, \quad x_2 = \varphi(0) = \frac{\nu-u}{\nu+u}, \quad x_3 = \varphi(1) = \frac{1-(\nu-u)}{1-(\nu+u)}.$$

The morphism $(x, y) \mapsto (x^2, y)$ maps $\mathcal{C}''$ onto the elliptic curve $\mathcal{E}$ of equation $y^2 = (x-1)(x-x_2^2)(x-x_3^2)$. The curve $\mathcal{E}$ has Legendre form $y^2 = x(x-1)(x-\Lambda)$, where

$$\Lambda = \frac{x_2^2 - x_3^2}{1 - x_3^2} = \frac{\mu}{\left(\nu \pm \sqrt{\nu(\nu - \lambda)}\right)^2}.$$

Computing the minimal polynomial of Λ proves the theorem. The conditions on λ, μ, ν show that none of the denominators vanishes, and that $\mathcal{E}$ is not singular. □

Corollary 5 *Let $\mathcal{C}$ be a curve whose moduli belong to $\mathcal{U}$, and (λ, μ, ν) defined as above. The j-invariants of the quotients of degree 2 of the Jacobian of $\mathcal{C}$ are the solutions of the equation $j^2 + c_1(\lambda, \nu)j + c_0(\lambda, \nu) = 0$, where (c_0, c_1) are rational functions.*

Proof. The previous theorem yields 2 elliptic curves that are quotients of the Jacobian of $\mathcal{C}$, and on the open set $\mathcal{U}$, they are the only ones. The polynomial equation giving j is obtained as the resultant of Equation 1 and the equation giving the j-invariant of an elliptic curve under Legendre form, $\Lambda^2(\Lambda - 1)^2 j = 2^8(\Lambda^2 - \Lambda + 1)^3$. □

We do not print the values of $c_0(\lambda, \nu)$ and $c_1(\lambda, \nu)$ for lack of space. Since the moduli (j_1, j_2, j_3) can be written in terms of λ and ν, an elimination procedure could give the coefficients c_0 and c_1 in terms of the moduli. Our approach is less direct, but yields lighter computations.

2.2 The Group Acting on Rosenhain Forms

In this section, we introduce two invariants that characterize the isomorphism classes of (2,2)-reducible curves.

Theorem 6 *Let $\mathcal{C}$ be a curve of genus 2 whose moduli belong to $\mathcal{H}_2$. There are 24 triples $(\lambda, \mu = \nu\frac{1-\lambda}{1-\nu}, \nu)$ for which the curve of equation $y^2 = x(x - 1)(x - \lambda)(x-\mu)(x-\nu)$ is isomorphic to $\mathcal{C}$. The unique subgroup of order 24 of $PGL(2, 5)$ acts transitively on the set of these triples.*

Proof. Theorem 4 yields a triple $(\lambda_1, \mu_1, \nu_1)$ that satisfies the condition, so from now on, we consider that $\mathcal{C}$ is the corresponding curve. Every curve isomorphic to $\mathcal{C}$ is given by a birational transformation $x \mapsto \frac{ax+b}{cx+d}$. Since this curve must be under Rosenhain form, the transformation must map 3 of the 6 Weierstraß points $(0, 1, \infty, \lambda_1, \mu_1, \nu_1)$ on the points $(0, 1, \infty)$. The corresponding homographic transformations form a group of order $6.5.4 = 120$, and an exhaustive search shows that only 24 of them satisfy the relation $\mu = \nu\frac{1-\lambda}{1-\nu}$ on the new values (λ, μ, ν). Let us denote by $(\lambda_i, \mu_i, \nu_i)_{i=1,\ldots,24}$ the corresponding triples. The exhaustive study shows that the curve of Rosenhain form $\{0, 1, \infty, \lambda_i, \mu_i, \nu_i\}$ is sent to the curve of Rosenhain form $\{0, 1, \infty, \lambda_j, \mu_j, \nu_j\}$ by successive applications on these 6 points of the maps $\sigma_1(x) = 1/x$, $\sigma_2(x) = 1 - x$, $\sigma_3(x) = \frac{x-\lambda}{1-\lambda}$,

$\sigma_4(x) = x/\mu$. These maps generate a group isomorphic to the unique subgroup of order 24 of $PGL(2,5)$, whose action on the triples (λ, μ, ν) is given by the following table. $\qquad\square$

map	σ_1	σ_2	σ_3	σ_4
λ	$\frac{1}{\nu}$	$1-\mu$	$\frac{\lambda}{\lambda-1}$	$\frac{\lambda}{\mu}$
μ	$\frac{1}{\mu}$	$1-\lambda$	$\frac{\mu-\lambda}{1-\lambda}$	$\frac{1}{\mu}$
ν	$\frac{1}{\lambda}$	$1-\nu$	$\frac{\nu-\lambda}{1-\lambda}$	$\frac{\nu}{\mu}$

The 24 triples $(\lambda_i, \mu_i, \nu_i)$ are explicitely given in Appendix. The symmetric functions in these triples are invariants of the isomorphism class of $\mathcal{C}$. We will now define two specific invariants that *characterize* these classes.

Definition 7 *Let $\mathcal{C}$ be a curve of genus 2 whose moduli belong to $\mathcal{H}_2$, and let $\{(\lambda_i, \mu_i, \nu_i)\}_{1 \leq i \leq 24}$ be the set of triples defined above. We denote by Ω and Υ the following functions:*
$$\Omega = \textstyle\sum_{i=1}^{24} \nu_i^2,$$
$$\Upsilon = \textstyle\sum_{i=1}^{24} \lambda_i \nu_i.$$

The following proposition shows that Ω and Υ characterize the isomorphism classes of such curves. It is straightforward to check all the following formulae, since (j_1, j_2, j_3), (c_0, c_1) and (Ω, Υ) can be written in terms of (λ, ν).

Proposition 8 *Let $\mathcal{C}$ be a curve of genus 2 whose moduli belong to $\mathcal{H}_2$, and (Ω, Υ) defined as above. If all terms are defined, then the following holds:*

$$j_1 = \frac{36(\Omega-2)\Upsilon^2}{(\Omega-8)(2\Upsilon-3\Omega)^2}, \quad j_2 = \frac{-216\Upsilon^2(\Omega\Upsilon+\Upsilon-27\Omega)}{(\Omega-8)(2\Upsilon-3\Omega)^3}, \quad j_3 = \frac{-243\Omega\Upsilon^4}{64(\Omega-8)^2(2\Upsilon-3\Omega)^5}.$$

The previous system can be solved for (j_1, j_2, j_3) only if the point (j_1, j_2, j_3) belongs to $\mathcal{H}_2$. In this case, Ω and Υ are given by the following proposition.

Proposition 9 *Let $\mathcal{C}$ be a curve of genus 2 whose moduli belong to $\mathcal{H}_2$, and (Ω, Υ) defined as above. If all terms are defined, then the following holds:*

$$
\begin{aligned}
\Upsilon = 3/4(162 j_1^4 &- 483729408 j_1 j_3 + 17199267840000 j_3^2 + 67184640 j_1^2 j_3 - 36 j_1^5 \\
&- 134369280 j_3 j_2 + 162 j_1 j_2^2 + 45 j_2^3 + 35251200 j_1 j_3 j_2 - 45 j_1^3 j_2 - 72 j_1^2 j_2^2 \\
&- 6912000 j_3 j_2^2 - 20 j_1 j_2^3 - 4 j_1^4 j_2)(349360128 j_1 j_3 - 29859840 j_3 j_2 \\
&+ 1911029760000 j_3^2 + 972 j_1^2 j_2 - 110730240 j_1^2 j_3 - 45 j_1 j_2^2 - 12441600 j_1 j_3 j_2 \\
&+ 6 j_2^3 + 45 j_1^4 - 330 j_1^3 j_2 - 56 j_1^2 j_2^2 - 16 j_1^5)/ \\
((27 j_1^4 &+ 161243136 j_1 j_3 + 1433272320000 j_3^2 - 53498880 j_1^2 j_3 - 9 j_1^5 \\
&+ 44789760 j_3 j_2 + 486 j_1^2 j_2 + 135 j_1 j_2^2 - 23846400 j_1 j_3 j_2 - 162 j_1^3 j_2 - 81 j_1^2 j_2^2 \\
&- 3456000 j_3 j_2^2 - 10 j_1 j_2^3 - 2 j_1^4 j_2)(-26873856 j_1 j_3 - 14929920 j_3 j_2 \\
&+ 955514880000 j_3^2 + 3732480 j_1^2 j_3 - 9 j_1 j_2^2 + 4147200 j_1 j_3 j_2 + 3 j_2^3 + 9 j_1^4 \\
&- 3 j_1^3 j_2 + 2 j_1^2 j_2^2 - 2 j_1^5)),
\end{aligned}
$$

$$
\begin{aligned}
\Omega = (349360128 j_1 j_3 &- 29859840 j_3 j_2 + 1911029760000 j_3^2 + 972 j_1^2 j_2 - 110730240 j_1^2 j_3 \\
&- 45 j_1 j_2^2 - 12441600 j_1 j_3 j_2 + 6 j_2^3 + 45 j_1^4 - 330 j_1^3 j_2 - 56 j_1^2 j_2^2 - 16 j_1^5)/ \\
(-26873856 j_1 j_3 &- 14929920 j_3 j_2 + 955514880000 j_3^2 + 3732480 j_1^2 j_3 - 9 j_1 j_2^2 \\
&+ 4147200 j_1 j_3 j_2 + 3 j_2^3 + 9 j_1^4 - 3 j_1^3 j_2 + 2 j_1^2 j_2^2 - 2 j_1^5).
\end{aligned}
$$

Remark The invariants (Ω, Υ) are rational functions defined on the variety $\mathcal{H}_2$. There may exist simpler formulae to express them.

We now give the coefficients of the minimal polynomial of the j-invariant in terms of Ω and Υ.

Proposition 10 *Let C be a curve of genus 2 whose moduli belong to the open set $\mathcal{U}$. The j-invariants of the elliptic quotients of degree 2 of its Jacobian are the solutions of the equation $j^2 + c_1 j + c_0$, where c_0 and c_1 are given below.*

$$c_0 = \frac{4096 \Upsilon^2 (\Omega - 32)^3}{\Omega^2 (\Omega - 8)}, \quad c_1 = \frac{-128 \Upsilon (\Omega^2 - 4\Omega\Upsilon + 56\Omega - 512)}{\Omega(\Omega - 8)}.$$

The previous two propositions lead to an expression of the form $j^2 + c_1(j_1, j_2, j_3) j + c_0(j_1, j_2, j_3) = 0$, where $c_1(j_1, j_2, j_3)$ and $c_0(j_1, j_2, j_3)$ are rational functions in (j_1, j_2, j_3). The denominators in these functions vanish on the two curves $\mathcal{D}$ and $\mathcal{V}$, and two additional curves. This last degeneracy is an artifact due to our choice of denominators; it is treated in Appendix.

Computational considerations. To derive the previous formulae, the first step is to obtain each of the functions $(c_0, c_1, j_1, j_2, j_3)$ in terms of Ω and Υ. Let us consider the case of, say, j_1. The indeterminates $(\lambda, \nu, j_1, \Omega, \Upsilon)$ are related by the system $\{\Omega = \Omega(\lambda, \nu), \Upsilon = \Upsilon(\lambda, \nu), j_1 = j_1(\lambda, \nu)\}$, where the right-hand sides are rational functions. The relation between (Ω, Υ, j_1) is the equation of the image of the corresponding rational function. Determining this relation is often called *implicitization*.

A well-known approach to this question relies on a Gröbner basis computation. The system can be rewritten as a polynomial system F_{j_1} in $(\lambda, \nu, j_1, \Omega, \Upsilon)$. The relation we seek is the intersection of the ideal generated by F_{j_1} and the additional equation $1 - ZD(\lambda, \nu)$ with $\mathbb{Q}[j_1, \Omega, \Upsilon]$, where Z is a new indeterminate, and D the lcm of the denominators [2, chapter 3.3]. The intersection can be computed by a Gröbner basis for an eliminating order. In our case, such computations take several hours, using Magma on a Alpha EV6 500 Mhz processor.

We followed another approach to treat this question. The system we consider defines a finite extension of the field $\mathbb{Q}(\Omega, \Upsilon)$, and the relation we seek is the minimal polynomial of j_1 in this extension. In [12], the second author proposes a probabilistic polynomial-time algorithm to compute this minimal polynomial; its Magma implementation solves the present question in a matter of minutes.

Finally, once j_1, j_2 and j_3 are obtained in terms of (Ω, Υ), we have to solve the system in Proposition 8 for (Ω, Υ). This system defines a finite extension of $\mathbb{Q}(j_1, j_2)$. Since Ω and Υ are know to be functions of (j_1, j_2, j_3), j_3 is a primitive element for this extension, and our question is reduced to compute Ω and Υ using this primitive element. The methods in [12] apply as well in this case, and give the formulae in Proposition 9.

3   The Curve $\mathcal{D}$

We now turn to the first special case, the curve $\mathcal{D}$ defined in the preliminaries, and prove Theorem 1 in this case. The computations turn out to be quite simpler, mainly because this variety has dimension only one. Our formulation also leads to additional results concerning the endomorphism ring of such Jacobians.

Theorem 11 *Let C be a curve of genus 2 whose moduli belong to $\mathcal{D}$. There are two elliptic curves that are quotients of degree 2 of $\mathrm{Jac}(C)$.*

Proof. As in the generic case, we start form a characterization of those curves due to Igusa [4].

Lemma 12 *Let C be a curve of genus 2. The reduced group of automorphisms of C is D_3 if and only if C is isomorphic to a curve of equation $y^2 = x(x - 1)(x - \lambda)(x - \mu)(x - \nu)$, $\mu = \frac{1}{1-\lambda}$ and $\nu = 1 - \frac{1}{\lambda}$, with λ different from 0, 1 and $(1 \pm \sqrt{3})/2$.*

If C is as above, its reduced group of automorphisms can be explicitly written. In the following table, u denotes $\pm\sqrt{\lambda^2 - \lambda + 1}$.

map	order
Id $(x, y) \mapsto (x, y)$	1
$\tau_1 \ (x, y) \mapsto \left(\frac{\lambda - x}{(\lambda-1)x+1}, \frac{u^3 y}{((\lambda-1)x+1)^3} \right)$	2
$\tau_2 \ (x, y) \mapsto \left(\frac{(\lambda-1)x+1}{\lambda x+1-\lambda}, \frac{u^3 y}{(\lambda x+1-\lambda)^3} \right)$	2
$\tau_3 \ (x, y) \mapsto \left(\frac{\lambda x+1-\lambda}{x-\lambda}, \frac{u^3 y}{(x-\lambda)^3} \right)$	2
$\rho_1 \ (x, y) \mapsto \left(1 - \frac{1}{x}, \frac{y}{x^3} \right)$	3
$\rho_2 \ (x, y) \mapsto \left(\frac{1}{1-x}, \frac{y}{(1-x)^3} \right)$	3

For each of the involutions τ_1, τ_2, τ_3, we repeat the construction done in the proof of Theorem 4: we associate to each τ_i a pair of elliptic curves.

To this effect, we determine an isomorphism φ from C to a curve where τ_i becomes $(x, y) \mapsto (-x, y)$, and denote by $x_1 = 1$, x_2 and x_3 the values taken by φ at $\{0, 1, \infty\}$. The means that the curve C is isomorphic to the curve $y^2 = (x^2 - 1)(x^2 - x_2^2)(x^2 - x_3^2)$, and the elliptic curves we look for are $y^2 = (x-1)(x-x_2^2)(x - x_3^2)$, whose Legendre forms is $y^2 = x(x-1)(x-\Lambda)$, where $\Lambda = (x_2^2 - x_3^2)/(1 - x_3^2)$. These computations are summarized in the following table.

involution	φ	x_2	x_3	Λ
τ_1	$x \mapsto \frac{(-1-u)x+\lambda}{(-1+u)x+\lambda}$	$\frac{-1-u+\lambda}{-1+u+\lambda}$	$\frac{u+1}{u-1}$	$\Lambda_1 = \lambda(\lambda - 1 - u)^2$
τ_2	$x \mapsto \frac{(\lambda-1-u)x+1}{(\lambda-1+u)x+1}$	$\frac{-1-u+\lambda}{-1+u+\lambda}$	$\frac{\lambda-u}{\lambda+u}$	$\Lambda_2 = \frac{1}{\lambda(\lambda-1+u)^2}$
τ_3	$x \mapsto \frac{x-\lambda+u}{x-\lambda-u}$	$\frac{-1-u+\lambda}{-1+u+\lambda}$	$\frac{\lambda-u}{\lambda+u}$	$\Lambda_3 = \frac{1}{\lambda(\lambda-1+u)^2}$

Let Λ_i' be the conjugate of Λ_i, obtained when u is replaced by $-u$. The elliptic curves corresponding to τ_i and $\tau_i\iota$ have Legendre parameters Λ_i and Λ_i', and we have $\Lambda_2 = \Lambda_3 = 1/\Lambda_1'$, $\Lambda_2' = \Lambda_3' = 1/\Lambda_1$. Since changing Λ_i to its inverse $1/\Lambda_i$ leaves the j-invariant unchanged, there are only 2 isomorphism classes of elliptic quotients.   □

We now give generators of the ideals defining the curves in $\mathcal{D}$, in terms of their moduli. We follow the Gröbner basis approach we already mentioned; Magma's Gröbner package takes about a minute to treat this simpler problem.

Lemma 12 gives the moduli in terms of λ, and these relations can be expressed by a polynomial system $F_{\mathcal{D}}$ in $\mathbb{Q}[j_1, j_2, j_3, \lambda]$. The ideal defining the curve $\mathcal{D}$ is obtained as the intersection of the ideal generated by $F_{\mathcal{D}}$ and $1 - ZD(\lambda)$ with $\mathbb{Q}[j_1, j_2, j_3]$, where Z is a new indeterminate, and $D(\lambda)$ the lcm of the denominators of (j_1, j_2, j_3) expressed in terms of λ:

$$7j_2^3 - 57600j_3j_1j_2 + 8991j_1j_2 + 2646j_2^2 - 34774272j_3j_1 - 22394880j_3j_2$$
$$-9953280000j_3^2 + 65610j_1 + 7290j_2 - 4901119488j_3 = 0,$$
$$j_1j_2^2 - 297j_1j_2 - 90j_2^2 - 725760j_1j_3 + 172800j_2j_3$$
$$-2187j_1 - 243j_2 + 169641216j_3 = 0,$$
$$-81j_1 + 21j_1^2 - 9j_2 + 5j_1j_2 + 864000j_3 = 0.$$

As in Section 2, the previous proof yields the minimal polynomial of the Legenbre parameters Λ, and then of j-invariants in terms of λ, under the form $j^2 + c_1(\lambda)j + c_0(\lambda)$. Eliminating λ is a simple task, which gives the formulae:

$$c_1 = \frac{3}{8} \frac{-85221j_1j_2 - 69228j_1^2 - 6621j_2^2 + 6054374400j_3j_1 + 692576000j_3j_2 - 5952061440j_3}{j_3(4705j_2 + 21492j_1 - 129816)},$$
$$c_0 = -81 \frac{2373j_1^2 + 1412j_1j_2 + 210j_2^2 + 33696000j_3j_1 + 4320000j_3j_2 - 246067200j_3}{j_3(5j_2 + 27j_1 - 108)}.$$

The points were a denominator vanishes must be treated separately, in Appendix. The previous results make the proof of the following corollaries easy.

Theorem 13 *Let $\mathcal{C}$ be curve of genus 2 whose moduli belong to $\mathcal{D}$. Its two elliptic quotients are 3-isogeneous.*

Proof. We use the same notation as in the previous proof. Let $\mathcal{E}_1$ be the elliptic curve associated to the involution τ_1, under the form $y^2 = x(x-1)(x-\Lambda_1)$. Its 3-division polynomial is $\psi_3(x) = 3x^4 + (-4\Lambda_1 - 4)x^3 + 6\Lambda_1 x^2 - \Lambda_1^2$. The linear form $S_3(x) = 3x + \lambda - 2(u+1)$ divides $\psi_3(x)$ and corresponds to a subgroup of $\mathcal{E}_1$ of order 3. Using Vélu's formulae [13], we can explicitly determine a curve 3-isogeneous to $\mathcal{E}_1$, of the form $y^2 = x^3 + a_2x^2 + a_4x + a_6$, where a_2, a_4, a_6 are defined by

$$a_2 = -(\Lambda_1 + 1), \quad a_4 = \Lambda_1 - 5t, \quad a_6 = 4(\Lambda_1 + 1)t - 14x_0(t - x_0^2 + \Lambda_1),$$

with $x_0 = (2(u+1) - \lambda)/3$, $t = 6x_0^2 - 4(\Lambda_1 + 1)x_0 + 2\Lambda_1$. It is straightforward to check that the j-invariant of this curve is Λ_1'. □

Corollary 14 *Let $\mathcal{C}$ be curve of genus 2 whose moduli are on $\mathcal{D}$. The endomorphism ring of the Jacobian of $\mathcal{C}$ contains an order in the quaternion algebra $\left(\frac{3,1}{\mathbb{Q}}\right)$. In particular, it admits a real multiplication by $\sqrt{3}$.*

Proof. The Jacobian of C is isogeneous to $\mathcal{E}_1 \times \mathcal{E}_2$, where $\mathcal{E}_1$ and $\mathcal{E}_2$ are 3-isogeneous elliptic curves. Let us denote by $\mathcal{I} : \mathcal{E}_1 \to \mathcal{E}_2$ a degree-3 isogeny, and $\hat{\mathcal{I}}$ its dual isogeny. Let $\mathcal{O}$ be the ring

$$\mathcal{O} = \left\{ \begin{pmatrix} a & \sqrt{3}b \\ \sqrt{3}c & d \end{pmatrix}, \text{ where } a, b, c, d \in \mathbb{Z} \right\}.$$

The map sending $\begin{pmatrix} a & \sqrt{3}b \\ \sqrt{3}c & d \end{pmatrix}$ to the endomorphism of $\mathcal{E}_1 \times \mathcal{E}_2$ given by $(P, Q) \mapsto$ $([a]P + [b]\hat{\mathcal{I}}Q, [c]\mathcal{I}P + [d]Q)$ is an injective ring homomorphism. Multiplication by $\sqrt{3}$ is for instance represented by the endomorphism $(P, Q) \mapsto (\hat{\mathcal{I}}Q, \mathcal{I}P)$. □

4   The Curve $\mathcal{V}$

This is the second special case; as previously, the study is based on a result due to Igusa.

Theorem 15 *Let C be a curve of genus 2 whose moduli belong to $\mathcal{V}$. There exist two elliptic curves $\mathcal{E}_1$ and $\mathcal{E}_2$ such that $\mathcal{V}$ is (2,2)-isogeneous to $\mathcal{E}_1 \times \mathcal{E}_1$ and $\mathcal{E}_2 \times \mathcal{E}_2$. These elliptic curves are 2-isogeneous.*

Proof. The following result is taken from [4].

Lemma 16 *Let C be a curve of genus 2. The reduced groups of automorphisms of C is V_4 if and only if C is isomorphic to the curve of equation $y^2 = x(x - 1)(x + 1)(x - \lambda)(x - 1/\lambda)$, where λ is different from 0, -1 and 1.*

If C is as above, its reduced automorphisms can be explicitly determined; in the following table, u denotes $\pm\sqrt{1 - \lambda^2}$ and $\bar{u}$ denotes $\pm\sqrt{\lambda^2 - 1}$.

map	order
Id $(x, y) \mapsto (x, y)$	1
τ_1 $(x, y) \mapsto \left(\frac{x-\lambda}{\lambda x - 1}, \frac{u^3 y}{(\lambda x - 1)^3} \right)$	2
τ_2 $(x, y) \mapsto \left(\frac{\lambda x - 1}{x - \lambda}, \frac{\bar{u}^3 y}{(x - \lambda)^3} \right)$	2
ρ $(x, y) \mapsto \left(\frac{1}{x}, \frac{iy}{x^3} \right)$	4

We follow the same method as in the proof of Theorem 11: for each τ_i, we make up an isomorphism φ from C to a curve where τ_i becomes $(x, y) \mapsto (-x, y)$. This curve is then isogeneous to the elliptic curve $y^2 = (x - 1)(x - x_2^2)(x - x_3^2)$, whose Legendre forms are $y^2 = x(x-1)(x-\Lambda)$. This leads to the following table.

involution	φ	x_2	x_3	Λ	j
τ_1	$x \mapsto \frac{(1-u)x-\lambda}{(1+u)x-\lambda}$	$\frac{\lambda+u-1}{\lambda-u-1}$	$\frac{1-u}{1+u}$	$\Lambda_1 = \frac{\lambda^2(1-\lambda)}{(\lambda-1-u)^2}$	$J_1 = 64\frac{(4-l^2)^3}{l^4}$
τ_2	$x \mapsto \frac{(-\lambda-\bar{u})x+1}{(-\lambda+\bar{u})x+1}$	$\frac{\lambda+\bar{u}-1}{\lambda-\bar{u}-1}$	$\frac{\lambda+\bar{u}}{\lambda-\bar{u}}$	$\Lambda_2 = \frac{\lambda-1}{\lambda(\lambda-1-\bar{u})^2}$	$J_2 = 64\frac{(4l^2-1)^3}{l^2}$

The invariants J_1 and J_2 do not depend on u. This implies that the Jacobian of $\mathcal{C}$ is (2,2)-isogeneous to the products $\mathcal{E}_1 \times \mathcal{E}_1$ and $\mathcal{E}_2 \times \mathcal{E}_2$, and consequently, also to $\mathcal{E}_1 \times \mathcal{E}_2$. Finally, the curves $\mathcal{E}_1$ and $\mathcal{E}_2$ are 2-isogeneous, since (J_1, J_2) cancels the modular equation of degree 2. $\square$

Following the same method as in the previous section, we obtain an ideal defining the moduli of such curves:

$$32j_1j_2^2 - 27j_1j_2 - 54j_2^2 + 4423680j_1j_3 + 14745600j_2j_3 - 13436928j_3 = 0,$$
$$64j_2^3 - 78643200j_1j_2j_3 + 243j_1j_2 - 378j_2^2 + 31850496j_1j_3 - 8847360j_2j_3$$
$$-36238786560000j_3^2 + 120932352j_3 = 0,$$
$$3j_1^2 - 10j_1j_2 + 18j_2 - 4608000j_3 = 0.$$

Their j-invariant are solution of the equation $j^2 + c_1j + c_0$, where c_0 and c_1 are given by the following formulae:

$$c_1 = \frac{9}{4} \frac{3j_1j_2 - 2j_2^2 + 1866240j_3 + 211200j_3j_1 + 64000j_3j_2}{j_3(-243 + 78j_1 + 20j_2)},$$

$$c_0 = 108 \frac{2560000j_3j_2 + 51j_1j_2 + 30j_2^2 + 768000j_3j_1 + 18662400j_3}{j_3(-243 + 78j_1 + 20j_2)}.$$

5   Examples

In this section, we present examples, mostly taken from the literature, that show the use of our results.

5.1   The Generic Case

Let $\mathcal{C}$ be the curve defined over $\mathbb{Q}$ by the equation $y^2 = x^6 - x^5 + x^4 - x^2 - x - 1$. Its moduli are

$$j_1 = \frac{2^3 \times 3^2 \times 5 \times 13}{37^2}, \quad j_2 = -\frac{2^3 \times 3^3 \times 11 \times 13}{37^3}, \quad j_3 = \frac{3^5 \times 53^2}{2^8 \times 37^5}.$$

They belong to the open set $\mathcal{U} \subset \mathcal{H}_2$, so $\mathrm{Jac}(\mathcal{C})$ is isogeneous to a product of two elliptic curves. On this example, finding these curves through a Rosenhain form requires to work in an extension of $\mathbb{Q}$ of degree 24. Propositions 9 and 10 directly give:

$$c_0 = \frac{2^{14} \times 5^6 \times 37^3}{53^2}, \quad c_1 = \frac{2^8 \times 3^4 \times 47}{53},$$

and the j-invariants of the elliptic curves are defined on $\mathbb{Q}(i)$ by

$$j = -\frac{2^7 \times 3^4 \times 47}{53} \pm \frac{2^8 \times 7 \times 11 \times 181}{53}i.$$

Notice that 53 divides the discriminant of the curve, it is no surprise to see it appear in the denominator of j.

5.2   The Curve $\mathcal{D}$

The following example is taken from [6], where Kulesz builds a curve admitting many rational points. Let $\mathcal{C}$ be the curve defined on $\mathbb{Q}$ by the equation

$$y^2 = 1412964(x^2 - x + 1)^3 - 8033507x^2(x-1)^2.$$

Its moduli are

$$j_1 = \frac{3^2 \times 149 \times 167 \times 239^2 \times 3618470803 \times 33613^2}{757^2 \times 76832154757^2},$$

$$j_2 = -\frac{3^3 \times 239^2 \times 33613^2 \times 195593 \times 31422316507485410373257}{757^3 \times 76832154757^3},$$

$$j_3 = -\frac{2^{22} \times 3^{17} \times 5^9 \times 7^6 \times 47^3 \times 89^3 \times 239^4 \times 33613^4}{757^5 \times 76832154757^5}.$$

We check that they belong to the curve $\mathcal{D}$, so the reduced group of automorphisms of $\mathcal{C}$ is D_3 (the construction of this curve in [6] already implies this result). Again, writing down a Rosenhain form for this curve requires to work in an algebraic extension of $\mathbb{Q}$. Our formulae readily give the j-invariants of the quotient elliptic curves:

$$-\frac{239 \times 33613 \times 84333563^3}{2^{24} \times 3^4 \times 5^9 \times 7^2 \times 47^3 \times 89} \quad \text{and} \quad \frac{19^3 \times 67^3 \times 239 \times 349^3 \times 33613}{2^8 \times 3^{12} \times 5^3 \times 7^6 \times 47 \times 89^3}.$$

5.3   The Curve $\mathcal{V}$

In the paper [9], Leprévost and Morain study the curve $\mathcal{C}_\theta$ defined on $\mathbb{Q}(\theta)$ by the equation $y^2 = x(x^4 - \theta x^2 + 1)$, with the purpose to study sums of characters. Its moduli are

$$j_1 = 144\frac{9\theta^2 - 20}{(3\theta^2 + 20)^2}, \quad j_2 = -3456\frac{27\theta^2 - 140}{(3\theta^2 + 20)^3}, \quad j_3 = 243\frac{(\theta^2 - 4)^2}{(3\theta^2 + 20)^5}.$$

We check that they belong to the curve $\mathcal{V}$, so the reduced group of automorphisms of $\mathcal{C}$ is V_4. This yields the j-invariants of the quotient elliptic curves:

$$j = 64\frac{(3\theta - 10)^3}{(\theta - 2)(\theta + 2)^2} \quad \text{and} \quad j' = 64\frac{(3\theta + 10)^3}{(\theta + 2)(\theta - 2)^2}.$$

Notice that the curves E_θ and E'_θ given in [9] $y^2 = x(x^2 \pm 4x + 2 - \theta)$, have the same invariants j'. The other quotient curves, with invariant j, admit the equation $y^2 = x(x^2 \pm 4x + 2 + \theta)$.

Appendix: Formulary

To complete the previous study, we give formulae describing the following cases:

- The reduced group of automorphisms $\mathcal{G}$ is neither D_3 nor V_4, nor $\mathbb{Z}/2\mathbb{Z}$: this is the case for the two points 2.(a) and 2.(b) below.
- A denominator vanishes. On the curve $\mathcal{D}$, this happens at a single point, treated in 2.(c); in the generic case, two curves must be studied in 2.(f) and 2.(g).
- The covariant A vanishes, so the moduli (j_1, j_2, j_3) are not adapted. We choose two other invariants and go through the same exhaustive process.

All these formulae are gathered as an algorithm, taking as input a curve of genus 2, with (2,2)-reducible Jacobian, that outputs the minimal polynomial of the j-invariants of the elliptic quotients.

1. Compute the covariants A, B, C, D, R of $\mathcal{C}$ given in [4], and check that $R = 0$.
2. If $A \neq 0$: compute j_1, j_2, j_3.
   (a) If $(j_1, j_2, j_3) = (\frac{81}{20}, -\frac{729}{200}, \frac{729}{25600000})$, then the reduced group of automorphisms is D_6; return $j(j - 54000)$.
   (b) If $(j_1, j_2, j_3) = (-\frac{36}{5}, \frac{1512}{25}, \frac{243}{200000})$, then the reduced group of automorphisms is $\mathfrak{S}_4$; return $j - 8000$.
   (c) If $(j_1, j_2, j_3) = (\frac{24297228}{885481}, -\frac{81449284536}{833237621}, -\frac{57798021931029}{47220229240364864})$, then the reduced group of automorphisms is D_3; return $j^2 + \frac{471690263168}{658503}j - \frac{8094076887461888}{57289761}$.
   (d) If (j_1, j_2, j_3) cancel the polynomials defining $\mathcal{D}$, then the reduced group of automorphisms is D_3; return j as computed in Section 3.
   (e) If (j_1, j_2, j_3) cancel the polynomial defining $\mathcal{V}$, then the reduced group of automorphisms is V_4; return j as computed in Section 4.
   (f) If (j_1, j_2, j_3) satisfy $331776 j_3 - j_2^2 - 24 j_1 j_2 - 144 j_1^2 = 0$ and $9 j_1 + j_2 = 0$, then the reduced group of automorphisms is $\mathbb{Z}/2\mathbb{Z}$; return $j^2 + \frac{150994944 j_3}{j_2 + 12 j_1} j - \frac{260919263232 j_3}{j_2 + 12 j_1}$.
   (g) If (j_1, j_2, j_3) satisfy $j_2^5 + 54 j_2^4 - 322486272 j_2^2 j_3 + 481469424205824 j_3^2 = 0$ and $18 j_1 + 5 j_2 = 0$, then the reduced group of automorphisms is $\mathbb{Z}/2\mathbb{Z}$; return $j^2 + c_1 j + c_0$, where

$$c_0 = -\frac{125}{9559130112} \frac{(-j_2^2 - 24 j_1 j_2 - 144 j_1^2 + 16257024 j_3)^2}{j_3^2},$$

$$c_1 = \frac{(16257024 j_3 - j_2^2 - 24 j_1 j_2 - 144 j_1^2)(2723051520 j_3 - 289 j_2^2 - 6936 j_1 j_2 - 41616 j_1^2)}{2064772104192 j_3^2}.$$

   (h) Else, we are in the generic case, and no denominator vanishes; return j as computed in Section 2.
3. The case $A = 0$
   (a) If $B = 0$ and $C^5 = 4050000 D^3$, the reduced group of automorphisms is $\mathbb{Z}/2\mathbb{Z}$; return $(j - 4800)(j - 8640)$.
   (b) If $C = 0$ and $B^5 = 3037500 D^2$, the reduced group of automorphisms is $\mathbb{Z}/2\mathbb{Z}$; return $(j - 160)(j + 21600)$.
   Compute the invariants $t_1 = \frac{3}{512} \frac{CD}{B^4}$ and $t_2 = 1536 \frac{BC}{D}$.
   (c) If $(t_1, t_2) = (1/576000, -460800)$, the reduced group of automorphisms is V_4; return $j^2 + 7200 j + 13824000$.
   (d) If $(t_1, t_2) = (-1/864000, -172800)$, the reduced group of automorphisms is D_3; return $j^2 + 55200 j - 69984000$.
   (e) The reduced group of automorphisms is $\mathbb{Z}/2\mathbb{Z}$. Compute

$$\Omega = -4 \frac{-238878720000 t_1 + 1555200 t_2 t_1 + 7 t_2^2 t_1 + 2 t_2}{477757440000 t_1 + 2073600 t_2 t_1 + t_2^2 t_1 - t_2} \quad \text{and} \quad \Upsilon = \frac{3}{2}\Omega,$$

   then c_0 and c_1 given in Section 2; return $j^2 + c_1 j + c_0$.

Appendix: The 24 Triples

The following table gives the full list of the triples defined in Theorem 6.

$$\left(\lambda, \tfrac{\lambda\nu-\nu}{\nu-1}, \nu\right) \qquad \left(\tfrac{-1}{\nu-1}, \tfrac{-\lambda+\nu}{\lambda\nu-\lambda-\nu+1}, \tfrac{-\lambda+\nu}{\lambda\nu-\lambda}\right) \quad \left(\tfrac{-\lambda\nu+2\nu-1}{\nu-1}, \tfrac{-\lambda+\nu}{\nu-1}, \tfrac{(-\lambda+\nu)}{\lambda\nu-\lambda}\right)$$

$$\left(\tfrac{\lambda\nu-\lambda}{\lambda\nu-\nu}, \tfrac{\nu-1}{\lambda\nu-\nu}, \tfrac{\nu-1}{\lambda-1}\right) \qquad \left(\lambda, \tfrac{-\lambda+\nu}{\nu-1}, \tfrac{\lambda-\nu}{\lambda\nu-2\nu+1}\right) \qquad \left(\tfrac{-\lambda\nu+2\nu-1}{\nu-1}, -\lambda+1, -\nu+1\right)$$

$$\left(\tfrac{\lambda\nu-\lambda}{\lambda\nu-\nu}, \tfrac{\nu-1}{\nu}, \tfrac{\lambda\nu-\lambda-\nu+1}{\lambda\nu-2\nu+1}\right) \quad \left(\tfrac{\lambda}{\lambda-1}, \tfrac{-\lambda+\nu}{\lambda\nu-\lambda-\nu+1}, \tfrac{\lambda-\nu}{\lambda-1}\right) \qquad \left(\tfrac{\lambda-1}{\nu-1}, \tfrac{\lambda\nu-\nu}{\nu-1}, \tfrac{\lambda\nu-\nu}{\lambda\nu-\lambda}\right)$$

$$\left(\tfrac{\lambda}{\lambda-1}, \tfrac{\nu}{\nu-1}, \tfrac{\lambda\nu-\nu}{\lambda\nu-2\nu+1}\right) \qquad \left(\tfrac{\lambda-1}{\lambda-\nu}, \tfrac{\lambda\nu-\nu}{\lambda-\nu}, \tfrac{\lambda\nu-\nu}{\lambda\nu-2\nu+1}\right) \qquad \left(\tfrac{\lambda\nu-2\nu+1}{\lambda\nu-\nu}, \tfrac{\nu-1}{\nu}, \tfrac{\lambda-1}{\lambda}\right)$$

$$\left(\tfrac{\lambda\nu-2\nu+1}{\lambda\nu-\lambda-\nu+1}, \tfrac{-1}{\lambda-1}, \tfrac{\nu-1}{\lambda-1}\right) \qquad \left(\tfrac{1}{\nu}, \tfrac{\lambda-\nu}{\lambda\nu-\nu}, \tfrac{\lambda-\nu}{\lambda\nu-2\nu+1}\right) \qquad \left(\tfrac{\lambda\nu-2\nu+1}{\lambda\nu-\lambda-\nu+1}, \tfrac{\nu}{\nu-1}, \tfrac{\lambda\nu-\nu}{\lambda\nu-\lambda}\right)$$

$$\left(\tfrac{-1}{\nu-1}, \tfrac{-1}{\lambda-1}, \tfrac{-\nu+1}{\lambda\nu-2\nu+1}\right) \qquad \left(\tfrac{\lambda\nu-2\nu+1}{\lambda\nu-\nu}, \tfrac{\lambda-\nu}{\lambda\nu-\nu}, \tfrac{\lambda-\nu}{\lambda-1}\right) \qquad \left(\tfrac{\lambda\nu-2\nu+1}{\lambda-\nu}, \tfrac{-\nu+1}{\lambda-\nu}, \tfrac{1}{\lambda}\right)$$

$$\left(\tfrac{\lambda-1}{\lambda-\nu}, \tfrac{-\lambda\nu+\lambda+\nu-1}{\lambda-\nu}, \tfrac{\lambda-1}{\lambda}\right) \qquad \left(\tfrac{\lambda\nu-2\nu+1}{\lambda-\nu}, \tfrac{\lambda\nu-\nu}{\lambda-\nu}, \nu\right) \qquad \left(\tfrac{1}{\nu}, \tfrac{\nu-1}{\lambda\nu-\nu}, \tfrac{1}{\lambda}\right)$$

$$\left(\tfrac{\lambda-1}{\nu-1}, -\lambda+1, \tfrac{\lambda\nu-\lambda-\nu+1}{\lambda\nu-2\nu+1}\right) \quad \left(\tfrac{-\lambda\nu+\lambda}{\lambda-\nu}, \tfrac{-\lambda\nu+\lambda+\nu-1}{\lambda-\nu}, -\nu+1\right) \quad \left(\tfrac{-\lambda\nu+\lambda}{\lambda-\nu}, \tfrac{-\nu+1}{\lambda-\nu}, \tfrac{-\nu+1}{\lambda\nu-2\nu+1}\right)$$

References

1. O. Bolza. On binary sextics with linear transformations onto themselves. *Amer. J. Math.*, 10:47–70, 1888.
2. D. Cox, J. Little, D. O'Shea Ideals, Varieties and Algorithms. Springer-Verlag, 1992.
3. E. Howe, F. Leprevost, and B. Poonen. Large torsion subgroups of split Jacobians of curves of genus 2 or 3. *Forum Math.*, 12:315–364, 2000.
4. J. Igusa. Arithmetic variety of moduli for genus 2. *Ann. of Math. (2)*, 72:612–649, 1960.
5. J. Igusa. On Siegel modular forms of genus 2. *Amer. J. Math.*, 84:175–200, 1962.
6. L. Kulesz. Courbes algébriques de genre 2 possédant de nombreux points rationnels. *C. R. Acad. Sci. Paris Sér. I Math.*, 321:91–94, 1995.
7. L. Kulesz. Application de la méthode de Dem'janenko-Manin à certaines familles de courbes de genre 2 et 3. *J. Number Theory*, 76:130–146, 1999.
8. H. Lange. Über die Modulvarietät der Kurven vom Geschlecht 2. *J. Reine Angew. Math.*, 281:80–96, 1976.
9. F. Leprévost and F. Morain. Revêtements de courbes elliptiques à multiplication complexe par des courbes hyperelliptiques et sommes de caractères. *J. Number Theory*, 64:165–182, 1997.
10. Magma. http://www.maths.usyd.edu.au:8000/u/magma/
11. J.-F. Mestre. Construction de courbes de genre 2 à partir de leurs modules. In T. Mora and C. Traverso, editors, *Effective methods in algebraic geometry*, volume 94 of *Progr. Math.*, pages 313–334. Birkhäuser, 1991.
12. É. Schost. Sur la résolution des systèmes polynomiaux à paramètres. PhD Thesis, École polytechnique, 2000.
13. J. Vélu. Isogénies entre courbes elliptiques. *C. R. Acad. Sci. Paris Sér. I Math.*, 273:238–241, July 1971. Série A.

Algebraic Constructions
for PSK Space-Time Coded Modulation

Andrew M. Guidi[1], Alex J. Grant[1], and Steven S. Pietrobon[2]

[1] Institute for Telecommunications Research
Mawson Lakes Boulevard
Mawson Lakes, SA, 5095, Australia
{aguidi,alex}@spri.levels.unisa.edu.au
[2] Small World Communications
6 First Avenue
Payneham South, SA, 5070 Australia
steven@sworld.com.au

Abstract. We consider the design of phase shift keyed space-time coded modulation for two antenna systems based on linear codes over rings. Design rules for constructing full diversity systematic space-time codes based on underlying existing algebraic codes were first presented by Hammons and El Gamal in 2000. We reformulate and simplify these design rules, resulting in the condition that the characteristic polynomial of the parity generation matrix must be irreducible. We further extend the results to non-systematic codes. These results yield a recursive construction based on the Schur determinant formula. The resulting block codes are guranteed to provide full diversity advantage. In addition, the code construction is such that the corresponding parity check matrix is sparse, enabling the use of the powerful Sum-Product algorithm for decoding.

1 Introduction

Wireless access to data networks such as the Internet is expected to be an area of rapid growth for mobile communications. High user densities will require very high speed low delay links in order to support emerging Internet applications such as voice and video. Even in the low mobility indoor environment, the deleterious effects of fading and the need for very low transmit power combine to cause problems for radio transmissions. Regardless of advanced coding techniques, channel capacity remains an unmovable barrier. Without changing the channel itself, not much can be done. Fortunately, increasing the number of antennas at both the base and mobile stations accomplishes exactly that, resulting in channels with higher capacity. Such systems can theoretically increase capacity by up to a factor equaling the number of transmit and receive antennas in the array [2,3,4,5,6]. There are currently two main approaches to realizing the capacity potential of these channels: coordinated space-time codes and layered space-time codes.

Coordinated space-time block codes [7,8,9] and trellis codes [10,11,12,13,14] are designed for coordinated use in space and time. The data is encoded using

S. Boztacs and I.E. Shparlinski (Eds.): AAECC-14, LNCS 2227, pp. 387–396, 2001.
© Springer-Verlag Berlin Heidelberg 2001

multi-dimensional codes that span the transmit array. Trellis codes are typically decoded using the Viterbi algorithm. Such codes are efficient for small arrays, and can achieve within 3 dB of the 90% outage capacity rate calculated in [3]. A serious obstacle to extension to larger arrays however is the rapid growth of decoder complexity with array size and data rate: the number of states in a full-diversity space-time trellis code for t transmit antennas with rate R is $2^{(t-1)R}$.

Another approach uses layered space-time codes [15,16], where the channel is decomposed into parallel single-input, single-output channels. The receiver successively decodes these layers by using antenna array techniques and linear or non-linear cancellation methods.

In this paper, we consider only the design of coordinated space-time codes achieving full space diversity over fading channels. Early work on space-time code design was based on considering the minimum rank of a difference matrix with entries found by determining all possible complex differences obtained between all valid codeword sequences [10]. One problem with constructing such codes in this manner was that the code design was not performed over a finite field, resulting in rather ad-hoc design methods, or codes found by exhaustive computer search. This problem was addressed in [1] whereby the minimum rank criteria was related back to a binary rank criteria such that codes satisfying the binary rank criteria were shown to provide maximum diversity advantage. Design rules for binary space-time codes based on phase shift keyed (PSK) modulation were presented in [1] and were further extended in [17] for quadrature amplitude modulation (QAM) constellations. In this paper, we expand on the ideas presented in [1] in designing binary systematic space-time codes which achieve full diversity advantage using binary-PSK (BPSK). The codes are based on ensuring that the characteristic polynomial of the parity generator matrix **P** is irreducible over $\mathbb{F}_2$. We give a recursive code design procedure to construct such matrices. The construction is shown to result in a sparse parity check matrix **H** similar to that of irregular low-density parity check (LDPC) codes. We then construct codes over $\mathbb{Z}_4$ based on the modulo 2 projection of the $\mathbb{Z}_4$ code matrix for use with quadrature-PSK (QPSK) modulation.

2   Preliminaries

2.1   Channel Model

We consider a single-user wireless communication link consisting of t transmit antennas and r receive antennas. The matched filtered signal $y_{jk} \in \mathbb{C}$ for receive antenna $j = 1, 2, \ldots, r$ at time $k = 1, 2, \ldots, N$ is given by

$$y_{jk} = \sum_{i=1}^{t} c_{ik} \gamma_{ji} + n_{jk} \tag{1}$$

where c_{ik} is the code symbol transmitted from antenna $i = 1, 2, \ldots, t$ at time k; γ_{ji} is the complex gain coefficient between transmit antenna i and receive antenna j; $n_{jk} \in \mathbb{C}$ is a discrete time circularly symmetric complex Gaussian noise process, independent over space and time, $\mathsf{E}[n_{jk}\overline{n}_{jk}] = \sigma^2$ and $\mathsf{E}[n_{jk}\overline{n}_{j'k'}] = 0$ for $j \neq j'$ and/or $k \neq k'$. $\mathsf{E}[X]$ denotes the expectation of a random variable X. Using matrix notation, the received $r \times N$ matrix $\mathbf{y} = (y_{jk})$ is given by $\mathbf{y} = \boldsymbol{\Gamma}\mathbf{c} + \mathbf{n}$ where we define the matrices $\mathbf{n} = (n_{jk})$ and $\boldsymbol{\Gamma} = (\gamma_{ji})$. The complex matrix $\mathbf{c} = (c_{ik})$ is referred to as the transmitted *space-time codeword*. Note that this implies that the channel matrix $\boldsymbol{\Gamma}$ is held constant for each codeword, but may vary from codeword to codeword. This is usually referred to as quasi-static flat-fading.

2.2 Design Rules for Space-Time Codes

Design rules for space-time codes were presented in [10]. Let $\mathbf{c}$ be a valid space-time codeword and let the erroneous codeword $\mathbf{e}$ be any other valid codeword that may be chosen at the receiver in preference to the transmitted codeword $\mathbf{c}$. Let $\mathbf{B}(\mathbf{c}, \mathbf{e}) = \mathbf{e} - \mathbf{c}$ be the codeword difference matrix. We define the matrix $\mathbf{A}(\mathbf{c}, \mathbf{e}) = \mathbf{B}(\mathbf{c}, \mathbf{e})\mathbf{B}^*(\mathbf{c}, \mathbf{e})$ where $*$ denotes the Hermitian (conjugate transpose) operation. Tarokh et al. [10] showed that for a quasi-static flat fading channel, the pairwise error probability is upper bounded as follows

$$P(\mathbf{c}, \mathbf{e}) \leq \left(\frac{1}{\prod_{i=1}^{t}(1 + \lambda_i E_s/4N_0)} \right)^r \tag{2}$$

where λ_i denotes the non-zero eigenvalues of the matrix $\mathbf{A}(\mathbf{c}, \mathbf{e})$. If $\mathbf{A}(\mathbf{c}, \mathbf{e})$ has rank s, then it has exactly s non-zero eigenvalues and (2) can be bounded by

$$P(\mathbf{c}, \mathbf{e}) \leq \left(\prod_{i=1}^{s} \lambda_i \right)^{-r} (E_s/4N_0)^{-sr}. \tag{3}$$

The resulting space-time code results in a *coding advantage* of $(\lambda_1\lambda_2 \ldots \lambda_s)^{1/s}$ and a *diversity advantage* of rs. The main gain to be obtained from a space-time code is the diversity advantage, which is generally optimised prior to considering the coding advantage [10]. The reason for this is that the diversity advantage governs the asymptotic slope of the bit error rate (BER) vs E_b/N_0 performance curve while the coding gain is responsible for shifts in the performance curve.

Criterion 1 (Rank). *In order to obtain maximum diversity advantage, the matrix $\mathbf{B}(\mathbf{c}, \mathbf{e})$ has to be of rank t (full rank) over all valid codeword pairs $\mathbf{c} \neq \mathbf{e}$.*

Criterion 2 (Determinant). *In order to obtain maximum coding advantage, the product $(\prod_{i=1}^{s} \lambda_i)^{1/s}$ has to be maximised over all valid codeword pairs $\mathbf{c} \neq \mathbf{e}$.*

These design rules were also extended for space-time codes operating in Rician channels and for fading channels with correlated coefficients [10].

2.3   The Stacking Construction

Let $M_{m \times n}(\mathbb{F})$ denote the set of all $m \times n$ matrices with elements from the field $< \mathbb{F}, \oplus, \otimes >$. Consider a (non space-time) rate $1/2$ binary systematic linear block code $\mathcal{C}$ with generator matrix

$$\mathbf{G} = [\mathbf{I}|\mathbf{P}] \in M_{k \times 2k}(\mathbb{F}_2), \tag{4}$$

where $\mathbf{I}$ is the $k \times k$ identity matrix and $\mathbf{P}$ is a $k \times k$ generator matrix for the parity bits only. For a given information (row) vector $\mathbf{x} = (x_1, x_2, \ldots, x_k)$, the output codeword consists of k information bits $\mathbf{xI}$ followed by k parity bits $\mathbf{xP}$.

A space-time code for a two-antenna system can be constructed from $\mathcal{C}$ in which the systematic bits and the parity bits are transmitted across the first and second antenna respectively. The resulting $2 \times k$ space-time codeword is

$$\mathbf{c} = \begin{bmatrix} \mathbf{xI} \\ \mathbf{xP} \end{bmatrix}. \tag{5}$$

This is the BPSK *stacking construction* from [1] (assuming $\mathbb{F}_2$ is mapped to an antipodal constellation for transmission). It was also shown in [1] that the stacking construction achieves full spatial diversity if and only if $\mathbf{P}$ and $\mathbf{I} \oplus \mathbf{P}$ are simultaneously full rank over $\mathbb{F}_2$. However no further guidance (other than manually checking these two binary rank conditions) was given for code design. The authors went on to present a binary code satisfying the above condition, in which the underlying systematic binary code was an expurgated and punctured version of the Golay code $\mathcal{G}_{23}$.

3   Eigentheory Based Design Rules

We now investigate conditions on the spectral properties of $\mathbf{P}$ such that $\mathbf{P}$ and $\mathbf{I} \oplus \mathbf{P}$ are both guaranteed to be full-rank. This will in turn allow us to give a new code construction. The design rules apply for codes which use BPSK modulation. In Sect. 5.2, we "lift" these properties to design QPSK space-time codes which also guarantee maximum diversity advantage. Using our approach we can easily extend our construction to non-systematic codes.

Let $\mathbf{P} \in M_{k \times k}(\mathbb{F})$ with spectrum $\sigma(\mathbf{P}) = \{\lambda_1, \lambda_2, \ldots\}$. By convention, we shall consider the eigenvalues $\lambda_i \in \overline{\mathbb{F}}$ where $\mathbb{F} \subseteq \overline{\mathbb{F}}$. $\overline{\mathbb{F}}$ is an algebraic closure of $\mathbb{F}^1$, thus $|\sigma(\mathbf{P})| = k$ counting multiplicities. Under our closure convention we can easily show that [18, Theorem 1.1.6 pp. 36] holds in the finite field case. For reference, we reproduce the theorem here as a lemma.

Lemma 1. *Let* $\mathbf{P} \in M_{k \times k}(\mathbb{F})$ *and* $\lambda \in \sigma(\mathbf{P})$. *Let* $f \in \mathbb{F}[x]$ *be a polynomial with coefficients in* $\mathbb{F}$. *Then* $f(\lambda) \in \sigma(f(\mathbf{P}))$.

---

[1] In fact, all we require is an extension field constructed by adjoining the roots of all polynomials of degree k. Without this convention, $\mathbf{P}$ may not have any eigenvalues at all.

Application of this lemma leads directly to our first main result.

Theorem 1 (Rank Criterion). *The binary matrices* $\mathbf{P}$ *and* $\mathbf{I} \oplus \mathbf{P}$ *are simultaneously full-rank if and only if the characteristic polynomial of* $\mathbf{P}$ *is irreducible over* $\mathbb{F}_2$.

Proof. Consider the function $f(x) = 1 + x$ with unique inverse $f^{-1}(y) = y - 1$. By Lemma 1, $\lambda \in \sigma(\mathbf{P})$ and $1 \oplus \lambda \in \sigma(\mathbf{I} \oplus \mathbf{P})$. Thus we have that $1 \in \sigma(\mathbf{P})$ iff $0 \in \sigma(\mathbf{I} \oplus \mathbf{P})$ and $1 \in \sigma(\mathbf{I} \oplus \mathbf{P})$ iff $0 \in \sigma(\mathbf{P})$. Thus the characteristic polynomials of $\mathbf{P}$ and $\mathbf{I} \oplus \mathbf{P}$ are irreducible, i.e., do not have any roots equal to 0 or 1, iff $\mathbf{P}$ and $\mathbf{I} \oplus \mathbf{P}$ have full rank (since a member of $M_{k \times k}(\mathbb{F})$ has full rank iff 0 is not a root of its characteristic polynomial). □

Theorem 1 can be illustrated via the following example. Table 1 lists some 2×2 binary matrices $\mathbf{P}$, $\mathbf{I} \oplus \mathbf{P}$ and their corresponding determinants. These four

Table 1. Binary 2×2 matrices and their determinants.

$\mathbf{P}$	$\mathbf{I} \oplus \mathbf{P}$	$(\det \mathbf{P}, \det(\mathbf{I} \oplus \mathbf{P}))$
$\begin{bmatrix} 1 & 1 \\ 0 & 1 \end{bmatrix}$	$\begin{bmatrix} 0 & 1 \\ 0 & 0 \end{bmatrix}$	(1,0)
$\begin{bmatrix} 1 & 1 \\ 1 & 1 \end{bmatrix}$	$\begin{bmatrix} 0 & 1 \\ 1 & 0 \end{bmatrix}$	(0,1)
$\begin{bmatrix} 0 & 0 \\ 1 & 1 \end{bmatrix}$	$\begin{bmatrix} 1 & 0 \\ 1 & 0 \end{bmatrix}$	(0,0)
$\begin{bmatrix} 1 & 1 \\ 1 & 0 \end{bmatrix}$	$\begin{bmatrix} 0 & 1 \\ 1 & 1 \end{bmatrix}$	(1,1)

scenarios cover all possible combinations for the determinants of $\mathbf{P}$ and $\mathbf{I} \oplus \mathbf{P}$. It can be seen that in the last example both determinants are nonzero and hence this case satisfies the conditions to ensure full spatial diversity. It is easily verified that the characteristic polynomial for both these matrices is $t^2 + t + 1$ which is irreducible over $\mathbb{F}_2$. In fact, these are the only two members of $M_{2 \times 2}(\mathbb{F}_2)$ with this property.

Of course we do not need to limit ourselves to $M_{2 \times 2}(\mathbb{F})$. In the general case, if $\mathbf{P} \in M_{k \times k}(\mathbb{F})$, then for the systematic space-time code to achieve maximum diversity, the characteristic polynomial $p(t)$ of $\mathbf{P}$ must be of the form

$$p(t) = \prod \phi_i(t) \tag{6}$$

where each ϕ_i is irreducible over $\mathbb{F}$. In fact, valid 3×3 matrices would have characteristic equation of degree three, again irreducible over GF(2) (i.e., $\lambda^3 + \lambda^2 + 1$ and $\lambda^3 + \lambda + 1$). The larger matrix size would result in a greater number of binary 3×3 matrices having irreducible characteristic equation over 2×2 matrices (2 possibilities).

4   Code Design

Since we want to be able to generate space-time block codes which guarantee full diversity for variable blocklengths, one approach is to design space-time block codes using 2×2 matrices given in the last row of Table 1. As these matrices satisfy Theorem 1 they can be used as building blocks for generating codes of longer blocklength. We use the Schur determinant formula [18] to recursively construct block upper triangular parity generator matrices $\mathbf{P}$ which satisfy the condition given by (6). Our construction is summarised in the following theorem,

Theorem 2 (Schur Construction). *Let $k \geq 2$ and $\mathbf{P} \in M_{k \times k}(\mathbb{F}_2)$ be a block upper-triangular parity generator matrix, defined as follows*

$$
\mathbf{P} = \begin{bmatrix} \mathbf{P}_1^* & & & \\ & \mathbf{P}_2^* & \mathbf{X} & \\ & \mathbf{0} & \ddots & \\ & & & \mathbf{P}_l^* \end{bmatrix}
\tag{7}
$$

where $\mathbf{0}$ denotes all-zero entries below the diagonal, $\mathbf{X}$ denotes arbitrary binary entries above the diagonal and the matrices $\mathbf{P}_1^*, \mathbf{P}_2^*, \ldots, \mathbf{P}_l^*$, $l < k$, are binary square matrices (not necessarily equal size) with irreducible characteristic polynomials. The resulting space-time code given by the stacking construction (5) achieves full spatial diversity.

Proof. Partition $\mathbf{P} \in M_{k \times k}(\mathbb{F})$ as follows

$$
\mathbf{P} = \begin{bmatrix} \mathbf{P}_{11} & \mathbf{P}_{12} \\ \mathbf{P}_{21} & \mathbf{P}_{22} \end{bmatrix}
\tag{8}
$$

such that for each submatrix we have $\mathbf{P}_{11}$ of size $p \times p$, $\mathbf{P}_{12}$ of size $p \times q$, $\mathbf{P}_{21}$ of size $q \times p$, and finally $\mathbf{P}_{22}$ of size $q \times q$ where $p = q = k/2$ for k even and $p = \lfloor k/2 \rfloor$, $q = p+1$ for k odd. The Schur determinant formula [18, Section 0.8.5] gives $\det \mathbf{P} = \det \mathbf{P}_{11} \det(\mathbf{P}_{22} - \mathbf{P}_{21}\mathbf{P}_{11}^{-1}\mathbf{P}_{12})$ (the term in brackets is the Schur complement). Thus, it is immediately clear that for $\mathbf{P}$ to be non-singular, we require $\mathbf{P}_{11}$ to be non-singular. We have some freedom in determining suitable sub-matrices to ensure that the Schur complement is also non-singular. Setting $\mathbf{P}_{21} = \mathbf{0}$ gives $\det \mathbf{P} = \det \mathbf{P}_{11} \det \mathbf{P}_{22}$ and therefore we are free to choose $\mathbf{P}_{12}$ provided $\mathbf{P}_{22}$ is non-singular. Thus we let

$$
\mathbf{P} = \begin{bmatrix} \mathbf{P}_{11} & \mathbf{X} \\ \mathbf{0} & \mathbf{P}_{22} \end{bmatrix}
\tag{9}
$$

where we require that $\mathbf{P}_{11}$ and $\mathbf{P}_{22}$ are non-singular. We now simply apply this procedure (9) recursively to the submatrices $\mathbf{P}_{11}$ and $\mathbf{P}_{22}$, until we obtain 2×2 or 3×3 sub-partitions. For the case where we obtain (2×2) $\mathbf{P}_{11}$ and $\mathbf{P}_{22}$ diagonal sub-partitions, we set these to one of the matrices with irreducible characteristic polynomials discussed earlier (last example in Table 1). Similarly, any 3×3 sub-partitions are set to any binary matrix having irreducible characteristic polynomial. This ensures that not only is $\mathbf{P}$ non-singular, but that $\mathbf{I} \oplus \mathbf{P}$ is also non-singular. □

As an example, the following 4×4 generator $\mathbf{P}$ constructed from 2×2 building blocks guarantees full diversity. Both the matrices $\mathbf{P}$ and $\mathbf{I} \oplus \mathbf{P}$ are full rank.

$$
\mathbf{P} = \begin{bmatrix} 1 & 1 & 1 & 0 \\ 1 & 0 & 0 & 0 \\ 0 & 0 & 0 & 1 \\ 0 & 0 & 1 & 1 \end{bmatrix}, \mathbf{I} \oplus \mathbf{P} = \begin{bmatrix} 0 & 1 & 1 & 0 \\ 1 & 1 & 0 & 0 \\ 0 & 0 & 1 & 1 \\ 0 & 0 & 1 & 0 \end{bmatrix}
$$

An example of a 6×6 generator $\mathbf{P}$ constructed from 2 unique 3×3 building blocks having irreducible characteristic equation is

$$
\mathbf{P} = \begin{bmatrix} 0 & 1 & 1 & 0 & 0 & 1 \\ 0 & 0 & 1 & 0 & 1 & 0 \\ 1 & 0 & 0 & 0 & 0 & 0 \\ 0 & 0 & 0 & 0 & 1 & 1 \\ 0 & 0 & 0 & 1 & 0 & 0 \\ 0 & 0 & 0 & 0 & 1 & 0 \end{bmatrix}.
$$

Again we can verify that $\mathbf{P}$ and $\mathbf{I} \oplus \mathbf{P}$ have full-rank. The recursive design discussed thus allows variable blocklength codes to be constructed. However, the code design is limited in the sense that it can only be applied to systematic BPSK modulated codes. In the next section, we extend the code design to deal with these limitations.

5   Further Extensions

5.1   Non-systematic Codes

Replacing the identity component $\mathbf{I}$ in (4) with some other $k \times k$ matrix $\mathbf{M}$, a non-systematic space-time code results. The corresponding space-time code is defined as follows

$$
\mathbf{c} = \begin{bmatrix} \mathbf{xM} \\ \mathbf{xP} \end{bmatrix}. \tag{10}
$$

From the BPSK stacking construction [1], $\mathbf{M}$, $\mathbf{P}$ and $\mathbf{M} \oplus \mathbf{P}$ are required to be simultaneously full rank in order for the space-time code to achieve maximum diversity advantage. We assume that $\mathbf{P}$ is constructed according to Theorem 2. For non-singular $\mathbf{M}$, we have $\mathbf{M} \oplus \mathbf{P} = (\mathbf{I} \oplus \mathbf{P}\mathbf{M}^{-1})\mathbf{M}$. Thus we require $\mathbf{I} \oplus \mathbf{P}\mathbf{M}^{-1}$ to be non-singular. Begin by restricting $\mathbf{M}^{-1}$ to be partitioned as follows

$$
\mathbf{M}^{-1} = \begin{bmatrix} \mathbf{I} & \mathbf{X}_2 \\ \mathbf{0} & \mathbf{I} \end{bmatrix}
$$

and similarly let $\mathbf{P}$ be partitioned

$$\mathbf{P} = \begin{bmatrix} \mathbf{P}_1 & \mathbf{X}_1 \\ \mathbf{0} & \mathbf{P}_2 \end{bmatrix}.$$

We then obtain

$$\mathbf{PM}^{-1} = \begin{bmatrix} \mathbf{P}_1 & \mathbf{X} \\ \mathbf{0} & \mathbf{P}_2 \end{bmatrix}$$

and by our assumption on $\mathbf{P}$, the resulting code achieves full-spatial diversity. In fact for this case, $\mathbf{M} = \mathbf{M}^{-1}$ for entries defined in $\mathbb{F}_2$. The non-systematic construction allows greater flexibility in the addition of non-zero terms within the parity check matrix $\mathbf{H}$ associated with the code resulting in possibly better performing codes (see Sect. 5.3). As an example, choosing $\mathbf{M}$ to have the following form

$$\mathbf{M} = \begin{bmatrix} 1 & 0 & 1 & 0 \\ 0 & 1 & 0 & 1 \\ 0 & 0 & 1 & 0 \\ 0 & 0 & 0 & 1 \end{bmatrix}$$

We can verify that if we choose $\mathbf{P}$ to be as defined in the previous example, then $\mathbf{M} \oplus \mathbf{P}$ is full-rank. The rank of $\mathbf{M} \oplus \mathbf{P}$ is in fact full irrespective of the selection of the top-right 2×2 sub-matrix belonging to $\mathbf{M}^2$.

5.2   QPSK Codes

In [1], a stacking construction for QPSK space-time codes is also presented. The natural discrete alphabet considered is the ring $\mathbb{Z}_4$ of integers modulo 4. Modulation is performed by mapping codeword symbols $c \in \mathbb{Z}_4$ to QPSK modulated symbols $x \in \{\pm 1, \pm i\}$ where the mapping is given by $x = i^c$ where $i = \sqrt{-1}$. For a $\mathbb{Z}_4$ matrix $\mathbf{c}$, the binary matrices $\alpha(\mathbf{c})$ and $\beta(\mathbf{c})$ are uniquely defined such that $\mathbf{c} = \alpha(\mathbf{c}) + 2\beta(\mathbf{c})$. If we define the $\mathbb{Z}_4$-valued matrix $\mathbf{c} = [\mathbf{c}_1^T \mathbf{c}_2^T \ldots \mathbf{c}_t^T]^T$ where $\mathbf{c}_i$ is the i-th row of the codeword matrix, then the row based indicant projection $\Xi(\mathbf{c})$ is defined as $\Xi(\mathbf{c}) = [\alpha(\mathbf{c}_1^T)\alpha(\mathbf{c}_2^T) \ldots \alpha(\mathbf{c}_t^T)]^T$. As a result of the QPSK stacking construction, the stacking of $\mathbb{Z}_4$-valued matrices will produce a QPSK based space-time code that achieves full spatial diversity provided that the stacking of the corresponding Ξ-projection matrices produces a BPSK based space-time code that also guarantees full-spatial diversity. This result allows the design of QPSK based spaced time codes to be effectively mapped back into the domain of BPSK space-time codes.

We can thus use the QPSK stacking construction along with the code construction technique presented in Sect. 4 to design QPSK based space-time codes. In fact, for QPSK modulation, we have a greater flexibility in choosing the diagonal 2×2 building blocks. For BPSK modulation, the number of possible 2×2 building blocks was limited to two. However for QPSK, provided the modulo 2

---

[2] Provided that $\mathbf{P}$ is full rank

projection results in a matrix with irreducible characteristic equation, full diversity is guaranteed. For example, the following 2×2 matrices with entries $\in \mathbb{Z}_4$ result in the same row-based indicant projection matrix Ξ

$$\begin{bmatrix} 3 & 3 \\ 3 & 0 \end{bmatrix}, \begin{bmatrix} 1 & 3 \\ 1 & 0 \end{bmatrix}, \begin{bmatrix} 3 & 1 \\ 1 & 2 \end{bmatrix}, \begin{bmatrix} 1 & 3 \\ 3 & 2 \end{bmatrix} \overset{\Xi}{\Rightarrow} \begin{bmatrix} 1 & 1 \\ 1 & 0 \end{bmatrix}.$$

5.3   Decoding Algorithm

Detailed discussion of the decoding algorithm and the resulting performance will be presented elsewhere. All that will be mentioned here is that Theorem 2 allows construction of low density parity check matrices [20,21], suitable for application of the Sum-Product algorithm [19] operating on the corresponding factor graph. In addition, the triangular structure of $\mathbf{P}$ leads to a low complexity encoding algorithm (which is an important feature for low-density parity check codes). Currently we are investigating row and column degree sequences for $\mathbf{P}$ resulting in the best possible BER vs E_b/N_0 performance.

6   Conclusions

We have presented a recursive construction for systematic and non-systematic space-time codes for two antenna systems. The codes are guaranteed to provide full diversity advantage. Performance results based on the code construction techniques discussed will appear in a subsequent publication.

References

1. Hammons, A. R., and El Gamal, H.: On the theory of space-time codes for PSK modulation. IEEE Trans. Inform. Theory Vol. **46** No. **2** 2000 524–542
2. Telatar, I. Emre.: Capacity of multi-antenna Gaussian channels. European Trans. Telecomm. Vol. **10** No. **6** 1999 585–595
3. Foschini, G. J., Gans, M. J.: On limits of wireless communications in a fading environment when using multiple antennas. Wireless Personal Communications Vol. **6** No **3** 1988 311–335
4. Winters, J.H., Salz, J., Gitlin, G.D.: The capacity of wireless communication systems can be substantially increased by the use of antenna diversity. Conf. Inform. Sci. Syst. Princeton, NJ, 1992
5. Winters, J.H., Salz, J., Gitlin, G.D.: The impact of antenna diversity on the capacity of wireless communication systems. IEEE Trans. Commun. Vol. **42** No. **2/3/4** 1994 1740–1751
6. Narula, A., Trott, M.D., Wornell, G.W.: Performance limits of coded diversity methods for transmitter antenna arrays. IEEE Trans. Inform. Theory Vol. **45** No. **7** 1999 2418–2433
7. Alamouti, S.M.: A simple transmit diversity technique for wireless communications. IEEE J. Selected Areas Commun. Vol. **16** No. **8** 1998 1451–1458
8. Tarokh, V., Jafarkhani, H., Calderbank, A.R.: Space-time block codes from orthogonal designs. IEEE Trans. Inform. Theory Vol. **45**, No. **5**, 1999 1456–1467

9. Tarokh, V., Jafarkhani, H., Calderbank, A.R.: Space-time block coding for wireless communications: Performance results. IEEE J. Selected Areas Commun. Vol. **17**, No. **3**, 1999 451–460

10. Tarokh, V., Seshadri, N., Calderbank, A.R.: Space-time codes for high data rate wireless communication: Performance criterion and code construction. IEEE Trans. Inform. Theory Vol. **44**, 1998 744–765

11. Naguib, A.F., Tarokh, V., Seshadri, N., and Calderbank, A.R.: A space-time coding modem for high-data-rate wireless communications. IEEE J. Selected Areas Commun. Vol. **16**, No. **8**, 1998 1459–1478

12. Agrawal, D., Tarokh, V., Naguib, A., Seshadri, N.: Space-time coded OFDM for high data-rate wireless communication over wideband channels. 48th IEEE Vehicular Technology Conference, Ottawa, Canada, Vol. **3**, 1998 2232–2236

13. Tarokh, V., Naguib, A., Seshadri, N., Calderbank, A.R.: Combined array processing and space-time coding. IEEE Trans. Inform. Theory Vol. **45**, No. **4**, 1999 1121–1128

14. Baro, S., Bauch, G., Hansmann, A.: Improved codes for space-time trellis-coded modulation. IEEE Commun. Lett. Vol. **4**, No. **1**, 2000 20–22

15. Foschini, G.J.: Layered space-time architecture for wireless communication in a fading environment when using multi-element antennas. Bell Labs Technical Journal. Vol. **1**, No. **2**, 1996 41–59

16. El Gamal, H., Hammons, A.R.: The layered space-time architecture: A new perspective. IEEE Trans. Inform. Theory submitted

17. Liu, Y., Fitz, M.P., Takeshita, O.: A rank criterion for QAM space-time codes," EEE Trans. Inform. Theory submitted

18. Horn R.A.,Johnson, C.R.: Matrix Analysis. Oxford University Press, Cambridge, 1985

19. Frey, B.: Graphical Models for Machine Learning and Computation. MIT Press, Cambridge, Massachusetts, 1998

20. Gallager, R.J.: Low-density parity-check codes. IEEE. Trans. Inform. Theory Vol. **8**, No. **1**, 1962 21-28

21. Richardson, T., Shokrollahi, A., Urbanke, R.: Design of provably good low-density parity-check codes. IEEE Trans. Inform. Theory submitted

Author Index

Lecture Notes in Computer Science

For information about Vols. 1–2148
please contact your bookseller or Springer-Verlag

Vol. 2187: U. Voges (Ed.), Computer Safety, Reliability and Security. Proceedings, 2001. XVI, 249 pages. 2001.

Vol. 2188: F. Bomarius, S. Komi-Sirviö (Eds.), Product Focused Software Process Improvement. Proceedings, 2001. XI, 382 pages. 2001.

Vol. 2189: F. Hoffmann, D.J. Hand, N. Adams, D. Fisher, G. Guimaraes (Eds.), Advances in Intelligent Data Analysis. Proceedings, 2001. XII, 384 pages. 2001.

Vol. 2190: A. de Antonio, R. Aylett, D. Ballin (Eds.), Intelligent Virtual Agents. Proceedings, 2001. VIII, 245 pages. 2001. (Subseries LNAI).

Vol. 2191: B. Radig, S. Florczyk (Eds.), Pattern Recognition. Proceedings, 2001. XVI, 452 pages. 2001.

Vol. 2192: A. Yonezawa, S. Matsuoka (Eds.), Metalevel Architectures and Separation of Crosscutting Concerns. Proceedings, 2001. XI, 283 pages. 2001.

Vol. 2193: F. Casati, D. Georgakopoulos, M.-C. Shan (Eds.), Technologies for E-Services. Proceedings, 2001. X, 213 pages. 2001.

Vol. 2194: A.K. Datta, T. Herman (Eds.), Self-Stabilizing Systems. Proceedings, 2001. VII, 229 pages. 2001.

Vol. 2195: H.-Y. Shum, M. Liao, S.-F. Chang (Eds.), Advances in Multimedia Information Processing – PCM 2001. Proceedings, 2001. XX, 1149 pages. 2001.

Vol. 2196: W. Taha (Ed.), Semantics, Applications, and Implementation of Program Generation. Proceedings, 2001. X, 219 pages. 2001.

Vol. 2197: O. Balet, G. Subsol, P. Torguet (Eds.), Virtual Storytelling. Proceedings, 2001. XI, 213 pages. 2001.

Vol. 2198: N. Zhong, Y. Yao, J. Liu, S. Ohsuga (Eds.), Web Intelligence: Research and Development. Proceedings, 2001. XVI, 615 pages. 2001. (Subseries LNAI).

Vol. 2199: J. Crespo, V. Maojo, F. Martin (Eds.), Medical Data Analysis. Proceedings, 2001. X, 311 pages. 2001.

Vol. 2200: G.I. Davida, Y. Frankel (Eds.), Information Security. Proceedings, 2001. XIII, 554 pages. 2001.

Vol. 2201: G.D. Abowd, B. Brumitt, S. Shafer (Eds.), Ubicomp 2001: Ubiquitous Computing. Proceedings, 2001. XIII, 372 pages. 2001.

Vol. 2202: A. Restivo, S. Ronchi Della Rocca, L. Roversi (Eds.), Theoretical Computer Science. Proceedings, 2001. XI, 440 pages. 2001.

Vol. 2204: A. Brandstädt, V.B. Le (Eds.), Graph-Theoretic Concepts in Computer Science. Proceedings, 2001. X, 329 pages. 2001.

Vol. 2205: D.R. Montello (Ed.), Spatial Information Theory. Proceedings, 2001. XIV, 503 pages. 2001.

Vol. 2206: B. Reusch (Ed.), Computational Intelligence. Proceedings, 2001. XVII, 1003 pages. 2001.

Vol. 2207: I.W. Marshall, S. Nettles, N. Wakamiya (Eds.), Active Networks. Proceedings, 2001. IX, 165 pages. 2001.

Vol. 2208: W.J. Niessen, M.A. Viergever (Eds.), Medical Image Computing and Computer-Assisted Intervention – MICCAI 2001. Proceedings, 2001. XXXV, 1446 pages. 2001.

Vol. 2209: W. Jonker (Ed.), Databases in Telecommunications II. Proceedings, 2001. VII, 179 pages. 2001.

Vol. 2210: Y. Liu, K. Tanaka, M. Iwata, T. Higuchi, M. Yasunaga (Eds.), Evolvable Systems: From Biology to Hardware. Proceedings, 2001. XI, 341 pages. 2001.

Vol. 2211: T.A. Henzinger, C.M. Kirsch (Eds.), Embedded Software. Proceedings, 2001. IX, 504 pages. 2001.

Vol. 2212: W. Lee, L. Mé, A. Wespi (Eds.), Recent Advances in Intrusion Detection. Proceedings, 2001. X, 205 pages. 2001.

Vol. 2213: M.J. van Sinderen, L.J.M. Nieuwenhuis (Eds.), Protocols for Multimedia Systems. Proceedings, 2001. XII, 239 pages. 2001.

Vol. 2214: O. Boldt, H. Jürgensen (Eds.), Automata Implementation. Proceedings, 1999. VIII, 183 pages. 2001.

Vol. 2215: N. Kobayashi, B.C. Pierce (Eds.), Theoretical Aspects of Computer Software. Proceedings, 2001. XV, 561 pages. 2001.

Vol. 2216: E.S. Al-Shaer, G. Pacifici (Eds.), Management of Multimedia on the Internet. Proceedings, 2001. XIV, 373 pages. 2001.

Vol. 2217: T. Gomi (Ed.), Evolutionary Robotics. Proceedings, 2001. XI, 139 pages. 2001.

Vol. 2218: R. Guerraoui (Ed.), Middleware 2001. Proceedings, 2001. XIII, 395 pages. 2001.

Vol. 2220: C. Johnson (Ed.), Interactive Systems. Proceedings, 2001. XII, 219 pages. 2001.

Vol. 2221: D.G. Feitelson, L. Rudolph (Eds.), Job Scheduling Strategies for Parallel Processing. Proceedings, 2001. VII, 207 pages. 2001.

Vol. 2224: H.S. Kunii, S. Jajodia, A. Sølvberg (Eds.), Conceptual Modeling – ER 2001. Proceedings, 2001. XIX, 614 pages. 2001.

Vol. 2225: N. Abe, R. Khardon, T. Zeugmann (Eds.), Algorithmic Learning Theory. Proceedings, 2001. XI, 379 pages. 2001. (Subseries LNAI).

Vol. 2227: S. Boztaş, I.E. Shparlinski (Eds.), Applied Algebra, Algebraic Algorithms and Error-Correcting Codes. Proceedings, 2001. XII, 398 pages. 2001.

Vol. 2229: S. Qing, T. Okamoto, J. Zhou (Eds.), Information and Communications Security. Proceedings, 2001. XIV, 504 pages. 2001.

Vol. 2230: T. Katila, I.E. Magnin, P. Clarysse, J. Montagnat, J. Nenonen (Eds.), Functional Imaging and Modeling of the Heart. Proceedings, 2001. XI, 158 pages. 2001.

Vol. 2232: L. Fiege, G. Mühl, U. Wilhelm (Eds.), Electronic Commerce. Proceedings, 2001. X, 233 pages. 2001.

Vol. 2233: J. Crowcroft, M. Hofmann (Eds.), Networked Group Communication. Proceedings, 2001. X, 205 pages. 2001.

Vol. 2234: L. Pacholski, P. Ružička (Eds.), SOFSEM 2001: Theory and Practice of Informatics. Proceedings, 2001. XI, 347 pages. 2001.

Vol. 2239: T. Walsh (Ed.), Principles and Practice of Constraint Programming – CP 2001. Proceedings, 2001. XIV, 788 pages. 2001.

Vol. 2241: M. Jünger, D. Naddef (Eds.), Computational Combinatorial Optimization. IX, 305 pages. 2001.